Atomic Symbols and Names

Symbol	Name	Symbol	Name	Symbol	Name
Ac	Actinium	Gd	Gadolinium	Po	Polonium
Ag	Silver	Ge	Germanium	Pr	Praseodymium
Al	Aluminum	H	Hydrogen	Pt	Platinum
Am	Americium	He	Helium	Pu	Plutonium
Ar	Argon	Hf	Hafnium	Ra	Radium
As	Arsenic	Hg	Mercury	Rb	Rubidium
At	Astatine	Ho	Holmium	Re	Rhenium
Au	Gold	Hs	Hassium	Rf	Rutherfordium
B	Boron	I	Iodine	Rg	Roentgenium
Ba	Barium	In	Indium	Rh	Rhodium
Be	Beryllium	Ir	Iridium	Rn	Radon
Bh	Bohrium	K	Potassium	Ru	Ruthenium
Bi	Bismuth	Kr	Krypton	S	Sulfur
Bk	Berkelium	La	Lanthanum	Sb	Antimony
Br	Bromine	Li	Lithium	Sc	Scandium
C	Carbon	Lr	Lawrencium	Se	Selenium
Ca	Calcium	Lu	Lutetium	Sg	Seaborgium
Cd	Cadmium	Md	Mendelevium	Si	Silicon
Ce	Cerium	Mg	Magnesium	Sm	Samarium
Cf	Californium	Mn	Manganese	Sn	Tin
Cl	Chlorine	Mo	Molybdenum	Sr	Strontium
Cm	Curium	Mt	Meitnerium	Ta	Tantalum
Cn	Copernicium	N	Nitrogen	Tb	Terbium
Co	Cobalt	Na	Sodium	Tc	Technetium
Cr	Chromium	Nb	Niobium	Te	Tellurium
Cs	Cesium	Nd	Neodymium	Th	Thorium
Cu	Copper	Ne	Neon	Ti	Titanium
Db	Dubnium	Ni	Nickel	Tl	Thallium
Ds	Darmstadtium	No	Nobelium	Tm	Thulium
Dy	Dysprosium	Np	Neptunium	U	Uranium
Er	Erbium	O	Oxygen	V	Vanadium
Es	Einsteinium	Os	Osmium	W	Tungsten
Eu	Europium	P	Phosphorus	Xe	Xenon
F	Fluorine	Pa	Protactinium	Y	Yttrium
Fe	Iron	Pb	Lead	Yb	Ytterbium
Fm	Fermium	Pd	Palladium	Zn	Zinc
Fr	Francium	Pm	Promethium	Zr	Zirconium
Ga	Gallium				

EXPLORING CHEMISTRY

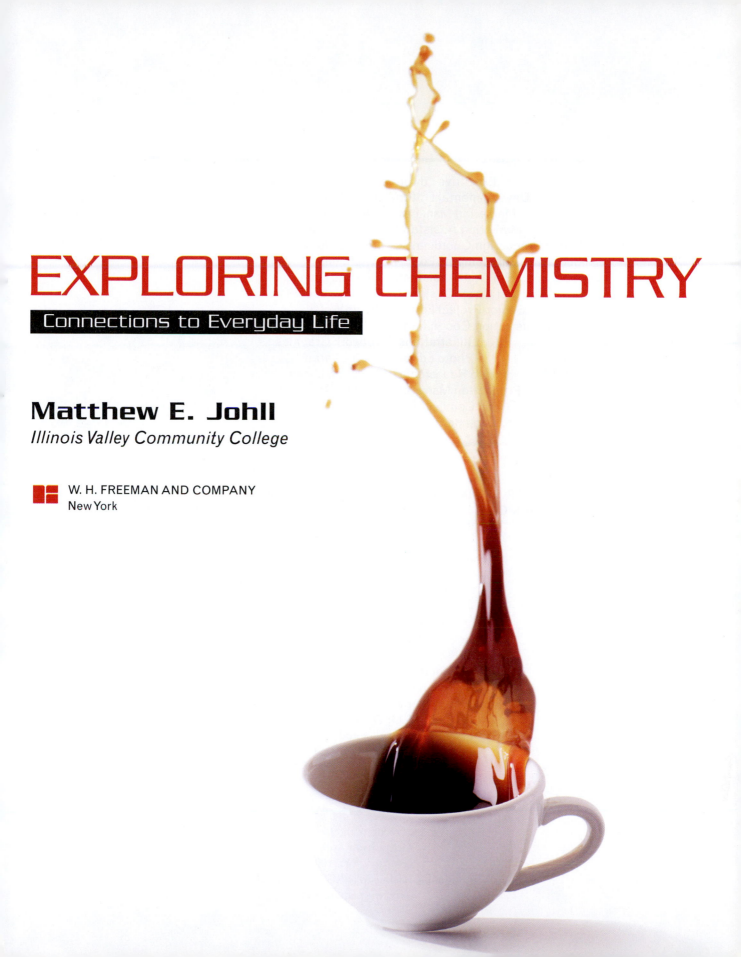

EXPLORING CHEMISTRY

Connections to Everyday Life

Matthew E. Johll
Illinois Valley Community College

W. H. FREEMAN AND COMPANY
New York

Executive Editor:	Anthony Palmiotto
Developmental Editor:	Donald Gecewicz
Marketing Manager:	Alicia Brady
Marketing Assistant:	Joanie Rothschild
Media Production Coordinator:	Jenny Chiu
Media and Supplements Editor:	Dave Quinn
Editorial Assistant:	Nicholas Ciani
Text and Cover Designer:	Vicki Tomaselli
Senior Project Editor:	Vivien Weiss
Senior Illustration Coordinator:	Bill Page
Illustrations:	Network Graphics
Photo Editor:	Cecilia Varas
Photo Researcher:	Christina Micek
Production Manager:	Julia DeRosa
Composition:	Aptara®, Inc.
Printing and Binding:	RR Donnelley

FRONT IMAGE (coffee spilling): © Radius Images/Corbis
BACK COVER (beans): Maria Toutoudaki/Stockbyte/Getty Images

Library of Congress Control Number: 2012933304

Hardcover: ISBN-10: 1-4641-0478-6
ISBN-13: 978-1-4641-0478-7
Loose-Leaf: ISBN-10: 1-4641-2781-6
ISBN-13: 978-4641-2781-6

ISBN for the SSM: ISBN-10: 1-4641-2519-8,
ISBN-13: 978-1-4641-2519-5
ISBN for the Computerized Test Bank: ISBN-10: 1-4641-2524-4,
ISBN-13: 978-1-4641-2524-9

First printing

W. H. Freeman and Company
41 Madison Avenue
New York, NY 10010
Houndmills, Basingstoke RG21 6XS, England
www.whfreeman.com

CONTENTS

PREFACE

Exploring Chemistry: One thing I have learned as a chemistry instructor is that chemistry has many great stories. These stories present themselves in a human context—showing the choices, efforts, strengths, and flaws of human beings. Too often these stories are relegated to boxed essays or the last chapters of a book that rarely get covered, and our students thereby lose the opportunity to see the impact chemistry has on their everyday lives. However, interesting stories in and of themselves are not sufficient. In the development of each case study, the stories are chosen to engage the students in a meaningful and thought-provoking manner, which leads to students *wanting* to read more. The concept of learning chemistry by means of narratives from other contexts and other disciplines has proven remarkably effective.

I wrote my first book, *Investigating Chemistry,* to address the tremendous learning opportunity that students have as they take a nonmajors chemistry course. That textbook combines the case-study teaching approach with the high-interest theme of forensic science. However, many of my colleagues from around North America expressed that, while they appreciated the role of forensic science within chemistry, they would like to further expand the students' horizons, taking the case-study approach into a wider array of subject areas.

That feedback was the kernel of the book in your hands, *Exploring Chemistry: Connecting Chemistry to Everyday Life.* This text relies on the same teaching method as its predecessor, and it exposes the students to the broad impact of chemistry. But rather than focusing only on forensic chemistry, the topics range from environmental issues, to cooking, to athletics, to genetics.

Each of these case studies, when properly examined, lends itself to an understanding of the scientific process. In the case studies, I focus on large-scale and small-scale observations, which lead to methodical processes designed to reach conclusions. The chapter themes allow us to demonstrate to students the importance of careful measurement, proper handling of substances, reliable data, and supportable conclusions. By relating chemistry to what are, in the students' view, unexpected topics, the text builds an appreciation for the practical, important, and accessible field of chemistry.

So, in writing this book, I have sought to bring together a wide variety of topics from everyday life and to integrate the fundamentals of chemistry in ways that are effective and accessible for students. With the help of reviewers, colleagues who provided their syllabi, chemical education research, and even the existing textbooks directed at the course, I identified the most widely taught concepts. Then I worked to apply the most relevant and interesting topics to the concept.

Consider molecular structure, gas laws, and organic chemistry—topics that can be quite challenging for nonmajors to connect with. In *Exploring Chemistry,* however, students first engage with these topics by connecting with how unexpected experimental results led to a new class of powerful chemotherapy drugs, how the Department of Homeland Security and the Transportation Security Administration utilize the gas laws to screen passengers for explosive devices, and how scientists are now creating biodegradable biopolymers. Students are motivated to understand the chemistry and engage in a meaningful learning experience, as they have made a connection between their everyday lives and the chemistry presented within the chapter.

The success of *Investigating Chemistry* demonstrates significant interest among instructors in teaching nonscience majors using the case-study approach. They report that students make more and better use of the textbook and increasingly retain information. In preparing *Exploring Chemistry,* then, I wanted to refine the explanation of the chemical concepts, which are the core of the book, as well as integrate the themes more thoroughly.

Organization and Teaching Method

Exploring Chemistry follows a mainstream organization, similar to many textbooks in the field. Its unique approach—describing the standard principles of chemistry within the context of high-interest cases—is presented with flexibility in mind.

To introduce chemical principles, every chapter begins with a case study that tells a story based on real events, such as how isotope ratios found in ancient ice cores reflect global climate. Through concept-oriented focusing questions that urge students to connect observations and information, the case then sets the stage for a detailed and supportive explanation. The case study is resolved at the end of the chapter, which provides students with an opportunity to review the major topics from within the chapter and better understand the chemistry connection to everyday life.

Within each chapter, particular attention is paid to two challenging areas in student preparation and understanding: mathematical skills and problem-solving capabilities.

- **Math is presented to enhance flexibility in teaching.** Each time the concepts lend themselves to a quantitative explanation, readers encounter a clearly marked "Mathematics of . . ." section, in which all of the appropriate mathematical coverage is found. These math sections provide a relatively detailed treatment of the topics. Furthermore, instructors have the freedom to choose which mathematics sections to cover to fulfill their own goals for the course.

- **Detailed worked examples demonstrate a consistent, step-by-step process of problem solving.** Examples appear frequently within the chapter, offering students as many opportunities as possible to practice and gain an understanding of the methods needed to solve chemistry problems. Each example has an embedded practice problem, closely modeled on the worked example. Students can reinforce their knowledge by reading the worked example and trying out the practice problem.

3.9 Mathematics of Light

Light travels through space in the form of waves, much as waves travel across the oceans. Waves can be described mathematically by their **wavelength**, symbolized by the Greek letter lambda (λ). The wavelength is the distance from one wave peak to another, as illustrated in Figure 3.13. The unit used to measure wavelength for light in the visible spectrum is typically the nanometer (nm). The prefix *nano* represents the multiplier 1×10^{-9}, or one billionth of a meter.

Waves are also mathematically described by their frequency, symbolized by the Greek letter nu (ν). **Frequency** is the number of times the wave peak passes a point in space within a specific time period. It is measured in units called *hertz* (Hz), and 1 Hz represents a wave that passes a point once per second (1 Hz = 1 s^{-1}). Mathematically, s^{-1} and 1/s are equivalent.

Wavelength and frequency are related to the **speed of light** (c), as shown in equation 2 on the facing page. The speed of light in a vacuum is a constant value equal to 3.00×10^8 m/s. Equation 2 is commonly rearranged to solve for the wavelength, as shown in equation 3, or for the frequency, as shown in equation 4. One of the most common mistakes made in using equations 2–4 is failing to make sure that the units of distance are the same. The speed of light is measured in meters per second (m/s), whereas the wavelength of light from the visible spectrum is measured in nanometers. Just remember to use the appropriate conversion factors for units!

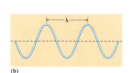

LEARNING OBJECTIVE
Examine and use the relevant mathematical equations to describe the behavior of light.

$$c = \lambda \times \nu \qquad (2)$$

$$\frac{c}{\nu} = \frac{\lambda \times \nu}{\nu} \Rightarrow \lambda = \frac{c}{\nu} \qquad (3)$$

$$\frac{c}{\lambda} = \frac{\lambda \times \nu}{\lambda} \Rightarrow \nu = \frac{c}{\lambda} \qquad (4)$$

(a)   (b)

FIGURE 3.13 Waves are characterized by their wavelength (the distance between the peaks of the wave) and their frequency (the time it takes for one wave to pass a given point in one second). (a) A series of ocean waves rolling toward a beach show the wavelengths highlighted by sunlight. The greater the frequency, the shorter the time interval between waves striking the beach. (b) Light travels through space as a wave, sharing the same characteristics (wavelength and frequency) as ocean waves. (a: The Irish Image Collection/Design Pics/Corbis)

Complete List and Explanations of Case Studies

I have introduced new case studies with exciting stories that illustrate the role of chemistry in our everyday lives and the world around us.

- **Chapter 1: Exploring the Impact Humans Have on the Environment** This case examines the unintended environmental consequences of a society that overuses and misuses prescription medication. The chemically induced modification of gender distribution among fish living downstream from a wastewater plant brings up issues of responsible use of resources, subtle impairment of habitat of certain species, and the scientific method.

- **Chapter 3: Exploring Historical Climate Change** Besides showing students that climate change occurs throughout the history of the earth, this case stresses how scientists can determine the way the earth's climate has changed by examining the stable isotope distribution in glacial ice cores as well as deposits of ancient shells.

- **Chapter 5: Exploring Chemotherapy Drugs** This case highlights the scientific method—but it also shows how keen observation leads scientists to apply unexpected results to create new, beneficial chemicals. Bonding matters in a new class of chemotherapy drugs, which also shows students that chemistry is a continually creative science.

- **Chapter 6: Exploring Chemistry in the Kitchen** We learn that a good cake, muffin, cookie, or brownie—staples of our cuisine—are also successful chemistry experiments. The key difference in these tasty experiments lies in the chemistry occurring in aqueous solutions.

■ **Chapter 7: Exploring Antibiotics and Drug-Resistant Infections** Antibiotics are commonly prescribed drugs, whose misuse has led to deadly drug-resistant infections that plague hospitals and make headlines in the news. Students must ponder whether our current model for drug development can outpace the evolutionary speed of the bacteria as those bacteria become increasingly resistant to multiple drugs.

■ **Chapter 8: Exploring Biodegradable Polymers** As consumers, students make choices that ultimately impact everything from product development to product packaging. This case study examines the development of a biodegradable biopolymer. Polymer chemistry and consumer recycling are used to illustrate the variety of organic molecules, the creativity of the scientific method, and the problems that modern society has to address by means of science.

■ **Chapter 10: Exploring Airport Security** Airport-security measures have evolved to meet the ever-changing threats posed by terrorism. The gas laws explain how explosive devices work, yet this case also shows students how the gas laws can be used to detect explosives, preventing their use. A discussion of the chemistry takes some of the mystery out of how security systems work.

■ **Chapter 11: Exploring Green Chemistry** Students are shown how the principles of green chemistry are challenges that have led scientific researchers to new levels of creativity. This case also stresses how "green" manufacturing processes help to produce a better product—the familiar prescription drug called Lipitor®—with less pollution.

■ **Chapter 12: Exploring Nuclear Power** This case highlights a dilemma for the well-informed citizen—is nuclear power still viable as a source of electricity? The case examines the extent to which nuclear power is used in the United States. Yet the March 2011 earthquake and tsunami in Japan demonstrated how fragile nuclear power plants can be.

■ **Chapter 13: Exploring Athletic Performance** How we feel during a workout is related to chemistry. This case considers the long-standing belief (almost an urban legend) that lactic acid makes muscles sore. By understanding chemical equilibrium, though, we can point to another culprit.

■ **Chapter 14: Exploring Genetically Modified Food** This case examines one of the biggest controversies and dilemmas of our times: What does it mean to modify food products genetically? The case details traditional modifications through agricultural practices and newer industrial methods of altering genes in food products.

I have also retained some forensic-chemistry cases when I thought those cases vividly introduced certain big ideas. (I have added new information to these cases for this book.)

■ **Chapter 2: Exploring Evidence from a Crime Scene** This case is a wonderful example of how keen observation and careful treatment of evidence played integral roles in catching a crime-mob serial killer. Chemists may not solve capital crimes every day, but observation and problem solving are part of daily life.

■ **Chapter 4: Investigating the Chemistry of a Poison** This case tells the story of murder by means of arsenic, but it also stresses that the dose makes the poison and that poisons can be detected through various tests.

■ **Chapter 9: Investigating the Chemistry of Fire and Arson** Most of the popular forensic-science-themed television shows trumpet the power of science to catch and convict the guilty suspect. But what happens when an innocent man is convicted and placed on death row by faulty evidence handling and analysis? Starting with the chemical basis of what a fire is, students also consider what arson evidence is and how evidence should be properly collected, analyzed, and introduced into court proceedings.

I also renewed each chapter by refining coverage, reworking figures and photos to be more information-rich, and replacing items within the end-of-chapter questions and problems. Adopters and reviewers of the first text asked for more case-study problems, and I have accommodated their requests in this book as well.

A Strong Connection Between the Thematic Topics and the Chemical Concepts

At the end of each chapter-opening case study, students will encounter a number of questions connecting the facts of the case to the chemical concepts in the chapter. These questions take advantage of the interest that students gain as they read the opening case, and seek to expand that interest to the chemical principles themselves. The questions serve to guide the student to look for the big ideas of chemistry and to connect those ideas to the case study and other applications.

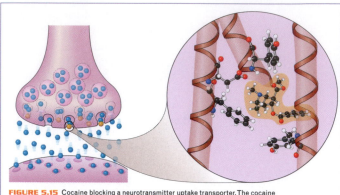

FIGURE 5.15 Cocaine blocking a neurotransmitter uptake transporter. The cocaine molecule is the correct size, shape, and polarity to wedge itself into the uptake channels of neurotransmitters. Once the uptake of neurotransmitters is blocked, the concentration of the neurotransmitters in the gap between neurons is greatly magnified, which results in the magnification of the signal being sent between neurons. The increased signal produces a pleasurable sensation, or "high." (Adapted from Dahl)

Emphasis on Bonding

A nonscience major should be conversant in what bonding means and how bonding relates to the creation of compounds, to their properties, and, by extension, to reactions. To highlight bonding, I aligned it with the coverage of chemical structures. Chapter 4, "Chemical Bonding and Reactions," gives an introduction to ionic and covalent bonding, as well as to the basics of reactions and to the concept of the mole. Chapter 5, "Chemistry of Bonding: Structure and Function of Drug Molecules," provides more detail about such important concepts as covalent bonding, Lewis structures, and VSEPR theory. Chapter 6 takes us into solutions chemistry, with a focus on aqueous solutions, the kind most familiar to our students. Chapter 7 delves into the details of solutions chemistry, especially intermolecular properties as evidenced by the freezing point and boiling point of solutions. In each of these chapters, the case study shows that bonding is crucial in creating distinctive chemical substances—as well as that bonding can be a life-or-death matter.

Enhanced Visuals Portray Chemical Processes over Time

Chemistry depends on visual observation. A textbook should build up students' skill at observation. Visuals in a text should reinforce student understanding and present concepts in an easily perceived way that complements the explanation in the main text.

After working carefully with the artists to produce dynamic and accurate art for all three editions of *Investigating Chemistry* and considering the challenges students have in grasping the material, I recognized a gap in student understanding. Students have difficulty recognizing changes and processes that occur over time, and most textbook art depicts a static moment in time, rather than representing the dynamic processes that are occurring.

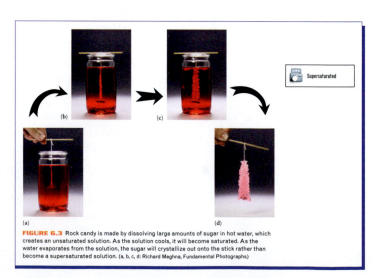

FIGURE 6.3 Rock candy is made by dissolving large amounts of sugar in hot water, which creates an unsaturated solution. As the solution cools, it will become saturated. As the water evaporates from the solution, the sugar will crystallize out onto the stick rather than become a supersaturated solution. (a, b, c, d: Richard Megna, Fundamental Photographs)

A significant amount of the new art in *Exploring Chemistry* is aimed directly at showing students how physical and chemical changes happen over a period of time. To do so, we have increased the scale of the artwork, striving for a realistic three-dimensional representation that includes motion and detailed explanations to guide students through the images.

Chapter Walkthrough

Case Studies introduce the students to chemical concepts in a context to which they can relate, and show the importance of a specific subject area in the students' everyday lives. As they read about the pharmaceutical development, new polymer materials, social issues, and products in everyday use, students become invested and motivated to learn the relevant chemistry. As mentioned below, the opening cases are resolved at the end of the chapter.

2.3 Mathematics of Unit Conversions

2.5 Mathematics of Significant Figure Calculations

3.7 Mathematics of Isotopic Abundance and Atomic Mass

6.5 Mathematics of Solutions: Concentration Calculations

6.10 Mathematics of Solutions: Calculating pH

Focusing Questions connect students' interest in the case to the chemistry in the chapter. These big-picture questions orient the students and help them identify what to look for as they read.

Learning Objectives start each section and draw students' attention to the key ideas they will encounter. The objectives are written to be straightforward and achievable.

Visuals highlight interesting aspects of forensic evidence and investigations. The layout of each page is designed to heighten the interaction between the written text and the many detailed and accurate figures and photos of chemical reactions, processes, equipment, and molecular models.

Detailed Worked Examples Paired with Practice Problems give students a helpful roadmap for solving. Reviewers have praised the step-by-step solutions, particularly the inclusion of the "simple" (often algebraic) steps left out of many textbooks. A practice problem follows each worked example, giving the students the immediate opportunity to check their understanding.

Flexible Mathematics Sections cover the mathematical background in depth to provide instructors with the freedom to customize mathematical coverage of their course. Set off clearly so they can be incorporated as needed, these math sections are placed so that readers encounter them as the related chemical topics are first discussed. Through the use of conceptual explanations, worked examples, and practice problems, students receive ample explanation and practice on the math topics.

In the Lab boxes illustrate the key investigative tools in chemistry, so students can see how chemists use modern equipment, laboratory techniques, and analytical methods to examine samples of interest. Considering the student audience, I do not go into too much depth in these boxes; rather, I show how certain procedures are tied to specific uses.

Scanning Electron Microscopy

The scanning electron microscope (SEM) functions like a traditional microscope except that the SEM uses electrons bouncing off a surface to form an image. The traditional microscope depends on photons of light bouncing off a surface. The main advantage to using electrons is that they are much smaller particles than photons of light, and therefore, they make it possible to obtain an image of much smaller particles. Scientists use scanning electron microscopes to obtained detailed images for such items as computer chips, insects, parasites, bacteria, viruses, and pollen. Figure 3.19 shows an image of the pollen from the passion flower (*Passiflora caerulea*), and Figure 3.20 shows a mixture of red and white blood cells.

In addition to creating amazingly detailed images of such small objects, it is possible to determine the elemental analysis of the particles at the same time. The SEM can image the particle using an instrument called an energy dispersive X-ray spectrometer (EDS), which is attached to the SEM. The combined analytical system is abbreviated SEM-EDS. To understand how the EDS system detects the presence of elements, Figure 3.21, showing a simplified Bohr model of an atom, can be of help. Although we know that electrons are not actually in orbits that resemble those of the planets, this simplified model, proposed by Niels Bohr, is a convenient device for demonstrating how the EDS system works.

In part (a), an electron from the SEM strikes an atom. Because the electron is so small, it can penetrate the atom and collide with enough energy to force an inner-core electron to leave the atom. Part (b) shows the atom with a vacancy in the inner core electrons. This condition is not stable, as electrons always go to the lowest possible energy level. Part (c) shows an electron from one of the outer shells, which has a higher energy, dropping down to fill the vacancy in the lower energy orbital and releasing excess energy as a photon. The photons emitted from atoms in this process are X-rays. The process involves the same basic concepts studied

FIGURE 3.19 SEM image of the pollen from the passion flower (*Passiflora caerulea*). (© Steve Gschmeissner/Science Photo Library/Corbis)

FIGURE 3.20 SEM image of red and white blood cells. (© Visuals Unlimited/Corbis)

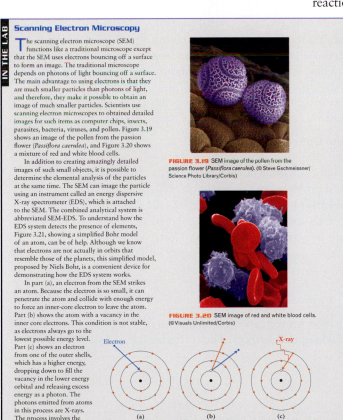

FIGURE 3.21 Basic principle of an energy-dispersive X-ray spectrometer (EDS).

Case Study Finales tie the chemistry from the chapter directly to the opening case study that piqued their interest. The finales provide an excellent opportunity to recap the major points of the chapter.

End-of-Chapter Summaries with "Takeaway Graphics" help students review the major concepts introduced in each chapter. Because chemistry examines many properties that are visual, students benefit from having a core concept from each chapter reinforced through a reprise of its related visual.

CHAPTER SUMMARY

• A solution is made up of a solvent (the major constituent) and a solute (the minor component). An aqueous solution specifically refers to the use of water as the solvent.

• Electrolytes are compounds that, when dissolved in water, can conduct electricity since they dissociate into component ions that are free to move between oppositely charged electrodes. Nonelectrolytes are compounds that will not conduct electricity because they do not dissociate into ions upon dissolving.

• Compounds that fully dissociate into ions in solution are called strong electrolytes; those that partially dissociate are called weak electrolytes.

• Saturated solutions contain the maximum concentration of dissolved solute that the solution can hold at a given temperature. Solutions with less than this concentration are termed unsaturated, and those unstable solutions containing more dissolved solute are termed supersaturated.

Making More Connections lists at the end of the chapter provide references for further research.

End-of-Chapter Questions and Problems are set up to provide maximum flexibility for instructor assignments and student practice.

- *Paired Chapter Review Problems* include answers for the odd-numbered questions, each followed by a related question for which the answer is not provided.

- *Case Study Problems* examine other situations and scenarios that apply the concepts presented in the case studies.

- *Clearly Labeled Section Numbers* are provided in parentheses following each problem, for efficient assignment. Other designations are "CP" for comprehensive problems that span concepts across chapters, and "ITL" for problems associated with the analysis of evidence in the optional In the Lab boxes.

Student Ancillary Support

Supplemental learning materials allow students to interact with concepts several times in a variety of scenarios. By reflecting on or exploring figures, reinforcing problem-solving methods, and drilling chapter objectives, students can obtain a practical understanding of the core concepts. With that in mind, W. H. Freeman has developed the most comprehensive student learning package available.

Printed Resources

Student Solutions Manual by Jason Powell, Ferrum College
ISBN: 1-4641-2519-8

The *Student Solutions Manual* includes worked-out solutions with detailed explanations to selected odd-numbered end-of-chapter questions and problems in the textbook. The walkthroughs are designed to provide a model for problem-solving techniques and teach students to avoid common pitfalls. The manual also includes integrated media icons, which point to selected problem-solving assets that can be accessed via the Book Companion Web site, the interactive eBook, and ChemPortal.

Free Media Resources

Book Companion Web Site

The *Exploring Chemistry* Book Companion Web site, www.whfreeman.com/johllexploring1e, provides a range of tools for problem solving and conceptual support. They include:

- Student self-quizzes

- Flashcards

- Chemistry demonstrations

- Access to select Multimedia Resources, which can be purchased for a nominal fee.

Premium Media Resources

The Premium Media Resources can be purchased directly from the Book Companion Web Site for a small fee, and are also embedded in the multimedia-enhanced eBook and the WebAssign online homework system.

ChemCasts are videos that replicate the face-to-face experience of watching an instructor work a problem. Using a virtual whiteboard, the ChemCasts tutors show students the steps involved in solving key worked examples, while explaining the concept along the way. The worked examples were chosen with the input of chemistry students and instructors across the country. ChemCasts can be viewed online or downloaded to a portable media device, such as an iPod.

Problem-Solving Tutorials are Web tutorials that offer support for skills critical to success in the introductory course. Based on worked examples and end-of-chapter problems from the text, these interactive walk-throughs reinforce the concepts presented in the classroom, providing students with self-paced explanation and practice. Each tutorial consists of the following four components:

1. *Conceptual Explanation:* An easy-to-follow, thorough explanation of the topic.
2. *Worked Example:* A step-by-step walkthrough of the problem-solving technique.
3. *Try It Yourself:* An interactive version of the worked example that prompts students to complete the problem and supplies answer-specific feedback.
4. *Practice Problems:* A series of problems designed to check understanding.

As they utilize the tutorials, students achieve a deeper understating of both the chemistry fundamentals and the problem-solving methods, which they can then apply to other problems and assessment.

Electronic Textbook Options

For students interested in digital textbooks, W. H. Freeman offers the complete text of *Exploring Chemistry*, First Edition, in two easy-to-use formats.

The Multimedia-Enhanced eBook

The Multimedia-Enhanced eBook contains the complete text with a wealth of helpful interactive functions. All student Premium Resources are linked directly from the eBook pages. Students are thus able to access supporting resources when they need them, taking advantage of the "teachable moment" as they read. Customization functions include instructor and student notes, highlighting, document linking, and editing capabilities. Access to the Multimedia-Enhanced eBook can be purchased from the Book Companion Web site.

The CourseSmart eTextbook

Though it does not include any Premium Resources, the CourseSmart eTextbook does provide the full digital text, along with tools to take notes, search, and highlight passages. A free app allows access to CourseSmart eTextbooks on Android and Apple devices, such as the iPad. They can also be downloaded to your computer and accessed without an Internet connection, removing any limitations for students when it comes to reading digital text. The CourseSmart eTextbook can be purchased at www.coursesmart.com.

Instructor Ancillary Support

For instructors using *Exploring Chemistry*, W. H. Freeman provides a complete suite of assessment tools and course materials for the taking.

Computerized Test Bank by Mark Benvenuto, University of Detroit-Mercy

ISBN: 1-4641-2524-4

The Test Bank offers over 2,500 multiple-choice questions, tackling core physics concepts as well as various life-science applications. While the Test Bank is also available in downloadable Word files off the companion Web site, the easy-to-use CD includes Windows and Macintosh versions of the widely used Diploma test generation software, allowing instructors to add, edit, and sequence questions to suit their testing needs.

Electronic Instructor Resources

Instructors can access valuable teaching tools through www.whfreeman.com/johllexploring1e. These password-protected resources are designed to enhance lecture presentations, and they include Textbook Images (available in JPEG and PowerPoint format), Clicker Questions, Lab Videos, Lecture PowerPoints, Instructor Solutions, and more.

Course Management System Cartridges

W. H. Freeman provides seamless integration of resources into your Course Management Systems. Four cartridges are available (Blackboard, WebCT, Desire2Learn, and Angel), and other select cartridges (Moodle, Sakai, etc.) can be produced upon request.

LabPartner Chemistry

W. H. Freeman's latest offering in custom lab manuals provides instructors with a diverse and extensive database of experiments published by W. H. Freeman and Hayden-McNeil Publishing—all in an easy-to-use, searchable online system. With the click of a button, instructors can choose from a variety of traditional and inquiry-based labs. LabPartner Chemistry sorts labs in a number of ways—from topic, title, and author, to page count, estimated completion time, and prerequisite knowledge level. Add content on lab techniques and safety, reorder the labs to fit your syllabus, and include your original experiments with ease. Wrap it all up in an array of bindings, formats, and designs. It's the next step in custom lab publishing—the perfect partner for your course.

LabPartner Chemistry specifically includes labs from *Investigating Chemistry for the Laboratory*—the widely adopted lab manual from David Collins (Brigham Young University–Idaho). These unique motivating labs employ forensic chemistry as their theme, while still focusing on the basics of chemistry and the scientific process.

Online Learning Environments

W. H. Freeman offers an online homework environment built specifically around your textbook.

ChemPortal

W. H. Freeman's ChemPortal is a flexible resource for instructors and students, providing multimedia learning and assessment tools that build on the teaching strategies presented in the textbook. For consistency and convenience, the ChemPortal offers a wide array of resources in one location, with one login. Within ChemPortal, you will find:

- Algorithmically generated problems: Students receive homework problems containing unique values for computation, encouraging them to work out the problems on their own.

- Complete access to the interactive eBook is available from a live table of contents, as well as from relevant problem statements.

- Links to select Multimedia Resources are provided as hints and feedback to ensure a clearer understanding of the problems and the concepts they reinforce.

To learn more, see a demo, or access ChemPortal, visit www.whfreeman.com/johllexploring1e.

CHAPTER 1

Understanding Our World with Chemistry

CASE STUDY: Exploring Our Water Supply

Water is one of the most basic resources needed for life. We largely take it for granted because it is plentiful and cheap. Consider that we use 349 billion gallons of freshwater per day in the United States for personal, agricultural, and industrial use. Approximately 44.2 billion gallons per day are used for drinking water. But what happens to all of that water after it leaves our kitchens and bathrooms?

The vast majority of the water will enter a wastewater treatment facility. The role of the wastewater treatment facility is to remove anything present in the wastewater that may be harmful and then return the treated water to the natural waters. Wastewater entering a facility is first sent through bar screens that remove trash and large debris. The water then enters sedimentation tanks where it is left to sit for several hours. As the water sits, heavy solids suspended in the water sink to the bottom and are removed. Oils and grease, however, will float to the surface where a large rotating skimmer can remove them.

The next stage of water treatment involves bubbling air through the water to add more oxygen to the water as well as to help the bacteria present in the tanks digest the organic material present in the wastewater. The water is then sent into additional tanks where the bacteria are separated out and returned to the previous tank, while the treated water proceeds into a disinfection system. Many water treatment plants use chlorine or powerful ultraviolet lights to sterilize the water of any potentially harmful organisms.

Finally, some water treatment plants use another set of biological treatment tanks that contain additional bacteria to remove phosphorus and nitrogen compounds from the water. These compounds are targeted because they are essential plant nutrients that can cause massive algae blooms if they are not removed. The final step is to discharge the treated water into nearby streams, rivers, lakes, or the ocean.

The treated water is mixed with the natural waters present in the watershed and, in many cases, is taken into drinking water treatment systems downstream. The water that is discharged from a water-treatment center is usually monitored to make sure that the major chemical pollutants have been removed and that it doesn't contain any harmful organisms. It is normally not monitored to see what other compounds it may contain.

As you read through the chapter, consider the following questions:

- **What happens to the chemicals you use to clean your bathroom?**
- **When you shower, what happens to the perfumes, colognes, and deodorants you previously used?**
- **When you are taking medication, does your body consume 100% of the drugs? If not, what happens to the remainder?**

Are there any unusual compounds that make it through the water treatment system, and if so, is there any reason for concern? This is a complex question that can vary from one location to another. There does appear to be reason for concern, though, as shown in one study conducted several years ago in Boulder, Colorado. Upstream from the water treatment facility, the male population of the white sucker fish (*Catostomus commersonii*) ranged from 36–46% with the remainder being female. The gender distribution shifted at the location of the waste-water discharge site with a decrease in the male population to 17–21%, with 18–22% of the fish showing intersex characteristics (both female and male genitals) and the remainder being clearly female.

MAKE THE CONNECTION

What is present in the water that could affect the gender of the wild fish of the river, and how did it get there?

1.1 Welcome

LEARNING OBJECTIVE

Explain how case studies can be used to learn chemistry.

Welcome to *Exploring Chemistry*. The book you are about to use takes a novel approach to teaching chemistry by focusing on how the fundamental principles of chemistry affect your everyday life and the world around you. At the beginning of every chapter, we start with a case study that highlights how chemistry applies to your life, but to fully understand the impact of the case study, you must learn the information presented within the chapter. Case studies help provide the context for how the information within a chapter is used and help increase a student's appreciation and interest in the material during the learning process. At the end of each chapter, we will revisit the case study to help you review the chemical principles and see how the material presented in the textbook really does affect the world around you.

Ocean-going cargo ships arrive daily at major shipping ports from Whittier, Alaska, to Miami, Florida bringing in goods from all regions of the world. (© Michael Wells/fstop/Corbis.)

In the case studies found within this book, you will learn how chemistry plays a role in solving crimes, how it uncovers our past in archeology excavations, how it enables blood to carry oxygen, and how it assists in detecting and combating cancer, to name just a few examples. We encourage you to think about how the chemistry that you are learning relates to your daily lives, your hobbies, and your future careers. Chemistry is everywhere, and the more you learn about it, the more you will appreciate the transforming power of chemistry. We hope you find this new case-study approach to learning chemistry engaging and interesting.

1.2 Chemistry, the Global Society, and You

By the time you reach your classroom today, you will already have had an impact on the global economy. As you dress yourself, the labels on your clothing might read Bangladesh, Honduras, Vietnam, or El Salvador. The dog food you pour into the bowl for your pets comes from China. The cup of coffee you brew originated in Cameroon, and the banana you eat on the way out the door came from Ecuador. You call a friend on your cell phone produced in Japan and make arrangements to meet for lunch. You drive to campus in a car assembled in the United States, but the steel originated in India, the electronics came from South Korea, and the car runs on gasoline refined from oil produced in Saudi Arabia. You meet your friend for lunch, and the lettuce in your salad came from Mexico while the sugar in the lemonade was from Brazil.

> **LEARNING OBJECTIVE**
>
> **Gain an appreciation for the global role of chemistry.**

The global economy has created ties that many people are not aware of. How we choose to spend our money has a real worldwide impact. When an *E. coli* outbreak occurs in Denver, it may trace back to poor food handling in Mexico. When a loved family pet dog dies from renal failure in New York, it could be linked to illegal doping of melamine into pet food in China. Our global economy creates the need for us to understand how we influence the world and how events occurring across the globe will impact our daily lives.

A chemist would look at your morning and see

—complex molecules used to color clothing that retain their intensity over repeated washings.
—dog-food ingredients chosen for their fat, carbohydrate, and protein levels.
—extraction of caffeine from coffee beans using supercritical carbon dioxide.
—proper cooking conditions that kill harmful bacteria found on raw food.
—the alloy process altering the physical properties of the iron metal making up the steel.
—altering the electronic properties of semiconducting elements to create computer chips.
—crude-oil refining based on vaporization of volatile organic compounds.

Your morning was made possible by an international economic and scientific coalition of business interests. Whether your future involves entering some aspect of international business or remaining mainly a consumer, our hope is that you will have a better understanding of your role within the global society after taking this course.

1.3 Matter and Its Forms

By now, you have already started to suspect that chemistry is not confined to a stereotypical laboratory but rather is embedded into every aspect of the world around us. **Chemistry** is defined as the study of the physical material of the universe and the changes it may undergo. This study may occur in a laboratory, a farm field, a manufacturing plant, the upper atmosphere, or even in the far reaches of outer space. The physical material that makes up the universe is referred to as **matter** and can also be defined as any substance that has mass and occupies space. To study chemistry, we must start with the physical forms of matter.

> **LEARNING OBJECTIVE**
>
> **Describe the three states of matter and distinguish elements from compounds.**

States of Matter

The three primary states of matter are the solid, liquid, and gaseous states, as illustrated in Figure 1.1. A **solid** is characterized by the particles that compose the solid being in close proximity to one another. The particles—which can be atoms, molecules, or ions—are held together by attractive forces that keep them in fixed positions. Because of this, solids do not flow, as liquids or gases do. Solids are very difficult to compress because the particles that make them up are already closely packed. The volume and shape of a solid are constant because of the fixed position of its particles and their inability to flow. When a solid is heated, the particles gain energy and the solid becomes a liquid.

Liquids are characterized by particles that are farther apart than those of a solid, but are still fairly close to one another. The greater distance between particles in a liquid allows them to move around freely, a property that is evidenced by the ability of liquids to flow. Liquids cannot easily be compressed because the particles remain in contact even as they move about. The volume of a liquid is constant, but because the liquid can flow, its shape changes to fit the container that holds it. If a liquid is cooled, it will form a solid; if heat is added, it will form a gas.

Gases are characterized by very large distances between particles that are moving at high speeds—approximately 300 meters per second! Gases can be compressed because of the large distances between particles, and gases readily flow because the particles are in continuous motion. Gases will always fill the volume and shape of the container in which they are placed.

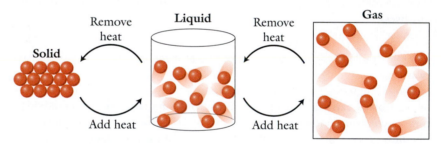

FIGURE 1.1 The physical states of matter are solid, liquid, and gas. The addition or removal of energy changes matter from one state to another. The volume, shape, and compressibility of matter depend on the physical state.

Pure Substances—Elements and Compounds

A **pure substance** is any form of matter that has a uniform composition and cannot be separated by physical methods such as filtration or evaporation into more than one component. For example, sodium chloride cannot be physically separated to form sodium metal and chlorine gas; therefore, it is a pure substance. Salt water is not a pure substance, however, because you can evaporate the water to recover the sodium chloride.

Elements and compounds are two subclasses of pure substances. An **element** is the simplest form of a pure substance. The smallest unit of an element that retains all of the properties of that element is called the **atom**. Each element has unique atoms that are unlike the atoms of any other element. One common example of an element is aluminum, as found in aluminum foil. A less familiar example is sodium, which is often found in clandestine drug labs where methamphetamine is manufactured. Sodium is very dangerous because it reacts violently with water—even water vapor in the air—and must be stored under a layer of oil to keep it from reacting. Gold is an element known since ancient

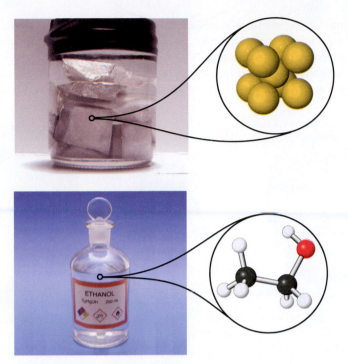

Pure substances can be found as pure elements (such as sodium, top) or as pure compounds (such as ethanol, bottom). (Top photo: Justin Urgitis; bottom photo: Tom Pantages)

times and has been valued for use in jewelry and coins from ancient to modern times. Our fascination with gold is reflected by Hollywood movies that come out almost yearly involving the theft of large amounts of the precious metal.

It is rare to find substances in elemental form because most elements react with other substances to form **compounds**, the second class of pure substances. A compound is a substance that is made up of atoms from two or more elements chemically bonded together.

There are two broad classes of compounds, **molecular compounds** and **ionic compounds**. A **molecule** is the simplest unit of a molecular compound while a **formula unit** is the simplest unit of an ionic compound. The properties of molecular and ionic compounds will be discussed in detail in future chapters, but for now it is important to know that the elements that make up either type of compound cannot be separated by physical methods but only by a chemical reaction.

Examples of compounds are sodium chloride, in which sodium is bonded to chlorine, or ethyl ether, in which carbon, hydrogen, and oxygen are bonded in a chemical combination. Advertisers sometimes make claims that a compound is all natural with the implication that it makes their product safe. However, many poisonous and toxic compounds are found in nature. Throughout history, arsenic compounds were used as rat poisons and pesticides, and the commercial availability of arsenic compounds led to their widespread use by those with murderous intent.

Mixtures

A *mixture* is a combination of two or more pure substances that are physically mixed together but not chemically bonded together. Matter is most commonly found as a mixture of two or more substances, and there are two subclasses of mixtures. A **homogeneous mixture** is one in which the substances that compose it are so evenly distributed that a sample from any part of the mixture will be chemically identical to a sample taken from any other part. A common example of a homogeneous mixture is a solution of sugar in water.

While many homogeneous mixtures are substances dissolved in a liquid, you can find gaseous and solid homogeneous mixtures, too. The air within a breath is an example of a homogeneous mixture of 78% nitrogen, 21% oxygen, and 1% carbon dioxide, helium, argon, and a few other gases. **Alloys** are an example of solid metallic elements forming a homogeneous mixture. Figure 1.2 shows several additional examples of homogeneous solutions.

(a)

(b)

(c)

(d)

FIGURE 1.2 Examples of homogeneous mixtures. (a) The bronze used in the sculpture is a homogeneous mixture of copper and tin metals. (b) Wine is a complex mixture that contains many compounds such as water, ethanol, sugar, and tannins, the molecules that give wine its color. (c) Gasoline is a complex mixture of several hundred compounds refined from crude oil. (d) Whipped cream is a homogeneous mixture of cream and air, both of which are homogeneous mixtures. (a: © Annebicque Bernard/Corbis Sygma; b: Ocean Corbis; c: ©Wave/Corbis; d: © Doable/amanaimages/Corbis.)

A **heterogeneous mixture** is one in which the composition varies from one region of a sample to another. For example, chocolate-chip ice cream is a heterogeneous mixture, as there are regions of many chips and regions of mostly just ice cream. Most mixtures are heterogeneous, as it is difficult to have a perfect homogeneous distribution of the components. Some examples of heterogeneous mixtures are shown in Figure 1.3.

Mixtures can be separated into their components by physical means such as evaporation or filtering. For example, if you wanted to separate salt water into salt and water, simply boil the water off, trap the vapor, and then condense the water vapor back to liquid water in a separate container by cooling it. Boiling works in this case because the solid salt, which is dissolved in the water, is not altered by heating water to its boiling point.

A summary of the many forms of matter can be found in Figure 1.4.

(a)

(b)

(c)

(d)

FIGURE 1.3 Examples of heterogeneous mixtures. (a) Soil contains sand, silt, and clay particles of different minerals and decaying vegetable matter. (b) Ranch dressing is a mixture of soybean oil, water, egg, sugar, salt, and spices. (c) Natural waters contain microorganisms, silt, algae, and aquatic invertebrates. (d) Mining depends on extracting a pure compound or element out of natural geological deposits containing many different mineral types. (a: ©Tetra Images/Corbis; b: Alamy; c: Jim Vecchi/Corbis; d: ©The Irish Image Collection/Design Pics/Corbis.)

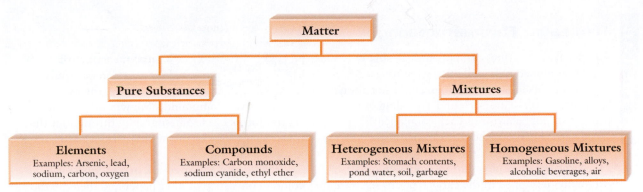

FIGURE 1.4 All matter, the physical material of the universe, can be classified as either a pure substance or a mixture. If found as a pure substance, matter must be in the form of a pure element or a pure compound. If found as a mixture, matter will be either a heterogeneous mixture with differing chemical composition throughout or a chemically uniform homogeneous mixture.

WORKED EXAMPLE 1

Label each of the containers below as either a pure substance or a mixture.

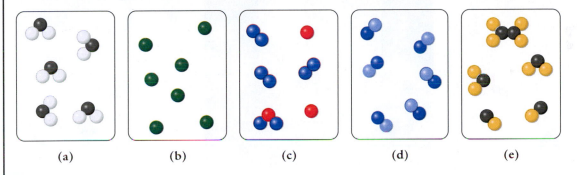

(a)   (b)   (c)   (d)   (e)

SOLUTION

(a) Pure substance   (b) Pure substance   (c) Mixture
(d) Pure substance   (e) Mixture

Practice 1.1

Label each container from Worked Example 1 as containing either elements, compounds, or both.

Answer

(a) Compounds   (b) Elements   (c) Both   (d) Compounds   (e) Compounds

1.4 The Periodic Table

One of the most widely used tools in science is the periodic table, copies of which are displayed on the inside covers of nearly all chemistry textbooks (including this one) and the walls of laboratories and science classrooms throughout the world. The **periodic table** shows all the known chemical elements, arranged in a specific pattern. The importance of the table stems from the fact that it places an extensive amount of information at our fingertips. For example, it allows us to predict chemical reactions, the formulas of many compounds, relative sizes of atoms and molecules, shapes of molecules, and whether compounds will dissolve in water or oil.

> **LEARNING OBJECTIVE**
> Trace the development of the periodic table.

Thin-Layer Chromatography

To make a mixture, components can simply be combined physically. But separating a mixture into its individual components is not always easy. Separation requires an understanding of the properties and behavior of each component.

One of the main methods of separating mixtures in the forensic science laboratory is called **chromatography**. Chromatography was originally developed in the early 1900s to separate the colored pigments in flowers. The name *chromatography* literally translates as "color writing" because the individual colors separate from the mixture. Chromatography exploits the fact that different compounds are attracted to or repelled by other compounds to varying degrees,

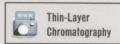
Thin-Layer
Chromatography

and the extent of attraction or repulsion is always the same for any two given compounds. We will talk about several forms of chromatography throughout this book, as it is one of the main methods used to analyze samples.

If you place a drop of ink or a mixture of food coloring on a strip of filter paper and then dip the strip into a liquid such as water or alcohol, allowing the liquid to move upward over the spot, you will see chromatography in action. Most inks consist of a mixture of colored compounds, as does a combination of food coloring. The compounds in the mixture have different levels of attraction for

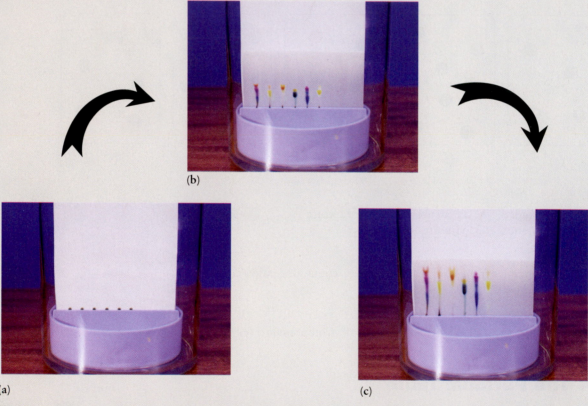

(b)

(a)

(c)

Water-soluble markers are often mixtures of several dyes that produce the desired colors. (a) Ink spots are placed at the bottom of a piece of paper. (b) As the mobile phase moves up the paper, some pigments in ink also move with it. (c) Each pigment moves at a different speed, separating out as it travels up the paper. (a, b, c: Photo Researchers, Inc.)

the paper (called the *stationary phase*) and for the liquid (called the *mobile phase*) passing over them. Chromatography allows the different compounds to be separated, so that you can see different colored spots at different distances from the starting point.

The simplest form of chromatography described above is called *paper chromatography*. In most laboratories, however, a variation of this technique, called **thin-layer chromatography** (**TLC**), is used. As with paper chromatography, there are always two parts to a TLC system, a mobile phase and a stationary phase. In TLC, the mobile phase is a liquid such as water; the stationary phase is a thin solid coating (such as silica gel) on a glass plate. The mixture to be separated is added onto the bottom of the stationary phase. The mobile phase is then allowed to move across the stationary phase. Each compound in the mixture is attracted to the liquid mobile phase and to the stationary solid phase to a different degree than are all of the other compounds in the mixture. If a compound is attracted only to the mobile phase, it will move as fast as the mobile phase. If a compound is attracted only to the stationary phase, it will not move at all. In most cases, a compound will be attracted somewhat to both the mobile phase and the stationary phase and it will move slowly across the stationary phase.

The key to chromatography is that each compound moves at a speed slightly different from all other compounds; this separates each compound in the mixture. We can tentatively identify the compounds present in a mixture by determining if they move at the same speed as a known compound. Thin-layer chromatography can be used to analyze inks, pigments, pharmaceuticals, and illegal drugs.

A very closely related method called *immunochromatography* is the basis for many drug-screening tests and home-pregnancy tests. A multiple-drug screening test typically used in hospital laboratories is shown in Figure 1.5. In this method the sample is drawn into a cartridge and starts to move upward through chromatography paper. The paper contains chemicals that react with the target compounds, producing the colored lines that indicate a positive or negative test. (More details on how this method works will be presented in Chapter 5.) However, it is possible for more than one compound to exhibit the same, or nearly the same, reactivity as the target compounds. This situation creates a false positive result. Confirming the identity of a compound requires further analysis, as explained in more detail in Chapter 9.

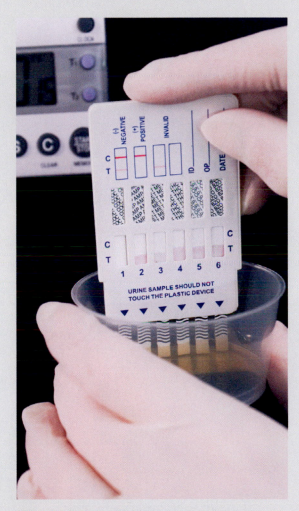

FIGURE 1.5 A multiple-drug screen test can screen for the presence of several illegal drugs at the same time. (Phototake)

The periodic table is so crucial to scientists that an international organization, the International Union of Pure and Applied Chemistry (IUPAC), maintains and updates any changes or additions to it. Although scientists long ago discovered all of the naturally occurring elements, scientists now create new elements that have to be characterized and named.

Development of the Periodic Table

One of the best examples to illustrate the power of the periodic table is the story of its creation in 1869 by Dmitri Mendeleev, a Russian chemist. The necessity to develop some kind of organizing scheme for chemical elements and their properties had become

(a) Mendeleev's original periodic table as published in 1871.

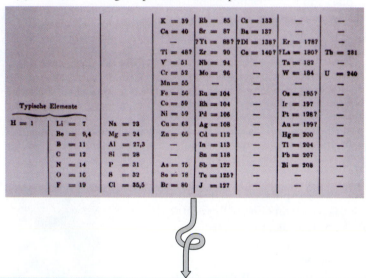

(b) The modern version of the periodic table.

(a: Photo Researchers)

evident around Mendeleev's time because of the success of chemists in discovering new elements. Some elements such as gold, silver, and lead had been known from ancient times, but most elements were unknown until the eighteenth and nineteenth centuries. (Remember that most elements exist in compounds rather than in pure form in nature.) More than 20 new elements were isolated in the 1700s; by the time Mendeleev started working on the periodic table in 1869, the number had grown by another 30, for a total of 63 known elements.

Mendeleev noticed that some elements have very similar reactions and properties. He also noticed that as the mass of the atoms increased, these properties seemed to repeat periodically. Mendeleev organized the elements according to their repeating, periodic properties, grouping into columns elements with similar properties. When the known elements were arranged this way, there were several gaps in the periodic table. Mendeleev boldly stated that the gaps were due to elements that existed but had not yet been discovered.

The first gap in his table is where we now find element number 31, gallium (Ga). Mendeleev called this *eka-aluminum*, which means "similar to aluminum." The second missing element was number 32, which today is called germanium (Ge), but Mendeleev called the undiscovered element *eka-silicon*. Table 1.1 illustrates the power of the periodic table as Mendeleev made predictions about each of the undiscovered elements. In the following chapters, we will discuss atomic mass, density, and formulas, but for now we focus on how closely Mendeleev was able to predict the values for the two undiscovered elements. Both gallium and germanium were discovered and isolated a few years after Mendeleev's predictions.

TABLE 1.1	Mendeleev's Predictions and Actual Values			
Property	Eka-aluminum Predictions	Actual Properties of Gallium	Eka-silicon Predictions	Actual Properties of Germanium
Atomic mass (amu)	About 68	69.7	About 72	72.6
Melting point (°C)	Low	29.8	—	—
Density (g/cm^3)	5.9	5.94	5.5	5.47
Oxide formula	X_2O_3	Ga_2O_3	XO_2	GeO_2
Chloride formula	XCl_3	$GaCl_3$	XCl_4	$GeCl_4$
Discovered (year)		1886		1875

1.5 Learning the Language of Chemistry

Throughout this book, we will be learning how to use the periodic table to determine the chemical formulas of compounds, what the various numbers represent, and how to predict both physical and chemical properties of the elements. However, it is important to start learning the names and symbols of the elements now so that the language of chemistry will be familiar to you when the periodic table is explored in more depth. The periodic table lists all of the elements by their symbols. The corresponding name and spelling of each element can be found in Table 1.2.

LEARNING OBJECTIVE

Interpret and use atomic symbols, the periodic table, and chemical formulas.

TABLE 1.2   Atomic Symbols and Names

Ac	Actinium	Gd	Gadolinium	Po	Polonium
Ag	Silver	Ge	Germanium	Pr	Praseodymium
Al	Aluminum	H	Hydrogen	Pt	Platinum
Am	Americium	He	Helium	Pu	Plutonium
Ar	Argon	Hf	Hafnium	Ra	Radium
As	Arsenic	Hg	Mercury	Rb	Rubidium
At	Astatine	Ho	Holmium	Re	Rhenium
Au	Gold	Hs	Hassium	Rf	Rutherfordium
B	Boron	I	Iodine	Rg	Roentgenium
Ba	Barium	In	Indium	Rh	Rhodium
Be	Beryllium	Ir	Iridium	Rn	Radon
Bh	Bohrium	K	Potassium	Ru	Ruthenium
Bi	Bismuth	Kr	Krypton	S	Sulfur
Bk	Berkelium	La	Lanthanum	Sb	Antimony
Br	Bromine	Li	Lithium	Sc	Scandium
C	Carbon	Lr	Lawrencium	Se	Selenium
Ca	Calcium	Lu	Lutetium	Sg	Seaborgium
Cd	Cadmium	Md	Mendelevium	Si	Silicon
Ce	Cerium	Mg	Magnesium	Sm	Samarium
Cf	Californium	Mn	Manganese	Sn	Tin
Cl	Chlorine	Mo	Molybdenum	Sr	Strontium
Cm	Curium	Mt	Meitnerium	Ta	Tantalum
Cn	Copernicium	N	Nitrogen	Tb	Terbium
Co	Cobalt	Na	Sodium	Tc	Technetium
Cr	Chromium	Nb	Niobium	Te	Tellurium
Cs	Cesium	Nd	Neodymium	Th	Thorium
Cu	Copper	Ne	Neon	Ti	Titanium
Db	Dubnium	Ni	Nickel	Tl	Thallium
Ds	Darmstadtium	No	Nobelium	Tm	Thulium
Dy	Dysprosium	Np	Neptunium	U	Uranium
Er	Erbium	O	Oxygen	V	Vanadium
Es	Einsteinium	Os	Osmium	W	Tungsten
Eu	Europium	P	Phosphorus	Xe	Xenon
F	Fluorine	Pa	Protactinium	Y	Yttrium
Fe	Iron	Pb	Lead	Yb	Ytterbium
Fm	Fermium	Pd	Palladium	Zn	Zinc
Fr	Francium	Pm	Promethium	Zr	Zirconium
Ga	Gallium				

The Periodic Table of Elements

1																	18
1 H 1.0079	2											13	14	15	16	17	2 He 4.003
3 Li 6.941	4 Be 9.012											5 B 10.811	6 C 12.011	7 N 14.007	8 O 15.999	9 F 18.998	10 Ne 20.180
11 Na 22.990	12 Mg 24.305	3	4	5	6	7	8	9	10	11	12	13 Al 26.982	14 Si 28.086	15 P 30.974	16 S 32.065	17 Cl 35.453	18 Ar 39.948
19 K 39.098	20 Ca 40.078	21 Sc 44.956	22 Ti 47.867	23 V 50.942	24 Cr 51.996	25 Mn 54.938	26 Fe 55.845	27 Co 58.933	28 Ni 58.693	29 Cu 63.546	30 Zn 65.409	31 Ga 69.723	32 Ge 72.64	33 As 74.922	34 Se 78.96	35 Br 79.904	36 Kr 83.798
37 Rb 85.468	38 Sr 87.62	39 Y 88.906	40 Zr 91.224	41 Nb 92.906	42 Mo 95.94	43 Tc (98)	44 Ru 101.07	45 Rh 102.906	46 Pd 106.42	47 Ag 107.868	48 Cd 112.411	49 In 114.818	50 Sn 118.710	51 Sb 121.760	52 Te 127.60	53 I 126.905	54 Xe 131.293
55 Cs 132.905	56 Ba 137.327	57 La 138.906	72 Hf 178.49	73 Ta 180.948	74 W 183.84	75 Re 186.207	76 Os 190.23	77 Ir 192.217	78 Pt 195.078	79 Au 196.967	80 Hg 200.59	81 Tl 204.383	82 Pb 207.2	83 Bi 208.980	84 Po (209)	85 At (210)	86 Rn (222)
87 Fr (223)	88 Ra (226)	89 Ac (227)	104 Rf (261)	105 Db (262)	106 Sg (266)	107 Bh (264)	108 Hs (277)	109 Mt (268)	110 Ds (271)	111 Rg (272)	112 Cn (283)						

58 Ce 140.116	59 Pr 140.908	60 Nd 144.24	61 Pm (145)	62 Sm 150.36	63 Eu 151.964	64 Gd 157.25	65 Tb 158.925	66 Dy 162.500	67 Ho 164.930	68 Er 167.259	69 Tm 168.934	70 Yb 173.04	71 Lu 174.967
90 Th 232.038	91 Pa 231.036	92 U 238.029	93 Np (237)	94 Pu (244)	95 Am (243)	96 Cm (247)	97 Bk (247)	98 Cf (251)	99 Es (252)	100 Fm (257)	101 Md (258)	102 No (259)	103 Lr (262)

Most **atomic symbols** of the elements are formed from the first letter of the element's name, and in many cases, either the second or the third letter of the name follows the first letter. For example, the symbol for helium is He, and for manganese, Mn. However, you will notice that some elements have very unusual symbols that do not derive from their contemporary names. Au is the symbol for gold and Na is the symbol for sodium. These symbols come from the Latin names of the elements: *aurum* for gold and *natrium* for sodium.

■ WORKED EXAMPLE 2

What are the correct names for the following elements?

(a) F   (b) Be   (c) Ni

SOLUTION

The elements listed in this example are some of those for which names are commonly misspelled by beginning chemistry students.

(a) Fluorine, commonly misspelled as flourine.

(b) Beryllium, commonly misspelled as berryllium.

(c) Nickel, commonly misspelled as nickle.

Practice 1.2

What are the correct names for the following elements that have their atomic symbols derived from their Latin names?

(a) Au   (b) K   (c) Pb

Answer

(a) Gold

(b) Potassium

(c) Lead

■ WORKED EXAMPLE 3

What are the atomic symbols for the elements listed below?

(a) Titanium and tin

(b) Arsenic and argon

(c) Nitrogen and nickel

SOLUTION

The elements listed above commonly have their atomic symbols confused.

(a) Ti and Sn

(b) As and Ar

(c) N and Ni

Practice 1.3

What are the correct atomic symbols for the following elements, which have their atomic symbols derived from their Latin names?

(a) Tungsten   (b) Copper   (c) Iron

Answer

(a) W   (b) Cu   (c) Fe

Regions of the Periodic Table

The periodic table can be divided into many subregions. It is often necessary to know if an element is a metal, nonmetal, or metalloid because that will influence the way a compound formula is written. Many elements are classified as metals and dominate the middle and left side of the periodic table, as shown in Figure 1.6. All **metals** share the following properties: They are good conductors of electricity and heat. They are solids at room temperature except for mercury, which is a liquid. The **melting point**, the temperature at which most metals melt to form a liquid, is generally very high. Metals are able to conduct electricity because the metal atoms form **metallic bonds** to each other in which the electrons are free to move from one atom to another; this model is called the *sea of electrons*.

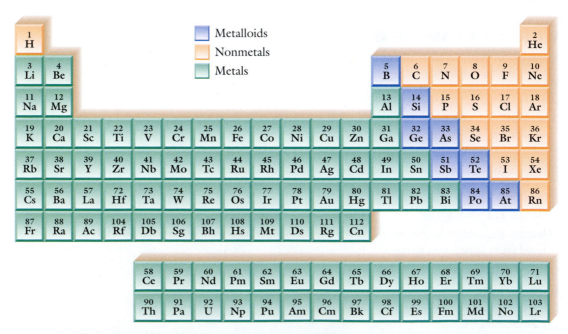

FIGURE 1.6 Regions of the periodic table: metals, metalloids, and nonmetals.

The nonmetals occupy the upper right corner of the periodic table. **Nonmetals** do not conduct electricity or heat very well and can be found as a solid, such as carbon, a liquid, such as bromine, or a gas, such as helium. In general, nonmetals have melting points much lower than metals.

There is a third class of elements known as metalloids. The **metalloids** have properties that are between the two extremes shown by metals and nonmetals. There is not a sharp line that distinguishes a metalloid from a metal or a nonmetal. Rather, there is a gradual change in the physical properties of the elements. Those elements generally considered to be metalloids are B, Si, Ge, As, Sb, Te, Po, and At. The location of metals, metalloids, and nonmetals on the periodic table is shown in Figure 1.6.

■ WORKED EXAMPLE 4

List the names and atomic symbols for five metals that could be found in a student dormitory. Include what object is made from the metal.

SOLUTION

Answers may vary, but some possibilities include:
   Gold, Au: jewelry
   Aluminum, Al: soda can
   Copper, Cu: power cord wires
   Iron, Fe: steel bed frames
   Titanium, Ti: lightweight bicycle parts

Practice 1.4

List the names and atomic symbols for two nonmetals that could be found in a student dormitory. Include what object is made from the nonmetal.

Answer

Answers may vary, but some examples include:
   Carbon, C: pencil graphite
   Neon, Ne: advertising sign
   Oxygen, O: air (as O_2)

Chemical Formulas

Formulas are a chemist's shorthand for showing the elements in a compound as well as the numbers of the atoms of each element that make up the compound. Using a chemical formula instead of writing out the name saves time and simplifies the writing process, provided that you know the symbols of the elements. Chemical formulas make it easier to communicate, in an exact manner, the composition of a substance. Notice that each **subscript** in the formulas refers only to the element directly preceding it. Metals and metalloids are listed before nonmetals when writing formulas.

■ WORKED EXAMPLE 5

The brilliant blue sapphire gemstone is composed of the mineral corundum, Al_2O_3, contaminated with small amounts of titanium and iron, whereas a ruby contains small amounts of chromium. What elements are present in corundum and how many atoms of each are there?

SOLUTION

The two elements in the compound are aluminum (Al) and oxygen (O). The subscript number 2 refers to two atoms of aluminum (Al), and the subscript number 3 refers to three atoms of oxygen (O).

Practice 1.5

How many atoms of each element are present in (a) $C_6H_{12}O_6$ and (b) H_2CO_3?

Answer

(a) Six carbon atoms, twelve hydrogen atoms, and six oxygen atoms

(b) Two hydrogen atoms, one carbon atom, and three oxygen atoms

■ WORKED EXAMPLE 6

Ludwig van Beethoven is believed to have died of lead poisoning. Foul play is not suspected because the dangers of lead and lead compounds were unknown at the time, and he could have been exposed accidentally. Write the formula for the compound lead sulfate that has one lead atom, one sulfur atom, and four oxygen atoms.

SOLUTION

When a compound contains only one atom of an element, we do not use a subscript (the subscript "1" is understood). Therefore, the formula has only a subscript number 4 after the oxygen atom: $PbSO_4$.

Practice 1.6

The compound mercury(II) chloride is extremely poisonous and has one mercury atom and two chlorine atoms. Write the correct formula for the compound.

Answer

$HgCl_2$

1.6 The Most Important Skill of a Scientist: Observation

LEARNING OBJECTIVE

Describe a critical skill scientists must learn.

"In the fields of observation, chance favors only the prepared minds."

—Louis Pasteur

The most valuable skill any scientist has is the ability to make accurate observations and, from those observations, use scientifically sound reasoning to understand what is occurring. Some of the most interesting discoveries occur when a scientist investigates unusual or unexpected results. History has recorded many such discoveries and often credits those discoveries to serendipity—the discovery made by accident or by luck. However, as Louis Pasteur so eloquently stated, "In the fields of observation, chance favors only the prepared minds."

Saccharin is an artificial sweetener that was discovered in 1879 by Constantin Fahlberg and is commonly used today to sweeten many diet products and as Sweet 'N Low ® sweetener. Fahlberg was conducting research on coal tar, a byproduct from processing coal. The discovery came one evening during supper when he noticed a sweet-tasting sensation on his hands, a result of failing to wash his hands after working with chemicals in the laboratory all day. While poor laboratory hygiene is not recommended, Fahlberg followed this observation and created the world's very first sugar substitute.

Another example of an unintended discovery is cisplatin, a chemotherapy drug discovered in 1965 by Barnett Rosenberg. He was examining the effect of electric fields on bacteria and came to determine that the electrode he was using was creating a compound that inhibits cell division. The electrode he was using was made out of platinum, a metal generally considered nonreactive and inert. This unintended discovery has saved countless lives, to the credit of Rosenberg for properly interpreting his unusual results.

Observation skills are critical to success in professions other than the sciences. Medical doctors, nurses, pharmacists, veterinarians, engineers, mechanics, stockbrokers, and teachers are all examples of professions that rely heavily on observation skills. Prepare your mind and be open to new possibilities.

1.7 Critical Thinking and the Scientific Method

The **scientific method** is a systematic process of solving problems driven by the collection and interpretation of data. A common misconception is that only laboratory scientists use the scientific method when, in truth, many people, from mechanics to medical doctors, commonly use it. You can probably think of a problem that you have solved in your own life using the scientific method. It is by making careful observations during an experiment and by interpreting those data using the scientific method that great discoveries are made. The scientific method is outlined stepwise below and graphically depicted in Figure 1.7.

LEARNING OBJECTIVE

Think like a scientist to address a problem.

The Scientific Method

1. Determine the nature of the problem. (Something is interfering with the gender of fish near a wastewater treatment plant.)
2. Collect and analyze all relevant data. (Consider how fish collection was done, age of the fish, etc.)
3. Form an educated guess, called a **hypothesis**, as to what happened. (Chemicals entering the water from residential homes may be altering the natural growth cycle of the fish.)
4. Test the hypothesis. (Collect and analyze water and fish tissue samples for traces of personal-care product chemicals, pharmaceuticals, and cleaning agents.)
5. If your hypothesis holds up to the testing, you are finished. If not, go back to step two.

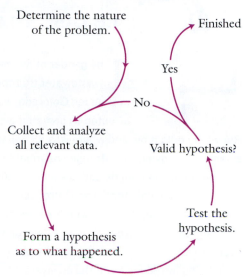

FIGURE 1.7 Graphical Depiction of the Scientific Method. The scientific method has been simplified into the basic components illustrated in the circular process shown here.

WORKED EXAMPLE 7

Discuss the steps a doctor might use to determine what is wrong with a patient who describes a shooting pain that developed in her arm after she fell down a flight of stairs.

SOLUTION

1. Is the arm broken or merely bruised? Hypothesis to test: The arm may be broken.
2. Ask the patient to describe how she landed. Feel the arm and apply slight pressure to it.
3. Consider the patient's reaction to pressure, the details of the accident description, and how the arm physically feels.
4. If the patient's reaction is not indicative of a broken arm, send her home with a pain reliever. If any doubt remains, order an X-ray to confirm a break.
5. Evaluate the X-ray and prescribe appropriate treatment.
6. If the patient returns within days still complaining of pain, repeat the procedure or refer her to a specialist.

A medical doctor collects information from a patient in order to form a hypothesis (diagnosis). Sometimes the hypothesis is verified by the testing (treatment) process. If not, the doctor will consider the new information in the formulation of a new hypothesis. (©Tetra Images/ Tetra Images/Corbis.)

Practice 1.7

The alternator in a vehicle supplies electricity to the car when the engine is running and recharges the car battery. If the alternator is malfunctioning, the car will drain the battery. Describe how a mechanic would use the scientific method to determine whether the problem is the alternator or the battery.

Answer

☐ Answers vary individually. Have a classmate read your answer to see if it seems reasonable.

1.8 CASE STUDY FINALE: Exploring Our Water Supply

The gender of fish near a wastewater treatment facility in Boulder, Colorado, was being affected by something in the water. But what could it be, and where was it coming from? The scientists studying this strange abnormality focused on a group of compounds called *endocrine disruptors,* which are chemicals that, when introduced into a biological system, interfere with the normal hormones that regulate cell reproduction and development. If their hypothesis was correct, endocrine disruptor chemicals were being introduced into the ecosystem through the wastewater treatment plant.

The analysis of the water coming from the wastewater treatment plant showed elevated levels of chemicals that behave like estrogen, the female hormone. When pharmaceutical drugs are ingested, a significant amount goes through the human body unchanged and passes out through the urinary system and into the wastewater treatment plant. The sewage that enters the treatment plant is a heterogeneous mixture of suspended solids in water with many dissolved compounds. As the solids settle out of the liquid phase, a homogeneous mixture of dissolved compounds forms. The homogeneous solution is mostly water, but it still contains compounds from the pharmaceutical drugs along with compounds from perfumes, detergents, soaps, and cleaning products. These compounds are not removed from the wastewater mixture.

The next step of the wastewater treatment plant involves chlorinating the water to sterilize it or,

alternatively, using very bright ultraviolet lights, which also will kill any harmful microorganisms present in the solution. This step does not remove the pharmaceutical drugs or other compounds from personal-care products. The water is then pumped out of the treatment plant into nearby rivers, streams, lakes, or oceans.

The water coming out of the wastewater treatment plant and entering the river in Boulder, Colorado, was collected and analyzed using high-performance liquid chromatography. (This is a more advanced form of chromatography that will be further covered in Chapter 7, but is based on the same principles as the thin-layer chromatography introduced in this chapter.) The analysis of the wastewater sample confirmed the hypothesis of the researchers investigating the gender disruption of the fish. The wastewater sample was a mixture that contained endocrine-disruptor chemicals. The water contained forms of estrogen present in hormone treatment drugs and birth-control pills. The water also contained endocrine-disrupting chemicals used to make polymers, fragrances, detergents, and flame retardants.

The water we use is a limited resource. Water is neither created nor destroyed, and we simply recycle a finite amount of water contained on this planet. Evidence such as this study shows that we have to pay closer attention to the effect our actions are having. Downstream from that wastewater treatment plant is undoubtedly an intake pipe for a drinking water treatment plant.

The contamination of our water system with pharmaceutical compounds is of great concern, and is a problem that you can make a direct impact on. The first step is to reduce pharmaceutical waste when possible by using all antibiotics when prescribed by your doctor. If you are prescribed a new medication, ask your doctor to prescribe just enough pills to determine whether the drug will work and then a number of refills, rather than a large prescription that will go to waste should it not work. The same applies for pet medications.

When shopping for over-the-counter medications, buy only the amount that you will use before the expiration date. Shopping for medications in bulk can be attractive for the price difference; but if any goes unused and requires disposal, the environmental costs can be considerable.

The next step you can take is to dispose of any expired or unused medication properly by contacting a local pharmacist or searching for a nearby hazardous materials drop-off location that accepts pharmaceuticals. Do not store expired or unused medications, as they can be accidentally ingested by children or pets.

Finally, the last and perhaps most important step you can take is to share this information with your friends and family.

THE SECRET TO SUCCESS

The secret to success in a chemistry course is very simple: You must practice solving problems long before an exam is ever given. Success in any aspect of life comes only after repeated practice. Professional athletes paid millions of dollars can be seen running the same practice drills as high school and college athletes. Why? Practice makes perfect! For the same reason, musicians will spend 90% of their time practicing an arrangement for a very few limited public performances. The end of each chapter of this book contains many problems for you to practice so that you will be able to succeed at exam time.

Each chapter of this book starts out with a case study that illustrates how the content of the chapter is related to real-world issues. Because the applications of the content are placed at the beginning of the chapter, you can see the relevancy of the information from the start. However, the end-of-chapter problems start with many examples of problems that are not directly tied into the case study. This was done so that you have the chance to practice a wide variety of problems and master the required skills and calculations. There are additional problems at the end of the problems section that return you to the case study to practice your newly mastered skills. Odd-numbered problems have answers provided at the end of the book so that you can check your work. If you feel that you need additional resources, you can purchase the *Student Solutions Manual*, which has the answers to all odd-numbered problems worked out in full, step-by-step detail. The even-numbered problems provide you with more practice and mirror the odd-numbered problems, but answers are not provided.

CHAPTER SUMMARY

• Chemistry has an impact on our daily lives and the world in which we live. Studying chemistry helps us better understand our role in the global society.

• Matter is the physical material of the universe and can exist in the solid, liquid, or gas states. Matter consists of either pure substances (elements and compounds) or mixtures.

• Elements are characterized by having atoms that are identical. Elements are represented by symbols that are usually derived from the first few letters of their modern or familiar name, although some are derived from Latin names.

• The periodic table is divided into the metals, nonmetals, and metalloids. All elements belonging to a class have similar properties. Metals have the ability to conduct electricity whereas nonmetals do not conduct electricity.

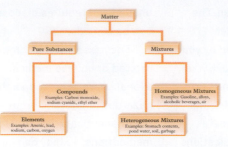

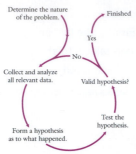

• Compounds are made up of two or more atoms of different elements bonded together. Chemical formulas indicate each element in the compound and the number of each type of atom.

• Mixtures can be classified as heterogeneous or homogeneous. A heterogeneous mixture is one that has a different chemical makeup from one region of a sample to another, whereas a homogeneous mixture is uniform throughout. Mixtures can be separated into their components by physical methods such as filtering, evaporation, or thin-layer chromatography.

• When investigating matter and the changes that it undergoes, scientists use a systematic, logical approach to problem solving called the scientific method. The scientific method consists of collecting data, making a hypothesis, and testing the hypothesis.

KEY TERMS

alloy (p. 8)
atom (p. 6)
atomic symbol (p. 15)
chemistry (p. 5)
chromatography (p. 10)
compound (p. 7)
element (p. 6)
formula (p. 17)
formula unit (p. 7)
gas (p. 6)

heterogeneous mixture (p. 8)
homogeneous mixture (p. 7)
hypothesis (p. 19)
ionic compound (p. 7)
liquid (p. 6)
matter (p. 5)
melting point (p. 16)
metallic bonds (p. 16)
metalloids (p. 17)
metals (p. 16)

molecular compound (p. 7)
molecule (p. 7)
nonmetals (p. 17)
periodic table (p. 9)
pure substance (p. 6)
scientific method (p. 19)
solid (p. 6)
subscript (p. 17)
thin-layer chromatography (p. 11)

MAKING MORE CONNECTIONS: Additional Readings, Resources, and References

For more information on water use in the United States: http://water.usgs.gov/watuse/

To see a real-time water quality map: http://waterwatch.usgs.gov/wqwatch/

For information on contamination and pollution of water: http://water.usgs.gov/owq/topics.html#cont

To find out more on pharmaceuticals in the environment: http://wastenotproject.org/securemedreturn_environmentalbackgrounder_112309.pdf

Vajda, A. M., Barber, L. B., Gray, J. L., Lopez, E. M., Woodling, J. D., and Norris, D. O. "Reproductive disruption in fish downstream from an estrogenic wastewater effluent," *Environmental Science & Technology* 42(9), 2008, pp. 3407–3414.

Woodling, J. D., Lopez, E. M., Maldonado, T. A., Norris, D. O., and Vajda, A. M. "Intersex and other reproductive disruption of fish in wastewater effluent dominated Colorado streams," *Comparative Biochemistry and Physiology Part C* 144, 2006, pp. 10–15.

For an interactive Web site about the case in Problem 46: www.crimelibrary.com/criminal_mind/forensics/mormon_forgeries

REVIEW QUESTIONS AND PROBLEMS

Questions

1. How does chemistry have an impact on your daily life? List several examples. (1.1)
2. What is the difference between compounds and elements? (1.3)
3. Explain how you can distinguish between a heterogeneous mixture and a homogeneous mixture. (1.3)
4. Illustrate the difference between a heterogeneous mixture and a homogeneous mixture by drawing two containers, each containing "●", "▲", and "■" particles. (1.3)
5. How could you determine whether a sample is a pure substance or a homogeneous mixture? (1.3)
6. Give an example of a homogeneous mixture not listed in the textbook. What could you add to the example you provided to make it a heterogeneous mixture? (1.3)
7. Explain how a mixture can be separated into its pure components. (1.3)
8. Why is the periodic table such a valuable tool? (1.4)
9. Why are elements rarely found as pure substances? (1.3)
10. What is the organization that is responsible for maintaining the periodic table? (1.4)
11. Why are some atomic symbols not based on the contemporary name of the element? (1.5)
12. Sketch the periodic table and indicate where metals, metalloids, and nonmetals are located. (1.5)
13. Why is the ability to make observations important to a scientist? (1.6)
14. Explain how the scientific method is used to solve problems. (1.7)
15. Is the failure to prove a hypothesis a failure of the scientific method? (1.7)
16. Why do many different professions use the scientific method? (1.7)

Problems

17. Identify the contents of the container as either a pure substance or a mixture, and determine whether it is made up of elements, compounds, or both. (1.3)

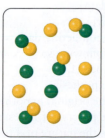

18. Identify the contents of the container as either a pure substance or a mixture, and determine whether it is made up of elements, compounds, or both. (1.3)

19. Identify each of the following substances as either a pure substance or a mixture. (1.3)
    (a) Gasoline
    (b) Air
    (c) Water
    (d) Steel
20. Identify each of the following substances as either a pure substance or a mixture. (1.3)
    (a) Ethanol
    (b) Soda
    (c) Soil
    (d) Brass
21. Identify each of the following substances as either an element or a compound. (1.3)
    (a) Silicon
    (b) Carbon dioxide
    (c) Arsenic
    (d) Water
22. Identify each of the following substances as either an element or a compound. (1.3)
    (a) Sugar
    (b) Carbon
    (c) Lead
    (d) Rust
23. Identify each of the following substances as either a heterogeneous or homogeneous mixture. (1.3)
    (a) Soil
    (b) Air
    (c) Diesel fuel
    (d) Concrete
24. Identify each of the following substances as either a heterogeneous or homogeneous mixture. (1.3)
    (a) River water   (c) Coffee
    (b) Brass alloy   (d) Whiskey

25. What are the names of the elements represented by the atomic symbols listed below? (1.5)
   (a) Mg   (c) P
   (b) Kr   (d) Ge
26. What are the names of the elements represented by the following atomic symbols? (1.5)
   (a) Cr   (c) Cl
   (b) S   (d) Br
27. What are the names of the elements represented by the atomic symbols listed below? (1.5)
   (a) Mn   (c) Cd
   (b) Be   (d) Rb
28. What are the names of the elements represented by the atomic symbols listed below? (1.5)
   (a) Si   (c) Sr
   (b) Ti   (d) Au
29. What are the atomic symbols of the elements listed below? (1.5)
   (a) Neon   (c) Rubidium
   (b) Zinc   (d) Iodine
30. What are the atomic symbols of the elements listed below? (1.5)
   (a) Sodium   (c) Vanadium
   (b) Aluminum   (d) Uranium
31. What are the atomic symbols of the elements listed below? (1.5)
   (a) Barium   (c) Silver
   (b) Cesium   (d) Iridium
32. What are the atomic symbols of the elements listed below? (1.5)
   (a) Zirconium   (c) Bismuth
   (b) Mercury   (d) Palladium
33. Examine each of the elements and symbols listed below and correct any spelling errors or incorrect atomic symbols. If there are no errors, indicate by writing *correct*. (1.5)
   (a) Cadmuim, Cd   (c) Flourine, Fl
   (b) Potasium, P   (d) Zinc, Zn
34. Examine each of the elements and symbols listed below and correct any spelling errors or incorrect atomic symbols. If there are no errors, indicate by writing *correct*. (1.5)
   (a) Sulfer, Su
   (b) Aluminum, Al
   (c) Manganese, Mn
   (d) Boron, Bo
35. Identify each of the following elements as a metal, metalloid, or nonmetal. (1.5)
   (a) Gallium
   (b) Phosphorus
   (c) Boron
   (d) Bismuth
36. Identify each of the following elements as a metal, metalloid, or nonmetal. (1.5)
   (a) Antimony
   (b) Tellurium
   (c) Thorium
   (d) Selenium
37. Identify each of the following elements as a metal, metalloid, or nonmetal. (1.5)
   (a) As   (c) Ge
   (b) C   (d) I
38. Identify each of the following elements as a metal, metalloid, or nonmetal. (1.5)
   (a) Cd   (c) Cl
   (b) Po   (d) Pb
39. Which formula below corresponds to the compound containing one magnesium atom, one sulfur atom, and four oxygen atoms? (1.5)
   (a) MnSO4   (c) $MgSuO_4$
   (b) $MgSO^4$ (d) $MgSO_4$
40. Which formula below corresponds to the compound containing three sodium atoms, one phosphorus atom, and four oxygen atoms? (1.5)
   (a) So_3PO_4 (c) Na^3PO^4
   (b) Na_3PO_4 (d) So3PO4
41. Write the formula for each compound, given the information below. (1.5)
   (a) 1 calcium atom, 2 fluorine atoms
   (b) 2 sodium atoms, 1 sulfur atom, 4 oxygen atoms
   (c) 2 hydrogen atoms, 1 oxygen atom
   (d) 3 magnesium atoms, 2 phosphorus atoms
42. Write the formula for each compound, given the information below. (1.5)
   (a) 1 lithium atom, 1 nitrogen atom, 3 oxygen atoms
   (b) 2 potassium atoms, 1 sulfur atom
   (c) 1 nitrogen atom, 4 hydrogen atoms, 1 chlorine atom
   (d) 2 carbon atoms and 6 hydrogen atoms
43. Write the formula for the mineral hematite, which is made of two iron atoms and three oxygen atoms. (1.5)
44. Which of the following steps is *not* part of the scientific method? (1.7)
   (a) Collect relevant data
   (b) Evaluate data
   (c) Create a hypothesis
   (d) Eliminate conflicting data
45. Which of the following steps is *not* part of the scientific method? (1.7)
   (a) Test the hypothesis
   (b) Validate the original hypothesis
   (c) Repeat the process, if necessary
   (d) Determine the nature of a problem

Case Study Problems

46. A document discovered in 1985, reportedly one of the very first documents printed on a printing press in the North American colonies in 1640, was being sold for $1.5 million. The Oath of a Freeman was a loyalty oath all citizens of the Massachusetts Bay Colony had to take. The age of the paper and ink was verified by carbon-14 dating, but an observant scientist determined it was a fraud. Printing presses of that period were based on arranging each letter individually on the page, inking the surface, and pressing the paper against it. All letters and individual characters were on blocks of exactly the same dimension. By looking at the document, can you determine why it is a forgery? (CP)

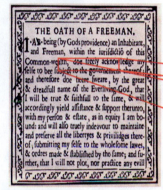

(Utah Lighthouse Ministry)

47. Discuss whether or not Sherlock Holmes is applying the scientific method correctly based on the following quotations: (1.7)
    (a) "We approached the case, you remember, with an absolutely blank mind, which is always an advantage. We had formed no theories. We were simply there to observe and to draw inferences from our observations."—*The Adventure of the Cardboard Box*
    (b) "It is a capital mistake to theorize before you have all the evidence. It biases the judgment."—*A Study in Scarlet*
    (c) "How often have I said to you that when you have eliminated the impossible, whatever remains, however improbable, must be the truth?"—*The Sign of Four*

48. The solid particles that settle out from wastewater treatment tanks, called sludge, are often spread out on farm fields as a fertilizer. What risks might be associated with this practice? Using the scientific method, identify the potential problem and outline how you would approach answering this question.

49. The levels of pharmaceuticals and personal-care product chemicals found in natural waters are typically very low (less than 1 part per billion). Using the scientific method, explain how you would determine whether these chemicals present a risk to aquatic wildlife in the river system.

50. The television shows that feature medical dramas and forensic-science crime teams use the scientific method to solve the medical mystery illness and crimes featured in the story line. Watch one episode of such a program and explain how the problems were solved using science. Make note of any key observations that were needed to solve the problems.

51. Evidence seized in a drug bust was analyzed using thin-layer chromatography (TLC), and the result is shown below. What illegal drug is identified by the TLC? Is this absolute proof that the person was in possession of the illegal drug? Explain why or why not. (ITL)

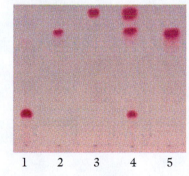

1 Cocaine
2 Methamphetamine
3 Heroin
4 Mixture of 1–3
5 Evidence

CHAPTER 2

Matter: Properties, Changes and Measurements

CASE STUDY: Exploring Evidence from a Crime Scene

In 1985, an assistant United States Attorney released the information that Wilfred "Willie Boy" Johnson had been serving as an FBI informant against John Gotti, the head of the Gambino crime family. This controversial strategy was used in an attempt to force Johnson to testify against Gotti and enter into the federal witness protection program. When this occurred, Johnson immediately feared for his life—and rightly so. He had long served as an informant but had always had the understanding that he would never testify against anyone in the family because betraying the family was a death sentence. He also knew that he didn't have to fear just the Gambino crew, but any member of the New York crime families and others looking to enhance their reputations by killing a perceived traitor.

The only chance Johnson had to escape the wrath of the families was to refuse to testify. By refusing to cooperate with the U.S. attorney, he was able to get a short reprieve from the Gambino family because the federal case against Gotti failed to secure a conviction. However, he had betrayed too many individuals, and ultimately, a contract was placed on his life.

Several members of the Bonanno crime family eventually carried out the murder in 1988. Two members of the family, Vincent "Kojak" Giattino and Thomas

"Tommy Karate" Pitera, took the contract. As they saw it, doing a favor for Gotti, the "boss of bosses," might be of benefit if they should ever need a favor in return. Both would stand trial for the murder of Johnson, but only Giattino was convicted. Pitera was acquitted and seemed to have escaped the justice system, to the dismay of the FBI and federal

prosecutors. It appeared that Pitera would return to his former life—that is, until a small sample of soil would betray Tommy Karate as a mass murderer.

MAKE THE CONNECTION

What information does a bit of soil on a shovel hold that might persuade a jury to convict and sentence a person to life in prison?

> **As you read through the chapter, consider the following questions:**
>
> • What properties of soil and its components could be measured to determine whether two soil samples are consistent with having a common origin?
>
> • How would you prove that a sample of soil is unique to one small geographic area?

2.1 Reactions, Properties, and Changes

LEARNING OBJECTIVE

Distinguish chemical changes and properties from physical changes and properties.

One of the requirements in the collection of physical evidence from a crime scene is to preserve the evidence in its original state. Specially trained police officers usually do the evidence collection, although in some police departments forensic laboratory personnel come to the crime scene. The evidence to be collected is first photographed in place and then collected for processing at a crime laboratory. Each item is packaged separately to prevent cross contamination.

Evidence is collected into clean containers, sealed, and labeled with information specific to the crime scene. If clothing is wet from blood or other liquids, it is collected in a temporary plastic container and repackaged after being air-dried in a controlled atmosphere. (Getty Images/Stockdisc Premium)

Amino acids in the sweat from your fingers react with a compound called ninhydrin to produce a visible fingerprint image. (Mauro Fermariello/Science Photo Library)

Chemical Change and Chemical Properties

When matter undergoes a chemical reaction and forms a new substance, we refer to this process as a **chemical change**. In some instances, it is important to prevent chemical change from taking place to preserve evidence from crime scenes. For example, bloody clothes collected at a crime scene are dried out to prevent the blood from decomposing, a process that involves chemical changes. In other instances, chemical changes are used in a beneficial way for the collection of evidence, as in the development of fingerprints.

Smokeless gunpowder, represented by the simplified formula $C_6H_6O_5(NO_2)_3$, undergoes a chemical reaction and produces gaseous compounds such as carbon dioxide (CO_2), water vapor (H_2O), nitrogen gas (N_2), hydrogen gas (H_2), and carbon monoxide (CO), which propel the bullet down the barrel. The carbon monoxide and hydrogen gas further react (ignite) with oxygen gas from the atmosphere to produce the muzzle flash. (Getty Images/Panoramic Images)

Latent fingerprints are invisible to the naked eye, but when treated with a variety of chemicals, images of the fingerprints become visible.

Chemical properties are used to describe the potential chemical reactions a substance can undergo. Examples of chemical properties include flammability (the potential of a substance to react rapidly with oxygen and burst into flames, releasing light and heat), rusting (the ability of iron to react with oxygen and form iron oxide—rust), and explosiveness (the tendency of compounds such as nitroglycerin to decompose violently, producing a mixture of gases that expand rapidly).

It is important to understand the chemical properties of the evidence being gathered so it can be collected in proper containers. Corrosive compounds such as battery acid are collected in either glass or plastic bottles. If battery acid were placed in a metal container, it would react with the metal, corroding it. Crime scene investigators use proper storage containers, refrigerate and freeze evidence, and separate incompatible types of evidence to preserve the evidence in its original state.

Physical Change and Physical Properties

A **physical change** occurs when matter is transformed in a way that does not alter its chemical identity, such as when matter changes between the solid, liquid, or gaseous states. One example of a physical change in evidence collection involves a sample collected from the suspected area of a fire's origin in an arson investigation. The area will often contain some of the flammable liquid used to start the fire. The liquid sample is collected into an airtight metal canister, but over time, the liquid evaporates into the air space inside the canister. However, the flammable vapor is still the same chemical compound; it has merely changed from the liquid state to the gaseous state.

Physical properties can be measured without altering the chemical identity of a substance. Some common physical properties are the color, melting point, boiling point, odor, refractive index, hardness, texture, solubility, electrical properties, and density of a substance. When laboratory personnel are presented with an unknown compound, they often begin their analysis by determining the melting point. Although the melting point does not specifically identify the compound, it allows all compounds with different melting points to be eliminated. The **refractive index** is a measure of how much light is bent when it passes through a transparent substance such as glass. Density, a property that we will discuss in depth later in this chapter, is a very useful tool in determining the identity of evidence—such as glass fragments from a window smashed by an intruder or from the headlights and rear-view mirrors left by a vehicle that flees a hit-and-run crime scene. Table 2.1 illustrates the physical properties of some common materials.

Water is being vaporized by the heat of the fire to form steam. However, the chemical makeup of water remains the same, H_2O. (Michael Salas/Getty Images)

TABLE 2.1	Physical Properties of Common Materials			
Substance	Appearance	Melting Point (°C)	Boiling Point (°C)	Density (g/mL)
Ethanol	Colorless liquid	−114	78.5	0.798
Ice	Colorless to white solid	0	—	0.917
Corn oil	Yellow viscous liquid	−18 to −10	—	0.920
Water	Colorless liquid	—	100	1.00
Sodium chloride	White, cubic crystal solid	804	—	2.17
Aluminum	Silver-white metal	660	2327	2.70
Iron	Silver to gray metal	1535	3000	7.86
Copper	Reddish metal	1083	2595	8.94
Lead	Silvery-gray metal	327	1740	11.3
Gold	Yellow metal	1065	2700	19.3

Source: The Merck Index, 11th edition, published by Merck & Co., Inc

2.2 Mass, Weight, and Units

LEARNING OBJECTIVE

Distinguish mass from weight and explain the importance of units of measurement.

Because evidence collected at a crime scene may have to be analyzed using several laboratory methods, it is important for investigators to know how large a sample to collect. If an insufficient sample is collected, the laboratory might not be able to completely analyze the evidence. On the other hand, forensic crime laboratories are often backlogged with evidence to analyze, which can create problems for storage if needlessly large samples are collected from crime scenes. Therefore, most crime laboratories publish guidelines for sample size.

The **mass** of a sample is a measure of how much matter is contained in the sample. The unit commonly used for reporting mass is the gram. In the laboratory, mass is measured on a balance, which compares the amount on the pan to a known mass and displays the value. Accuracy in determining the mass of a sample is important for many types of evidence but particularly when a drug arrest is made. The charge that will be filed—"intent to deliver" versus "intent for personal use"—is defined in terms of the mass of drugs recovered.

The mass of a sample is not to be confused with its weight. **Weight** is a measure of how strongly gravity is pulling on matter. If an object is moved to a location where the pull of gravity is lower, the weight of the object decreases. An astronaut on the moon weighs only 1/6 as much as on earth because the moon's gravity is 1/6 of the earth's. However, the astronaut's mass remains the same. The mass of an object does not depend on gravity and, therefore, does not change from one location to another.

Weight is commonly reported in units of pounds and is measured with a scale. A scale has a spring that compresses as the force of gravity acts on an object placed on the scale. A dial attached to the spring indicates the weight.

Collecting sufficient evidence at a crime scene is critical so that the forensic crime lab can run multiple tests on the sample without consuming the entire sample of evidence. (Brand X Pictures/Alamy)

TABLE 2.2	Standard Units of Measurement	
Measurement	Units	Abbreviation
Mass	grams	g
Volume	liters	L
Distance	meters	m
Time	seconds	s

The standard units are used to measure experimental variables to facilitate communication between scientists.

It should be pointed out that in the laboratory we use *only* the metric system for measurements. However, because many other agencies, from police departments to engineering firms, use the older English-based measurement units such as feet and pounds, a scientist should be able to convert from English units to metric units. Appendix A lists the metric–English equivalents for converting between the two sets of units. Table 2.2 summarizes the units used in a laboratory to take measurements.

Units of measurement are often modified to reflect the size of the sample being measured. When police seize a major shipment of drugs, they commonly report how many "kilos" were seized. The term *kilo* is a slang expression for kilogram (kg), which is equal to 1000 grams. It is easier to communicate the size of a large sample by saying "100 kilos" rather than "100,000 grams." On the other hand, when police officers search a vehicle during a routine traffic stop and come across a small quantity of illegal narcotics, they measure how many grams were seized rather than kilos. It is simpler to describe the mass of a small sample as 5 grams rather than 0.005 kilogram. Other units commonly modified with prefixes are milliliters and centimeters. The prefix modifiers often used in the laboratory are summarized in Table 2.3.

The **unit** of a measurement contains critical information about what system of measurement is used and whether the base unit is modified with a prefix. Therefore, it is

Evidence Collection Guidelines for Missouri State Highway Patrol

Liquids: 1 ounce
Drugs: Small seizure, entire sample
Drugs: Large seizure, representative sample
Blood: 2 vials, 10 mL each
Urine: 50 mL
Stomach contents: 50 mL
Hair: 50 head, 25 pubic
Clothing: Dried, packaged separately

TABLE 2.3	Common Metric Prefixes		
Prefix	Symbol	Decimal Equivalent	Exponential Value
tera	T	1,000,000,000,000	10^{12}
giga	G	1,000,000,000	10^{9}
mega	M	1,000,000	10^{6}
kilo	k	1,000	10^{3}
deci	d	0.1	10^{-1}
centi	c	0.01	10^{-2}
milli	m	0.001	10^{-3}
micro	μ	0.000001	10^{-6}
nano	n	0.000000001	10^{-9}
pico	p	0.000000000001	10^{-12}

Prefix modifiers are added to convey large or small quantities in a clear and concise manner.

always important to state the unit along with the number. If a police officer were to write that an amount of cocaine seized in an arrest was "10," the information would be useless. It is very important to know whether the officer meant 10 grams, 10 kilograms, or 10 milligrams.

2.3 Mathematics of Unit Conversions

LEARNING OBJECTIVE
Use conversion factors properly in solving mathematical problems.

Solving problems in chemistry often requires converting from one unit of measurement to another by means of a **conversion factor**. A conversion factor is simply an equality by which a quantity is multiplied to convert the original units of the quantity to the new units that are desired. To use a conversion factor, divide by the units you wish to cancel and multiply by the units you want in the final answer. Because you will be doing conversions throughout this course, it is advisable for you to master this skill now.

■ WORKED EXAMPLE 1

How many meters are in 328 mm?

SOLUTION

The equality 1 mm = 0.001 m provides the information needed for a conversion factor between millimeters and meters.

$$328 \text{ mm} \times \frac{0.001 \text{ m}}{1 \text{ mm}} = 0.328 \text{ m}$$

Notice that when you divide millimeters by millimeters, the units cancel out, leaving the meter as the remaining unit.

Practice 2.1

How many millimeters are in 4.73 cm?

Answer

❏ 47.3 mm

■ WORKED EXAMPLE 2

How many millimeters are in 0.671 m?

SOLUTION

$$0.671 \text{ m} \times \frac{1 \text{ mm}}{0.001 \text{ m}} = 671 \text{ mm}$$

The equality needed to solve this is 1 mm = 0.001 m. But how do we know whether to use 1 mm/0.001 m or 0.001 m/1 mm as the conversion factor? In this case, the question asks us to find *millimeters*. Therefore, we want *meters* to cancel out in the calculation, and we use 1 mm/0.001 m as the conversion factor.

Practice 2.2

How many kilometers are in 218 meters?

Answer

❏ 0.218 km

It is often necessary to use multiple conversion factors to get from the given information to the desired units. For example, how many centimeters are there in 2.000 feet? You probably do not know the conversions for centimeters to feet. However, you know that 1 foot = 12 inches, and from Appendix A you can find that 2.54 centimeters = 1 inch.

■ **WORKED EXAMPLE 3**

How many centimeters are in 5.80 ft.?

SOLUTION

$$5.80 \text{ ft.} \times \frac{12 \text{ in.}}{1 \text{ ft.}} \times \frac{2.54 \text{ cm}}{1 \text{ in.}} = 177 \text{ cm}$$

Practice 2.3

How many centimeters are in 0.829 yd.?

Answer

❑ 75.8 cm

■ **WORKED EXAMPLE 4**

A bar of gold has a mass of approximately 14.1 kg. Calculate the mass in pounds, given the conversion factor 1 lb = 454 g.

SOLUTION

$$14.1 \text{ kg} \times \frac{1000 \text{ g}}{1 \text{ kg}} \times \frac{1 \text{ lb}}{454 \text{ g}} = 31.1 \text{ lb}$$

Practice 2.4

If the average adult has a weight of 154 lb, calculate the mass of the average adult in kg.

Answer

❑ 69.9 kg

You are likely to encounter several more conversion problem formats, such as the conversion of an area (distance^2) or volume (distance^3). For example, if the volume of a container is 0.005 m^3, what is the volume expressed in cm^3? To answer this, we start with the conversion 1 cm = 0.01 m. Then the units *and* the corresponding values are raised to the desired power, 1^3 cm^3 = 0.01^3 m^3, which is simplified to 1 cm^3 = 0.000001 m^3.

$$0.004 \text{ m}^3 \times \frac{1 \text{ cm}^3}{0.000001 \text{ m}^3} = 4000 \text{ cm}^3$$

The final conversion format corresponds to the conversion of rates (distance/time). Recall that English–metric conversions can be found in Appendix A. In these problems, care must be taken to arrange the conversion factors so that the unit being canceled is always placed in the appropriate location. For example, note the arrangement in this conversion of 65 miles per hour to kilometers per minute:

$$\frac{65 \text{ mi.}}{\text{hr}} \times \frac{1.61 \text{ km}}{1 \text{ mi.}} \times \frac{1 \text{ hr}}{60 \text{ min}} = 1.7 \text{ km/min}$$

■ WORKED EXAMPLE 5

Calculate the volume in liters of a 3.5 m³ container, given that 1 L = 1000 cm³.

SOLUTION

$$3.5 \ \cancel{m^3} \times \frac{1 \ \cancel{cm^3}}{(0.01)^3 \ \cancel{m^3}} \times \frac{1 \ L}{1000 \ \cancel{cm^3}} = 3500 \ L$$

Practice 2.5

Calculate the area in m² of a room that is 18.0 yd.²

Answer

❏ 15.1 m²

■ WORKED EXAMPLE 6

If a car is going 31.0 km/hr, calculate the speed in ft./s, given that 5280 ft. = 1 mi.

SOLUTION

$$\frac{31.0 \ km}{hr} \times \frac{0.621 \ mi.}{km} \times \frac{5280 \ ft.}{1 \ mi.} \times \frac{1 \ hr}{60 \ min} \times \frac{1 \ min}{60 \ s} = 34.0 \ \frac{ft.}{s}$$

Practice 2.6

A vehicle is rated to get 26 mpg (miles per gallon). Convert this to km/L.

Answer

❏ 11 km/L

2.4 Errors and Estimates in Laboratory Measurements: Significant Figures

LEARNING OBJECTIVE

Identify significant figures in measurements and know how to determine them.

In making measurements for scientific purposes, we intentionally include one number that is an estimate in every measurement. It seems odd, but by including one digit that is an estimate (which means it may contain error), the measurement is actually more accurate than if we used only the digits that are exactly known. Consider the shoeprint shown in Figure 2.1. How would an investigator report the width of the print? It is greater than 11 cm but less than 12 cm. Depending on the person, estimates of the width might range from 11.4 cm to 11.6 cm, all of those estimates are acceptable. Because the last decimal place is an estimate, it will vary from one person's observation to another's. The rule in the laboratory is that we keep all digits that are known exactly, plus one digit that is an estimate and contains some error. Collectively, these digits are called **significant figures**.

■ WORKED EXAMPLE 7

A buret is a piece of scientific glassware used to accurately measure out solutions during experiments. What is the volume reading on the buret shown to the left? (*Hint:* The solution forms a concave shape, called a *meniscus*. Read the volume at the lowest point of the meniscus.)

SOLUTION

The volume reading of the buret is more than 24.0 mL but less than 25.0 mL, so an estimate of 24.2 mL would be reasonable.

FIGURE 2.1 When making measurements, it is necessary to include an estimated digit in the answer to obtain the most accurate value. When taking a photograph of evidence, a ruler is always included for later reference. It is critical that the camera be at a 90-degree angle to the ruler. Otherwise, the readings will be distorted. (Alamy)

Practice 2.7

Using a metric ruler, measure the height of a can of soda in units of centimeters.

Answer

12.2 cm

It is often necessary to use measured quantities in calculations. When this is done, it is necessary to know how many significant figures are in each number involved in the calculation, as the answer must reflect the proper number of significant figures. There are guidelines that are helpful in telling which figures are significant.

Zero and Nonzero Numbers in Measurements

The first rule for determining significant figures is that all nonzero numbers in a measurement are significant. For example, the number 459.61 cm contains five significant figures.

The rules for determining significant figures get more complex when there is a zero in the number. The rules for zeros are summarized below.

Zero Rules

1. Zeros located between nonzero numbers are significant.
   Example: 101 has three significant figures.

2. Zeros at the beginning of a number containing a decimal point are *not* significant.
   Example: 0.015 has two significant figures.

3. Zeros at the end of a number containing a decimal point are significant.
   Example: 25.20 has four significant figures.

4. Zeros at the end of a number not containing a decimal point are ambiguous.
   Example: 100 contains at least one significant figure but could contain up to three significant figures, depending on whether the estimate is in the tens or the ones place.

How should the number 100 be written to indicate the correct number of significant figures? The solution is to use **scientific notation** because only those zeros that are considered significant are included in the number, as shown below. There are some older methods, such as underlining the significant zeros or placing a decimal at the end to indicate significant digits. These antiquated methods are *not* used in modern laboratories and are not considered proper methods for indicating significant figures.

Scientific Notation and Significant Figures

100 written with 1 significant digit is 1×10^2

100 written with 2 significant digits is 1.0×10^2

100 written with 3 significant digits is 1.00×10^2

The procedure for writing numbers in scientific notation is:

1. Count the number of places that the decimal point will have to move to get one nonzero digit to the left side of the decimal place.

2. The number of places you moved the decimal point is the number used as the exponent.

    (a) If you moved the decimal to the right, you make the exponent a negative value.

    (b) If you moved the decimal to the left, you make the exponent a positive value.

■ WORKED EXAMPLE 8

Write the number 12,000 with two significant figures.

SOLUTION

1. Count the number of places the decimal point is shifted to the left until you have one nonzero digit to the left of the decimal point:

$$12,000$$
$$\leftarrow \leftarrow\leftarrow\leftarrow$$

2. Write the number in scientific notation with the exponent equal to the number of shifts. The exponent is positive because you shifted to the left: 1.2×10^4.

Practice 2.8

Write the number 25,000 using three significant figures.

Answer

☐ 2.50×10^4

■ WORKED EXAMPLE 9

Write the number 0.000120 in scientific notation. *Note:* This number contains three significant digits with no ambiguity. However, we commonly write very small or very large numbers in scientific notation form.

SOLUTION

1. Count the number of places the decimal point is shifted to the right until you have one nonzero digit to the left of the decimal point:

$$0.000120$$
$$\rightarrow\rightarrow\rightarrow\rightarrow$$

2. Write the number in scientific notation with the exponent equal to the number of shifts. The exponent is negative because the decimal point was shifted to the right: 1.20×10^{-4}.

Practice 2.9

How would you write the number 0.05370 using scientific notation?

Answer

5.37×10^{-2}

2.5 Mathematics of Significant Figure Calculations

Some measurements used in a calculation may have more significant figures than others. In such a case, how do we determine the number of significant figures the answer should have? The last digit, which contains the estimate in a measurement, affects any number by which it is multiplied, divided, added to, or subtracted from. The rules for counting significant figures in mathematical operations are listed below. The digit containing the estimate is colored in blue, and any number it affects is also colored blue to help you visualize why the rules apply as they do.

Rules for Mathematical Operations of Significant Figures

1. Addition and Subtraction: The answer can have only as many decimal places as the number with the fewest digits after the decimal point.

2. Multiplication and Division: The answer can have only as many significant figures as the number with the fewest significant figures.

When writing a number with the correct number of significant figures from a mathematical problem, it will often need to be rounded up or down according to the rules below.

Rules of Rounding

1. If the number being dropped is greater than 5, increase the last saved digit by 1.

2. If the number being dropped is less than 5, leave the last saved digit as it is.

3. If the number being dropped is exactly 5, flip a coin.*

*This rule changes from one source to another, as there is not a standard procedure for this situation. A truly random coin flip prevents a bias from being introduced.

WORKED EXAMPLE 10

What is the answer to 145.056 + 7.01 + 22.0261?

SOLUTION

```
  145.056
    7.01   ← fewest digits after decimal point
+  22.0261
 ─────────
  174.0921
```

The answer is 174.09 since there are two decimal places in 7.01.

Practice 2.10

What is the answer to 61.83 − 59.241?

Answer

The answer is 2.58. Notice that there are three significant digits even though the 61.83 has four significant digits and 59.241 has five significant digits. The answer is limited by the number with the fewest digits after the decimal point, not the total number of significant digits.

■ WORKED EXAMPLE 11

What is the answer to: 10.31×2.5?

SOLUTION

$$
\begin{array}{r}
10.31 \\
\times \quad 2.5 \leftarrow \text{fewest significant digits} \\
\hline
5155 \\
+ \quad 2062 \\
\hline
25.775
\end{array}
$$

The answer is 26 because there are two significant digits in 2.5.

Practice 2.11

What is the answer to 0.03010/0.20?

Answer

0.15

There are some problems that require multiplication or division and addition or subtraction in the same problem. For these mixed operation problems, always follow the rules for order of operations giving priority to any mathematical steps in parentheses first, followed by multiplication/division, and finally addition/subtraction. When a problem includes multiple steps, *never* round a number until the calculation is completely finished. To determine the correct number of significant digits in a number, examine each step in the order completed and use the rules appropriately for each step.

■ WORKED EXAMPLE 12

What is the answer to $\dfrac{(14.01 - 1.025)}{0.0120}$?

SOLUTION

Answer is significant to two decimal places   Answer contains five numbers, but only four are significant digits

$$
\frac{(14.01 - 1.025)}{0.0120} = \frac{12.985}{0.0120} = 1082.083333 \Rightarrow 1.08 \times 10^3
$$

Contains three significant digits   Calculator would provide this number   Answer rounded to three significant digits

Practice 2.12

What is the answer to the problem $\dfrac{(2.160 + 20.0)}{0.0030}$?

Answer

7.5×10^3

2.6 Experimental Results: Accuracy and Precision

LEARNING OBJECTIVE

Distinguish between accuracy and precision in experimental results.

The forensic scientists who analyze evidence are often called to court to explain the results of their analysis. There are two aspects of the results generated by the scientists that are critical for both the prosecution and defense to understand.

Accuracy

The first aspect of importance is how close the experimental results are to the true or real value for the quantity being measured. This represents the **accuracy** of the results. For example, if a cocaine sample were 50.0% pure and the lab reported back a value of 50.1%, the analysis would be considered accurate because the experimental value was close to the true value. If the lab had reported a value of 45%, the results would be inaccurate. How is the true or real value of a sample determined? Many times, it is not possible to know the exact value. However, there are analytical methods and advanced statistics that enable scientists to predict how close the results are to the real or true value.

Precision

Another important aspect of the experimental results is the **precision** of a measurement. Precision refers to how reproducible a measurement is if the same sample is measured multiple times. From the cocaine example, if the scientists reported that the cocaine sample was analyzed three times and the results were 45.0%, 44.9%, and 45.2%, those results are numerically close and the measurement would be considered precise. However, because the measurements are not close to the true value of 50%, they are said to be inaccurate. For reported results to be considered valid, measurements must be both precise and accurate. This would be the case if the cocaine analysis gave results of 49.9%, 50.0%, and 50.2%. The concepts of accuracy and precision are also illustrated in Figure 2.2.

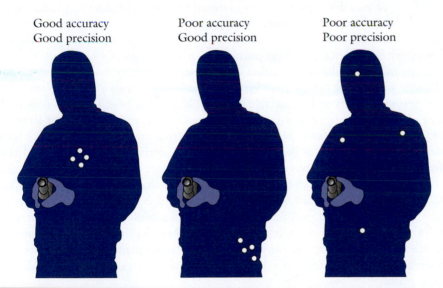

Good accuracy
Good precision

Poor accuracy
Good precision

Poor accuracy
Poor precision

FIGURE 2.2 Law enforcement officers must train using human-shaped targets because the visual difference between a traditional bull's-eye target and a human target can affect both their accuracy and precision.

◼ WORKED EXAMPLE 13

Label each of the sets of data below as accurate, precise, inaccurate, or imprecise, as appropriate. In each case, the true value of the measurement is 15.27.

(a) 14.02, 12.56, 17.22

(b) 15.24, 15.21, 15.28

(c) 14.34, 14.21, 14.29

SOLUTION

(a) Scattered data = imprecise. Not close to true value = inaccurate.

(b) Close to true value = accurate.

(c) Not close to true value = inaccurate.

Practice 2.13

When forensic chemists testify at trial, it is quite common for the defense attorney to question them about the procedures, methods, and chemicals used to calibrate the instrument used for the analysis. Explain how this line of inquiry is used to question the accuracy of the results.

Answer

The calibration of an instrument ensures that the instrument is providing accurate results. If the instrument were calibrated incorrectly, then all of the results would be inaccurate.

2.7 Density Measurements

> **LEARNING OBJECTIVE**
>
> **Use density as another physical property to investigate chemical evidence.**

Earlier in this chapter, we introduced some physical properties of matter—such as melting point, boiling point, and density—that can aid in the identification of unknown substances collected as evidence. Physical properties of the evidence are compared with physical properties of a series of known materials. This comparison allows investigators to eliminate from the list of possible substances any material that does not match. A tentative identification of the evidence can be made if its physical properties match those of a known substance. More extensive experiments need to be done to positively identify the material, but the job is much easier if the possibilities have been greatly narrowed.

One of the physical properties that can be used to identify evidence is density, defined as the ratio of the mass of an object to its volume. Density is often confused with weight or mass. This is understandable because objects that have a high density, such as cement or iron, are those that are often described as "heavy." Objects that have a low density, such as foam or feathers, are typically ones that are considered "light." We can therefore think of density as the "heaviness" of a material, but density should not be confused with either the mass or the weight of an object. The density of water can be determined by measuring both the mass and volume of the water sample. If the mass of a 1.00-mL sample of water is 1.00 gram, the density is 1.00 g/mL. Any object that has a density less than 1.00 g/mL will float; any object with a density greater than water's will sink. Table 2.1 lists some common materials and their densities. Glass evidence lends itself to analysis by the examination of density, as there are many different types of glass, each with a different density. Listed in Table 2.4 are some physical properties of various types of glass.

Both obsidian and pumice are volcanic rocks created from cooling lava. The pumice (*D* = 0.60 g/ml) floats in water (*D* = 1.00 g/mL) as it is less dense than water. However, obsidian (*D* = 2.60 g/mL) is denser than water, and therefore sinks. (Sheila Terry/Science Photo Library)

■ WORKED EXAMPLE 14

If the density of a glass sample is determined to be 2.26 ± 0.02 g/mL, what are the possible types of glass the sample could be? The ± sign signifies that any glass that has a density within 0.02 units of 2.26 g/mL would be a possibility.

SOLUTION

The range of acceptable answers is 2.24 to 2.28 g/mL. The following glass samples all fall within this range: borosilicate, alkali barium borosilicate, soda borosilicate, and alkali strontium.

Practice 2.14

If the refractive index for the same glass sample from Worked Example 14 is determined to be 1.48 ± 0.01, can you further reduce the number of possible types of glass? If so, what are the remaining types?

Answer

Borosilicate, alkali barium borosilicate, or soda borosilicate

TABLE 2.4 Physical Properties of Glass			
Type	Softening Point[1] (°C)	Density (g/mL)	Refractive Index[2]
Alkali barium	646	2.64	1.511
Alkali barium (optical)	647	2.60	1.512
Alkali barium borosilicate	712	2.27	1.484
Alkali borosilicate	718	2.29	1.486
Alkali strontium	688	2.26	1.519
Alkali zinc borosilicate	720	2.57	1.523
Borosilicate	720	2.28	1.490
Borosilicate (Pyrex®)	821	2.23	1.473
Lead borosilicate	447	5.46	1.860
Potash borosilicate	820	2.16	1.465
Potash soda lead	630	3.05	1.560
Soda borosilicate	808	2.27	1.476
Soda-lime	696	2.47	1.510

[1]The softening point is the temperature at which heated glass starts to deform under its own weight.
[2]The refractive index of all samples is measured at a wavelength of 589.3 nm.

2.8 Mathematics of Density Measurements

Take a look at the equations below and on the next page. Equation 1 can be used to determine the density of an object, provided that the mass and volume of the object are known. Furthermore, it is possible to rearrange equation 1 so that either mass or volume can be calculated from the data. This is shown in equations 2 and 3.

LEARNING OBJECTIVE
Reinforce the use of density measurement as a method of evidence analysis.

$$\text{Density} = \frac{\text{mass}}{\text{volume}} \text{ or } D = \frac{m}{V} \tag{1}$$

Solve for volume

Step 1: Multiply both sides by V
$$D \times V = \frac{m}{V} \times V$$

Step 2: Divide both sides by D
$$\frac{D \times V}{D} = \frac{m}{D}$$

Step 3: Final equation:
$$V = \frac{m}{D} \tag{2}$$

Solve for mass

Step 1: Multiply both sides by V

$$D \times V = \frac{m}{\cancel{V}} \times \cancel{V}$$

Step 2: Final equation:

$$D \times V = m \quad \text{or} \quad m = D \times V \tag{3}$$

The units of density are most commonly g/mL, g/cc, or g/cm^3 ("cc" is an older abbreviation for a cubic centimeter, cm^3). Because 1 mL = 1 cc = 1 cm^3, all of the previously listed density units are interchangeable. You will occasionally hear the term *cc* instead of *milliliter* on a TV medical drama, and *cc* is also commonly used in engineering and material science literature. Therefore, you should be aware of its usage and meaning.

■ **WORKED EXAMPLE 15**

A shattered glass jar was found at a burglary scene and was determined to be made of soda-lime glass. A glass fragment recovered from a suspect's home had a mass of 4.652 g. When the glass fragment was placed into a graduated cylinder with 20.00 mL of water, the level of the water rose to 21.53 mL. Does the glass fragment link the suspect to the crime scene? Consult Table 2.4.

SOLUTION

$$D = \frac{m}{V} = \frac{4.652 \text{ g}}{(21.53 - 20.00)} = \frac{4.652 \text{ g}}{1.53 \text{ mL}} = 3.04 \text{ g/mL}$$

No, the glass samples do not match. Soda-lime glass has a density of 2.47 g/mL. The fragment recovered is most likely potash soda lead glass, which has a density of 3.05 g/mL.

Practice 2.15

Determine the identity of a 12.471-g sample of glass that displaces 2.33 mL of water.

Answer

❏ D = 5.35 g/mL, which is closest to lead borosilicate glass.

■ **WORKED EXAMPLE 16**

The typical density range for urine samples recovered from an autopsy is 1.002 to 1.028 g/cm^3. Density values higher than this range can indicate that the victim suffered from vomiting, excessive sweating, or dehydration shortly before death. Density values lower than this range can indicate that the victim was diabetic or had acute renal failure due to exposure to certain metals or solvents. Calculate the acceptable mass range that a 25.00-mL urine sample could have and still be considered normal.

SOLUTION

Two calculations are necessary. The first is based on the low density value of 1.002 g/cm^3, and the second is based on the high density value of 1.028 g/cm^3.

$$\overbrace{1 \text{ mL} = 1 \text{ cm}^3}$$

$$\text{Low density: } m = D \times V \Rightarrow m = 1.002\frac{\text{g}}{\text{cm}^3} \times 25.00 \text{ } \cancel{\text{cm}^3} = 25.05 \text{ g}$$

$$\text{High density: } m = D \times V \Rightarrow m = 1.028\frac{\text{g}}{\text{cm}^3} \times 25.00 \text{ } \cancel{\text{cm}^3} = 25.70 \text{ g}$$

Practice 2.16

Calculate the mass of 75.00 mL of urine with a density of 1.018 g/cm^3.

Answer

☐ 76.35 g

■ WORKED EXAMPLE 17

Cerebral edema is swelling of the brain tissue caused by the absorption of water and by increased water content in the brain cavity, which can be caused by a blunt force trauma. The average human brain is 1.30 × 10^3 g, and the density of the normal human brain tissue is 1.05 g/mL. Calculate the initial volume the brain tissue occupies and then the volume it would occupy after absorption of water decreases the brain tissue density to 1.01 g/mL.

SOLUTION

$$\text{Initial: } V = \frac{m}{D} \Rightarrow V = \frac{1.30 \times 10^3 \cancel{g}}{1.05 \cancel{g}/\text{mL}} = 1.24 \times 10^3 \text{ mL}$$

$$\text{Final: } V = \frac{m}{D} \Rightarrow V = \frac{1.30 \times 10^3 \cancel{g}}{1.01 \cancel{g}/\text{mL}} = 1.29 \times 10^3 \text{ mL}$$

Practice 2.17

The density of PEK-1 plastic explosive is 1.45 g/cc. A single cartridge of PEK-1 has a mass of 1.00 × 10^2 g. What is the volume of the PEK-1 cartridge?

Answer

☐ 69.0 cc

2.9 How to Analyze Glass and Soil: Using Physical Properties

Glass fragments and soil samples are two types of evidence that have historically been analyzed by measuring their physical properties, especially density. Today there are more modern instrumental methods for analyzing the exact elemental makeup of evidence, and we will discuss these methods in the next chapter. However, density is still used today for an initial evaluation of glass and soil.

> **LEARNING OBJECTIVE**
>
> Show how a physical property such as density is used to analyze evidence.

> [Sherlock Holmes] Tells at a glance different soils from each other. After walks has shown me splashes upon his trousers, and told me by their colour and consistence in what part of London he had received them.
>
> —*A Study in Scarlet,* Sir Arthur Conan Doyle

Glass Evidence

There are many types of glass and glasslike materials such as Plexiglas, which is used for car windows and headlights. When glass material is collected at a crime scene, density can be measured to help identify the material and to see if it matches evidence collected from a suspect. The density measurement is often made by the **sink-float method**, in which the glass fragment is placed into a solution of known density. If the glass has a higher density than the liquid, it sinks; if it has a lower density than the liquid, it floats. If the glass has the same density as the liquid, it will stay suspended in the solution.

Soil Evidence

Soil evidence is analyzed in a slightly different manner. Soil is a heterogeneous mixture containing a variety of components such as minerals, dust, organic materials, pollen, clay, pebbles, and so on. Each component has a different density, and ideally, the components are a unique mixture found only around the crime scene area. If this is true and matching soil is recovered from the suspect, an evidence trail has been established. Remember that the soil evidence does not prove guilt, merely that the person was present at that location. More investigative work is needed to prove whether a suspect was, in fact, at the location at the time of the crime and committed the criminal act.

The method used historically to analyze soil is called the **density gradient method**. A tall cylinder contains a solution that has a high density at the bottom of the cylinder and a low density near the top of the cylinder. When the soil sample is introduced to the top of the cylinder, the particles begin to sink down through the lighter density liquid until they reach a place where the density of the liquid matches the density of the particles. The particles remain suspended at the level where their density and the density of the liquid are equal. The result is that the soil forms bands throughout the cylinder, as illustrated in Figure 2.3.

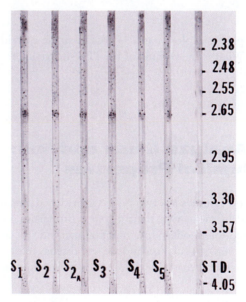

FIGURE 2.3 Soil evidence from a crime (S_1) is compared to soil samples recovered from the clothing of the victim (S_2 and S_{2A}) and the suspect (S_3, S_4, S_5). The final tube contains samples of standard materials with known densities. (Petraco, N and Kubic, T., Journal of Forensic Science, 2000; 45(4): 872-873)

When soil evidence is obtained from other sources, such as a suspect's shovel, matching the density bands allows investigators to make a preliminary assumption as to whether the new soil sample came from the same area as a previous one, such as a crime scene. Matching the soil density bands works best when the soil components significantly change from one area to another, making the soil near the crime scene unique. One problem with using a gradient tube is that many different minerals have nearly identical density values; thus, a unique band structure might not be obtained even though the samples are different.

Modern forensic geologists will use a host of physical properties to determine whether the soil sample from a crime scene matches one obtained from a suspect. The first step in the analysis of soil is to compare the color of the two soil samples and determine whether they match. The human eye is one of the best methods for detecting even slight differences in colors and is an accepted method for determining whether soils match. In order for soil colors to be compared properly, it is important that both soil samples have the same moisture content, because moisture will change the color of the soil. Moisture level is commonly adjusted by heating the samples to dryness or, less often, by adding excess moisture to both samples. If the colors of the soil samples match, the forensic geologist will next compare the textures of the soil samples.

> Observation tells me that you have a little reddish mould adhering to your instep. Just opposite the Wigmore Street Office they have taken up the pavement and thrown up some earth, which lies in such a way that it is difficult to avoid treading in it in entering. The earth is of this peculiar reddish tint which is found, as far as I know, nowhere else in the neighborhood.
>
> —*The Sign of Four*, Sir Arthur Conan Doyle

The soil texture depends on how much sand, silt, and clay are present; the texture can change dramatically if the ratio of these components changes. The terms *sand*, *silt*, and *clay* refer only to the physical sizes of the particles, not to their chemical identities. Two sand samples can have a different chemical makeup. However, if two soil samples come from a common point of origin, the ratio of sand to silt to clay should be constant. The particles of soil also demonstrate another physical property used to characterize the soil—the shape of the particles. The three shape descriptions a forensic geologist uses are angular, sub-rounded, and well-rounded. Figure 2.4 illustrates physical properties used to analyze soils.

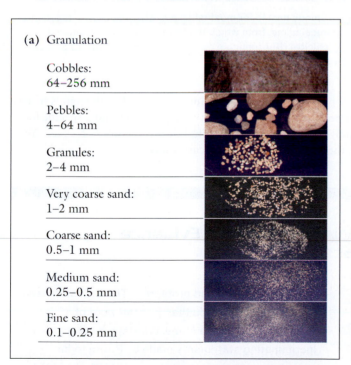

(a) Granulation

Cobbles: 64–256 mm
Pebbles: 4–64 mm
Granules: 2–4 mm
Very coarse sand: 1–2 mm
Coarse sand: 0.5–1 mm
Medium sand: 0.25–0.5 mm
Fine sand: 0.1–0.25 mm

(b)

FIGURE 2.4 The size and shape of soil components dictate the overall texture and appearance of soil samples. The size of the particle, not its chemical makeup, determines whether a soil particle is classified as sand, silt, or clay. (a, b: Courtesy Sally Johll)

The final step forensic geologists can perform is analyzing the mineral composition of the soil. Each mineral is a unique chemical compound. For example, the mineral quartz has the chemical formula SiO_2 and the mineral calcite has the formula $CaCO_3$. Mineral identification can be based on physical appearance under a microscope because the shape and color of a mineral can be quite distinctive. Further identification is also possible using advanced laboratory instrumentation. Figure 2.5 illustrates the distinctive shapes and appearances of minerals.

(a)

(b)

(c)

FIGURE 2.5 Geologists give unique names for most minerals, whereas chemists tend to refer to each mineral by the chemical name, from which the formula can be determined. (a) Calcium carbonate, $CaCO_3$ (calcite) (b) Aluminum oxide, Al_2O_3 (corundum) (c) Silicon dioxide, SiO_2 (quartz) (Courtesy Sally Johll)

The analysis of physical properties of soils is an important tool in the hands of law enforcement because it can provide a physical link between a suspect and a specific location. It was just such an analysis that led to the downfall of one member of a New York crime family. Now back to the Tommy Karate Pitera case study . . .

2.10 CASE STUDY FINALE: Exploring Evidence from a Crime Scene

Tommy Karate Pitera had evaded the justice system for the murder of Willie Boy Johnson, but his luck was running out. An associate of Pitera who had assisted him on several of his murderous drug deals confessed his role to police and agreed to testify against Pitera in exchange for a lighter sentence and protection. The U.S. attorney wanted more evidence than just the word of a co-conspirator to the crime. Wiretaps had recorded incriminating statements made by Pitera to his associates, but the prosecutors wanted to be able to physically link Pitera to the crimes. Soil evidence would provide that positive link from Pitera to a mass grave site containing suitcases filled with dismembered body parts that were discovered outside of the William R. Davis Wildlife Sanctuary located on Staten Island, New York.

The suitcases contained the remains of seven people who, it was later determined, had the misfortune of crossing Pitera. Pitera was in charge of the drug dealing aspects of the Bonanno crime family, and his main work entailed stealing drug shipments and money from other drug dealers. If the drug dealers protested the loss of their money and drugs, they were murdered. Pitera's other victims were those whom he suspected of being informants or who had insulted him.

The detectives at the mass grave site decided to take soil samples at the location of the graves, hoping that the soil at the grave site would provide a color, texture, and composition that could be linked to Pitera. The value of such evidence lies in its uniqueness. If the soil all over Staten Island were uniform, there would be no way to link a suspect to the grave site rather than to any other location on the island. To prove that the soil around the graves was unlike soil found elsewhere on Staten Island, detectives took samples not only at the site but also at various distances from the grave location to show that the soil changes in composition as one goes away from the site.

The soil samples from Staten Island were sent to Special Agent Bruce W. Hall, supervisor of the forensic mineralogy unit at FBI headquarters. He was also given one other sample: soil that was stuck in a shovel obtained from a home on East 12th Street in the Gravesend section of Brooklyn, the home of Pitera. Shovels have a small loop on top for the foot to apply pressure, as shown in the image. This area works as an excellent soil sampler since it takes a sample and then, because it is filled, does not allow future soil to mix with the sample.

Hall was able to examine the color, texture, and composition of the soil found in the shovel and link that shovel to the burial site. When defense attorneys argued that the shovel had been used for gardening and the soil could have come from other locations, Hall had samples taken from the alibi locations and was able to show that the soils were vastly different from the soil on the shovel. Did this mean the shovel was not used at the alibi locations? Not at all—it simply meant that no soil from those sites was found on the shovel, but soil from the burial site was.

Tommy Karate Pitera had escaped conviction for the murder of Willie Boy Johnson, but his conviction for the murders of the victims whose bodies were found at the mass grave site on Staten Island resulted in a life sentence in prison. A sample of soil stuck to a shovel and a scientist who could show how the physical properties of that soil connected it to a burial site contributed to getting a mass murderer off the streets.

CHAPTER SUMMARY

• Physical and chemical properties of matter are important in evidence collection and analysis. Physical changes do not alter the chemical identity of a substance, whereas chemical changes result in the formation of a new substance.

• A measurement consists of a numerical value and a unit describing the system of measurement. Units can be modified with prefixes to indicate the magnitude of the measurement.

• The mass of a sample is a measure of the amount of matter in the sample. This differs from weight, which is a measure of gravity pulling on the sample.

• The reproducibility of an experimental measurement is an indication of the precision of the measurement. The accuracy of measurements depends on how close the results are to the true value. Both precision and accuracy are important in the interpretation of experimental results.

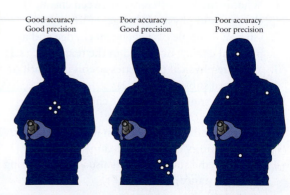

Good accuracy
Good precision

Poor accuracy
Good precision

Poor accuracy
Poor precision

• There is always error in a measurement because numbers resulting from measurements consist of all known digits plus one digit that contains an estimate. These digits are collectively called the significant figures of a number.

• Density, the ratio of mass to volume of a substance, is a physical property that can aid in the identification of glasses and soils. Color and texture of soils are also used for identification, as are the distinctive shapes of mineral crystals in the soil.

KEY TERMS

accuracy (p. 39)
chemical change (p. 28)
chemical properties (p. 29)
conversion factors (p. 32)
density (p. 40)
density gradient method (p. 44)

mass (p. 30)
physical change (p. 29)
physical properties (p. 29)
precision (p. 39)
refractive index (p. 29)

scientific notation (p. 36)
significant figures (p. 34)
sink-float method (p. 43)
units (p. 31)
weight (p. 30)

MAKING MORE CONNECTIONS: Additional Readings, Resources, and References

Baumann, E., and O'Brien, J. *Murder Next Door: How Police Tracked Down 18 Brutal Killers,* Chicago: Bonus Books, 1991.

Doyle, A. C. *Sherlock Holmes: The Complete Novels and Stories,* vol. 1, New York: Random House, 2003.

Lubasch, A. H. "Reputed Mobster Guilty in Six Narcotics Murders; Death Penalty Possible Under Federal Law," *New York Times,* June 26, 1992, p. B3.

Missouri State Highway Patrol Forensic Laboratory. *Forensic Evidence Handbook,* Jefferson City: Missouri State Highway Patrol, 2003.

Murray, R. C., and Tedrow, J. C. F. *Forensic Geology,* Englewood Cliffs: Prentice Hall, 1998.

Petraco, N., and Kubic, T. A density gradient technique for use in forensic soil analysis. *Journal of Forensic Science* 45(4), July 2000.

For more information about the properties of glass and ceramics: www.mindrum.com/tech.html

For more information on FBI procedures for the analysis of glass by the sink-float or density gradient methods: www.fbi.gov/about-us/lab/forensic-science-communications/fsc/oct2009/review/2009_04_review01.htm/

REVIEW QUESTIONS AND PROBLEMS

Questions

1. What is the key difference between chemical changes and physical changes? (2.1)
2. Give an example of a chemical change and a physical change not listed in the textbook. (2.1)
3. Why are physical properties useful in identifying unknown substances? (2.1)
4. What are some examples of chemical properties? (2.1)
5. What is the difference between mass and weight? (2.2)
6. Why are units important in a measurement? (2.2)
7. What are the standard units used for measuring mass? Distance? (2.2)

8. What are the two other common units that are interchangeable with the milliliter (mL)? (2.2)
9. What are the common prefix modifiers and their abbreviations for the metric system? (2.2)
10. Explain how conversion factors are used to change units on a number. (2.3)
11. What is the conversion factor between centimeters and inches? (2.3)
12. Why is there always error involved in making a measurement? (2.4)
13. How can estimating the last digit on a measurement provide you with a more accurate value? (2.4)

14. Sketch a dartboard and illustrate, using "X" marks, the concepts of accuracy and precision. (2.6)

Problems

15. Identify each of the following as either a chemical change or a physical change. (2.1)
    (a) Evaporation of gasoline
    (b) Toasting a marshmallow
    (c) Filtering a pond water sample
    (d) Burning documents

16. Identify each of the following as either a chemical change or a physical change. (2.1)
    (a) Formation of clouds and rain
    (b) Freezing biological samples for storage
    (c) Baking bread dough
    (d) Detonation of TNT

17. Identify each of the following as either a chemical property or a physical property. (2.1)
    (a) Hardness
    (b) Corrosiveness
    (c) Flammability
    (d) Color

18. Identify each of the following as either a chemical property or a physical property. (2.1)
    (a) Refractive index
    (b) Boiling point
    (c) Explosiveness
    (d) Inertness

19. Identify whether each of the following phrases relates to mass or weight. (2.2)
    (a) Depends on gravity
    (b) Measured in grams
    (c) Measures the amount of matter
    (d) Measured with a scale

20. Identify each of the following phrases as part of the definition for mass or weight. (2.2)
    (a) Measured in pounds
    (b) Measures gravity pulling on matter
    (c) Measured with a balance
    (d) Independent of gravity

21. Write the symbol and decimal equivalent for the following metric prefixes. (2.2)
    (a) micro   (c) centi
    (b) kilo   (d) nano

22. Write the symbol and decimal equivalent for the following metric prefixes. (2.2)
    (a) mega   (c) milli
    (b) deci   (d) giga

23. Write the metric prefix that corresponds to the decimal equivalent below. (2.2)
    (a) 0.001   (c) 0.1
    (b) 1000   (d) 0.000000001

24. Write the metric prefix that corresponds to the decimal equivalent below. (2.2)
    (a) 0.01   (c) 0.000001
    (b) 1,000,000   (d) 1,000,000,000

25. Convert each of the following numbers to the units indicated below. (2.3)
    (a) 0.025 g = _____ mg
    (b) 3525 mL = _____ L
    (c) 0.78 m = _____ dm
    (d) 433 cm = _____ m

26. Convert each of the following quantities to the units indicated below. (2.3)
    (a) 2.3×10^6 g = _____ Mg
    (b) 3.64 m = _____ cm
    (c) 7.4×10^{-4} s = _____ ms
    (d) 0.314 g = _____ mg

27. Convert each of the following quantities to the units indicated below. (2.3)
    (a) 54.0 in. = _____ ft.
    (b) 36.0 ft. = _____ yd.
    (c) 0.820 m = _____ mm
    (d) 9.40 in. = _____ cm

28. Convert each of the following quantities to the units indicated below. (2.3)
    (a) 27.5 ft. = _____ cm
    (b) 146 cm = _____ yd.
    (c) 34.8 in. = _____ yd.
    (d) 1.27 yd. = _____ m

29. Convert each of the following quantities to the units indicated below. (2.4)
    (a) 32 ft.^{2} = _____ m^2
    (b) 49 in.^{2} = _____ ft.^{2}
    (c) 224 cm^3 = _____ dm^3
    (d) 45 ft.^{3} = _____ cm^3

30. Convert each of the following quantities to the units indicated below. (2.4)
    (a) 27 yd.^{2} = _____ m^2
    (b) 166 km^2 = _____ cm^2
    (c) 15 m^3 = _____ cm^3
    (d) 38 in.^{3} = _____ mm^3

31. Convert each of the following quantities to the units indicated below. (2.4)
    (a) 554 mm/s = _____ ft./hr
    (b) 73 kg/hr = _____ lb/day
    (c) 26 km/hr = _____ ft./s
    (d) 47 mi./gal = _____ km/L

32. Convert each of the following quantities to the units indicated below. (2.4)
    (a) 63 μL/min = _____ mL/hr
    (b) 41 gal/day = _____ L/s
    (c) 217 ft./s = _____ km/hr
    (d) 66 mg/mL = _____ kg/L

33. How many significant figures are in each of the numbers below? (2.4)
    (a) 3007
    (b) 0.00250
    (c) 0.01410
    (d) 3000

34. How many significant figures are in each of the numbers below? (2.4)
    (a) 5.00×10^3
    (b) 2004
    (c) 0.02010
    (d) 6.00

35. Write the following numbers in scientific notation. (2.4)
    (a) 2300
    (b) 0.0010
    (c) 17,500
    (d) 0.0000240

36. Write the following numbers in scientific notation. (2.4)
    (a) 400
    (b) 0.000045
    (c) 71
    (d) 0.00925

37. Write the following numbers in regular decimal notation. (2.4)
    (a) 6.14×10^{-3}
    (b) 2.59×10^5
    (c) 1.0025×10^4
    (d) 2.226×10^{-2}

38. Write the following numbers in regular decimal notation. (2.4)
    (a) 3.50×10^{-3}
    (b) 5.95×10^3
    (c) 7.8×10^1
    (d) 2.510×10^2

39. Record the value (mL) of the buret readings below with the proper number of significant figures. (2.4)

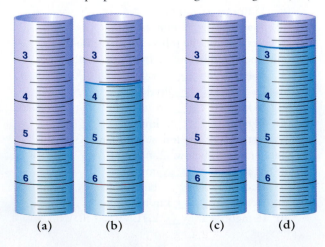

(a)   (b)   (c)   (d)

40. Record with the proper number of significant figures the value of the temperatures shown in the following thermometers. (2.4)

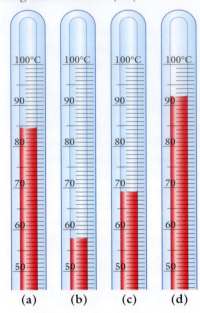

(a)   (b)   (c)   (d)

41. Which of the following numbers contains three significant figures? If a number does not contain three significant figures, alter the number so it will contain exactly three significant figures. (2.4)
    (a) 0.0401   (c) 0.05
    (b) 1.2×10^3 (d) 250.0

42. Which of the following numbers contains three significant figures? If a number does not contain three significant figures, alter the number so it will contain exactly three significant figures. (2.4)
    (a) 3,000
    (b) 0.0501
    (c) 212.0
    (d) 3.85×10^5

43. Rewrite the following numbers in scientific notation. Round the answer to contain the number of significant figures indicated in parentheses. (2.4)
    (a) 35,200 (4)
    (b) 0.008705 (3)
    (c) 1,937 (2)
    (d) 0.0219 (2)

44. Rewrite the following numbers in scientific notation. Round the answer to contain the number of significant figures indicated in parentheses. (2.4)
    (a) 32.160 (3)
    (b) 2,843.53 (4)
    (c) 31.56 (3)
    (d) 0.000263 (1)

45. Rewrite the following numbers in scientific notation. Round the answer to contain the number of significant figures indicated in parentheses. (2.4)
   (a) 50.96 (3)
   (b) 78.16 (3)
   (c) 0.3341 (2)
   (d) 247.23 (2)

46. Rewrite the following numbers in scientific notation. Round the answer to contain the number of significant figures indicated in parentheses. (2.4)
   (a) 0.001342 (3)
   (b) 47,716 (3)
   (c) 6,750,025 (4)
   (d) 0.01990 (2)

47. Write the answer to the mathematical problems below with the correct number of significant figures. (2.5)
   (a) $101.34 - 92.1 - 1.793 =$ _____
   (b) $345.3 + 12.12 + 16.10 =$ _____
   (c) $14.5 + 12.34 - 8.991 =$ _____
   (d) $33.9 - 15.60 + 12 =$ _____

48. Write the answer to the mathematical problems below with the correct number of significant figures. (2.5)
   (a) $45.5 + 0.0023 + 17 =$ _____
   (b) $34.4 - 7.92 - 0.0731 =$ _____
   (c) $56 - 17.98 + 0.02 =$ _____
   (d) $1.45 + 101 - 12.02 =$ _____

49. Write the answer to the mathematical problems below with the correct number of significant figures. (2.5)
   (a) $12.2 \div 3.4 \div 0.0127 =$ _____
   (b) $14.9 \div 12.29 \times 0.020 =$ _____
   (c) $3.0 \times 2.34 \times 329 =$ _____
   (d) $76.3 \div 875.023 \times 31.1 =$ _____

50. Write the answer to the mathematical problems below with the correct number of significant figures. (2.5)
   (a) $642 \div 32.90 \div 100.0 =$ _____
   (b) $47 \times 23.3 \times 10.1 =$ _____
   (c) $82.901 \div 26.8 \times 3.33 =$ _____
   (d) $3967 \times 0.022 \div 9.09 =$ _____

51. Write the answer to the mathematical problems below with the correct number of significant figures. (2.5)
   (a) $(19.83 \times 2.3) + 4.100 =$ _____
   (b) $(14.3 - 2.3) \div 2.0 =$ _____
   (c) $0.020 \times 211.2 - 40.0 =$ _____
   (d) $12.11 \times (2.8 - 13.3) =$ _____

52. Write the answer to the mathematical problems below with the correct number of significant figures. (2.5)
   (a) $76.3 - 23.345 \div 16.0 =$ _____
   (b) $8.240 \times 37.2 - 119.00 =$ _____
   (c) $(1.003 \times 23.0) + 173.90 =$ _____
   (d) $56.2 \div 2.300 + 9 =$ _____

53. Which set of numbers below represents measurements that have good precision and good accuracy? The true value of the measurement is 4.75. (2.6)
   (a) 3.64, 3.82, 3.74, 3.98
   (b) 4.71, 3.98, 5.78, 3.03
   (c) 4.79, 4.68, 4.81, 4.83
   (d) 4.32, 4.93, 4.05, 4.11

54. Which set of numbers below represents measurements that have good precision but poor accuracy? The true value of the measurement is 18.44. (2.6)
   (a) 17.81, 17.10, 19.99, 18.43
   (b) 18.46, 18.39, 18.52, 18.48
   (c) 18.10, 19.21, 17.44, 17.99
   (d) 17.35, 17.41, 17.29, 17.38

55. Calculate the density of an object given the masses and volumes below. (2.8)
   (a) Mass = 14.45 g, volume = 10.0 cc
   (b) Mass = 12.2 g, volume = 3.43 mL
   (c) Mass = 9.02 g, volume = 6.23 cm^3
   (d) Mass = 7.02 g, volume = 8.29 mL

56. Calculate the density of an object given the masses and volumes below. (2.8)
   (a) Mass = 12.82 g, volume = 13.28 cc
   (b) Mass = 2.34 g, volume = 1.11 mL
   (c) Mass = 9.23 g, volume = 6.67 cm^3
   (d) Mass = 4.73 g, volume = 5.72 mL

57. Calculate the volume of an object given the densities and masses below. (2.8)
   (a) Density = 0.982 g/mL, mass = 14.45 g
   (b) Density = 3.231 g/mL, mass = 10.0 g
   (c) Density = 1.34 g/cc, mass = 4.71 g
   (d) Density = 2.90 g/cm^3, mass = 11.67 g

58. Calculate the volume of an object given the densities and masses below. (2.8)
   (a) Density = 0.864 g/mL, mass = 44.99 g
   (b) Density = 2.77 g/cc, mass = 21.4 g
   (c) Density = 11.0 g/mL, mass = 5.76 g
   (d) Density = 8.76 g/cm^3, mass = 2.003 g

59. Calculate the mass of an object given the densities and volumes below. (2.8)
   (a) Density = 0.935 g/mL, volume = 23.30 mL
   (b) Density = 1.45 g/cc, volume = 12.22 cc
   (c) Density = 13.6 g/mL, volume = 9.32 mL
   (d) Density = 2.25 g/cm^3, volume = 5.60 cm^3

60. Calculate the mass of an object given the densities and volumes below. (2.8)
    (a) Density = 3.91 g/mL, volume = 9.44 mL
    (b) Density = 0.791 g/mL, volume = 10.9 mL
    (c) Density = 2.34 g/cc, volume = 8.45 cc
    (d) Density = 7.44 g/cm^3, volume = 11.08 cm^3

Case Study Problems

61. Describe how the scientific method was used in the soil analysis done by Special Agent Bruce Hall in the Tommy Karate case study. (CS)

62. Soil analysis is a comparative science, which means that two samples are compared to determine whether, in fact, they come from the same source. Describe what physical properties are used to compare soils. If a sample from a suspect matches that from a crime scene, what else must be proven to show that the suspect was at the crime scene location? (2.9)

63. Why would it be more difficult to compare soil recovered from the blade of a shovel at a crime scene than soil from the rounded-over flange of the shovel? (CS)

64. Most states are adopting as the legal limit for driving under the influence of alcohol a blood alcohol content (BAC) of 0.08 g/dL. Many clinical and forensic laboratories will report BAC in units of mg/mL, g/L, or mg/dL. Convert 0.08 g/dL to each of the alternate units for BAC. (2.3)

65. Diazodinitrophenol (DDNP) is an explosive with a density of 1.63 g/mL. What volume (cm^3) would a 1.00-pound sample of DDNP occupy? (1 lb = 453.5 g) (2.8)

66. One complication of examining a cadaver for BAC is that the decomposition process can produce alcohol. Postmortem alcohol production is usually less than 70 mg/dL, but it can be as high as 100 mg/dL. Would it be possible to mistakenly report that a sober victim had been legally drunk at the time of death? Assume that the legal BAC level is 0.08 g/dL. (CP)

67. Alcohol is eliminated from the human body at an approximate rate of 15 mg/(dL · hr). What is the rate of alcohol elimination in units of g/(dL · min)? (2.3)

68. A piece of glass with a mass of 7.89 g was recovered from a crime scene. When submerged in a graduated cylinder, the volume of water in the cylinder rose from 24.00 mL to 27.54 mL. Identify from Table 2.4 the type of glass found at the crime scene. (2.8)

69. If a 17.84-g sample of alkali barium glass is submerged into a cylinder containing 30.00 mL, what is the final level of the water in the cylinder? (2.8)

70. Does a chemical or physical change force the bullet out of a gun? Is the smoke coming out the end of this gun a homogeneous or heterogeneous mixture? Explain your answer. (2.1)

(Index Stock Photos/Fotosearch)

71. If a suspected arson sample sent to the laboratory comes back negative for petroleum-based accelerants, does that prove that the fire was not deliberately set? If the sample comes back positive, does that automatically mean the fire was deliberately set? Explain how an investigator might use this information (whether affirmative or negative laboratory results) in determining the chain of events leading up to the fire. (CP)

72. A glass sample has a softening point in the range of 718°C to 723°C. List the possible glass types (consult Table 2.4). What are the possibilities if the density is determined to be 2.6 g/cm^3? (2.7)

73. A glass sample has a softening point in the range of 819°C to 833°C. List the possible glass types (consult Table 2.4). What are the possibilities if the density is determined to be 2.23 g/cm^3? (2.7)

74. A bullet recovered from a crime scene has a mass of 9.68 g. A suspect was arrested for possession of a recently fired .357 Magnum revolver. The masses of bullets are traditionally measured using the unit *grain*. Given that 1 grain = 0.0648 g, is the bullet consistent with coming from a .357 Magnum revolver, which is available with bullets that range from 125 to 180 grains?

75. The velocity of a 180-grain bullet shot from a .357 Magnum revolver is approximately 1000 ft./s. Convert this to m/s. How many meters can the bullet travel in the first 15 s after being fired?

76. The toxicity of a compound is often reported as the LD_{50}, which is the lethal dose for 50% of a population. The LD_{50} is typically measured using mice and is reported in units of milligrams of the compound per kilogram of the body mass of the population. The LD_{50} (mice) for caffeine is

192 mg/kg, and for nicotine it is 0.3 mg/kg. Assuming the LD_{50} for humans is similar to that measured for mice (or that you have a mouse the size of an average adult weighing 154 lb), determine the number of grams of caffeine and nicotine that would be lethal to 50% of the human (giant mouse) population.

77. Based on your answer to Problem 76, calculate the number of cigarettes that would have to be smoked in a single sitting to reach the LD_{50}, given that 1.89 mg nicotine is typically absorbed per cigarette. Calculate how many cans of Red Bull energy drink it would take to reach the LD_{50} of caffeine, given that there are 9.76 mg/oz and each can is 8.2 oz.

78. Search and rescue efforts are often complicated by adverse weather conditions and large areas to search. Given that the average surface area of a human is 1.8 m², estimate the percentage of area visible if the search and rescue team is looking in a 10-sq.-mi. area. (Your answer should also take into account that not all of the surface area of a body would be visible in your estimation.)

79. A graduated cylinder is carefully filled with four different solutions, each with a different density. Next, a mixture of soil containing four primary minerals is placed into the cylinder. Sketch the appearance of the cylinder; identify each layer and the location of each mineral in the cylinder, assuming perfect layering of solutions.

      Solution 1: 50 mL, D = 4.25 g/mL
      Solution 2: 25 mL, D = 2.45 g/mL
      Solution 3: 50 mL, D = 3.34 g/mL
      Solution 4: 25 mL, D = 1.80 g/mL
      Mineral 1: oldhamite, D = 2.58 g/mL
      Mineral 2: realgar, D = 3.56 g/mL
      Mineral 3: haycockite, D = 4.35 g/mL
      Mineral 4: graphite, D = 2.16 g/mL

80. A Molotov cocktail is a fuel-filled glass bottle that has a cloth rag acting as both a plug and a fuse. When thrown at a target, the glass breaks and releases the fuel, which is immediately ignited by the fuse. An arsonist used such a device to burn down a property in Dallas, Texas. The suspect was a professional arsonist from the Kansas City, Kansas, region. The only evidence from the crime scene was a shard of glass. How could a forensic scientist link the shard of glass found in Texas to a suspect from Kansas? No DNA or fingerprints were found on the glass. Use the scientific method and remember from the Tommy Pitera example that it is often as important to exclude possibilities as it is to find matching samples. (CP)

81. An eyewitness positively identified an individual as the person seen fleeing the scene of a homicide. There were two blood types found at the scene, that of the victim and another blood type matching the suspect. The suspect had no known connections to the victim and claimed he had spent the evening at home watching TV by himself. The suspect was convicted and sentenced to life in prison. Fifteen years later, DNA analysis of evidence from the crime scene proved the convicted man innocent. Based on the limited information within this problem, was it reasonable to convict the suspect? Was the scientific method properly applied in the original case and subsequent acquittal? (CP)

82. A high-speed chase captured on a surveillance camera showed the suspect running a red light at an intersection. The suspect vehicle collided with a vehicle that had just turned into the path of his car. The tape shows that it took the car 0.546 s to cross the intersection, which was measured to be 41.2 ft. The posted speed limit was 25 mph. Determine whether the suspect was speeding at the moment he collided with the second vehicle. (CP)

83. A fired cartridge case pictured (to scale) below was recovered at a crime scene with a deformed but relatively complete lead bullet. The mass of the bullet was determined to be 2.236 g. Two suspects, each with a different gun, are in custody. One was carrying a .22 LR pistol and the other a .25 automatic. A .22 LR pistol is manufactured to accommodate bullet cases that are 5.7 mm in diameter and 15.0 mm in length, with bullet masses from 30 grains (most common) up to 60 grains for specialty ammo. The 0.25 automatic loads bullet cases that are 6.4 mm in diameter, 16.0 mm in length, with bullet masses that range from 35 to 50 grains. Determine which gun fired the cartridge found at the crime scene and whether the recovered bullet and case are consistent with either one of the guns, both of the guns, or neither gun. (CP)

(© Christina Micek/www.christinamicek.com)

CHAPTER 3

Understanding Atoms

CASE STUDY: Exploring Historical Climate Change

Global climate change becomes evident when we look at the historical data. Great seas once covered regions that are now deserts, and glaciers covered large portions of North America, Europe, and Asia, although they have now retreated to the northernmost regions. As scientists study global climate change, we are starting to realize that change is a natural part of the earth's history. However, a new variable has been introduced—the effect humans have on the natural process of climate change.

Our ever-increasing hunger for energy and power has created unprecedented levels of carbon dioxide in the atmosphere through the burning of fossil fuels. The problem with carbon dioxide is that, as a greenhouse gas, it absorbs energy and traps energy in the earth's atmosphere rather than allowing the energy to pass out of the earth's atmosphere. In the last 400,000 years, the highest recorded level of carbon dioxide was 300 parts per million. We surpassed that historical level in the 1950s and are now at 390.92 parts per million—an unprecedented level of carbon dioxide. As the atmosphere's concentration of carbon dioxide increases, so too does average global temperature.

Global climate change is a major international and national political topic that influences decisions about national fuel standards, international climate treaties, and sources of energy (fossil, nuclear, or alternative), among others. The momentous nature of this topic makes it critically important for people to educate themselves on the facts of global climate change. Seeking the facts can pose a challenge to those with little exposure to the scientific methods used to research global climate change.

One of the standard methods used to research the history of climate is to examine changes recorded in the glaciers of the Arctic and Antarctic ice sheets. Each year, a new layer of snow and ice is deposited on top of the previous year's snow, compacting the layers beneath it. Glacial geologists then cut out a cylindrical shaft of ice from the glaciers, revealing samples of ice that are

As you read through the chapter, consider the following questions:

• What is the structure of an atom, and what properties distinguish atoms from one another?

• How can the differences between atoms of the same element affect the physical properties of compounds?

tens of thousands of years old or more. Carbon dioxide dissolves in water, something you have experienced every time you drink a carbonated beverage. The amount of carbon dioxide dissolved in the ice is directly related to the amount of carbon dioxide in the atmosphere. The analysis of historical carbon dioxide levels is fairly straightforward, but how do scientists determine what the average global temperature was thousands of years ago?

MAKE THE CONNECTION

The temperature of the ice is below zero, but the temperature of the earth is recorded in the water molecules themselves . . .

3.1 Origins of the Atomic Theory: Ancient Greek Philosophers

LEARNING OBJECTIVE

Distinguish the two early schools of thought on the nature of the atom.

In the past decade, high-performance polymer composites have revolutionized production of airplanes. Advances in semiconductor technology have exponentially increased computing power while allowing for the technology to be miniaturized. Today, scientists are exploring the potential uses of nanotechnology by manipulating chemical structures that are 1 to 100 billionth of a meter in size. The foundation of all these advances lies in our understanding of the fundamental building block of all matter, the atom. The gradual elucidation of the nature and structure of atoms constitutes one of the greatest detective stories in science. It took over 2300 years to solve the case! In fact, the case was closed only about seventy years ago. Moreover, the first image of an atom was taken only in 1983, when IBM researchers invented an instrument called the scanning tunneling microscope (STM), which was capable of sensing the presence of an individual atom and creating an image. Figure 3.1 shows an STM image of silicon atoms. The STM image is not a traditional picture; it is a computer-generated image depicting the forces between the electrons in an atom and the STM instrument.

FIGURE 3.1 Silicon atoms imaged by an STM. The colors in the image are computer generated. (Science VU/IBMRL/Visuals Unlimited)

The mystery began in the year 440 B.C. in ancient Greece. Two philosophers, Leucippus and his student Democritus, started a revolution of thought by claiming that matter was made of small, hard, indivisible particles they called *atoms*. It is said that Leucippus and Democritus originated their idea during a walk on the beach. From a distance, the sand looks like one large continuous object, but upon closer examination, one can see that it is really comprised of countless small grains of sand. According to their theory, atoms came in various sizes, shapes, and weights, were in constant motion, and combined to make up all the various forms of matter. The observable properties of matter could be directly related to the types of atoms it contained. Wherever atoms did not exist, there was a void or vacuum.

This theory of Leucippus and Democritus set the stage for a later confrontation with the philosopher Aristotle, who believed in a continuous model of matter. Aristotle taught that matter could be infinitely divided, which meant that no indivisible particles (atoms) existed. He convinced the majority of his colleagues that the atomic view of matter was illogical.

The argument for the existence of atoms was not advanced for the next 1900 years. In fact, teachings contrary to Aristotle, including the atomic theory, were against the law in some parts of Europe! To make matters worse, the atomic theory was associated with atheistic beliefs.

(a)   (b)

(a) A beach seen from a distance appears to be a solid object that stretches along the coast. (b) Upon closer examination, one can see that the beach consists of a vast number of tiny sand particles. (a: Julian Nieman/ Alamy; b: Blackout Concepts/Alamy)

3.2 Foundations of a Modern Atomic Theory

In the 1600s, Pierre Gassendi, a priest, philosopher, and scientist, reopened the atomic debate by publishing a work that defended the atomic theory and argued that the principles of atomic theory were not contradictory to Christian beliefs. He modified the original principles of Leucippus and Democritus to state that all atoms were actually created by God and that their motion was a gift of God. Current atomic theory is neutral on the existence of a supreme being, but Gassendi's work was important because it allowed people to engage in open debate about atomic theory without fear of retribution.

Even though the atomic theory could now be debated and discussed in academic settings, there were no significant developments until the late 1700s. Experiments that laid the groundwork for a better understanding of matter and chemical change were carried out around that time and became necessary precursors to the further development of atomic theory.

Antoine Lavoisier, who is considered the founder of modern chemistry, made a discovery in 1785 that was to become known as the law of conservation of mass. A scientific law states how a system behaves, usually in mathematical terms, but does not attempt to explain why it behaves in the stated manner. The **law of conservation of mass** states that mass is neither created nor destroyed in a chemical reaction but merely changes form. For many centuries, scientists had pondered what happened to matter during chemical reactions. Perhaps even our earliest ancestors who mastered the art of fire wondered what happened to the wood that was burning in their campfires. The fire seemed to consume the wood, since the only tangible evidence left from the fire was a very small amount of ash.

The key to unraveling this mystery was the awareness that matter undergoing a chemical change (such as burning wood) in an open environment only appears to lose mass because one or more of the products of the reaction escapes into the surroundings as a gas. Lavoisier conducted experiments in closed systems in which the substances undergoing a transformation could not exchange matter with the surroundings, as shown in Figure 3.2.

Pierre Gassendi, a seventeenth-century priest, philosopher, and scientist. Besides his writings on theology, he investigated many topics in physics, chemistry, and geology. (Science Photo Library/Photo Researchers, Inc.)

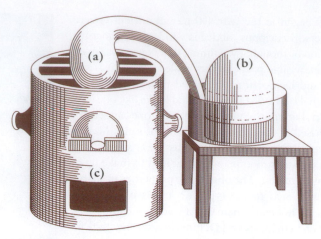

FIGURE 3.2 Lavoisier's reaction vessel (a) could be heated by the furnace (c), and any gases that formed during the reaction would be collected in the jar (b).

Lavoisier heated mercury(II) oxide, a red powdery solid, in a sealed system to form elemental mercury, a silver-colored, liquid metal, and colorless oxygen gas, as shown in Figure 3.3. Taking careful quantitative measurements, he found that the mass of the mercury(II) oxide before heating equaled the combined masses of the liquid mercury and oxygen gas that formed after heating. Thus, there was no overall change in mass as long as the reaction occurred in a sealed system.

The ashes left by countless campfires could finally be explained as the solid remains of a combustion reaction that also produced carbon dioxide gas and water vapor. The carbon dioxide and water vapor escape into the surrounding air.

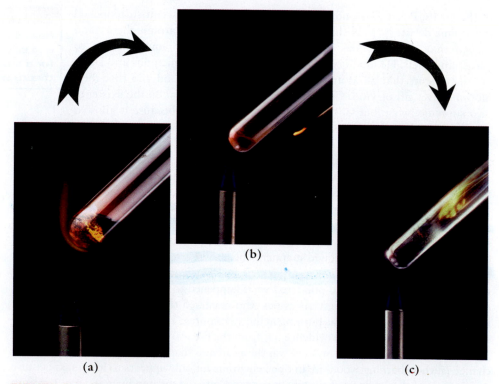

FIGURE 3.3 Mercury(II) oxide is a red powder (a) that decomposes (b) to form a silver-colored liquid metal (c) and oxygen gas that causes a glowing splint to burst into flames. In Lavoisier's experiment, this visually dramatic reaction was performed in a closed system to illustrate the law of conservation of mass. (a, b, c: © Joel Gordon)

WORKED EXAMPLE 1

Lavoisier studied the production of water from hydrogen and oxygen gases. If he produced 72.0 g of water starting with 8.0 g of hydrogen, how many grams of oxygen must have reacted with the hydrogen?

SOLUTION

According to the law of conservation of mass, the total mass of reactants must equal the total mass of products:

$$\text{grams hydrogen} + \text{grams oxygen} = \text{grams water}$$
$$8.0 \text{ g} + \text{g oxygen} = 72.0 \text{ g}$$
$$\text{g oxygen} = 72.0 \text{ g} - 8.0 \text{ g} = 64.0 \text{ g oxygen}$$

Practice 3.1

Alcohol in our digestive system reacts with oxygen gas to form carbon dioxide and water. How many grams of carbon dioxide are released if a 10.00-g alcohol sample reacts with 20.85 g of oxygen gas and produces 11.77 g of water?

Answer

❑ 19.08 g CO_2

Another series of important experiments, which were carried out by Joseph Louis Proust in 1797, led to a principle that became known as the law of definite proportions (also called the law of constant composition). The **law of definite proportions** states that a compound is always made up of the same relative masses of the elements that compose it. At the time Proust was investigating compounds, some scientists believed that a compound could contain any ratio of its elements and still be the same substance. Their argument was that only the type of elements in the compound mattered, not necessarily the amounts of each element. Proust was the first to show that analysis of a carefully prepared and purified compound always showed the same mass ratio of elements—even if different methods were used to prepare the compound—from one experiment to another.

The key to Proust's work was that he did it very carefully so that he was able to obtain both accurate and precise results. The work by earlier scientists had not been carried out as carefully as Proust's; therefore, earlier scientists based their conclusions on inaccurate data.

3.3 Dalton's Atomic Theory

John Dalton is credited with giving the world the first modern atomic theory. The word *theory* in the sciences is the term for the best current explanation of a phenomenon. It is *not* synonymous with an opinion and does *not* change as a result of political or personal agendas. A **theory** is accepted by the scientific community until such time that new data contradict the theory. A theory can be viewed as a hypothesis that has been validated by many experiments and can be used to make accurate predictions of future observations.

Dalton's passion was studying meteorology and weather patterns; this led him to study how gas mixtures behave. As a result of his studies, he developed an atomic theory that explained the behavior of gases and the previous findings of Lavoisier and Proust. Dalton's atomic theory of matter, published in 1803, had four basic tenets:

Dalton's Atomic Theory

1. All matter is made up of tiny, indivisible particles called atoms.
2. Atoms cannot be created, destroyed, or transformed into other atoms in a chemical reaction.
3. All atoms of a given element are identical.
4. Atoms combine in simple, whole-number ratios to form compounds.

With **Dalton's atomic theory**, the law of conservation of mass could now be explained. When a chemical reaction occurs, the atoms of each compound are rearranged to form new compounds, as shown in Figure 3.4, but the number of atoms remains the same. Thus, no mass is lost or gained in the process.

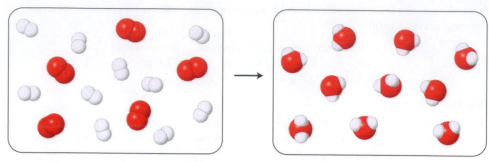

Hydrogen (⚬) reacts with oxygen (🔴) to form water (🔴)

FIGURE 3.4 The law of conservation of mass explains a chemical reaction as the rearrangement of atoms. The number of atoms before or after the reaction remains constant (10 oxygen and 20 hydrogen); the arrangement of chemical bonds changes.

Dalton's atomic theory could also explain the law of definite proportions. According to the atomic theory, compounds are made up of atoms in simple ratios. Pure water always has two hydrogen atoms to every oxygen atom, regardless of where the water comes from or how it is created. This constant atomic ratio provides the reason the mass ratio of elements in the compound is also constant.

Finally, Dalton's atomic theory predicted the law of multiple proportions formulated by Dalton himself. The **law of multiple proportions** states that any time two or more elements combine in different ratios, different compounds are formed. For example, laughing gas, which has the chemical formula N_2O, is quite different from the gas NO_2, a byproduct of explosions and burning diesel fuel. Figure 3.5 shows various compounds of nitrogen and oxygen.

In spite of the success of Dalton's theory in explaining the behavior of matter and the widespread support the theory received, it took almost another hundred years before the atomic theory of matter was completely accepted. In fact, there were still a few major scientists who disbelieved the atomic theory into the early 1900s. It would take more than logical reasoning and circumstantial evidence to get a unanimous decision by a jury of scientists.

Nitrogen (🔵)

Oxygen (🔴)

FIGURE 3.5 Nitrogen atoms can combine with oxygen atoms in a variety of simple ratios. Each combination represents a unique compound, illustrating the law of multiple proportions.

WORKED EXAMPLE 2

Label each of the following drawings as illustrating the law of conservation of mass (LCM), the law of multiple proportions (LMP), the law of definite proportions (LDP), or more than one law, if applicable.

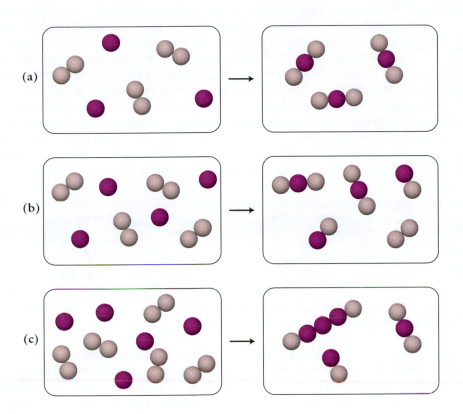

SOLUTION

(a) LCM, LDP. The reaction illustrates the law of conservation of mass because both reactants and products have an identical number of atoms. The law of definite proportions is also illustrated because a single compound has been created with a constant formula.

(b) LCM and LMP. The reaction illustrates the law of conservation of mass because both reactants and products have an identical number of atoms. The law of multiple proportions is also illustrated because several compounds have been created, each with a unique formula.

(c) LMP. The law of multiple proportions is illustrated because several compounds have been created, each with a unique formula.

Practice 3.2

Write the chemical formulas for all the nitrogen-oxygen compounds shown in Figure 3.5. Explain how the figure illustrates the law of multiple proportions.

Answer

NO, NO_2, N_2O, N_2O_3, N_2O_4, N_2O_5. The law of multiple proportions states that elements can combine in multiple whole-number ratios of their elements (1:1, 1:2, 2:1, and so forth), but each new ratio of atoms represents a new compound.

3.4 Atomic Structure: Subatomic Particles

Some of the most convincing evidence for the existence of atoms came from experiments that actually contradicted the first principle of Dalton's atomic theory. Experimental evidence began to show that atoms were not indivisible; in fact, there were three major components to the atom. These subatomic particles are the positively charged **proton**, the negatively charged **electron**, and the **neutron**, which has no electrical charge and is thus electrically neutral.

Experiments that led to the discovery of the electron depended on the invention in the 1850s of the cathode ray tube, the direct ancestor of the conventional television picture tube and computer monitor. A **cathode ray tube** is a sealed glass tube from which nearly all gases inside have been removed to create a vacuum. In the cathode ray tube are two metal electrodes called the *cathode* and the *anode* to which a large voltage is applied. When this process occurs, rays can be seen emanating from the cathode. The rays can be deflected by the poles of a magnet, moving away from the north-seeking (N) pole and toward the south-seeking (S) pole, as shown in Figure 3.6.

(a) (b) (c)

FIGURE 3.6 A beam of cathode rays is (b) deflected by the north-seeking pole and (c) attracted to the south-seeking pole of a magnet.

In 1897, the British physicist J. J. Thomson was able to measure the deflection of the cathode rays from magnetic and electric fields. Using the data from his deflection experiments, he calculated that the mass-to-charge ratio of the cathode ray particles was less than 1/1000 of the value calculated for any other known element. Thomson thus discovered that whatever made up the cathode rays was smaller than the smallest atom! Thomson correctly concluded that the cathode rays consisted of negatively charged subatomic particles called *electrons*. However, the actual electronic charge of the electron was not determined until 1909, when Robert Millikan measured the rate at which charged droplets of oil fell in an electric field.

The discovery of the electron presented a problem for scientists. They knew that atoms are electrically neutral, yet now they had evidence that atoms also contain negative particles. A positive charge has to balance the negative charge, but where in the atom is it? One popular theory at the time was that the electron particles were randomly stuck into a ball of positive charge. This model was commonly called the *plum pudding model*, referring to the way the fruit is distributed throughout a plum pudding. A more familiar model today might be a chocolate chip muffin.

Plum pudding   Chocolate chip muffin   Plum pudding model

The plum pudding model consisted of a positively charged matrix in which the negative electrons are embedded into the positive region. (Left: Foodfolio/Alamy; center: Index Stock)

In 1909, Ernest Rutherford set out to investigate the plum pudding model, which he thought was correct, by taking an extremely thin layer of gold foil and exposing it to a radioactive element. **Radioactivity** is the spontaneous emission of particles from unstable elements. The alpha particle, which is emitted from some radioactive elements, has a positive electrical charge and a mass four times that of a hydrogen atom. In Rutherford's experiment, the alpha particle emitter was aimed at the target placed inside a detector that would show a flash of light when an alpha particle struck it (Figure 3.7).

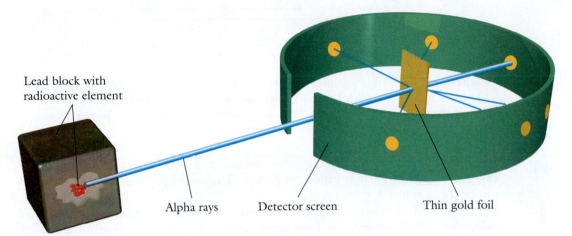

Lead block with radioactive element

Alpha rays   Detector screen   Thin gold foil

FIGURE 3.7 In Rutherford's gold foil scattering experiment, alpha rays from a radioactive element are directed out of a lead container toward a gold foil target surrounded by a detector screen that flashes at any location struck by an alpha particle.

The three major experimental observations made during the gold foil scattering experiment, with the conclusions Rutherford made from each observation, are as follows.

Observation 1: The vast majority of alpha particles passed directly through the solid gold foil.

Conclusion 1: The atom must consist mostly of empty space for the alpha particles to go directly through.

Observation 2: Occasionally an alpha particle would veer from a straight-line path and hit the detector on the side.

Conclusion 2: (a) The positively charged alpha particle was pushed off course when it came close to a positive region of the atom because like charges repel. (b) The positive region had to be small because only a few alpha particles ever came close enough to be repelled.

Observation 3: Rarely, an alpha particle would bounce directly back toward the alpha particle source after striking the gold foil.

Conclusion 3: (a) In these instances, the alpha particle made a direct hit on a very dense object that caused it to bounce back. (b) Since the direct hits were a rare event, the dense object must be very small and must be positively charged to repel the alpha particle.

The final result stunned Rutherford. He would later compare it to firing an artillery shell at a piece of paper and having it bounce back at you! The dense positive region Rutherford discovered in the atom is called the **nucleus**. Rutherford continued to explore the structure of the atom and would later discover that the positively

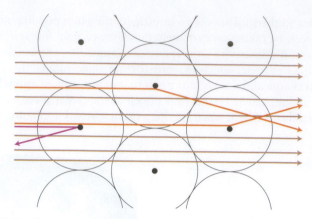

FIGURE 3.8 Rutherford's model of the atom consisted almost entirely of empty space in which the protons were located in a very small, dense nucleus. This model could account for the three experimental observations: (1) Most rays pass through (gray). (2) Some rays are deflected (orange). (3) Rarely, an alpha ray is returned (purple).

charged hydrogen ion has the simplest nucleus consisting of only one positively charged subatomic particle called the *proton*. The model Rutherford put forward is illustrated in Figure 3.8.

The task of determining the structure and components of an atom was complete except for an explanation for the fact that protons made up only a portion of the mass of most atoms. The difference in mass could not be explained by the presence of electrons, because they were so much lower in mass that they contributed very little to the overall mass of the atom. The search for the missing mass continued until 1932, when James Chadwick correctly interpreted several nuclear experiments that were producing heavy neutral particles called *neutrons*. A neutron has a mass equal to that of a proton but no electronic charge. Neutrons and protons are themselves made up of even smaller particles such as the *quarks* and *leptons*. However, we will limit our discussion to electrons, protons, and neutrons.

Neutrons were later linked to the violation of another principle of Dalton's original atomic theory. Dalton had postulated that all atoms of an element are identical, but, in fact, some atoms of the same element have different numbers of neutrons than others. These atoms are called **isotopes**. Isotopes of an element behave and react chemically like all other atoms of that element, but they differ in their atomic masses.

Because the masses of single atoms and subatomic particles are so small, using units of grams to express mass does not make sense. Instead, mass is expressed in a unit called the **atomic mass unit** (**amu**), defined as 1/12 the mass of a carbon atom that contains 6 protons and 6 neutrons. The properties of the subatomic particles are summarized in Table 3.1. Note how small the mass of the electron is compared with the masses of the proton and the neutron. The electron's mass is only 1/1838 (or 0.0005486) the mass of a proton or neutron. For most applications, the mass of the proton and neutron can be rounded to 1 amu, and the mass of the electron to 0 amu.

TABLE 3.1	Subatomic Particles		
Particle	Charge	Mass (amu)	Symbol
Electron	−1	0.0005486	e^-
Proton	+1	1.0073	p or p^-
Neutron	0	1.0087	n

3.5 Nature's Detectives: Isotopes

Isotopes are used quite often for solving chemical and medical mysteries. For example, radioactive isotopes are used routinely at most hospitals to investigate whether a patient has properly functioning organs or to locate traces of cancer. Isotopes are also commonly used in the laboratory to investigate how each step of a chemical reaction occurs. A closer look at the periodic table and the nature of isotopes is warranted because of the important role isotopes play in science and medicine.

Isotopes differ only in the masses of their atoms, so they have the same chemical symbol. How then do we represent the different isotopes of an element symbolically to distinguish one from another? The goal of any symbolic form used in chemistry is to clearly relay the relevant information in an abbreviated form.

For example, the element hydrogen has three isotopes, as shown in Table 3.2. The **mass number (A)** is the mass of the isotope measured in atomic mass units (amu). The mass of the isotope is the mass of the protons plus the mass of the neutrons; each particle has a mass of 1 amu. The mass of an electron is so small that it does not affect the mass of the atom to any important extent. It would take the mass of almost 2000 electrons to equal the mass of just one proton or neutron.

The number of protons in each isotope of an element is the same. In this example, each isotope of hydrogen has one proton. All atoms of an element, regardless of the number of neutrons, have the same number of protons. The periodic table is organized by ascending number of protons, shown in the upper center or upper right corner of the block for each element. The number of protons in an element is called the **atomic number (Z)**. Figure 3.9 shows the pattern that is used to represent specific isotopes.

TABLE 3.2	Isotopes of Hydrogen				
Isotope Name	Number of Neutrons	Number of Protons	Number of Electrons	Mass Number	Model Protons • Neutrons •
Hydrogen-1 (Protium)	0	1	1	1	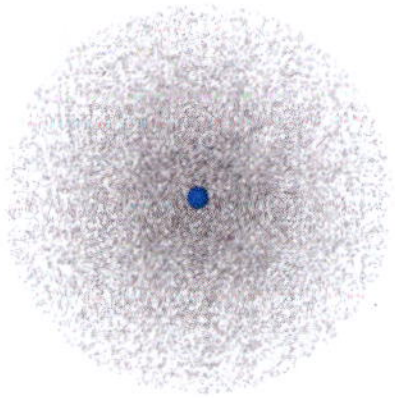
Hydrogen-2 (Deuterium)	1	1	1	2	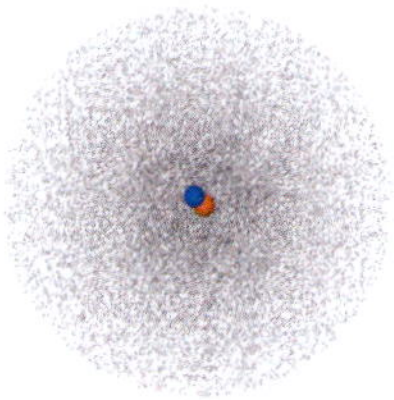
Hydrogen-3 (Tritium)	2	1	1	3	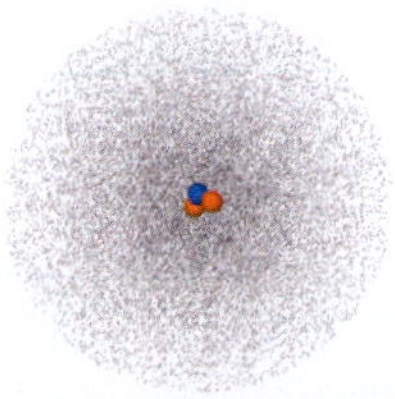

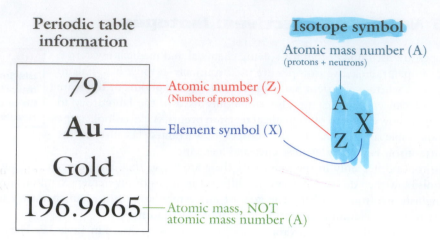

Periodic table information

Isotope symbol

FIGURE 3.9 Relationship between the information in the periodic table and the information in the isotope symbol.

The isotopes of hydrogen can then be written $_1^1H$ for protium, $_1^2H$ for deuterium, and $_1^3H$ for tritium. This system makes clear the distinction between isotopes without the need for memorization of names or formulas. One common modification of this method is the omission of the atomic number. The three isotopes of hydrogen could just as easily be written ^1H, ^2H, and ^3H. This shortcut is valid because all isotopes of hydrogen have the same atomic number, which can be determined from the periodic table.

Giving each isotope a distinct name would be impractical. Therefore, an isotope is simply given the name of the element and the atomic number (Z). For example, ^3H is written *hydrogen-3* and is properly pronounced "hydrogen three." This terminology may be familiar to you: The technique of carbon-14 dating of artifacts is commonly mentioned in the popular media in discussions of archeological discoveries.

WORKED EXAMPLE 3

The carbon and nitrogen atoms found in protein molecules come from the food a person consumes and consist of several isotopes. The ratio of the isotopes present can distinguish between a vegetarian and a meat-protein-rich diet. Write the isotope symbol for each of the following isotopes:

(a) Carbon atom containing 6 neutrons

(b) Nitrogen atom containing 8 neutrons

(c) Carbon atom containing 7 neutrons

(d) Nitrogen atom containing 7 neutrons

SOLUTION

(a) Carbon's atomic number is 6 (also the number of protons). The mass number is equal to protons + neutrons: $6 + 6 = 12$. The isotope symbol is $_6^{12}C$.

(b) Nitrogen's atomic number is 7 (also the number of protons). The mass number is equal to protons + neutrons: $7 + 8 = 15$. The isotope symbol is $_7^{15}N$.

(c) Carbon's atomic number is 6 (also the number of protons). The mass number is equal to protons + neutrons: $6 + 7 = 13$. The isotope symbol is $_6^{13}C$.

(d) Nitrogen's atomic number is 7 (also the number of protons). The mass number is equal to protons + neutrons: $7 + 7 = 14$. The isotope symbol is $_7^{14}N$.

Practice 3.3

Write the isotope symbols for the following:

(a) Atom containing 26 protons and 30 neutrons

(b) Atom containing 14 protons and 14 neutrons

(c) Atom containing 16 protons and 17 neutrons

Answer

(a) $^{56}_{26}Fe$ (b) $^{28}_{14}Si$ (c) $^{33}_{16}S$

Given the isotope symbol, it is possible to determine the number of protons, neutrons, and electrons of any atom using simple mathematics. The mass number of an atom is the sum of protons and neutrons. Subtracting the number of protons (the atomic number) from the mass number yields the number of neutrons. The number of electrons present in an atom is always equal to the number of protons, as long as the atom does not have a positive or negative charge.

■ WORKED EXAMPLE 4

Determine the number of protons, neutrons, and electrons found in each of the following isotopes:

(a) $^{22}_{10}Ne$ (b) $^{136}_{56}Ba$ (c) $^{106}_{46}Pd$

SOLUTION

(a) $^{22}_{10}Ne \Rightarrow ^{p+n=22}_{p=10}Ne$ (b) $^{136}_{56}Ba \Rightarrow ^{p+n=136}_{p=56}Ba$ (c) $^{106}_{46}Pd \Rightarrow ^{p+n=106}_{p=46}Pd$

  p = 10   p = 56   p = 46

  n = 22 − 10 = 12   n = 136 − 56 = 80   n = 106 − 46 = 60

  e^- = p = 10 e^- = p = 56 e^- = p = 46

Practice 3.4

Determine the number of protons, neutrons, and electrons in each of the following isotopes:

(a) Krypton-78   (b) Tin-126   (c) Zinc-68

Answer

(a) 36 p, 42 n, 36 e^- (b) 50 p, 76 n, 50 e^- (c) 30 p, 38 n, 30 e^-

Some elements, such as beryllium and fluorine, do not have any naturally occurring isotopes. Other elements, such as mercury and osmium, have seven naturally occurring, stable (nonradioactive) isotopes. When an element has isotopes, the proportion of each isotope is generally constant in natural samples. For example, 92.23% of silicon atoms are silicon-28, 4.67% of silicon atoms are silicon-29, and the remaining 3.10% are silicon-30. Distribution of isotopes and percentages is unique to each element.

3.6 Atomic Mass: Isotopic Abundance and the Periodic Table

What is the mass of a single atom? If you're thinking not very much, you're right. A single atom of hydrogen has a mass of 1.67×10^{-24} g. For most purposes, scientists do not use grams to communicate atomic masses. They use the atomic mass unit (amu), introduced earlier in this chapter. One amu is defined as 1/12 of the mass of a carbon-12 atom. The masses of the rest of the elements are calculated by the ratio of the mass of an element to that

LEARNING OBJECTIVE

Determine the atomic mass of an element with isotopes.

of carbon-12. For example, if the mass of an atom were twice that of carbon-12, the mass would be near 24 amu.

The existence of isotopes complicates the matter of reporting the atomic weights of the elements in the periodic table. How do we determine the atomic mass of an element with isotopes? To explore this problem, consider the isotopes of silver, ^{107}Ag and ^{109}Ag. The natural abundance of ^{107}Ag is 51.8% and that of ^{109}Ag is 48.2%. The atomic mass reported on the periodic table should reflect that a little more than 50% of all silver atoms have a mass of 107 amu, and a little less than 50% of all silver atoms have a mass of 109 amu. It should be no surprise that the atomic mass of silver is reported as 107.865 amu, just under 108 amu. No single atom of silver will ever have a mass of exactly 107.865 amu, but on average, it is the correct value. This method of determining atomic mass by natural abundance is called a **weighted average**.

■ WORKED EXAMPLE 5

The oxygen atoms found in protein molecules consist of several isotopes that come from consumed water. The ratio of the isotopes present varies by geographic location. Oxygen has three isotopes: oxygen-16, oxygen-17, and oxygen-18. Which of these isotopes has the highest natural abundance?

SOLUTION

The atomic mass of oxygen in the periodic table is very close to 16 amu (its actual value is 15.999 amu), which indicates that the vast majority of oxygen atoms exist as the oxygen-16 isotope.

Practice 3.5

Bromine has two isotopes: bromine-79 and bromine-81. Which of these isotopes has the highest natural abundance?

Answer

❑ Bromine-79 has a slightly higher abundance.

3.7 Mathematics of Isotopic Abundance and Atomic Mass

> **LEARNING OBJECTIVE**
>
> **Show how the distribution of isotopes relates to the atomic masses listed in the periodic table.**

The atomic mass reported in the periodic table is a weighted average that accounts for the natural distribution of isotopes of various elements across the entire earth.

You may notice that some elements, such as polonium, have their mass reported as a whole number in parentheses. These elements are radioactive, that is, the mass of the elements cannot be calculated because the atoms are undergoing radioactive decay that transforms them into new elements with different masses. The number indicated in parentheses is the mass number of the most stable isotope of the element. However, there are several radioactive elements, such as uranium, that have their atomic masses calculated as the weighted average of their isotopes, much like the stable elements. These elements, which also include thorium and protactinium, have radioactive isotopes that can take millions to billions of years to be transformed to a new element. Therefore, the weighted average of their isotopes will not significantly change, and their atomic mass is calculated.

The equation for calculating the atomic mass is the sum of the mass of each isotope multiplied by its abundance, as shown:

Atomic mass = (mass of isotope 1)(% abundance) + (mass of isotope 2)(% abundance) + ...   (1)

83	84		92
Bi	**Po**		**U**
208.980	(209)		238.029

For example, carbon is found in nature with two stable isotopes: 98.93% is found as carbon-12 and 1.07% is found as carbon-13. The mass of carbon-12 is 12.000 and the mass of carbon-13 is 13.003. The atomic mass is therefore:

$$\text{Atomic mass} = (12.000)(0.9893) + (13.003)(0.0107) = 12.011$$

■ WORKED EXAMPLE 6

Chlorine has two stable isotopes. Cl-35 makes up 75.78% of a sample with a mass of 34.969 amu; the remaining 24.22% is Cl-37 with a mass of 36.966 amu. Calculate the atomic mass of chlorine as reported in the periodic table.

SOLUTION

Using equation 1:

$$\text{Atomic mass} = (34.969)(0.7578) + (36.966)(0.2422) = 35.453 \text{ amu}$$

Practice 3.6

Potassium has three stable isotopes. Calculate the atomic mass of potassium as reported in the periodic table, given the following information: K-39 is 93.258% at 38.9637 amu; K-40 is 0.01171% at 39.9640; and K-41 is 6.7302% at 40.9618 amu.

Answer

❏ 39.0983 amu

3.8 Atomic Structure: Electrons and Emission Spectra

> **LEARNING OBJECTIVE**
>
> **Assess data from clues that the emission spectra of elements provide about the location of electrons within an atom.**

To complete the understanding of atomic structure, scientists had to investigate where electrons are located in an atom. Do electrons orbit the nucleus the way planets orbit the sun, or do the electrons move around the nucleus in random paths? The clues needed to solve this atomic mystery came from experiments involving light and its relationship to the behavior of electrons in atoms.

When white light passes through a prism, all the colors of the rainbow appear in a **continuous spectrum**, as shown in Figure 3.10. It is extremely difficult to look at a continuous spectrum and pick an exact point where one color starts and another stops. Instead, the colors gradually transition from one to another. A musical analogy of a continuous spectrum is the unbroken progression of sound made by a trombone as the trombone player pushes the slide all the way out and back in one smooth motion.

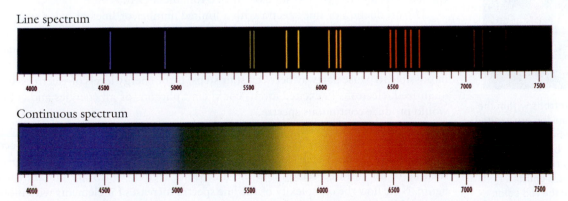

FIGURE 3.10 A continuous spectrum has no clear distinction between colors, whereas a line spectrum has clear separation between colors. (Wabash Instrument Corp./Fundamental Photographs)

White light is being separated into its continuous spectrum by a prism. (Paul Silverman/Fundamental Photographs)

Spectra Glasses, Neon Tubes, Ions

The opposite of a continuous spectrum is a **line spectrum**, the single lines of various colors that are clearly separated when certain types of light pass through a prism. A musical analogy to a line spectrum is a scale of individual notes played on an instrument.

Experimentally it had been shown that atoms produce light when placed in a flame or when electricity is passed through a gas vapor of the element. The colors of fireworks are an example of light produced by atoms in a flame, whereas neon light is an example of light that results from electricity flowing through a gas vapor. If the light emitted from excited atoms (such as those in a flame) passes through a prism, it appears as the thin bands of a line spectrum. What is especially interesting is that every element produces a different line spectrum. The unique spectrum of each element helps scientists determine what elements are present in a sample. One example would be the analysis of water for toxic heavy metals.

The discovery of line spectra posed a scientific mystery: Why do they occur? The answer turned out to provide a strong clue to how electrons are arranged in the atom. When atoms are at room temperature, the electrons are located in the lowest possible energy level, called the **ground state** (Figure 3.11a). When an atom absorbs energy, the electrons are pushed into higher energy levels, called **excited states** (Figure 3.11b). The excited state of an atom is unstable, and electrons tend to go back to the lower energy ground state. For the electron to go back to the ground state, the excess energy that it absorbed must be released (Figure 3.11c–e).

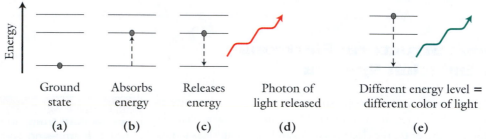

Ground state	Absorbs energy	Releases energy	Photon of light released	Different energy level = different color of light
(a)	**(b)**	**(c)**	**(d)**	**(e)**

FIGURE 3.11 (a) The ground-state electron (b) absorbs energy and then (c) spontaneously releases excess gained energy as (d) a photon of light. If the electron were to absorb a different amount of energy corresponding to a higher excited state, the photon of light produced as the electron releases the excess energy would be of a different color (e).

The electrons within the strontium ion are excited by the energy from the Bunsen burner flame. As they relax, photons of light are emitted that produce the red color. (David Taylor/Photo Researchers, Inc.)

The electron emits the excess energy in the form of a **photon**—a single "particle" or bundle of light. Multiple colors of light, corresponding to photons of different energies, are produced when electrons make the transition from higher energy levels back to the ground state. The energy of the photon, which equals the difference in energy between an excited state and the ground state of an electron, dictates the color.

The production of line spectra with a limited number of lines of color suggests that electrons in atoms are found in a few fixed energy levels. In the terminology of science, we say that the electrons are found in **quantized energy levels**. An analogy for quantized energy levels would be stair steps: The allowed states are a fixed distance apart, and it is impossible to pause partway between stair steps. If the energy levels of electrons were not quantized, electrons in excited atoms could emit photons of all energies and colors and would produce continuous spectra.

Each line in a spectrum is the result of an electron releasing energy from a higher quantized excited state to the ground state. Figure 3.12 shows the visible line spectra for several of the Group 1 elements. Many lines are produced outside the visible spectrum, such as in the infrared region and the ultraviolet region, which are not shown in the figure. Note how the complexity of the line spectra increases for elements with more electrons. The complex spectra result because a greater number of electrons are available for excitation, and there are more possible excited energy levels for the electrons to occupy.

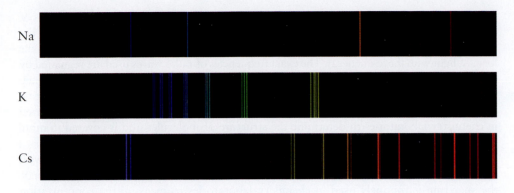

Na

K

Cs

3.9 Mathematics of Light

Light travels through space in the form of waves, much as waves travel across the oceans. Waves can be described mathematically by their **wavelength**, symbolized by the Greek letter lambda (λ). The wavelength is the distance from one wave peak to another, as illustrated in Figure 3.13. The unit used to measure wavelength for light in the visible spectrum is typically the nanometer (nm). The prefix *nano* represents the multiplier 1×10^{-9}, or one billionth of a meter.

LEARNING OBJECTIVE
Examine and use the relevant mathematical equations to describe the behavior of light.

Waves are also mathematically described by their frequency, symbolized by the Greek letter nu (ν). **Frequency** is the number of times the wave peak passes a point in space within a specific time period. It is measured in units called *hertz* (Hz), and 1 Hz represents a wave that passes a point once per second (1 Hz = 1 s^{-1}). Mathematically, s^{-1} and 1/s are equivalent.

Wavelength and frequency are related to the **speed of light** (c), as shown in equation 2 on the facing page. The speed of light in a vacuum is a constant value equal to 3.00×10^8 m/s. Equation 2 is commonly rearranged to solve for the wavelength, as shown in equation 3, or for the frequency, as shown in equation 4. One of the most common mistakes made in using equations 2–4 is failing to make sure that the units of distance are the same. The speed of light is measured in meters per second (m/s), whereas the wavelength of light from the visible spectrum is measured in nanometers. Just remember to use the appropriate conversion factors for units!

$$c = \lambda \times \nu \tag{2}$$

$$\frac{c}{\nu} = \frac{\lambda \times \nu}{\nu} \Rightarrow \lambda = \frac{c}{\nu} \tag{3}$$

$$\frac{c}{\lambda} = \frac{\lambda \times \nu}{\lambda} \Rightarrow \nu = \frac{c}{\lambda} \tag{4}$$

(a)

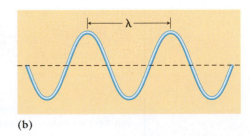

(b)

FIGURE 3.13 Waves are characterized by their wavelength (the distance between the peaks of the wave) and their frequency (the time it takes for one wave to pass a given point in one second). (a) A series of ocean waves rolling toward a beach show the wavelengths highlighted by sunlight. The greater the frequency, the shorter the time interval between waves striking the beach. (b) Light travels through space as a wave, sharing the same characteristics (wavelength and frequency) as ocean waves. (a: The Irish Image Collection/Design Pics/Corbis)

Visible light is only a small portion of the **electromagnetic spectrum**, which consists of all forms of light, as illustrated in Figure 3.14. The vast majority of light cannot be detected by the human eye. Radio waves, which are used for communication signals, are the form of light with the least energy. The exact frequencies of radio waves used are regulated by the government, and certain frequencies are reserved for police and emergency personnel. Continuing along the spectrum is the microwave region, perhaps best known for heating food in microwave ovens. The infrared region, also called the IR region, is commonly used in chemistry laboratories to determine the structure of molecules and is covered in more detail in Chapter 8.

The visible region of the spectrum is next highest in energy and consists of light with wavelengths ranging from 750 nm (red) to 380 nm (violet). The ultraviolet (UV) region of the spectrum is routinely used in chemistry laboratories to measure the concentration of compounds in solution. This method is covered in more detail in Chapter 4. The X-ray region is routinely used in hospitals and dental offices for imaging. Finally, gamma rays, the most energetic form of light, are generated during the decay process of radioactive elements.

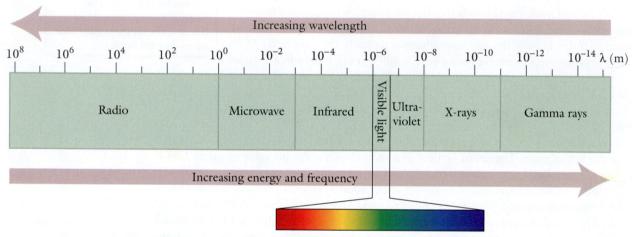

FIGURE 3.14 The electromagnetic spectrum shows all forms of light from the low-energy, large-wavelength radio waves to the high-energy, small-wavelength gamma rays. The visible region of the spectrum is a small section between 750 nm and 380 nm.

■ WORKED EXAMPLE 7

Arsenic is a toxic element that can sometimes be found in well water from naturally occurring minerals. When arsenic is in an excited state, it will emit light at a wavelength of 193.7 nm. Calculate the frequency (Hz) of light that corresponds to the wavelength of arsenic.

SOLUTION

Using equation 4,

$$\nu = \frac{c}{\lambda}$$

but first we must convert 193.7 nm to meters:

$$193.7 \ \cancel{nm} \times \frac{1 \times 10^{-9} \ m}{1 \ \cancel{nm}} = 1.937 \times 10^{-7} \ m$$

$$\nu = \frac{c}{\lambda} = \frac{3.00 \times 10^{8} \ m/s}{1.937 \times 10^{-7} \ m} = 1.549 \times 10^{15} \ 1/s = 1.55 \times 10^{15} \ Hz$$

Practice 3.7

Lead is a toxic metal that was historically used as an additive in paints. Old homes that contain lead paints can be a source of accidental lead poisoning to children, as they are more likely to ingest paint chips and lead-containing dust. The detection of lead is accomplished by measuring emitted light that corresponds to a wavelength of 220.4 nm. What is the frequency (Hz) of this wavelength of light?

Answer

❑ 1.36×10^{15} Hz

■ WORKED EXAMPLE 8

Determine the wavelength (nm) of light used for measuring cobalt, given that the frequency of the light is 1.2976×10^{15} Hz.

SOLUTION

Using equation 3,

$$\lambda = \frac{c}{\nu}$$

$$\lambda = \frac{c}{\nu} = \frac{3.00 \times 10^8 \text{ m/s}}{1.298 \times 10^{15} \text{ 1/s}} = 2.3112 \times 10^{-7} \text{ m}$$

$$2.3112 \times 10^{-7} \text{ m} \times \frac{1 \text{ nm}}{1 \times 10^{-9} \text{ m}} = 231.12 \text{ nm} \Rightarrow 231 \text{ nm}$$

Practice 3.8

Calculate the wavelength of light that lead atoms in an excited state will emit, given that the frequency of light is 1.361×10^{15} Hz.

Answer

❑ 2.20×10^2 nm

Recall that the color of each line present in a spectrum represents the transition from an excited-state energy level to a lower ground-state energy level. The difference in energy between these two levels can be calculated by the frequency of light according to equation 5 below, where E is energy with units of joules (J) and h is *Planck's constant*, which has a value of 6.626×10^{-34} J · s.

$$E = h \times \nu \tag{5}$$

Combine (5) and (4)

$$E = h \times \nu \text{ and } \nu \times \frac{c}{\lambda} \Rightarrow E = \frac{h \times c}{\lambda} \tag{6}$$

■ WORKED EXAMPLE 9

Determine the energy emitted from an arsenic atom as it releases a photon of light with a wavelength of 193.7 nm.

SOLUTION

$$E = \frac{h \times c}{\lambda} = \frac{(6.626 \times 10^{-34} \text{ J} \cdot \text{s}) \times (3.00 \times 10^8 \text{ m/s})}{193.7 \text{ nm} \times \frac{1 \times 10^{-9} \text{ m}}{1 \text{ nm}}}$$

$$= \frac{1.9908 \times 10^{-25} \text{ J}}{1.937 \times 10^{-7}} = 1.03 \times 10^{-18} \text{ J}$$

Practice 3.9

Determine the energy emitted from a mercury atom as it releases a photon of light with a wavelength of 545.2 nm.

Answer

❏ 3.65×10^{-19} J

■ **WORKED EXAMPLE 10**

Determine the energy emitted from a cobalt atom as it releases a photon of light with a frequency of 1.2976×10^{15} Hz.

SOLUTION

$E = h \times \nu = (6.626 \times 10^{-34}$ J $\cdot \cancel{s})(1.2976 \times 10^{15}$ 1/$\cancel{s}) = 8.598 \times 10^{-19}$ J

Practice 3.10

Determine the energy emitted from a lead atom as it releases a photon of light with a frequency of 1.361×10^{15} Hz.

Answer

❏ 9.018×10^{-19} J

3.10 Atomic Structure: Electron Orbitals

LEARNING OBJECTIVE

Describe the modern model of the atom and contrast it to earlier, obsolete models.

The quantized energy levels in which electrons exist led to speculation about where these levels were located around the nucleus. Early models showed electrons orbiting the nucleus as planets orbit the sun. In fact, we still draw a similar model when discussing the emission of photons from an excited energy level. However, you should recognize that what is fixed are the energy levels of electrons, *not* the locations of the electrons. Scientists often use models that are known to be oversimplified, but these tools help us understand a topic, and we also recognize that the model is not perfect.

One challenge that arises when we attempt to locate the exact position of an electron is that we cannot do so without disturbing the energy of that electron. If we attempt to measure the energy of an electron, the corresponding location is altered. Therefore, finding the exact location of electrons that have a specific, fixed energy level is impossible. This physical limitation to the measurement of electrons is summarized in the **Heisenberg uncertainty principle:**

> The more precisely the position is determined, the less precisely the momentum is known in this instant, and vice versa.
>
> —Werner Heisenberg, 1927

The energy of a particle is related to the momentum to which Heisenberg refers in his statement.

The best that scientists can do is to predict the highest probability of the location of an electron of a given energy around the nucleus of an atom. How these predictions are made is detailed in an area of study known as quantum mechanics, which requires an extensive background in calculus, chemistry, and physics. However, a full understanding of quantum mechanics is not necessary to comprehend and use the results fully. Just as we do not need an extensive knowledge of computer chip technology to use a computer, we do not have to dwell on the details of quantum mechanics to gain some understanding of electrons in atoms.

Examine the campus map shown in Figure 3.15. Although you would not be able to predict the exact location of a campus administrator, science, or music student, you have the highest probability of locating them within specific regions of campus. A similar process happens when trying to locate the electrons around an atom.

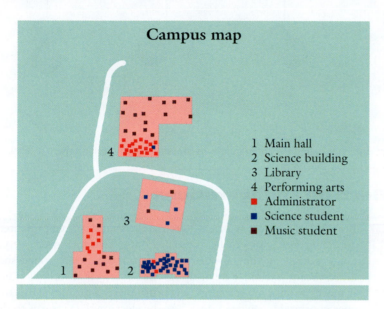

FIGURE 3.15 This campus map shows a high density of science students found around the science building. The greatest probability of finding a science student would be near the science building.

Through the mathematics of quantum mechanics, density maps for electrons of a fixed energy level have been determined. The **orbital** of an electron is a three-dimensional region of space in which an electron has the highest probability of being located. Much like the map example, an area of highest probability is defined and shown in Figure 3.16. Figure 3.16a shows the cross-sectional (two-dimensional) electron density map for locating the single electron of the hydrogen atom. It is apparent from the figure that the greatest probability of locating the electron is in a region surrounding the center of the atom. The probability of finding the electron decreases at distances farther away from the nucleus. Figure 3.16b is the three-dimensional representation of the orbital, which has a spherical shape. A sphere-shaped orbital is known as an *s* **orbital**. (The *s* actually stands for "sharp," not "spherical," as one might predict. The term *sharp* was used because it describes the emission lines observed when the electrons in the atom are excited.)

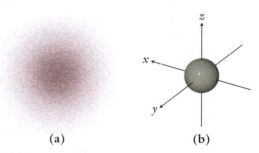

FIGURE 3.16 (a) Cross-sectional model and (b) spherical three-dimensional model of the *s* orbital.

The shape of the electron orbital changes as the energy level of an electron increases. Beyond the s orbital, there are three other orbitals, starting with the ***p* orbital**, shown as a probability density map in Figure 3.17a. The *p* orbital has a dumbbell shape: Electrons are located on either side of the nucleus in a teardrop-shaped lobe. One major difference between *p* and *s* orbitals is that there are three of the *p* orbitals, all having identical energies. As you can see in Figure 3.17b, one of the *p* orbitals is lined up on the *x* axis (p_x), one on the *y* axis (p_y), and the third on the *z* axis (p_z).

The two other electron orbital types, the *d* and *f* **orbitals**, are provided in Appendix B for completeness; this book does not delve further into the orbital discussion.

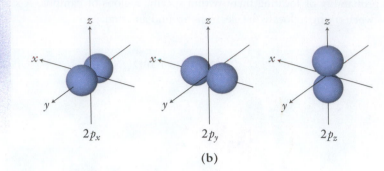

$2p_x$ orbitals

(a)

(b)

FIGURE 3.17 (a) Cross-sectional model and (b) spherical three-dimensional models of the *p* orbital.

3.11 Electron Configurations

Let's take a moment to briefly summarize what has been presented thus far about the behavior of electrons in atoms. The line spectra from excited atoms come from the transitions of electrons between a ground-state orbital and a vacant, excited-state orbital. When energy is absorbed by an electron, it moves from a lower energy level (usually the ground state) to a higher energy orbital (an excited state). As the electron then returns to the ground state, it emits a photon. The line emission spectra for atoms with fewer electrons (the lighter elements) are simpler than those of heavier elements. Heavier elements have a greater number of electrons and a larger variety of possible excited states.

For atoms with multiple electrons, a question then arises: Exactly which electron is being excited to a higher excited state? To answer this question, a system is needed to identify each electron in an atom. This system is called the *electron configuration* of the element. The periodic table is a necessary tool for understanding electron configurations.

Two rules govern the distribution of electrons in orbitals:

Rules for Electron Orbitals

1. A single orbital can contain a maximum of two electrons.

2. When filling electrons in the *p*, *d*, and *f* subsets, each orbital of the subset gets a single electron before any orbital of the subset receives a second electron.

Electron configurations can be thought of in much the same way as house numbers in your street address. Each set of orbitals is like a city block. Figure 3.18 is a color-coded map of the orbital blocks. Remember that the periodic table is arranged by increasing number of protons. For each proton that is gained, so is one electron. The electrons go into the orbital designated by the color-coded blocks in the periodic table.

Let's examine which orbitals the electrons occupy as we increase the atomic number.

Element	Number of e⁻	Orbital
H	1	First *s* orbital
He	2	Both e⁻ located in first *s* orbital
Li	3	2 e⁻ located in first *s* orbital, 1 e⁻ in second *s* orbital
Be	4	2 e⁻ located in first *s* orbital, 2 e⁻ in second *s* orbital

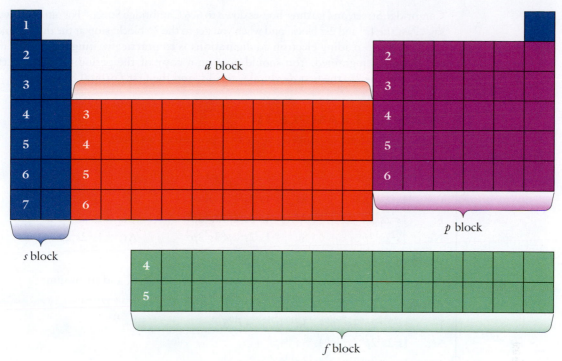

FIGURE 3.18 In examining the elements by increasing atomic number, it is apparent that electrons first occupy the 1s orbital. The second orbital to fill is the 2s followed by the 2p, and so on.

Writing the electron location in this fashion is cumbersome and time-consuming. Therefore, a shorthand method has been developed:

Number of electrons in orbital

$$\text{Be: } 1s^2\ 2s^2$$

Orbital type

Orbital number

Rewriting and expanding on our list of electron configurations:

Element	Number of e⁻	Orbital
H	1	$1s^1$
He	2	$1s^2$
Li	3	$1s^2 2s^1$
Be	4	$1s^2 2s^2$
B	5	$1s^2 2s^2 2p^1$ (Note that the first *p* orbital is in the second energy level.)
C	6	$1s^2 2s^2 2p^2$
N	7	$1s^2 2s^2 2p^3$

Let's take a closer look at reading the electron configuration for nitrogen directly from the periodic table. First, starting at the beginning of the periodic table in Figure 3.18, read each row until you get to the element of interest. Doing so is much the same as giving someone driving directions such as, "Pass the 100 and 200 block of Elm Street, turn left on

Cambridge Street, and go three houses down to 606 Cambridge Street." For nitrogen, we would say, "Pass the $1s^2$ and $2s^2$ block, and when you get to the $2p$ block, stop at the third element."

The key to reading electron configurations is to practice writing them until the pattern becomes ingrained. You should note on a copy of the periodic table that the first p orbital is the $2p$, the first d orbital is the $3d$, and the first f orbital is the $4f$.

■ WORKED EXAMPLE 11

Write the electron configurations for nitrogen through argon.

SOLUTION

N:	$1s^22s^22p^3$	Na:	$1s^22s^22p^63s^1$	P:	$1s^22s^22p^63s^23p^3$
O:	$1s^22s^22p^4$	Mg:	$1s^22s^22p^63s^2$	S:	$1s^22s^22p^63s^23p^4$
F:	$1s^22s^22p^5$	Al:	$1s^22s^22p^63s^23p^1$	Cl:	$1s^22s^22p^63s^23p^5$
Ne:	$1s^22s^22p^6$	Si:	$1s^22s^22p^63s^23p^2$	Ar:	$1s^22s^22p^63s^23p^6$

Practice 3.11

Write the full electron configurations for iodine, selenium, and strontium.

Answer

I: $1s^22s^22p^63s^23p^64s^23d^{10}4p^65s^24d^{10}5p^5$

Se: $1s^22s^22p^63s^23p^64s^23d^{10}4p^4$

Sr: $1s^22s^22p^63s^23p^64s^23d^{10}4p^65s^2$

Writing the entire electron configuration for the heavier elements can become time-consuming and repetitive. Furthermore, the electrons that are of interest to chemists are the outermost s and p orbital electrons. These electrons are collectively called the **valence shell electrons** and are of interest because chemical reactions involve the gaining, losing, and sharing of valence shell electrons.

The inner electrons that make up the beginning of the electron configuration are called the **core electrons**. Core electrons are not involved in chemical reactions except in rare cases. The separation of electrons into those that participate in reactions and those that do not leads to a method of abbreviating and shortening the electron configurations. When you provide directions to someone, you usually don't list every single house they will pass. Instead, you provide a reference point to start the direction such as, "Go down to the Y mart, turn left, go two blocks, and stop at the sixth house."

The reference point for electron configurations is the noble gas elements. Always start at the noble gas element that directly precedes the element of interest. For example, to write the electron configuration of sodium, we would start at neon, the noble gas directly before sodium. The newly abbreviated electron configuration of sodium is $[Ne]3s^1$.

■ WORKED EXAMPLE 12

Write out the abbreviated electron configurations for cobalt, xenon, sulfur, nickel, tin, and calcium.

SOLUTION

Co: $[Ar]4s^23d^7$

Xe: $[Kr]5s^24d^{10}5p^6$

S: $[Ne]3s^23p^4$

Ni: $[Ar]4s^23d^8$

Sn: $[Kr]5s^24d^{10}5p^2$

Ca: $[Ar]4s^2$

Practice 3.12

Which elements have the following electron configurations?

$[Ar]4s^23d^5$ $[Kr]5s^24d^{10}5p^3$ $[Ne]3s^23p^5$ $[Ar]4s^23d^{10}4p^2$

Answer

❑ Manganese   Antimony   Chlorine   Germanium

Table 3.3 lists the full and abbreviated electron configurations of the first 54 elements, for your reference. An excellent use for this table is in practicing writing both the full and abbreviated electron configurations of all the elements: Check your work against the table. Several elements do not follow the rules for writing electron configurations; these elements are highlighted in orange.

TABLE 3.3 Electron Configuration of the Elements Hydrogen through Xenon

	Full	Abbreviation		Full	Abbreviation
H	$1s^1$	—	Ni	$1s^22s^22p^63s^23p^64s^23d^8$	$[Ar]4s^23d^8$
He	$1s^2$	—	Cu	$1s^22s^22p^63s^23p^64s^13d^{10}$	$[Ar]4s^13d^{10}$
Li	$1s^22s^1$	$[He]2s^1$	Zn	$1s^22s^22p^63s^23p^64s^23d^{10}$	$[Ar]4s^23d^{10}$
Be	$1s^22s^2$	$[He]2s^2$	Ga	$1s^22s^22p^63s^23p^64s^23d^{10}4p^1$	$[Ar]4s^23d^{10}4p^1$
B	$1s^22s^22p^1$	$[He]2s^22p^1$	Ge	$1s^22s^22p^63s^23p^64s^23d^{10}4p^2$	$[Ar]4s^23d^{10}4p^2$
C	$1s^22s^22p^2$	$[He]2s^22p^2$	As	$1s^22s^22p^63s^23p^64s^23d^{10}4p^3$	$[Ar]4s^23d^{10}4p^3$
N	$1s^22s^22p^3$	$[He]2s^22p^3$	Se	$1s^22s^22p^63s^23p^64s^23d^{10}4p^4$	$[Ar]4s^23d^{10}4p^4$
O	$1s^22s^22p^4$	$[He]2s^22p^4$	Br	$1s^22s^22p^63s^23p^64s^23d^{10}4p^5$	$[Ar]4s^23d^{10}4p^5$
F	$1s^22s^22p^5$	$[He]2s^22p^5$	Kr	$1s^22s^22p^63s^23p^64s^23d^{10}4p^6$	$[Ar]4s^23d^{10}4p^6$
Ne	$1s^22s^22p^6$	$[He]2s^22p^6$	Rb	$1s^22s^22p^63s^23p^64s^23d^{10}4p^65s^1$	$[Kr]5s^1$
Na	$1s^22s^22p^63s^1$	$[Ne]3s^1$	Sr	$1s^22s^22p^63s^23p^64s^23d^{10}4p^65s^2$	$[Kr]5s^2$
Mg	$1s^22s^22p^63s^2$	$[Ne]3s^2$	Y	$1s^22s^22p^63s^23p^64s^23d^{10}4p^65s^24d^1$	$[Kr]5s^24d^1$
Al	$1s^22s^22p^63s^23p^1$	$[Ne]3s^23p^1$	Zr	$1s^22s^22p^63s^23p^64s^23d^{10}4p^65s^24d^2$	$[Kr]5s^24d^2$
Si	$1s^22s^22p^63s^23p^2$	$[Ne]3s^23p^2$	Nb	$1s^22s^22p^63s^23p^64s^23d^{10}4p^65s^24d^3$	$[Kr]5s^24d^3$
P	$1s^22s^22p^63s^23p^3$	$[Ne]3s^23p^3$	Mo	$1s^22s^22p^63s^23p^64s^23d^{10}4p^65s^14d^5$	$[Kr]5s^14d^5$
S	$1s^22s^22p^63s^23p^4$	$[Ne]3s^23p^4$	Tc	$1s^22s^22p^63s^23p^64s^23d^{10}4p^65s^24d^5$	$[Kr]5s^24d^5$
Cl	$1s^22s^22p^63s^23p^5$	$[Ne]3s^23p^5$	Ru	$1s^22s^22p^63s^23p^64s^23d^{10}4p^65s^14d^7$	$[Kr]5s^14d^7$
Ar	$1s^22s^22p^63s^23p^6$	$[Ne]3s^23p^6$	Rh	$1s^22s^22p^63s^23p^64s^23d^{10}4p^65s^14d^8$	$[Kr]5s^14d^8$
K	$1s^22s^22p^63s^23p^64s^1$	$[Ar]4s^1$	Pd	$1s^22s^22p^63s^23p^64s^23d^{10}4p^64d^{10}$	$[Kr]4d^{10}$
Ca	$1s^22s^22p^63s^23p^64s^2$	$[Ar]4s^2$	Ag	$1s^22s^22p^63s^23p^64s^23d^{10}4p^65s^14d^{10}$	$[Kr]5s^14d^{10}$
Sc	$1s^22s^22p^63s^23p^64s^23d^1$	$[Ar]4s^23d^1$	Cd	$1s^22s^22p^63s^23p^64s^23d^{10}4p^65s^24d^{10}$	$[Kr]5s^24d^{10}$
Ti	$1s^22s^22p^63s^23p^64s^23d^2$	$[Ar]4s^23d^2$	In	$1s^22s^22p^63s^23p^64s^23d^{10}4p^65s^24d^{10}5p^1$	$[Kr]5s^24d^{10}5p^1$
V	$1s^22s^22p^63s^23p^64s^23d^3$	$[Ar]4s^23d^3$	Sn	$1s^22s^22p^63s^23p^64s^23d^{10}4p^65s^24d^{10}5p^2$	$[Kr]5s^24d^{10}5p^2$
Cr	$1s^22s^22p^63s^23p^64s^13d^5$	$[Ar]4s^13d^5$	Sb	$1s^22s^22p^63s^23p^64s^23d^{10}4p^65s^24d^{10}5p^3$	$[Kr]5s^24d^{10}5p^3$
Mn	$1s^22s^22p^63s^23p^64s^23d^5$	$[Ar]4s^23d^5$	Te	$1s^22s^22p^63s^23p^64s^23d^{10}4p^65s^24d^{10}5p^4$	$[Kr]5s^24d^{10}5p^4$
Fe	$1s^22s^22p^63s^23p^64s^23d^6$	$[Ar]4s^23d^6$	I	$1s^22s^22p^63s^23p^64s^23d^{10}4p^65s^24d^{10}5p^5$	$[Kr]5s^24d^{10}5p^5$
Co	$1s^22s^22p^63s^23p^64s^23d^7$	$[Ar]4s^23d^7$	Xe	$1s^22s^22p^63s^23p^64s^23d^{10}4p^65s^24d^{10}5p^6$	$[Kr]5s^24d^{10}5p^6$

Scanning Electron Microscopy

The scanning electron microscope (SEM) functions like a traditional microscope except that the SEM uses electrons bouncing off a surface to form an image. The traditional microscope depends on photons of light bouncing off a surface. The main advantage to using electrons is that they are much smaller particles than photons of light, and therefore, they make it possible to obtain an image of much smaller particles. Scientists use scanning electron microscopes to obtained detailed images for such items as computer chips, insects, parasites, bacteria, viruses, and pollen. Figure 3.19 shows an image of the pollen from the passion flower (*Passiflora caerulea*), and Figure 3.20 shows a mixture of red and white blood cells.

In addition to creating amazingly detailed images of such small objects, it is possible to determine the elemental analysis of the particles at the same time. The SEM can image the particle using an instrument called an energy dispersive X-ray spectrometer (EDS), which is attached to the SEM. The combined analytical system is abbreviated SEM-EDS. To understand how the EDS system detects the presence of elements, Figure 3.21, showing a simplified Bohr model of an atom, can be of help. Although we know that electrons are not actually in orbits that resemble those of the planets, this simplified model, proposed by Niels Bohr, is a convenient device for demonstrating how the EDS system works.

In part (a), an electron from the SEM strikes an atom. Because the electron is so small, it can penetrate the atom and collide with enough energy to force an inner-core electron to leave the atom. Part (b) shows the atom with a vacancy in the inner core electrons. This condition is not stable, as electrons always go to the lowest possible energy level. Part (c) shows an electron from one of the outer shells, which has a higher energy, dropping down to fill the vacancy in the lower energy orbital and releasing excess energy as a photon. The photons emitted from atoms in this process are X-rays. The process involves the same basic concepts studied

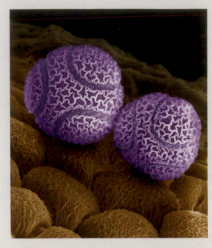

FIGURE 3.19 SEM image of the pollen from the passion flower (*Passiflora caerulea*). (© Steve Gschmeissner/ Science Photo Library/Corbis)

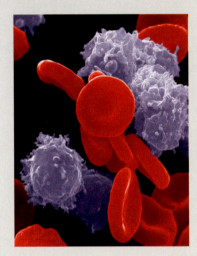

FIGURE 3.20 SEM image of red and white blood cells. (© Visuals Unlimited/Corbis)

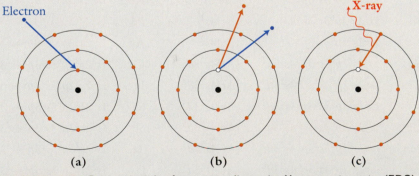

FIGURE 3.21 Basic principle of an energy-dispersive X-ray spectrometer (EDS).

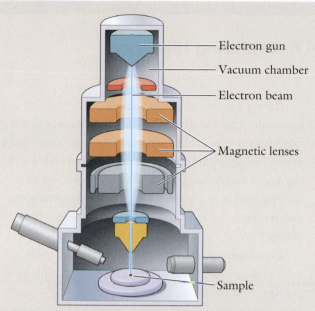

Scanning electron microscope (SEM) schematic.

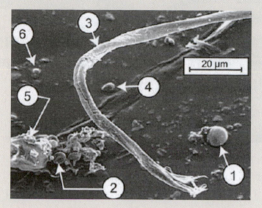

FIGURE 3.22 Six distinct particles collected from skin near the burn wound of the victim. (Kosanke et al., 2003)

earlier for understanding the creation of line spectra from energetically excited electrons relaxing to a lower energy level. Just as the line spectra of excited electrons in atoms can fingerprint an element, each element detected in an EDS emits X-rays that have a unique energy for that element. An SEM-EDS system can be used to investigate any material in which the combination of high magnification and elemental analysis would be useful to the researchers.

Consider an investigation into an accidental explosion of consumer fireworks. A person claimed to have been severely burned by the negligent use of fireworks by neighbors. The neighbors admitted to having launched fireworks but disputed the claim that they were responsible for causing the burns. If the victim's claim were true, residue from the explosives in the fireworks should be found in the vicinity of the burn. Figure 3.22 is an SEM image of the trace evidence residue obtained from skin near the burn. The next step was to determine whether any of the particles were in fact from the firework itself. The forensic scientists obtained unused fireworks of the same type used on the evening in question and compared the elemental makeup of the particles shown in Figure 3.22 to the unused fireworks. Two recovered particles were determined to be residue from an explosive material and had the same elemental composition, as shown in Figure 3.23.

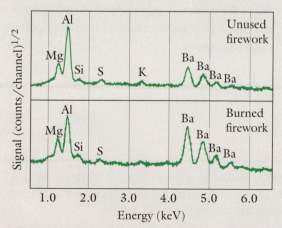

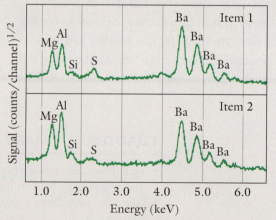

FIGURE 3.23 EDS spectra. Elemental analysis of the explosive particles in an unused firework and a carefully burned firework are compared with particles found on the victim and suspected to be from the firework.

3.12 CASE STUDY FINALE: Exploring Historical Climate Change

The ice cores record the dissolved carbon dioxide levels each year, as the new layer of snow forms. The layer also traps dust samples and air bubbles that can give more data about the climate. Some of the most interesting information about global temperature, however, comes from the molecules of water themselves.

You learned in the chapter that there are isotopes of each element that differ only by the number of neutrons found within the nucleus. Water is comprised of hydrogen, which has two main isotopes, ^1H (99.98%) and ^2H (0.016%), and oxygen, which also has two main isotopes, ^{16}O (99.76%) and ^{18}O (0.21%). A quick examination of the isotopes shows that the vast majority of water will be comprised of ^1H and ^{16}O, resulting in water molecules that have a mass of 18.0 amu. However, a very small percent of water will consist of isotopes that produce a slightly heavier water molecule.

The mass of water molecules present in the ice core layers is the key to unlocking the historical global temperature record. The oceans occupy approximately 71% of the surface of our planet, and as the water evaporates, it will move with the wind currents over the continents and come down as precipitation. The rate at which light water evaporates is greater than that of heavier water. This occurs because heavier molecules have a greater attraction for one another, and more energy is required to evaporate heavy water molecules.

During a period of global cooling, the amount of light water containing the ^{16}O isotope increases in the rain water, as the light water molecule evaporates more easily. The light water then is precipitated as snow over the Arctic and Antarctic glaciers leaving behind an isotopic temperature scale. The cooler the climate, the greater the enrichment of light water molecules that occurs.

Conversely, in a period of global warming, there is more energy available to evaporate a greater amount of the heavier water molecules. The resulting snow layer shows an increase in the relative amount of ^{18}O isotope in the ice core. Glacial geologists and other climate-change scientists cut cylindrical ice cores, which are then cataloged and stored in freezers awaiting analysis of the snow layers.

The oxygen isotopes can provide a second independent method of measuring temperature by examining carbonate containing shells and corals. When the ^{16}O isotope is being preferentially trapped on the glaciers during cool periods, there is an increasing concentration of the ^{18}O in the ocean water, which is then reflected in the calcium carbonate-based shells of shellfish. When the climate warms, a portion of the water in glaciers melts, and the water returns the ^{16}O to the oceans. The amount of ^{16}O in the main ocean increases, as does the amount found in the calcium carbonate shells.

CHAPTER SUMMARY

• Leucippus and Democritus first proposed an atomic theory of matter in ancient Greece. In the 1700s, work by Antoine Lavoisier and Joseph Proust provided sufficient experimental data for John Dalton to develop his atomic theory of matter.

• Experiments to prove the atomic theory produced results that seemed to contradict the indivisible nature of the atom. The experiments would lead to the discovery of the sub-atomic particles: electrons, protons, and neutrons.

• In Rutherford's model of the atom, most of the atom was empty space. The protons were in an extremely small region called the nucleus. Later experiments showed that the nucleus also contained neutrons.

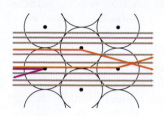

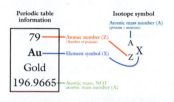

Periodic table information

79	Atomic number (Z) (Number of protons)
Au	Element symbol (X)
Gold	
196.9665	Atomic mass, NOT atomic mass number (A)

Isotope symbol

Atomic mass number (A) (protons + neutrons)

$^A_Z X$

• Isotopes are atoms of the same element that have a different number of neutrons. The isotopes undergo identical chemical reactions, which allow them to be used as medical and investigative indicators. The atomic weight of an element listed in the periodic table is actually a weighted average of all the isotopes, taking into consideration the natural abundance of each isotope.

• When energy is absorbed by an atom, the electrons can be promoted to higher quantized energy levels. The excited atoms then release photons of light as the electrons relax back to the ground state, forming a unique line spectrum that can be used to identify the element.

• Determining the exact location and the energy of an electron is impossible, according to the Heisenberg uncertainty principle, because measuring one variable inevitably alters the other. Modern atomic theory uses quantum mechanics to determine the region in space with the highest probability of containing an electron. There are four distinct orbital types, the s, p, d, and f orbitals. The s orbital is spherical in shape and the p orbital is shaped like a dumbbell. The d and f orbitals are more complex.

• Electron configurations provide a method for identifying each electron within an atom. The configuration lists each orbital in the atom and the number of electrons located in the orbitals.

• The SEM-EDS is a powerful investigative tool. It uses electrons scattered from the surface of an object to provide magnification beyond that available using a traditional light microscope, easily viewing objects that are only 10 μm in size. As the electrons strike the surface, core electrons are ejected. As a higher energy valence electron assumes the place of the missing core electron, it releases a photon of light corresponding to the X-ray region of the spectrum. The energy of the X-ray is unique to the element producing it and provides the chemical makeup of the particles being viewed.

KEY TERMS

atomic mass unit (amu) (p. 64)
atomic number (Z) (p. 65)
cathode ray tube (p. 62)
continuous spectrum (p. 69)
core electrons (p. 78)
Dalton's atomic theory (p. 60)
d orbital (p. 75)
electromagnetic spectrum (p. 72)
electron (p. 62)
excited state (p. 70)
f orbital (p. 75)
frequency (p. 71)
ground state (p. 70)

Heisenberg uncertainty
   principle (p. 74)
isotopes (p. 64)
law of conservation of
   mass (p. 57)
law of definite proportions (law of
   constant composition) (p. 59)
law of multiple proportions (p. 60)
line spectrum (p. 70)
mass number (A) (p. 65)
neutron (p. 62)
nucleus (p. 63)

orbital (p. 75)
photon (p. 70)
p orbital (p. 75)
proton (p. 62)
quantized energy level (p. 70)
radioactivity (p. 63)
s orbital (p. 75)
speed of light (p. 71)
theory (p. 57)
valence shell electrons (p. 78)
wavelength (p. 71)
weighted average (p. 68)

MAKING MORE CONNECTIONS: Additional Readings, Resources, and References

Beard, B. L., and Johnson, C. M. Strontium isotope composition of skeletal material can determine the birthplace and geographic mobility of humans and animals. *Journal of Forensic Science* 45(5), September 2003.

Emsley, J. *Nature's Building Blocks: An A–Z Guide to the Elements*, Oxford University Press, New York, 2001.

Kosanke, K. L., Dujay, R. C., and Kosanke, B. Characterization of pyrotechnic reaction residue particles by SEM/EDS. *Journal of Forensic Science* 48(3), May 2003.

Meier-Augenstein, W., and Fraser, I. Forensic isotope analysis leads to identification of a mutilated murder victim. *Science and Justice* 48, 153–159, 2008.

For more information about global climate change: http://climate.nasa.gov/

REVIEW QUESTIONS AND PROBLEMS

Questions

1. Describe atoms according to the model of Leucippus and Democritus. (3.1)

2. What did Leucippus and Democritus propose that existed where atoms did not? (3.1)

3. How did Aristotle view the nature of matter? (3.1)

4. Although Gassendi's work is not part of the modern atomic theory, why was it important at the time he published it? (3.2)

5. What does the law of conservation of mass state? (3.2)

6. How was the law of conservation of mass important to the development of modern atomic theory? (3.2)

7. What did Lavoisier do differently from other scientists that led to his success? (3.2)

8. What does the law of definite proportions state? (3.2)

9. How was the law of definite proportions important to the development of modern atomic theory? (3.2)

10. Why did Joseph Proust succeed with his experiments where others had failed? (3.2)

11. How is the term *theory* used differently in the sciences and in popular usage? (3.3)

12. Can a theory be disproved? (3.3)

13. What are the four basic principles of Dalton's atomic theory? (3.3)

14. Explain the law of conservation of mass in terms of Dalton's atomic theory. (3.3)

15. Explain the law of definite proportions in terms of Dalton's atomic theory. (3.3)

16. Explain how the law of multiple proportions came from Dalton's atomic theory. (3.3)

17. Sketch a reaction between five O_2 molecules and five C atoms that follows the law of conservation of mass. (3.3)

18. Using ● as a carbon atom and ○ as an oxygen atom, illustrate the concept of the law of definite proportions for CO_2. (3.3)

19. Using ● as a carbon atom and ○ as an oxygen atom, illustrate the concept of the law of multiple proportions for CO and CO_2. (3.3)

20. Describe a cathode ray tube and explain what the results of cathode ray tube experiments meant in terms of the properties of electrons. (3.4)

21. What are the three subatomic particles discussed in this chapter? (3.4)

22. What were the three major observations Rutherford made in his gold foil experiment? (3.4)

23. What conclusions about atomic structure did Rutherford reach after the gold foil experiment? (3.4)

24. Sketch Rutherford's atomic model. (3.4)

25. What is the main difference between isotopes of the same element? (3.4)

26. Explain how isotopes are used to investigate chemical mysteries. (3.5)

27. Why is the atomic mass for einsteinium (Es) listed as (252)? (3.7)

28. Why is the atomic mass of thorium (Th) reported as 232.038 if it is known that thorium is a radioactive element that is transformed into a different element? (3.7)

29. What type of light produces a continuous spectrum? A line spectrum? (3.8)

30. Describe what happens to the electrons in an atom when they are excited by electricity or a flame. (3.8)

31. Why is the line spectrum produced by excited cesium atoms more complex than the line spectrum produced by excited lithium atoms? (3.8)

32. Which form of electromagnetic radiation has the greatest energy? The least energy? (3.9)

33. Which form of electromagnetic radiation has the longest wavelength? The shortest wavelength? (3.9)

34. What happens to the frequency of light as the energy of the light increases? As it decreases? (3.9)

35. What happens to the energy of light as the wavelength increases? As it decreases? (3.9)

36. What is the Heisenberg uncertainty principle? (3.10)

37. What happens to the energy of an electron when the location of the electron is determined? (3.10)
38. Sketch the shape of the s and p orbitals. (3.10)
39. Sketch and label the regions of the periodic table that correspond to the s, p, d, and f electron orbitals. (3.11)
40. Describe how an X-ray photon is generated by a SEM-EDS. (ITL)

Problems

41. Place the following scientists in chronological order of their contributions to atomic theory. (3.1–3.4)
   (a) Lavoisier (c) Dalton
   (b) Democritus (d) J. J. Thomson
42. Place the following scientists in chronological order of their contributions to atomic theory. (3.1–3.4)
   (a) Gassendi (c) Chadwick
   (b) Rutherford (d) Proust
43. Which of the following statements about the experiments that led to the law of conservation of mass is false? (3.2)
   (a) The chemical reactions were studied in closed systems.
   (b) The chemical reactions neither gained nor lost mass.
   (c) The chemical reactions could exchange mass with the surroundings.
   (d) The chemical reactions could be heated externally from the surroundings.
44. Which of the following statements about the experiments that led to the law of definite proportions is false? (3.2)
   (a) A compound is made up of two or more elements.
   (b) A compound produced in the laboratory is identical to the same compound found in nature.
   (c) A compound is made up of differing relative ratios of the same elements.
   (d) A compound must be properly purified and carefully analyzed to obtain accurate results.
45. What is the mass of the missing reactant or product in each of the following reactions, given that the reaction goes to completion? (3.2)
   (a) 13.5 g water + 33.0 g carbon dioxide → _____ g carbonic acid
   (b) _____ g sodium hydroxide + 30.0 g hydrofluoric acid → 27.0 g water + 63.0 g sodium fluoride
   (c) 50.0 g calcium carbonate + 40.0 g sodium hydroxide → 53.0 g sodium carbonate + _____ g calcium hydroxide
   (d) 8.0 g sodium hydroxide + _____ g carbon dioxide → 16.8 g sodium hydrogen carbonate

46. What is the mass of the missing reactant in the following reactions, given that the reaction goes to completion? (3.2)
   (a) 2.7 g water + _____ g carbon dioxide → 9.3 g carbonic acid
   (b) 32.0 g sodium hydroxide + 16.0 g hydrofluoric acid → _____ g water + 33.6 g sodium fluoride
   (c) 0.60 g calcium carbonate + 0.48 g sodium hydroxide → _____ g sodium carbonate + 0.45 g calcium hydroxide
   (d) 0.53 g sodium hydroxide + 0.37 g carbon dioxide → _____ g sodium hydrogen carbonate
47. In Rutherford's gold foil experiment, the vast majority of alpha particles passed directly through the gold foil. This observation leads to which conclusion? (3.4)
   (a) The positive region of the atom has to be small.
   (b) The majority of the atom must consist of empty space.
   (c) The alpha particle makes a direct hit on the positive region.
   (d) The positive region of the atom is very dense.
48. In Rutherford's gold foil experiment, occasionally the alpha particle veered from a straight-line path. This observation leads to which conclusion? (3.4)
   (a) The positive region of the atom has to be small.
   (b) The majority of the atom must consist of empty space.
   (c) The alpha particle makes a direct hit on the positive region.
   (d) The positive region of the atom is very dense.
49. Fill in the missing information in the following table of subatomic particles. (3.4)

Particle	Charge	Mass (amu)	Symbol
_____	−1	_____	_____
_____	_____	1	_____
_____	_____	_____	n

50. Fill in the missing information in the following table of subatomic particles. (3.4)

Particle	Charge	Mass (amu)	Symbol
_____	_____	1.0073	_____
_____	0	_____	_____
_____	_____	0.0005486	_____

51. Using a periodic table, fill in the missing information in the following table: (3.5)

Protons	Neutrons	Electrons	Isotope
			$^{58}_{28}\text{Ni}$
22	25		
			^{12}C
18	20	18	

52. Using a periodic table, fill in the missing information in the following table: (3.5)

Protons	Neutrons	Electrons	Isotope
	143		$^{235}_{92}\text{U}$
82	126	82	
			^{64}Cu
14	14		

53. Write the isotope symbol for each of the following isotopes. (3.5)
    (a) Z = 12, neutrons = 12
    (b) Z = 19, neutrons = 22
    (c) Z = 34, neutrons = 40
    (d) Z = 56, neutrons = 80
54. Write the isotope symbol for each of the following isotopes. (3.5)
    (a) Z = 18, neutrons = 20
    (b) Z = 38, neutrons = 50
    (c) Z = 50, neutrons = 65
    (d) Z = 48, neutrons = 64
55. Which isotope of magnesium is most likely to be in the highest percentage? (3.6)
    (a) Mg-24   (b) Mg-25   (c) Mg-26
56. Which isotope of argon is most likely to be in the highest percentage? (3.6)
    (a) Ar-36   (b) Ar-38   (c) Ar-40
57. Given the isotopic abundance and mass for each element, calculate the atomic mass of each element as reported in the periodic table. (3.7)

Element	Isotope	Mass (amu)	Abundance (%)
Boron	B-10	10.013	19.9
	B-11	11.009	80.1
Bromine	Br-79	78.918	50.69
	Br-81	80.916	49.31
Rubidium	Rb-85	84.912	72.17
	Rb-87	86.909	27.83
Antimony	Sb-121	120.904	57.21
	Sb-123	122.904	42.79

58. Given the isotopic abundance and mass for each element, calculate the atomic mass of each element as reported in the periodic table. (3.7)

Element	Isotope	Mass (amu)	Abundance (%)
Lithium	Li-6	6.015	7.59
	Li-7	7.016	92.41
Copper	Cu-63	62.930	69.17
	Cu-65	64.928	30.83
Gallium	Ga-69	68.926	60.108
	Ga-71	70.925	39.892
Indium	In-113	112.904	4.29
	In-115	114.904	95.71

59. Which of the following statements about the production of line spectra is false? (3.8)
    (a) Electrons can absorb energy from being heated in a flame.
    (b) Electrons are stable in excited states.
    (c) Electrons release energy as they relax to the ground state.
    (d) Photons are produced when electrons release energy.
60. Which of the following statements about the production of line spectra is false? (3.8)
    (a) The energy of the photon can have any value.
    (b) The energy of the photon is related to the energy levels of the electron orbitals.
    (c) Electrons can be excited by passing electricity through a gaseous vapor.
    (d) Electrons always relax back to the ground state.
61. Calculate the frequency (Hz) of photons with the following wavelengths. (3.9)
    (a) 425.0 nm   (c) 850.0 nm
    (b) 682.0 nm   (d) 282.0 nm
62. Calculate the frequency (Hz) of photons with the following wavelengths. (3.9)
    (a) 376.0 nm   (c) 1153.0 nm
    (b) 686.0 nm   (d) 771.0 nm
63. Calculate the wavelength (nm) of photons with the following frequencies. (3.9)
    (a) 2.35×10^{15} Hz   (c) 8.41×10^{15} Hz
    (b) 5.68×10^{15} Hz   (d) 3.03×10^{15} Hz
64. Calculate the wavelength (nm) of photons with the following frequencies. (3.9)
    (a) 4.11×10^{15} Hz   (c) 6.00×10^{15} Hz
    (b) 7.76×10^{15} Hz   (d) 9.51×10^{15} Hz
65. Determine the energy of each photon in Problem 61. (3.9)
66. Determine the energy of each photon in Problem 62. (3.9)

67. Determine the energy of each photon in Problem 63. (3.9)
68. Determine the energy of each photon in Problem 64. (3.9)
69. Write the full electron configuration for an atom of each of the following elements. (3.11)
    (a) Ni   (b) O   (c) Cl   (d) Mn
70. Write the full electron configuration for an atom of each of the following elements. (3.11)
    (a) Fe   (b) Si   (c) Na   (d) V
71. Write the abbreviated electron configuration for an atom of each of the following elements. (3.11)
    (a) C   (b) Ge   (c) Sb   (d) As
72. Write the abbreviated electron configuration for an atom of each of the following elements. (3.11)
    (a) Rb   (b) Pb   (c) Ag   (d) P
73. Which elements have the following electron configurations? (3.11)
    (a) $1s^2 2s^2 2p^6 3s^2 3p^6 4s^2 3d^{10} 4p^6 5s^2 4d^7$
    (b) $1s^2 2s^2 2p^6 3s^2 3p^3$
    (c) $1s^2 2s^2 2p^6 3s^2 3p^6 4s^2 3d^8$
    (d) $1s^2 2s^2 2p^6 3s^2 3p^6 4s^2 3d^3$
74. Which elements have the following electron configurations? (3.11)
    (a) $1s^2 2s^2 2p^6 3s^2 3p^6 4s^2 3d^{10} 4p^6 5s^2 4d^4$
    (b) $1s^2 2s^2 2p^6 3s^2 3p^6 4s^1$
    (c) $1s^2 2s^2 2p^6 3s^2 3p^6 4s^2 3d^{10} 4p^6 5s^2$
    (d) $1s^2 2s^2 2p^2$
75. Which elements have the following electron configurations? (3.11)
    (a) $[Ne]3s^2 3p^3$ (c) $[Ne]3s^1$
    (b) $[Kr]5s^2 4d^{10} 5p^4$ (d) $[Kr]5s^2 4d^{10} 5p^5$
76. Which elements have the following electron configurations? (3.11)
    (a) $[Ar]4s^2 3d^8$ (c) $[He]2s^2 2p^6$
    (b) $[Ar]4s^2 3d^{10} 4p^3$ (d) $[He]2s^2$
77. The electron configuration for an atom of each of the following elements is incorrect. Correct the errors. (3.11)
    (a) Y: $[Ar]5s^2 4d^1$ (c) Fe: $1s^2 2s^2 3p^6 4s^2 5p^6 6s^2 7d^6$
    (b) Sc: $[Ar]4s^2 4d^1$ (d) F: $1s^2 2s^2 2p^6$
78. The electron configuration for an atom of each of the following elements is incorrect. Correct the errors. (3.11)
    (a) Ca: $1s^2 s^2 p^3 s^3 p^4 s$
    (b) P: $[Na]3s^2 3p^3$
    (c) Ge: $[Ar]4s^2 4d^{10} 4p^2$
    (d) Sn: $1s^2 2s^2 2p^5 3s^2 3p^5 4s^2 3d^{10} 4p^5 5s^2 4d^{10} 5p^2$

Case Study Problems

79. Strontium, a commonly found trace element in food, is incorporated into human bone structure because of its similarity in properties to calcium. Four stable isotopes of strontium are found in nature: ^{88}Sr, ^{87}Sr, ^{86}Sr, and ^{84}Sr. Determine the number of protons, neutrons, and electrons in each isotope of strontium. (3.5)
80. Using isotope abundance information, determine the atomic mass of strontium listed in the periodic table. (3.7)

Isotope	Mass (amu)	Abundance (%)
Sr-84	83.913	0.56
Sr-86	85.909	9.86
Sr-87	86.908	7.00
Sr-88	87.905	82.58

81. Below is the emission spectrum obtained from strontium. Determine the wavelength of the major lines and then calculate the energy and frequency for each of the main emission lines. (3.9)
82. Discuss how line emission spectra are generated. Explain why line emission spectra cannot be used to determine the isotope distribution in strontium. (3.8)
83. Strontium levels in teeth are established during the childhood years as the permanent teeth grow. However, bone material is continually replaced during our lifetime. Given these facts, explain how analyzing the isotopes of strontium in human remains might be useful to an archeologist.
84. In the latter portion of the In the Lab: case study (about the victim burned by fireworks), why was it important to analyze the pyrotechnical mixture from both an unburned firework and one that was burned in the laboratory? Could a conclusion have been drawn from an analysis of only the evidence recovered from the burn? Use your knowledge of the scientific method to help answer this question. (CP)
85. In the In the Lab: case study, the evidence recovered from the burn victim matched the elemental profile of the explosives used by the neighbors. Does this fact prove the neighbors guilty? Given the facts, provide an alternative explanation of the data that would exonerate the neighbors, if true. (CP)

Figure for Problem 81. (Wabash Instrument Corporation)

CHAPTER 4

Bonding and Reactions

CASE STUDY: Exploring the Chemistry of a Poison

George Trepal was a quiet, shy person of extraordinary intellect. He was quite knowledgeable about computer programming as well as a member of Mensa, a society for those with a measured IQ in the top 2% of the nation. He and his wife would host "murder mystery" weekends for the members of their Mensa chapter at a local hotel. The only problem that George was not able to figure out was what to do about his troublesome neighbors in the small town of Alturas, Florida.

George and his neighbors, the Carr family, had a series of run-ins over the years. The disputes ranged from such matters as the Carrs' dog—which would come into his yard—to the sons of the Carr family—who would tear up his beach with their all-terrain vehicles, play their radios too loud, and drive too fast in the neighborhood. George would confront the children and then try to talk with their parents, Peggy and Parealyn Carr. The situation never seemed to get any better.

In late October 1988, however, the Carrs abruptly moved out of their home—except for Peggy, who had unexpectedly slipped into a coma. She had earlier gone to the hospital complaining of a burning sensation in her feet and hands. Soon, her boys would complain of similar symptoms. Doctors quickly diagnosed the cause as thallium poisoning. (Thallium compounds are extremely poisonous and, in the past, were used as rat poisons.) Although her sons would recover, Peggy never emerged from her coma and passed away in March 1989.

Police started their investigation of Peggy's murder by interviewing the neighbors to see if they might provide information about Peggy or her family, or any lead that might answer the question that Peggy surely asked herself before slipping into a coma: "Why?" No one could think of anybody who would want to kill her for any reason—that is, no one except George. He ventured a guess about why someone might poison the Carr family: "To get them to move out, like they did."

The mind game George was now playing with the detectives had started several months earlier with the Carr family. They had received an anonymous typed note saying, "You and all your so-called family have two weeks to move out of Florida forever or else you all die. This is no joke." Detectives thought it more than a

> **As you read through the chapter, consider the following questions:**
>
> • **Does the type of bonding between atoms lead to different physical and chemical properties for a compound?**
>
> • **What are the different types of chemical reactions?**
>
> • **Do all chemical compounds undergo each type of reaction, and can this information be useful to a criminal investigation?**

coincidence that George would offer such an unusual reason with so little prompting. The investigation thus quickly focused on him and resulted in several key findings.

First of all, George Trepal had no record of employment for the last 15 years, although he falsely claimed to go with his wife, a physician, to her office each day. George also had been convicted of manufacturing methamphetamine in the 1970s as one of the lead chemists at a large clandestine drug laboratory. This conviction was of special interest to the authorities, as thallium compounds are often used to make one of the raw materials needed to produce methamphetamine. Finally, after George rented his home to an undercover police officer posing as a fellow member of Mensa, a bottle of thallium(I) nitrate was discovered in his garage.

The proof that the thallium in Trepal's garage was used to poison the Carr family was not without some controversy, and a flawed laboratory notebook would almost set him free on a technicality.

MAKE THE CONNECTION

But first we must investigate chemicals as evidence and follow the trail of thallium. . . .

4.1 Regions of the Periodic Table

> **LEARNING OBJECTIVE**
>
> **Describe the various regions of the periodic table and the terminology used to discuss the elements.**

This chapter will extend the discussion of chemical compounds and formulas that began in Chapter 1. Here we will explore two broad classes of compounds called *covalent* and *ionic* compounds, their formulas and names, and the kinds of reactions they undergo. But to understand these topics, more information on the periodic table is needed.

As we saw in Chapter 1, the periodic table is arranged so that elements with common properties and reactions are in the same column. Each of the columns is referred to as a **group** or **family.** Sodium and potassium, two elements in the same group, are both soft metals and form similar reaction products. Reactions 4.1 and 4.2 illustrate how sodium and potassium, respectively, undergo similar reactions with water to form hydrogen gas (H_2) and metal hydroxides (NaOH or KOH). Reactions 4.3 and 4.4 illustrate how magnesium and calcium, respectively, undergo similar reactions with water to form hydrogen gas and the resulting metal hydroxides ($Mg(OH)_2$ or $Ca(OH)_2$).

The elements within a group share many common traits, as illustrated previously with sodium and potassium. There are also differences between members of a group, such as the number of protons, neutrons, electrons, atomic mass, and isotopes of the elements.

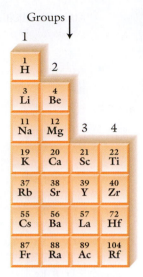

Groups

$2Na + 2H_2O \rightarrow NaOH + H_2$
(Reaction 4.1)

$2K + 2H_2O \rightarrow 2KOH + H_2$
(Reaction 4.2)

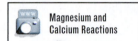

Magnesium and Calcium Reactions

$Mg + 2H_2O \rightarrow Mg(OH)_2 + H_2$
(Reaction 4.3)

$Ca + 2H_2O \rightarrow Ca(OH)_2 + H_2$
(Reaction 4.4)

(Reactions 4.1, 4.2: Charles D. Winters/Photo Researchers, Inc.; Reaction 4.3: Martin F. Chillmaid/ Science Photo Library/Photo Researchers, Inc.; Reaction 4.4: Andrew Lambert Photography/ Science Photo Library/Photo Researchers, Inc.)

There are three different methods for numbering the groups of the periodic table. This book uses the simplest version, which is also the method used by the International Union of Pure and Applied Chemistry (IUPAC), the body that governs changes to the periodic table. This method numbers the columns from 1 to 18, starting from the left and proceeding to the right, as shown in Figure 4.1.

Another method for identifying elements is to use group names. Except for hydrogen, all the elements of Group 1 of the periodic table (Li to Fr) are often referred to as the **alkali metals.** The elements of Group 2, Be to Ra, are called the **alkaline earth metals.** Not all groups are named individually. The elements of Groups 3 to 12 are collectively known as the **transition metals.** The elements of Groups 13 to 16 do not have a common name. Continuing across the periodic table, Group 17 elements are known as the **halogens,** and finally, Group 18 elements are designated the **noble gases.** Figure 4.1 (see page 92) shows a color-coded periodic table representing the members of each group with group names indicated.

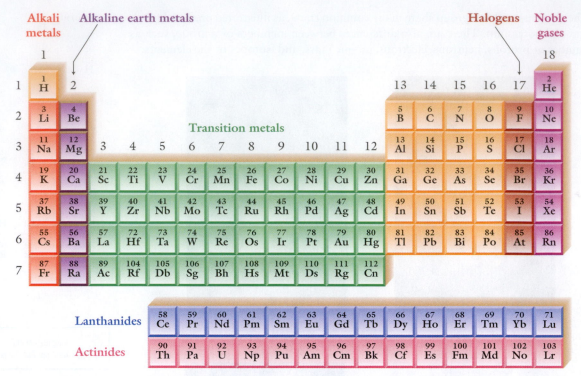

FIGURE 4.1 Regions of the periodic table.

The rows of the periodic table are referred to as **periods**, simply numbered from 1 to 7, as shown on the left side of Figure 4.1. The case study focuses on a compound of thallium, which is located in group 13 and period 6. It should be noted that the two unnumbered rows at the bottom of the table actually belong with the transition metals and are called the **inner transition metals.** The elements with atomic numbers 58–71 belong after lanthanum (La), which has atomic number 57. For this reason, they are called the *lanthanides.* Similarly, the elements with atomic numbers 90–103 would fit after actinium (Ac, atomic number 89) and are called the *actinides.* However, inserting these elements would make the periodic table awkward to use, so they are set apart at the bottom.

■ **WORKED EXAMPLE 1**

Identify the group number, group name (if any), and period for lead, arsenic, antimony, barium, and copper.

SOLUTION

Locate each element on the periodic table and then locate the column number at the top and period number at the side of the table. The common names can be identified from Figure 4.1.

Lead (Pb): Group 14, no common group name, Period 6
Arsenic (As): Group 15, no common group name, Period 4
Antimony (Sb): Group 15, no common group name, Period 5
Barium (Ba): Group 2, alkaline earth metals, Period 6
Copper (Cu): Group 11, transition metals, Period 4

Practice 4.1

The following elements are part of the electrolytes lost in sweat: Na, Ca, Cu, and Cl. Identify the group number, common group name, and period number for each element.

Answer

Sodium (Na): Group 1, alkali metals, Period 3

Calcium (Ca): Group 2, alkaline earth metals, Period 4

Copper (Cu): Group 11, transition metal, Period 4

Chlorine (Cl): Group 17, halogens, Period 3

WORKED EXAMPLE 2

Identify the following elements by their position on the periodic table.

(a) The element used to provide thermal shock resistance in glass is located in Group 13, Period 2.

(b) The element that is typically found in high concentrations in common window glass is located with the alkaline earth metals, Period 4.

(c) A toxic element, historically used as a rodent poison, which is located in Group 13, Period 6.

SOLUTION

(a) Count over 13 columns and down 2 rows on the periodic table to get to boron. Note: The periods are numbered on the left side of the periodic table. Boron is in Period 2, despite being the first element of Group 13.

(b) Alkaline earth metals are Group 2. The fourth row down is calcium.

(c) Counting over 13 columns and down 6 rows gives the element thallium.

Practice 4.2

Identify the following elements, from the information provided.

(a) Group 13, Period 3 element, a metal used to make high-strength, lightweight alloys.

(b) Alkaline earth metal in Period 3, a critical component in chlorophyll.

(c) Group 14, Period 3 element, used to make computer processors and memory chips.

Answer

(a) Aluminum

(b) Magnesium

(c) Silicon

4.2 Types of Compounds: Covalent Compounds

In Chapter 1, a compound was defined as a pure substance that is composed of two or more elements bonded together. We will now consider compounds in more detail. There are two types of chemical bonds—covalent and ionic—that form between atoms within a compound. Each bonding type imparts different physical and chemical properties to the compound. The two types of

> **LEARNING OBJECTIVE**
> Write the names and formulas of covalent compounds.

compounds must then be handled differently when being collected and analyzed as evidence. Each type also has its own distinct set of rules for naming compounds.

The first form of bonding in compounds that we will study occurs when two atoms share electrons. This sharing is referred to as a **covalent bond**; compounds that have this type of bonding are called *covalent compounds*. Covalent bonding occurs when the atoms from two nonmetal elements form a bond, such as H_2S, CO_2, and NH_3. The covalent compounds discussed in this chapter will be limited to binary (two-element) compounds. A much larger class of covalent compounds consists of the organic compounds, which are covered in depth in Chapter 8.

How exactly are electrons shared between two atoms? Recall that the electrons are located in orbitals that describe the area of highest probability of finding the electron. If atoms are to share electrons, then their orbitals must overlap with one another. In the reaction shown in Figure 4.2a, a covalent bond forms between the two hydrogen atoms by the overlap of s orbitals to form hydrogen gas (H_2). In a similar fashion, the reaction shown in Figure 4.2b illustrates the overlap of a p orbital of fluorine with an s orbital of hydrogen to form HF.

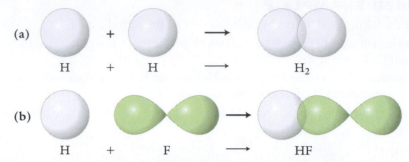

FIGURE 4.2 The formation of covalent bonds occurs through the overlap of electron orbitals from each of the atoms participating in the bond.

Names and Formulas of Covalent Compounds

The name of a compound must be unique to the compound so that confusion does not occur between differing compounds. The name must also provide sufficient information such that the formula for the exact compound can be determined. This is accomplished by including in the compound name the number of atoms of each element and the name of the element. As an example, consider the covalent compound carbon dioxide, CO_2. The name of the first element (carbon) is given first, followed by the name of the second element (oxygen) with an *-ide* ending and a prefix that indicates the number of atoms of each element. The prefixes used are listed in Table 4.1. Thus, CO_2 is named *carbon dioxide*. It is not named *monocarbon dioxide*, as it is customary to omit the prefix *mono-* if the first element in the formula has only one atom. The prefix identifying the number of oxygen atoms in the compound is needed because carbon can combine with oxygen to form carbon monoxide (CO) and the name must clearly distinguish one from the other.

TABLE 4.1	Prefixes for Covalent Compounds
Prefix	**Number**
mono-	1
di-	2
tri-	3
tetra-	4
penta-	5
hexa-	6
hepta-	7
octa-	8
nona-	9
deca-	10

If the name of an element starts with a vowel, omit the letters *a* or *o* from the end of the prefix for ease of pronunciation (as was shown above for carbon monoxide). However, do not omit the letter *i* in the prefixes *tri-* or *di-*, as in carbon dioxide.

WORKED EXAMPLE 3

Predict the formula for the following compounds:

(a) Carbon tetrachloride, used to make the blob in a lava lamp.

(b) Sulfur dioxide, used to prevent spoilage of wine.

(c) Dihydrogen monosulfide, responsible for the smell of rotten eggs.

SOLUTION

(a) One carbon atom and four chlorine atoms form CCl_4.

(b) One sulfur atom and two oxygen atoms form SO_2.

(c) Two hydrogen atoms and one sulfur atom form H_2S.

Practice 4.3

Name the following compounds:

(a) PF_5

(b) SO_3

(c) SF_2

(d) ICl

Answer

(a) Phosphorus pentafluoride

(b) Sulfur trioxide

(c) Sulfur difluoride

(d) Iodine monochloride

WORKED EXAMPLE 4

In Section 3.3, the law of multiple proportions was illustrated using various compounds composed of nitrogen and oxygen. Name each of the following nitrogen/oxygen compounds.

(a) NO

(b) N_2O

(c) N_2O_3

SOLUTION

(a) *Mono-* represents one atom, but *mono-* is not used before the first element in a compound's name. Use *mono-* only in front of the second element. The name of NO is nitrogen monoxide.

(b) *Di-* represents two atoms, and *mono-* represents one atom. The name of N_2O is dinitrogen monoxide.

(c) *Di-* represents two atoms, and *tri-* represents three atoms. The name of N_2O_3 is dinitrogen trioxide.

Practice 4.4

Name each of the following nitrogen/oxygen compounds.

(a) NO_2

(b) N_2O_5

(c) N_2O_4

Answer

(a) Nitrogen dioxide

(b) Dinitrogen pentoxide

(c) Dinitrogen tetroxide

Properties of Covalent Compounds

When a covalent compound is in the solid state, the attractive forces between molecules are weak. The addition of only a small amount of energy is needed to melt the solid. For this reason, the melting point of most covalent compounds is typically low, as compared with that of ionic compounds. For example, the covalent compound HF has a melting point of −35°C, whereas the ionic compound NaF has a melting point of 995°C. A similar pattern exists with the boiling points of covalent and ionic compounds.

4.3 Types of Compounds: Ionic Compounds

LEARNING OBJECTIVE

Write the names and determine the formulas of ionic compounds.

If the bond between two atoms forms by one atom transferring an electron (or electrons) to the other atom, the result is an **ionic compound**. When an atom loses or gains electrons during the formation of an ionic compound, it becomes electrically charged and is referred to as an **ion**. More specifically, the loss of an electron leaves an atom with a +1 charge (recall that electrons have a −1 charge), thus forming a positively charged ion called a **cation**. The gain of an electron gives an atom a −1 charge, forming a negatively charged ion called an **anion**. An **ionic bond** is the attractive force between the positively charged cation and the negatively charged anion.

Iron(III) sulfate   Iron(II) sulfate

Copper(II) carbonate   Copper(II) sulfate   Sodium chloride

(a)

(b)

(a) Ionic compounds are found as solid powders with a variety of colors. (b) The white ionic powder shown here is actually the deadly compound thallium(I) nitrate, believed to be the poison used to kill Peggy Carr. (a: Tom Pantages; b: Courtesy of American Elements)

Naming Ionic Compounds

Ionic compounds always contain at least one cation and one anion. The name of an ionic compound begins with the name of the cation followed by the name of the anion. The rule for naming a cation is quite simple: Write the name of the element followed by the word *ion*. The reaction below shows the formation of the sodium ion from elemental sodium. Notice that the charge of the sodium ion is indicated as a superscript following the atomic symbol, as shown below.

$$Na \rightarrow Na^+ + e^- \qquad \textbf{(Reaction 4.5)}$$

The rule for naming an anion is to change the ending of the element name with the suffix *-ide* and add the word *ion*. The next reaction shows the formation of the oxide ion from atomic oxygen. For any ion with a charge magnitude greater than 1, the number appearing before the positive (+) or negative (−) charge indicates the magnitude of the charge.

$$O + 2e^- \rightarrow O^{2-} \qquad \textbf{(Reaction 4.6)}$$

■ WORKED EXAMPLE 5

Name each of the following ions, which are major components in commercial mineral water.

(a) K^+

(b) F^-

(c) Ca^{2+}

(d) H^+

SOLUTION

(a) Cations have the same name as the element but with *ion* added: potassium ion.

(b) Anions have the ending of the element name modified with *-ide*: fluoride ion.

(c) Cations have the same name as the element but with *ion* added: calcium ion.

(d) Cations have the same name as the element but with *ion* added: hydrogen ion.

Practice 4.5

Name the following physiologically important ions:

(a) Cl^-

(b) Mg^{2+}

(c) Na^+

Answer

(a) Chloride ion

(b) Magnesium ion

(c) Sodium ion

Determining Formulas of Ionic Compounds

The periodic table provides assistance in determining whether an atom will form a cation or anion and, to some degree, what the charge of the ion will be. Metals tend to lose electrons and form cations, while nonmetals tend to gain electrons and form anions. (The metals and nonmetals can be distinguished by their positions on the periodic table in Figure 1.6.)

The charges of ions for some groups of elements follow a predictable pattern. For example, the alkali metals (Group 1) form cations that have a charge of +1. Cations formed by the alkaline earth metals (Group 2) have a +2 charge. The Group 16 elements typically form −2 anions, and the halogens (Group 17) form −1 anions. The noble gases do not form ions.

As for other elements with charges that tend to be consistent, silver in Group 11 always forms a +1 cation, zinc and cadmium in Group 12 form +2 cations, and aluminum in Group 13 forms a +3 cation. Finally, nitrogen and phosphorus in Group 15 tend to form −3 anions. It may be helpful for you to write the common charges of the ions on a copy of the periodic table, as shown in Figure 4.3, to refer to while doing homework. The remaining elements not listed above can have several different charges and will be discussed later in the chapter.

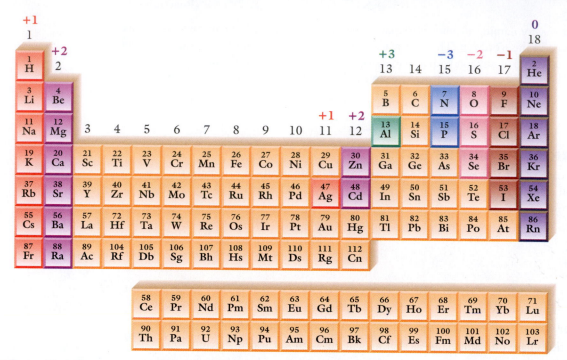

FIGURE 4.3 The charges of ions can be predicted based on the location of the elements in the periodic table. The noble gas elements do not form ions because of the stability of having filled the outermost *s* and *p* orbitals with eight electrons. The stability of having eight valence electrons is called the *octet rule* and explains why the ionic charges of the elements follow trends. All halogens tend to gain one electron, forming a −1 anion, to fill the outer *s* and *p* orbitals. For the alkali metals, it is easier to lose one electron than to gain seven electrons to reach the octet. The octet rule will be studied in greater depth in Chapter 5.

■ **WORKED EXAMPLE 6**

Predict the charge on the following ions, which can be found in the analysis of glass fragments recovered from a crime scene.

(a) Barium ion

(b) Aluminum ion

(c) Lithium ion

(d) Oxide ion

SOLUTION

(a) Ba is a member of Group 2, alkaline earth metals, and forms ions with a +2 charge.

(b) Al is a member of Group 13 and forms ions with a +3 charge.

(c) Li is a member of Group 1, alkali metals, and forms ions with a +1 charge.

(d) O is a member of Group 16 and forms ions with a −2 charge.

Practice 4.6

Predict the charge on the following ions:

(a) Zinc ion

(b) Phosphide ion

(c) Fluoride ion

(d) Strontium ion

Answer

(a) $+2$

(b) -3

(c) -1

(d) $+2$

The formulas of ionic compounds are determined by examining the charges of the cation and the anion and remembering that the total positive charge must be equal to the total negative charge so that the final compound has a zero (neutral) charge. For example, the formula for sodium fluoride, used to provide fluoride in drinking water, can be determined by looking at the periodic table and recalling the rules for charges of each group. Sodium forms the cation Na^+ and fluorine forms the anion F^-. Because the $+1$ and -1 charges sum to zero, the formula for sodium fluoride is NaF. Remember that both elements exist as charged ions within the compound, which is neutral overall. To determine the formula for calcium chloride, a common component in fingerprint residue, note that calcium belongs to the Group 2 elements, which form $+2$ cations, and that chlorine forms a -1 anion. The final formula must have a zero charge. Therefore, the formula must have two chloride ions for each calcium ion, forming $CaCl_2$.

To determine the formula for ionic compounds when the charges are not equal, it is necessary to find the lowest common multiple of the ionic charges. For example, Ca^{2+} and P^{3-} have a lowest common multiple of 6 ($2 \times 3 = 6$). Therefore, the formula would require three Ca^{2+} ions and two P^{3-} ions to form the neutral compound Ca_3P_2. If one carefully examines the magnitude of the ion charges, those values indicate the number of atoms of the opposite ion needed, as shown below for Ca_3P_2. Always verify that the formula produces a neutral (zero charge) compound and that the least common multiple is used.

$$Ca^{2+} \quad P^{3-} \rightarrow Ca_3P_2 \qquad \text{Because:} \qquad \begin{array}{l} 3 \times Ca^{2+} = +6 \\ \underline{2 \times P^{3-} \; = -6} \\ \qquad\qquad 0 \end{array}$$

■ WORKED EXAMPLE 7

Predict the formulas for the following ionic compounds formed from the elements indicated below.

(a) A compound formed with K and Br, used by veterinarians to stop canine seizures.

(b) A compound formed with Ca and O, commonly called lime and historically used for stage lighting: "the limelight."

(c) A compound formed with Al and O, the primary elements composing rubies and sapphires, although their colors come from trace impurities.

SOLUTION

(a) KBr: The +1 charge of K^+ and the −1 charge of Br^- balance to zero.

(b) CaO: The +2 charge of Ca^{2+} and the −2 charge of O^{2-} balance to zero.

(c) Al_2O_3: The +3 charge of Al^{3+} and the −2 charge of O^{2-} must balance to zero; this requires two Al^{3+} and three O^{2-}.

Practice 4.7

Predict the formula for the ionic compound formed from the elements indicated below.

(a) A compound formed with Al and Cl and used industrially to prepare clothing dyes.

(b) A compound formed with Ca and Cl and used to melt ice in the winter.

(c) A compound formed with K and I, a micronutrient added to common table salt to prevent the formation of goiters.

Answer

(a) $AlCl_3$

(b) $CaCl_2$

(c) KI

Drawing upon your personal experiences, you probably already know the rule for naming ionic compounds. What is the chemical name and formula for table salt? The answer is, of course, sodium chloride, or NaCl, which consists of the sodium ion and the chloride ion. Simply combine the names of the cation and the anion, omitting the word *ion* from each.

■ WORKED EXAMPLE 8

Name the compounds from Worked Example 7.

SOLUTION

(a) Potassium ion and bromide ion form the compound potassium bromide.

(b) Calcium ion and oxide ion form the compound calcium oxide.

(c) Aluminum ion and oxide ion form the compound aluminum oxide.

Practice 4.8

Name the compounds from Practice 4.7.

Answer

(a) Aluminum chloride

(b) Calcium chloride

(c) Potassium iodide

Cations with Multiple Charges and Polyatomic Ions

As shown in Figure 4.3, the majority of transition metals as well as the metals of Group 13 through Group 15 did not have a common charge. These elements will form cations. The charges of the ions cannot be easily predicted, as many of the elements can form cations with different charge magnitudes.

Different ions of the same element can have greatly different properties. For example, copper can form either a +2 ion or a +1 ion. However, the Cr^{3+} ion is an essential nutrient in our diets. In the case against George Trepal, it was specifically the thallium(I) ion and not the thallium(III) ion that was used to poison the Carr family. The names of compounds containing transition metal cations, or cations from the metals found in Groups 13 through 15 that do not follow a pattern, should indicate which ion of the element is present in the compound. Table 4.2 lists some of the common elements that can have multiple charges.

The method for distinguishing the ions from each other is to place the charge of the ion in Roman numerals after naming the metal element. For example, Cr^{3+} is named chromium(III) ion, and the Cr^{6+} ion is named chromium(VI) ion. When naming a compound, the Roman numeral charge is included as part of the name. Thus, the name of $CrCl_3$ is chromium(III) chloride.

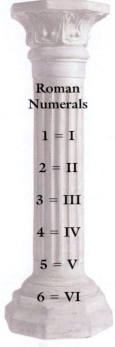

Roman Numerals

1 = I
2 = II
3 = III
4 = IV
5 = V
6 = VI

(Getty Images)

TABLE 4.2 Cations with Multiple Common Charges			
Element	Common Ionic Forms	Stock System Name	Latin System
Copper	Cu^{1+}	Copper(I) ion	Cuprous ion
	Cu^{2+}	Copper(II) ion	Cupric ion
Iron	Fe^{2+}	Iron(II) ion	Ferrous ion
	Fe^{3+}	Iron(III) ion	Ferric ion
Cobalt	Co^{2+}	Cobalt(II) ion	Cobaltous ion
	Co^{3+}	Cobalt(III) ion	Cobaltic ion
Chromium	Cr^{2+}	Chromium(II) ion	Chromous ion
	Cr^{3+}	Chromium(III) ion	Chromic ion
Lead	Pb^{2+}	Lead(II) ion	Plumbous ion
	Pb^{4+}	Lead(IV) ion	Plumbic ion
Tin	Sn^{2+}	Tin(II) ion	Stannous ion
	Sn^{4+}	Tin(IV) ion	Stannic ion
Mercury	Hg^{1+}	Mercury(I)	Mercurous ion
	Hg^{2+}	Mercury(II)	Mercuric ion
Gold	Au^{1+}	Gold(I)	Aurous ion
	Au^{3+}	Gold(III)	Auric ion

WORKED EXAMPLE 9

Predict the formula and name the compound formed from the following ions:

(a) Au^{3+} and Cl^-, a fixative used to create bronze-colored photographic images.

(b) Hg^{2+} and Cl^-, a historical treatment for syphilis.

(c) Fe^{3+} and Cl^- removes phosphorus in wastewater treatment, which, if not done, can lead to algae blooms.

SOLUTION

(a) Three chloride ions are needed to balance the +3 charge on the gold(III) ion: $AuCl_3$, gold(III) chloride.

(b) Two chloride ions are needed to balance the +2 charge on the mercury(II) ion: $HgCl_2$, mercury(II) chloride.

(c) Three chloride ions are needed to balance the +3 charge on the iron(III) ion: $FeCl_3$, iron(III) chloride.

Practice 4.9

Predict the formula and name the compound formed from the following ions:

(a) Cu^{2+} and Cl^-, a compound used to provide blue-green color to fireworks.

(b) Ti^{4+} and O^{2-}, a compound used as a pigment for white paint that wasn't made until the 1800s. One method of detecting forged artwork is to look for this compound on works of art supposedly from earlier periods.

(c) Pb^{2+} and Cl^-, a compound used as a pigment in white paints before the 1800s.

Answer

(a) $CuCl_2$, copper(II) chloride

(b) TiO_2, titanium(IV) oxide

(c) $PbCl_2$, lead(II) chloride

Thus far, we have discussed only ions that are formed when one element gains or loses electrons. These are commonly referred to as **monatomic ions**, meaning ions formed from a single atom. There is another group of ions called **polyatomic ions**, which consist of two or more atoms covalently bonded to each other that also have a net ionic charge. Polyatomic ions can be found as either cations or anions, as illustrated in Table 4.3.

One of the most infamous polyatomic ions is the cyanide ion, which has the formula CN^-. The cyanide ion consists of a carbon atom tightly bonded to a nitrogen atom in a unit that has an overall −1 charge. It is important to note that the carbon and nitrogen atoms stay bonded together under typical conditions in compounds containing the cyanide ion. Several polyatomic ions were featured in the investigation of George Trepal. The case against George Trepal was based on evidence that the compound found in his garage, thallium(I) nitrate, was used to poison the Carr family by spiking their soft drinks. The analysis of various Coca-Cola samples from the Carr home included tests for the presence of the sulfate ion. Table 4.3 has a list of common polyatomic ions.

TABLE 4.3	Polyatomic Ions		
Polyatomic Ion	**Symbol**	**Polyatomic Ion**	**Symbol**
Ammonium ion	NH_4^+	Chlorate ion	ClO_3^-
Nitrate ion	NO_3^-	Perchlorate ion	ClO_4^-
Nitrite ion	NO_2^-	Hydrogen carbonate ion	HCO_3^-
Hydroxide ion	OH^-	Carbonate ion	CO_3^{2-}
Acetate ion	$C_2H_3O_2^-$	Sulfate ion	SO_4^{2-}
Cyanide ion	CN^-	Chromate ion	CrO_4^{2-}
Permanganate ion	MnO_4^-	Dihydrogen phosphate ion	$H_2PO_4^-$
Hypochlorite ion	ClO^-	Hydrogen phosphate ion	HPO_4^{2-}
Chlorite ion	ClO_2^-	Phosphate ion	PO_4^{3-}

◼ WORKED EXAMPLE 10

Predict the formula for the compounds consisting of the following ions:

(a) K^+ and NO_3^- form a compound that is a common component in fertilizers.

(b) Fe^{2+} and NO_3^- form a compound used by jewelers to etch designs into sterling silver.

(c) Na^+ and CN^- form an extremely deadly compound that is used to extract gold and precious metals from ores in mining.

SOLUTION

(a) KNO_3: The +1 and −1 ion charges balance to zero.

(b) $Fe(NO_3)_2$: The Fe^{2+} requires two NO_3^- ions to balance the charge to zero.

(c) NaCN: The +1 and −1 ion charges balance to zero.

Practice 4.10

Name the compounds shown in the solution to Worked Example 10.

Answer

(a) Potassium nitrate

(b) Iron(II) nitrate

(c) Sodium cyanide

■ WORKED EXAMPLE 11

Predict the formulas for the following compounds:

(a) Calcium sulfate, a cheap whitening agent

(b) Iron(II) cyanide, a blue pigment

(c) Zinc phosphate, a corrosion-resistant white pigment

SOLUTION

(a) $CaSO_4$: Calcium always forms +2 ions and sulfate is SO_4^{2-}.

(b) $Fe(CN)_2$: Iron(II) is Fe^{2+} and cyanide is CN^-.

(c) $Zn_3(PO_4)_2$: Zn always forms +2 ions and phosphate is PO_4^{3-}. To achieve a formula with a neutral charge, a lowest common multiple of 6 is needed. This requires three Zn^{2+} ions and two PO_4^{3-} ions.

Practice 4.11

Determine the charge on each ion within the following compounds.

(a) $FeCO_3$

(b) $CuNO_3$

(c) Al_2S_3

Answer

(a) Fe^{2+}, CO_3^{2-}

(b) Cu^+, NO_3^-

(c) Al^{3+}, S^{2-}

Properties of Ionic Bonds

The strength of an ionic bond comes from the attraction between the opposite charges of the cation and the anion. The positive cation is attracted to the negative anion, just as the south pole of one magnet is attracted to the north pole of another. The ions are held together in fixed positions within a crystal structure. The crystal structure has a three-dimensional, repeating pattern called the **crystal lattice** in which the cations and anions are arranged. Shown in Figure 4.4 is the crystal lattice for KBr. Note that the crystal is made up of many potassium ions and many bromide ions. There is not actually a single KBr unit comparable to a molecule in a covalent compound. (The term *formula unit* rather than *molecule* is used to describe the simplest ratio of the ions that make up an ionic compound.) Each ion is effectively held in place by the attractive forces of the other ions surrounding it on all sides. The crystal lattice imparts the properties of hardness and rigidity that ionic compounds share.

The **melting point** for an ionic compound is the temperature at which the ions have sufficient energy to escape the attractive forces holding them in the lattice structure. When the melting point is reached, the crystal lattice collapses and a liquid is formed in which the ions are able to move past one another. In general, the melting point of an ionic compound is very high and increases with the magnitude of the charge on the ions.

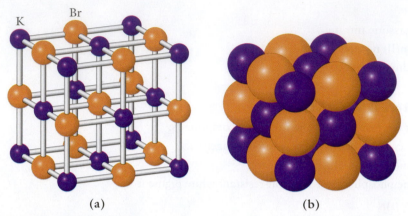

K   Br

(a)   (b)

FIGURE 4.4 (a) The lattice structure of KBr shows the three-dimensional repeating units of a crystal. (b) The space-filling model of KBr illustrates the actual spacing of atoms within the crystalline compound.

■ WORKED EXAMPLE 12

The melting point of CaO is 2572°C and that of KBr is 730°C. Explain why there is such a large difference in melting points.

SOLUTION

Because CaO has the higher melting point, the Ca^{2+} ions must be attracted more strongly to the O^{2-} ions than the K^+ ions are attracted to the Br^- ions. This makes sense because the attractive force between a $+2$ charge and a -2 charge would be stronger than the force between a $+1$ charge and a -1 charge.

Practice 4.12

Place the following compounds in order from lowest melting point to highest melting point:

(a) SrS

(b) FeN

(c) KCl

Answer

❏ KCl < SrS < FeN

Summary of Rules for Naming Ionic and Covalent Compounds

It is quite common for beginning students to confuse the rules for naming ionic compounds with the rules for naming covalent compounds, especially under the stress of an exam. The best way to prevent this is, of course, to practice! The second is to remember the rules and follow a systematic method for determining the correct name. Listed in the flowchart are some helpful questions to ask while naming compounds.

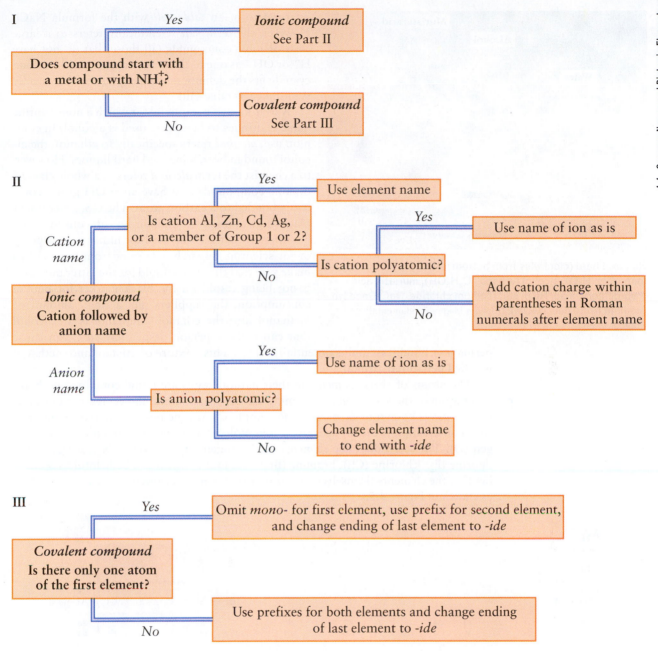

4.4 Common Names and Diatomic Elements

Historically, many compounds have been given names that were not as sys-
tematically determined as those described in the previous sections. Despite
the implementation of the standard rules for naming compounds, some of
these older names are still used in industry and laboratories around the
world. The best example is water, which has the formula H_2O. The systematic name for water is dihydrogen monoxide, which is not used because it would cause
more confusion than clarity. Another name commonly used is *salt*, which most people

LEARNING OBJECTIVE
Designate chemical
compounds when
common names are used.

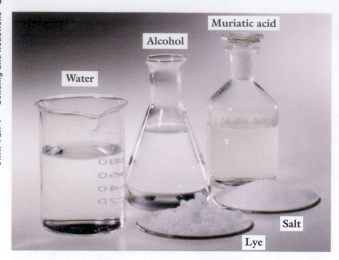

Pictured here (clockwise from bottom left): water (dihydrogen monoxide, H_2O), alcohol (ethanol, C_2H_5OH), muriatic acid (hydrochloric acid, HCl), salt (sodium chloride, NaCl), and lye (sodium hydroxide, NaOH). (Richard Megna/Fundamental Photographs)

interpret to mean *table salt* with the formula NaCl. However, the term **salt** scientifically refers to a large class of ionic compounds (all those that do not have H^+ or OH^- as one of the ions), not just NaCl. Sodium cyanide fits the definition of a salt and could be deadly if mistaken as table salt!

A well-known compound for which a nonscientific name continues to be widely used is alcohol. In common use, *alcohol* refers specifically to ethanol, the alcohol found in beer, wine, and hard liquors. However, to a chemist the term *alcohol* refers to a whole class of organic compounds that have an —OH group covalently bonded to a carbon atom. The U.S. government places high taxes on ethanol for human consumption, but ethanol that is to be used for industrial purposes or for scientific research is exempt from this tax. To make sure that the ethanol sold for the latter purposes is not being resold on the black market for human consumption, the suppliers mix a small amount of methanol into the ethanol. Methanol is an alcohol that can make a person desperately sick, can cause permanent blindness, and is potentially lethal. This mixture of ethanol and methanol is called **denatured alcohol**.

The atoms of seven elements in their natural states are found combined with another atom of the same element, forming diatomic (two-atom) molecules. For example, oxygen gas is known to exist as O_2, not simply O. Anyone referring to oxygen in ordinary conversation is usually speaking about molecular oxygen (O_2) and not atomic oxygen (O). The seven diatomic elements are hydrogen (H_2), nitrogen (N_2), oxygen (O_2), fluorine (F_2), chlorine (Cl_2), bromine (Br_2), and iodine (I_2). It can be helpful to remember that the elements themselves appear to form the number seven on the periodic table, as shown in Figure 4.5.

FIGURE 4.5 The seven diatomic elements (shown in violet).

4.5 Basics of Chemical Reactions

One of the key methods for identifying the type of evidence at a crime scene or in a laboratory is the use of chemical reactions. For example, when detectives are attempting to determine whether a white powder found on a kitchen table is an illegal narcotic or an innocent baking ingredient, they can conduct a field test for the presence of illegal drugs. A field test requires a small amount of the suspected drug, which is mixed with chemicals that are known to react with illegal drugs. The chemical reactions will produce colored compounds with illegal drugs but not with most other substances.

Contrary to what you may see in Hollywood movies, detectives do not taste a substance to identify it. This practice would endanger their lives and would not provide reliable or usable evidence. It is standard practice in laboratories to forbid not only the tasting of chemicals but also the consumption of any food or beverages within the laboratory because these items could become contaminated with laboratory chemicals.

Chemical reactions are represented symbolically by writing a chemical equation. The compounds that react with each other are collectively known as the **reactants** and are always listed on the left side of the chemical equation. The compounds that are formed as a result of the reaction are called the **products** and appear on the right side of the equation. An arrow separates the reactants from the products.

The chemical equation for the detonation of gunpowder is shown in Reaction 4.7. After the formula for each compound, abbreviations for the physical state may be, but are not always, included: (s) = solid, (l) = liquid, (g) = gas.

Officers can make a preliminary identification of illegal drugs by placing a small sample in a field test reaction vial. A positive reaction is indicated by the formation of a product compound with a specific color. Confirmatory tests are then conducted by forensic chemists in the laboratory. (Mistral Security)

$$4KNO_3(s) + 7C(s) + S(s) \rightarrow 3CO_2(g) + 3CO(g) + 2N_2(g) + K_2CO_3(s) + K_2S(s)$$

(Reaction 4.7)

Reactants · Products

Written in words rather than chemical symbols, Reaction 4.7 would be stated: "Four potassium nitrate formula units react with seven carbon atoms and one sulfur atom to produce three molecules of carbon dioxide, three molecules of carbon monoxide, two molecules of nitrogen gas, one formula unit of potassium carbonate, and one formula unit of potassium sulfide."

Recall that the law of conservation of mass states that mass is neither created nor destroyed during a chemical reaction. Also recall that Dalton described a chemical reaction as the rearrangement of atoms to form new compounds. If you carefully examine Reaction 4.7, you will see that each atom that makes up the reactants is accounted for in the products. The key to writing proper chemical equations is to make sure the reaction follows the law of conservation of mass. This means that chemical equations must be *balanced*.

4.6 Balancing Chemical Equations

Balancing a chemical equation involves manipulating the **coefficients** in front of each compound. The equation cannot be balanced by altering the subscripts in the formulas of compounds because this would change the identity of the compounds and create an incorrect representation of the reaction. The balancing of a chemical equation can be done in a systematic process outlined below.

1. List each element found in the reactants in a column beneath the reactant side of the equation. If oxygen and hydrogen are present, always place them at the bottom of the list.

2. Copy the identical list of elements in a column beneath the product side.

3. Record the number of atoms of each element found on the reactant side. Repeat this process on the product side.

4. Starting at the top of the list, attempt to balance the equation by placing coefficients in front of compounds to increase the number of atoms. Note: The coefficient modifies the number of *all* atoms in a compound; therefore, in your list, readjust the number of each atom in the compound.

5. Repeat until all elements are balanced in the equation.

■ WORKED EXAMPLE 13

Arsenic compounds have been used as poisons for most of recorded history. An early method for the detection of arsenic was the reaction of As_2O_3 with soot (carbon) to produce metallic arsenic and carbon monoxide. Balance the following equation:

$$\underline{}As_2O_3 + \underline{}C \rightarrow \underline{}As + \underline{}CO$$

SOLUTION

Step 1. List each element beneath each side of the reaction. Record the number of atoms of each element on both sides of the reaction.

$$\underline{}As_2O_3 + \underline{}C \rightarrow \underline{} As + \underline{}CO$$

As: 2	As: 1
C: 1	C: 1
O: 3	O: 1

Step 2. Balance As by placing the coefficient 2 in front of the product As.

$$\underline{}As_2O_3 + \underline{}C \rightarrow 2As + \underline{}CO$$

As: 2	As: ~~1~~ 2
C: 1	C: 1
O: 3	O: 1

Step 3. Balance O by placing the coefficient 3 in front of the product CO. Note: This changes the number of carbon atoms also.

$$\underline{}As_2O_3 + \underline{}C \rightarrow 2As + 3CO$$

As: 2	As: ~~1~~ 2
C: 1	C: ~~1~~ 3
O: 3	O: ~~1~~ 3

Step 4. Balance C by placing the coefficient 3 in front of the reactant C.

$$\underline{}As_2O_3 + 3C \rightarrow 2As + 3CO$$

As: 2	As: ~~1~~ 2
C: ~~1~~ 3	C: ~~1~~ 3
O: 3	O: ~~1~~ 3

Practice 4.13

The Marsh test for arsenic was based on the production of arsine gas (AsH_3) through the following chemical reaction: Hydrogen gas and arsenic(III) oxide react to produce arsine gas and water. Write the equation for this reaction and balance it.

Answer

$$6H_2 + As_2O_3 \rightarrow 2AsH_3 + 3H_2O$$

WORKED EXAMPLE 14

Accidental poisonings can occur with many household products that contain toxic substances. For example, toluene (C_7H_8) is a toxic and flammable compound found in some fingernail polish removers. Balance the equation for the combustion reaction of toluene with oxygen gas to produce carbon dioxide and water.

SOLUTION

Step 1. List each element beneath each side of the equation. Record the number of atoms of each element on both sides of the equation.

$$\_\_\_\_C_7H_8 + \_\_\_\_O_2 \rightarrow \_\_\_\_CO_2 + \_\_\_\_H_2O$$

C: 7   C: 1
H: 8   H: 2
O: 2   O: 3

Step 2. Balance C by placing the coefficient 7 in front of the product CO_2.

$$\_\_\_\_C_7H_8 + \_\_\_\_O_2 \rightarrow 7CO_2 + \_\_\_\_H_2O$$

C: 7   C: ~~1~~ 7
H: 8   H: 2
O: 2   O: ~~3~~ 15

Step 3. Balance H by placing the coefficient 4 in front of the product H_2O.

$$\_\_\_\_C_7H_8 + \_\_\_\_O_2 \rightarrow 7CO_2 + 4H_2O$$

C: 7   C: ~~1~~ 7
H: 8   H: ~~2~~ 8
O: 2   O: ~~3~~ ~~15~~ 18

Step 4. Balance O by placing the coefficient 9 in front of the reactant O_2.

$$C_7H_8 + 9O_2 \rightarrow 7CO_2 + 4H_2O$$

C: 7   C: ~~1~~ 7
H: 8   H: ~~2~~ 8
O: ~~2~~ 18   O: ~~3~~ ~~15~~ 18

Practice 4.14

Sodium hydrogen carbonate, $NaHCO_3$, decomposes upon heating to form sodium carbonate, water, and carbon dioxide. Write and balance the equation.

Answer

$$2NaHCO_3 \rightarrow Na_2CO_3 + H_2O + CO_2$$

The final balanced equation should have coefficients that are integer values in their simplest ratio. While using coefficients that are fractions or integer values that are multiples of the simplest ratio does provide a mathematically balanced equation, it is not commonplace to do so and should be avoided.

4.7 Mathematics of Chemical Reactions: Mole Calculations

A balanced chemical reaction can be viewed as a recipe. It lists the ingredients to make a set of products. Just as a recipe requires that we measure out a certain quantity of each ingredient, in the laboratory it is necessary to know how to measure out appropriate quantities of reactants. However, the individual atoms and molecules that are represented in the chemical formulas of an equation are far too small to measure on a laboratory balance. We need a system for measuring manageable quantities of atoms and molecules that will also tell us how many atoms or molecules we have in our sample.

Iodine, I_2
254 g

Sulfur, S
32.1 g

Sodium carbonate, Na_2CO_3
106 g

Lead(IV) oxide, PbO_2
239 g

The mass of one mole of each compound is determined by adding together the molar mass of each element found in the compound. The total mass of each compound pictured here varies, yet each contains the same number of molecules or formula units—Avogadro's number (6.022×10^{23}), which represents one mole of the compound. (Tom Schultz)

A simple system has been devised for relating a measured mass of an element or compound to the number of atoms or molecules contained in the mass. The system is based on the fact that the atomic mass of an element, expressed in grams, contains a known quantity of atoms: 6.022×10^{23}. This number is called **Avogadro's number** (N_A) (named for the scientist who discovered the principle), and it represents one **mole** (mol) of a substance. Just as a dozen represents a quantity of 12, a mole represents a quantity of 6.022×10^{23}. For example, one carbon atom has an atomic mass of 12.011 amu. If we measure 12.011 grams of carbon on a balance, this amount is one mole of carbon and contains 6.022×10^{23} carbon atoms.

The mass of one mole of any substance (element or compound) is called the **molar mass** (M). The molar mass of an element is equal to the mass listed on the periodic table for that element, with the unit being grams/mole. The molar mass of a compound is calculated by adding the mass of each atom found in the compound formula. For example, the molar mass for helium is 4.003 g/mol. The molar mass of NaCl is determined by adding the molar mass of Na to the molar mass of Cl:

$$\underbrace{22.990 \text{ g/mol}}_{\text{Molar mass Na}} + \underbrace{35.453 \text{ g/mol}}_{\text{Molar mass Cl}} = \underbrace{58.443 \text{ g/mol}}_{\text{Molar mass NaCl}}$$

WORKED EXAMPLE 15

What is the molar mass of the following substances?

(a) Lead

(b) Zinc fluoride

(c) Oxygen

(d) Sulfur trioxide

SOLUTION

(a) The molar mass of an element is determined by expressing the atomic mass from the periodic table in units of grams per mole. Lead is element number 82, symbol Pb, and has a molar mass of 207.2 g/mol.

(b) The molar mass of a compound is determined by adding the atomic mass of each atom in the compound together in units of grams per mole. Zinc fluoride has the formula ZnF_2, and the molar mass is

$$65.409 \text{ g/mol} + 2(18.998 \text{ g/mol}) = 103.405 \text{ g/mol } ZnF_2$$

(c) The molar mass of an element is determined by expressing the atomic mass from the periodic table in units of grams per mole. However, oxygen is a diatomic element (as are hydrogen, nitrogen, fluorine, chlorine, bromine, and iodine). Oxygen, with a formula of O_2, has a molar mass totaling $2 \times (15.999 \text{ g/mol}) = 31.998 \text{ g/mol}$.

(d) The molar mass of a compound is determined by adding the atomic mass of each atom in the compound together in units of grams per mole. Sulfur trioxide has the formula SO_3, and the molar mass is

$$32.065 \text{ g/mol} + 3(15.999 \text{ g/mol}) = 80.062 \text{ g/mol } SO_3$$

Practice 4.15

What is the molar mass of the following substances?

(a) Iodine

(b) Nitrogen triiodide

(c) Chromium

(d) Lithium phosphide

Answer

(a) Molar mass of I_2 = 253.810 g/mol

(b) Molar mass of NI_3 = 394.722 g/mol

(c) Molar mass of Cr = 51.996 g/mol

(d) Molar mass of Li_3P = 51.797 g/mol

How many significant digits should be kept when determining the molar mass values for use in calculations? The rule is that, when possible, the molar mass should have at least one more significant digit than the experimental values given in the problem.

Gram–Mole Conversions

In Figure 4.6, a beaker contains 30.00 g of water, an amount we want to convert to moles of water. The first step involved in the conversion of grams of a substance to moles of a substance is to calculate the molar mass of the substance, i.e., the molar mass serves as our conversion factor between grams and moles.

$30.00 \text{ g } H_2O = ? \text{ mol } H_2O$

FIGURE 4.6 A beaker containing 30.00 g of water. (Courtesy of Sally Johll)

Step 1. Calculate the molar mass of water, H_2O:

$$(2 \times 1.0079 \text{ g/mol H}) + 15.999 \text{ g/mol O} = 18.015 \text{ g/mol } H_2O$$

Step 2. Use the molar mass of water to convert from grams to moles:

$$1 \text{ mol } H_2O = 18.015 \text{ g } H_2O$$

$$30.00 \text{ g } H_2O \times \frac{1 \text{ mol } H_2O}{18.015 \text{ g } H_2O} = 1.665 \text{ mol } H_2O$$

The molar mass also serves as the conversion factor for converting from moles of a substance to grams of a substance; for example, what is the mass of 2.75 mol of H_2O?

Step 1. Calculate molar mass of water, H_2O:

$$(2 \times 1.0079 \text{ g/mol H}) + 15.999 \text{ g/mol O} = 18.015 \text{ g/mol } H_2O$$

Step 2. Use the molar mass of water to convert from grams to moles:

$$1 \text{ mol } H_2O = 18.015 \text{ g } H_2O$$

$$2.75 \text{ mol } H_2O \times \frac{18.015 \text{ g } H_2O}{1 \text{ mol } H_2O} = 49.5 \text{ g } H_2O$$

■ WORKED EXAMPLE 16

Calculate the mass of each of the following samples.

(a) 3.30 mol of Fe

(b) 2.49 mol of NaCN

(c) 0.500 mol of Ca

SOLUTION

(a) $3.30 \text{ mol Fe} \times \overbrace{\dfrac{55.847 \text{ g Fe}}{1 \text{ mol Fe}}}^{\text{Molar mass Fe}} = 184 \text{ g Fe}$

(b) $2.49 \text{ mol NaCN} \times \overbrace{\dfrac{49.008 \text{ g NaCN}}{1 \text{ mol NaCN}}}^{\text{Molar mass NaCN}} = 122 \text{ g NaCN}$

(c) $0.500 \text{ mol Ca} \times \overbrace{\dfrac{40.078 \text{ g Ca}}{1 \text{ mol Ca}}}^{\text{Molar mass Ca}} = 20.0 \text{ g Ca}$

Practice 4.16

Calculate the mass of each of the following samples.

(a) 1.50 mol of Cu

(b) 2.41 mol of Br_2

(c) 0.660 mol of Al

Answer

(a) 95.3 g of Cu

(b) 385 g of Br_2

(c) 17.8 g of Al

WORKED EXAMPLE 17

Calculate the number of moles of each substance found in the samples below.

(a) 23.4 g of As_2O_3
(b) 66.1 g of C_7H_8
(c) 116 g of KNO_3

SOLUTION

(a) **Step 1:** Determine the molar mass of As_2O_3:

$$(2 \times 74.922 \text{ g/mol As}) + (3 \times 15.999 \text{ g/mol O}) = 197.841 \text{ g/mol As}_2O_3$$

Step 2: Calculate the number of moles of As_2O_3:

$$23.4 \text{ g As}_2O_3 \times \frac{1 \text{ mol As}_2O_3}{197.841 \text{ g}} = 0.118 \text{ mol As}_2O_3$$

(b) **Step 1:** Determine the molar mass of C_7H_8:

$$(7 \times 12.011 \text{ g/mol C}) + (8 \times 1.0079 \text{ g/mol H}) = 92.141 \text{ g/mol C}_7H_8$$

Step 2: Calculate the number of moles of C_7H_8:

$$66.1 \text{ g C}_7H_8 \times \frac{1 \text{ mol C}_7H_8}{92.141 \text{ g}} = 0.717 \text{ mol C}_7H_8$$

(c) **Step 1:** Determine the molar mass of KNO_3:

$$(1 \times 39.098 \text{ g/mol K}) + (1 \times 14.007 \text{ g/mol N})$$
$$+ (3 \times 15.999 \text{ g/mol O}) = 101.10 \text{ g/mol KNO}_3$$

Step 2: Calculate the number of moles of KNO_3:

$$116 \text{ g KNO}_3 \times \frac{1 \text{ mol KNO}_3}{101.10 \text{ g}} = 1.15 \text{ mol KNO}_3$$

Practice 4.17

Calculate the number of moles in each substance below.

(a) 85.6 g of N_2
(b) 71.8 g of K_2CO_3
(c) 185 g of K_2S

Answer

(a) 3.06 mol of N_2
(b) 0.520 mol of K_2CO_3
(c) 1.68 mol of K_2S

WORKED EXAMPLE 18

Calculate how many moles of each element or compound is present, given the masses below.

(a) 42.0 g of C
(b) 0.469 g of As_2O_3
(c) 59.7 g of NaCN

SOLUTION

(a) $42.0 \text{ g C} \times \dfrac{1 \text{ mol C}}{12.011 \text{ g}} = 3.50 \text{ mol C}$

(b) $0.469 \text{ g As}_2\text{O}_3 \times \dfrac{1 \text{ mol As}_2\text{O}_3}{197.841 \text{ g}} = 0.00237 \text{ mol As}_2\text{O}_3$

$\underbrace{\phantom{197.841 \text{ g}}}$
$2\text{As} \times 74.922 \text{ g/mol}$
$+ 3\text{O} \times 15.999 \text{ g/mol}$

(c) $59.7 \text{ g NaCN} \times \dfrac{1 \text{ mol NaCN}}{49.008 \text{ g}} = 1.22 \text{ mol NaCN}$

$\underbrace{\phantom{49.008 \text{ g}}}$
$1\text{Na} \times 22.990 \text{ g/mol}$
$+1\text{C} \times 12.011 \text{ g/mol}$
$+1\text{N} \times 14.007 \text{ g/mol}$

Practice 4.18

Calculate how many moles of each element or compound are present, given the masses below.

(a) 32.0 g of CaO   (b) 70.1 g of KBr   (c) 89.2 g of $Zn_3(PO_4)_2$

Answer

(a) 0.571 mol of CaO

(b) 0.589 mol of KBr

(c) 0.231 mol of $Zn_3(PO_4)_2$

4.8 Mathematics of Chemical Reactions: Stoichiometry Calculations

> **LEARNING OBJECTIVE**
>
> **Use balanced chemical equations to determine the relationship between quantities of reactants and products.**

Stoichiometry is the study and use of balanced chemical equations for quantifying the amount of reactants and products in a given chemical reaction under a specific set of conditions. In combination with a balanced chemical equation, mole calculations make it possible to determine the maximum amount of a compound that can be produced from particular quantities of reactants. Recall the chemical reaction that takes place when gunpowder is detonated:

$$4KNO_3(s) + 7C(s) + S(s) \rightarrow 3CO_2(g) + 3CO(g) + 2N_2(g) + K_2CO_3(s) + K_2S(s) \qquad \textbf{(Reaction 4.7)}$$

The coefficients in this equation represent how many molecules, formula units, or atoms of reactants and products are involved in the reaction. But the coefficients can also represent the number of moles of each reactant and product. Reaction 4.7 can be interpreted as, "Four moles of potassium nitrate react with seven moles of carbon and one mole of sulfur to produce three moles of carbon dioxide, three moles of carbon monoxide, two moles of nitrogen gas, one mole of potassium carbonate, and one mole of potassium sulfide."

Mole–Mole Calculations

How many moles of nitrogen gas will be created if 2.50 mol of KNO_3 completely react according to Reaction 4.7? The conversion factor comes from the balanced chemical equation.

$$\text{Conversion factor: } 4 \text{ mol KNO}_3 = 2 \text{ mol N}_2$$

$$2.50 \text{ mol KNO}_3 \times \dfrac{2 \text{ mol N}_2}{4 \text{ mol KNO}_3} = 1.25 \text{ mol N}_2$$

WORKED EXAMPLE 19

If 3.75 mol of carbon completely react according to Reaction 4.7, calculate the following quantities:

(a) How many moles of sulfur reacted?

(b) How many moles of potassium nitrate reacted?

(c) How many moles of carbon monoxide were formed?

SOLUTION

(a) Conversion factor: 7 mol C = 1 mol S

$$3.75 \text{ mol C} \times \frac{1 \text{ mol S}}{7 \text{ mol C}} = 0.536 \text{ mol S}$$

(b) Conversion factor: 7 mol C = 4 mol KNO_3

$$3.75 \text{ mol C} \times \frac{4 \text{ mol } KNO_3}{7 \text{ mol C}} = 2.14 \text{ mol } KNO_3$$

(c) Conversion factor: 7 mol C = 3 mol CO

$$3.75 \text{ mol C} \times \frac{3 \text{ mol CO}}{7 \text{ mol C}} = 1.61 \text{ mol CO}$$

Practice 4.19

Calculate the number of moles of sulfur, according to Reaction 4.7, needed to form:

(a) 4.50 mol of carbon dioxide

(b) 17.2 mol of nitrogen

(c) 16.4 mol of potassium sulfide

Answer

(a) 1.50 mol of S   (b) 8.60 mol of S   (c) 16.4 mol of S

Two calculations often used in the laboratory are: (1) calculation of the number of grams of product that can be obtained from a certain number of grams of reactant; and (2) calculation of how many grams of reactant are needed to produce a given amount of product. These calculations are a combination of the gram–mole calculations covered in Section 4.7 and the mole–mole calculations discussed earlier in Section 4.8. The method for solving both of these problem types is outlined below.

1. Convert grams to moles of the given compound using the molar mass.

2. Use the balanced chemical reaction to find the conversion factor between the given compound and the desired compound.

3. Multiply by the conversion factor to get moles of desired compound.

4. Convert from moles to grams of desired compound using molar mass.

This procedure is also summarized in Figure 4.7.

Converting mass compound A to mass compound B

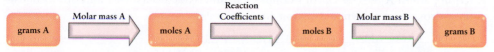

FIGURE 4.7 Mathematical map for calculations relating grams of reactant A to grams of product B.

■ WORKED EXAMPLE 20

Calculate how many grams of carbon monoxide are formed when 15.0 g of KNO_3 are consumed in Reaction 4.7.

SOLUTION

$$15.0 \text{ g } KNO_3 \xrightarrow{\text{Molar mass } KNO_3} \text{mol } KNO_3 \xrightarrow{4:3} \text{mol } CO \xrightarrow{\text{Molar mass } CO} \text{g } CO$$

$$15.0 \text{ g } KNO_3 \times \frac{1 \text{ mol } KNO_3}{101.1 \text{ g } KNO_3} \times \frac{3 \text{ mol } CO}{4 \text{ mol } KNO_3} \times \frac{28.01 \text{ g } CO}{1 \text{ mol } CO} = 3.12 \text{ g } CO$$

Practice 4.20

Using Reaction 4.7, calculate how many grams of sulfur are needed to form 11.3 g of K_2S.

Answer

❑ 3.29 g of S

■ WORKED EXAMPLE 21

Calculate how many grams of carbon must react to produce 3.20 g of arsenic according to the reaction:

$$As_2O_3 + 3C \rightarrow 2As + 3CO$$

SOLUTION

$$3.20 \text{ g } As \xrightarrow{\text{Molar mass } As} \text{mol } As \xrightarrow{2:3} \text{mol } C \xrightarrow{\text{Molar mass } C} \text{g } C$$

$$3.20 \text{ g } As \times \frac{1 \text{ mol } As}{74.922 \text{ g } As} \times \frac{3 \text{ mol } C}{2 \text{ mol } As} \times \frac{12.011 \text{ g } C}{1 \text{ mol } C} = 0.770 \text{ g } C$$

Practice 4.21

Calculate the number of grams of water produced in the following Marsh test, assuming the reaction consumes 2.72 g of As_2O_3.

$$6H_2 + As_2O_3 \rightarrow 2AsH_3 + 3H_2O$$

Answer

❑ 0.743 g of H_2O

4.9 Types of Reactions

Chemical reactions can be classified into several broad classes, according to the patterns in their reactivity. Being able to recognize and identify the various reaction types is useful because it then allows one to predict the product of other reactions within the same class. Several of the reaction types have subclasses, all of which are summarized in this section.

The **synthesis reaction**, also called a *combination reaction*, occurs when two compounds react together to form a single product. A synthesis reaction follows the generic reaction pattern A + B → C. For example, calcium oxide reacts with water to form calcium hydroxide, as shown in Reaction 4.8.

$$CaO(s) + H_2O(l) \rightarrow Ca(OH)_2(s) \qquad \textbf{(Reaction 4.8)}$$

The opposite process, when a single reactant breaks down to form several smaller compounds, is called a **decomposition reaction.** The generic reaction for a decomposition reaction is: C → A + B. Baking soda decomposes when heated, as shown in Reaction 4.9, to produce carbon dioxide, sodium carbonate, and water.

$$2NaHCO_3(s) \rightarrow Na_2CO_3(s) + CO_2(g) + H_2O(l) \quad \textbf{(Reaction 4.9)}$$

The **metathesis reactions** are sometimes called **double displacement reactions,** a term that describes the pattern of reactivity observed. A segment of each compound in the reaction is exchanged with the other reactant, leading to the formation of two new compounds as shown in the generic reaction: AB + CD → AD + CB. There are two distinct types of metathesis reactions, the first being the precipitation reaction. **Precipitation reactions** can be distinguished by the formation of a solid compound from two aqueous reactants. The abbreviation *aq*, which stands for *aqueous*, is used in equations to designate reactants that are dissolved in water. Shown here is the precipitation of solid silver chloride from the reaction of aqueous sodium chloride and silver nitrate.

$$NaCl(aq) + AgNO_3(aq) \rightarrow NaNO_3(aq) + AgCl(s) \quad \textbf{(Reaction 4.10)}$$

The precipitation of the chloride ion by the addition of silver nitrate was conducted on the evidence recovered from the Carr home. If thallium(I) chloride had been present, a precipitate would indicate the possible presence of chloride ions. If thallium(I) sulfate had been used as the poison, the possible presence of sulfate ions could be determined by the addition of barium nitrate, triggering the precipitation of barium sulfate.

Neutralization reactions are the second type of metathesis reactions and are characterized by the reaction of an acid with a base to produce a salt and water. A simple definition of an **acid** is any compound that can release the hydrogen ion, H^+, when dissolved in water. In common acids, hydrogen is usually listed first in the formula. A simple definition of a **base** is any compound that produces the hydroxide ion, OH^-, in solution. The common bases are identified by the presence of the hydroxide ion at the end of the chemical formula. The following reaction is representative of acid-base neutralizations:

$$HCl(aq) + NaOH(aq) \rightarrow NaCl(aq) + H_2O(l) \quad \textbf{(Reaction 4.11)}$$

Another class of reactions is **redox reactions.** Redox is an abbreviation for reduction-oxidation, which involve the gain and loss of electrons between atoms in the reacting compounds. **Oxidation** is the loss of electrons. **Reduction** is the gain of electrons. Oxidation and reduction processes happen at the same time.

The **single displacement** reaction is a class of redox reactions involving the replacement of part of a compound with a more active metal element as shown in the generic reaction pattern: A + BC → AB + C. An example is the reaction of a metal with an acid, as illustrated in Reaction 4.12. The hydrochloric acid reacts with iron to produce Fe^{2+} ions and hydrogen gas. The Fe(s) loses two electrons in the process of becoming Fe^{2+} ions, and the hydrogen ions from the acid gain those electrons in the process of forming hydrogen gas.

$$Fe(s) + 2HCl(aq) \rightarrow FeCl_2(aq) + H_2(g) \quad \textbf{(Reaction 4.12)}$$

A second class of redox reactions is the **combustion reactions.** The most familiar combustion reactions are distinguished by the presence of oxygen gas reacting with a fuel, an organic compound (a compound containing carbon and hydrogen), to produce carbon dioxide and water as illustrated with the generic reaction: Fuel + O_2 → CO_2 + H_2O. One of the components of gasoline is octane (C_8H_{18}), which reacts with oxygen gas in the combustion reaction shown in Reaction 4.13.

$$2C_8H_{18}(l) + 25O_2(g) \rightarrow 18H_2O(g) + 16CO_2(g) \quad \textbf{(Reaction 4.13)}$$

A summary of all the classes of chemical reactions is found in Table 4.4.

TABLE 4.4	Classes of Chemical Reactions	
Class (alias)	Generic Reaction	Example Reaction
Synthesis (combination)	A + B → C	$CaO(s) + H_2O(l) \rightarrow Ca(OH)_2(s)$
Decomposition	C → A + B	$2C_7H_5N_3O_6(l) \rightarrow 7CO(g) + 7C(s) + 5H_2O(g) + 3N_2(g)$
Metathesis (double displacement)	AB + CD → AD + CB	
Precipitation		$NaCl(aq) + AgNO_3(s) \rightarrow AgCl(s) + NaNO_3(aq)$
Neutralization		$NaOH(aq) + HCl(aq) \rightarrow H_2O(l) + NaCl(aq)$
Redox		
Single displacement	A + BC → AB + C	$Mg(s) + 2HCl(aq) \rightarrow MgCl_2(aq) + H_2(g)$
Combustion	Fuel + O_2 → CO_2 + H_2O	$CH_4(g) + O_2(g) \rightarrow CO_2(g) + H_2O(l)$

WORKED EXAMPLE 22

List the key features that distinguish each of the reaction types: synthesis, decomposition, metathesis-precipitation, metathesis-neutralization, redox-single displacement, redox-combustion.

SOLUTION

Synthesis: Two compounds or elements forming a single product

Decomposition: A single reactant compound forming smaller product compounds

Metathesis-precipitation: Solid forming from two aqueous reactants

Metathesis-neutralization: Acid reacting with base

Redox-single displacement: Reaction of a single element with a compound in which electrons are transferred from one reactant to the other

Redox-combustion: Organic (carbon and hydrogen) compound reacting with oxygen gas to produce carbon dioxide and water

Practice 4.22

Classify each of the following reactions as synthesis, decomposition, metathesis-precipitation, metathesis-neutralization, redox-single displacement, redox-combustion. More than one reaction type may apply.

(a) Test for chloride presence: $BaCl_2(aq) + Na_2SO_4(aq)$ $2NaCl(aq) + BaSO_4(s)$

(b) Gas line explosion: $C_3H_8(g) + 5O_2(g)$ $3CO_2(g) + 4H_2O(g)$

(c) Battery acid and zinc metal: $H_2SO_4(aq) + Zn(s)$ $H_2(g) + ZnSO_4(aq)$

Answer

(a) Metathesis-precipitation   (b) Redox-combustion   (c) Redox-single displacement

4.10 Mathematics of Chemical Reactions: Limiting Reactants and Theoretical Yields

LEARNING OBJECTIVE

Apply stoichiometry calculations in solving limiting reactant problems.

In many laboratory experiments, it is usual for at least one of the reactants to be present in an excess amount—beyond the amount that could be completely consumed by the other reactants. An alternate perspective is that one of the reactants will be completely consumed before the others are used up, at which point the reaction will come to a stop. The reactant that is completely consumed is called the **limiting reactant** (or limiting reagent). It is critical to know which reactant is the limiting reactant because the maximum amount of product that can be theoretically produced, the **theoretical yield**, is dictated by the limiting reagent.

Firefighters use this same idea when approaching a blaze. The goal of the firefighter is to prevent either further fuel or further oxygen from reaching the flames. When firefighters are successful, the fire will stop once the existing fuel or oxygen is consumed.

Consider Figure 4.8, which represents the reaction of carbon with oxygen gas to produce carbon dioxide. In the first diagram, there are six carbon atoms and four oxygen molecules that contain a total of eight oxygen atoms. The balanced chemical reaction for the formation of carbon dioxide is $C(s) + O_2(g) \rightarrow CO_2(g)$. If one carbon atom reacts with one oxygen molecule, then there will be two carbon atoms that do not react because there is an insufficient amount of oxygen in the container. The oxygen is the limiting reactant, whereas the carbon is present in excess.

Limiting reagent calculations are modifications of the stoichiometry problems that were introduced earlier in this chapter. The essence of the limiting reagent calculation is to determine whether the amount of one reactant present is enough to react completely with the other reactant(s) present in the reaction. Once the limiting reagent is identified, it is common to calculate the theoretical yield of the reaction for a given product. There are many approaches that can be used to solve limiting reactant problems. The method presented below combines determining the limiting reagent with determining the theoretical yield of the reaction, using the same stoichiometry calculations learned in Section 4.8.

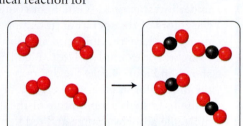

A Limiting Reagent

FIGURE 4.8 Reaction between carbon atoms (black) and oxygen gas (red) (a) before the reaction starts and (b) at completion. Diagram (b) shows the excess carbon present as the excess reagent. Oxygen gas, by its absence in container (b), indicates that it has completely reacted and, therefore, is the limiting reactant.

1. Determine the product of interest in the problem.

2. Convert grams of each reactant to grams of the product of interest.
   (a) Grams reactant 1 → moles reactant 1 → moles product → grams product
   (b) Grams reactant 2 → moles reactant 2 → moles product → grams product

3. The smaller amount calculated previously is the theoretical yield of the product produced by the limiting reactant.

For example, if you have 25.0 g of carbon reacting with 36.0 g of oxygen gas, what will be the theoretical yield of carbon dioxide, given the reaction $C(s) + O_2(g) \rightarrow CO_2(g)$?

Step 1: Product of interest is $CO_2(g)$.

Step 2: Convert grams of each reactant to grams $CO_2(g)$.

$$25.0 \text{ g C} \times \frac{1 \text{ mol C}}{12.011 \text{ g C}} \times \frac{1 \text{ mol CO}_2}{1 \text{ mol C}} \times \frac{44.01 \text{ g CO}_2}{1 \text{ mol CO}_2} = 91.6 \text{ g CO}_2$$

$$36.0 \text{ g O}_2 \times \frac{1 \text{ mol O}_2}{32.00 \text{ g O}_2} \times \frac{1 \text{ mol CO}_2}{1 \text{ mol O}_2} \times \frac{44.01 \text{ g CO}_2}{1 \text{ mol CO}_2} = 49.5 \text{ g CO}_2$$

Step 3: Compare the amounts of products from step 2: The oxygen gas produces the smaller amount of carbon dioxide. Therefore, it is the limiting reagent and the theoretical yield of carbon dioxide is 49.5 g. The excess reagent is carbon, as there is not enough oxygen present to make the full 91.6 g of carbon dioxide that would result if carbon fully reacted.

It is worth noting in the example above that while there was a greater mass of oxygen gas in the reaction as compared to carbon, it was still the limiting reagent. One *cannot* determine the limiting reagent by comparing mass of reactants. Only by doing the stoichiometric calculations can the limiting reagent be determined.

Spectrophotometry

One method for detecting the presence and quantity of compounds in a sample is a versatile laboratory technique called **spectrophotometry**. It is based on the absorption of light that occurs when light is passed through a solution containing molecules of the substance under investigation. White light contains all of the colors of the spectrum (the rainbow). Some molecules have the ability to absorb certain colors of light. For example, FD&C Blue Dye No. 1, shown in Figure 4.9, is the dye responsible for the blue color in blue raspberry Kool-Aid. When white light strikes a molecule of the dye, the molecule absorbs orange and red light. The remaining light passes through the solution and creates the blue color you observe. The absorption of light is due to the electrons in the dye molecule absorbing the energy from certain frequencies of light and going from a lower energy level to a higher energy level.

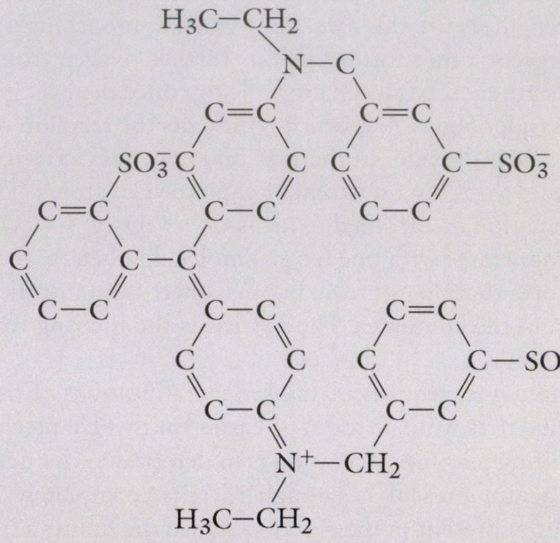

An unknown substance can tentatively be identified by measuring the wavelengths of light that are absorbed, called an *absorption spectrum*, when a solution containing the unknown compound is placed in a spectrophotometer. Shown in Figure 4.10 is the absorption spectrum for FD&C Blue Dye No. 1 and FD&C Yellow Dye No. 5. The two dyes clearly absorb different wavelengths of light and are clearly different substances.

FIGURE 4.9 FD&C Blue Dye No. 1.

A spectrophotometer measures the amount of light absorbed by a sample for each wavelength of light in the spectrum. This is accomplished by dispersing white light generated by a light bulb into the rainbow of colors and then selecting the specific wavelength of light (color) by turning the prism. That wavelength can then pass through a small slit in the instrument and enter the sample chamber, as shown in Figure 4.11.

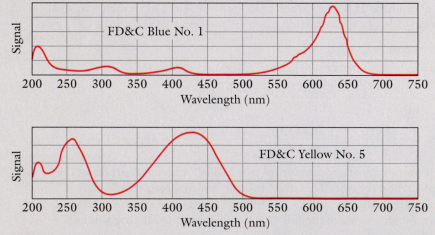

FIGURE 4.10 Absorption spectra for FD&C Blue Dye No. 1 and FD&C Yellow Dye No. 5.

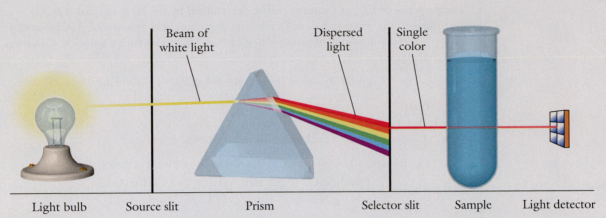

FIGURE 4.11 The amount of light absorbed by a sample is determined by first generating a complete white-light spectrum from a light bulb. Then, by using a prism to disperse the light, a single wavelength of light can pass through the sample.

The spectra in Figure 4.10 show the absorption of a sample over the entire ultraviolet and visible spectrum ranges by sequentially measuring each wavelength of light. A location on the spectrum that shows a maximum absorbance can then be used for measuring the amount of the compound in solution. For example, the FD&C Blue Dye No. 1 shown in Figure 4.10 has a maximum absorbance of light at approximately 630 nm.

To measure the concentration of a colored compound, a range of **standard solutions**—those that have a known concentration of the compound—are prepared; then the absorbance of each standard is measured. A graph of absorbance versus concentration is shown in Figure 4.12. The data points (green dots) form a straight line. A solution that contains an unknown amount of the compound (red dot) can then be measured by first determining its absorbance. The line defined by the standard solutions is used to determine the concentration of the unknown solution that corresponds to the experimental absorbance value.

Compounds that do not absorb light in the UV-VIS range can still be analyzed, provided that some additional steps are taken. For example, the nitrate ion does not absorb light in the UV-VIS range. However, upon the addition of cadmium metal, the nitrate ion undergoes a redox reaction

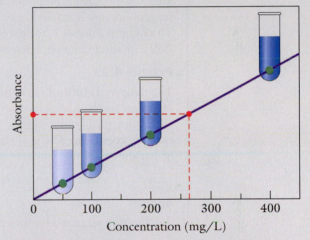

FIGURE 4.12 The measurement of a solution containing an unknown concentration of FD&C Blue Dye No. 1.

to form the nitrite ion. The nitrite ion is then mixed with several larger molecules that absorb light (sulfanilic acid and chromotropic acid), and the resulting compound absorbs light in the UV-VIS range. The concentration of nitrate can then be determined by the procedure outlined above. The nitrate ion, and the failure to accurately measure the amount of nitrate in the tampered Coke from the murder of Peggy Carr, will become an issue during the trial of George Trepal.

■ WORKED EXAMPLE 23

The presence of Cl^- in a sample can be determined by the formation of $AgCl(s)$ according to the reaction below. Determine the theoretical amount of AgCl that can be formed if a solution containing 30.0 g of NaCl is mixed with a solution containing 30.0 g of $AgNO_3$.

$$AgNO_3(aq) + NaCl(aq) \rightarrow AgCl(s) + NaNO_3(aq)$$

SOLUTION

Step 1: AgCl is the product of interest.

Step 2: Convert grams of each reactant to product of interest.

Reactant 1: 0.10 g NaCl → mol NaCl → mol AgCl → g AgCl

$$30.0 \text{ g NaCl} \times \frac{1 \text{ mol NaCl}}{58.443 \text{ g NaCl}} \times \frac{1 \text{ mol AgCl}}{1 \text{ mol NaCl}} \times \frac{143.321 \text{ g AgCl}}{1 \text{ mol AgCl}} = 7.50 \text{ g AgCl}$$

Reactant 2: 0.10 g $AgNO_3$ → mol $AgNO_3$ → mol AgCl → g AgCl

$$30.0 \text{ g AgNO}_3 \times \frac{1 \text{ mol AgNO}_3}{169.872 \text{ g AgNO}_3} \times \frac{1 \text{ mol AgCl}}{1 \text{ mol AgNO}_3} \times \frac{143.321 \text{ g AgCl}}{1 \text{ mol AgCl}} = 2.52 \text{ g AgCl}$$

Step 3: Compare the amounts of products from step 2. The silver nitrate is the limiting reagent because it would be completely used up by producing 2.52 g of silver chloride. The sodium chloride is the excess reagent because there is enough present to make up to 7.50 g of silver chloride; however, there is not sufficient silver nitrate.

Practice 4.23

The tampered bottle of Coke from the Trepal–Carr case was analyzed for the presence of sulfate ion by adding barium nitrate to trigger the precipitation of barium sulfate. Determine the limiting reagent and the theoretical yield of barium sulfate given that a solution containing 32.0 g of sodium sulfate is mixed with a solution containing 48.0 g of barium nitrate.

$$Na_2SO_4(aq) + Ba(NO_3)_2(aq) \rightarrow BaSO_4(s) + 2NaNO_3(aq)$$

Answer

❑ The limiting reagent is barium nitrate and the theoretical yield is 42.8 g of $BaSO_4$.

■ WORKED EXAMPLE 24

Determine how many grams of lead(II) iodide would be formed if 0.250 g of lead(II) nitrate were added to a solution containing 0.42 g of potassium iodide, given the reaction $Pb(NO_3)_2(aq) + 2KI\ (aq) \rightarrow PbI_2(s) + 2KNO_3(aq)$.

SOLUTION

Step 1: Product of interest is $PbI_2(s)$.

Step 2: Convert grams of each reactant to grams of $PbI_2(s)$.

$$0.250 \text{ g Pb(NO}_3)_2 \times \frac{1 \text{ mol Pb(NO}_3)_2}{331.2 \text{ g Pb(NO}_3)_2} \times \frac{1 \text{ mol PbI}_2}{1 \text{ mol Pb(NO}_3)_2} \times \frac{461.0 \text{ g PbI}_2}{1 \text{ mol PbI}_2}$$
$$= 0.348 \text{ g PbI}_2$$

$$0.42 \text{ g KI} \times \frac{1 \text{ mol KI}}{166.0 \text{ g KI}} \times \frac{1 \text{ mol PbI}_2}{2 \text{ mol KI}} \times \frac{461.0 \text{ g PbI}_2}{1 \text{ mol PbI}_2} = 0.584 \text{ g PbI}_2$$

Step 3: Compare the amounts of products from step 2: The theoretical yield of lead(II) iodide is 0.348 g produced by the complete reaction of the limiting reagent, lead(II) nitrate.

Practice 4.24

Determine the limiting reagent and theoretical yield of calcium phosphate when a solution containing 14.86 g of sodium phosphate is mixed with 14.86 g of calcium nitrate, given the reaction

$$2Na_3PO_4(aq) + 3Ca(NO_3)_2(aq) \rightarrow Ca_3(PO_4)_2(s) + 6NaNO_3(aq)$$

Answer

The limiting reagent is calcium nitrate and the theoretical yield is 9.360 g of calcium phosphate.

4.11 CASE STUDY FINALE: Exploring the Chemistry of a Poison

The detectives investigating Peggy Carr's death realized that to get George Trepal, they would not be able to use traditional interrogation methods—he was too good at playing mind games. Instead, undercover detective Susan Goreck posed as a member of the local Mensa group to befriend him and gain his trust. The detectives were worried that George himself might spot their game, but George seemed to enjoy the attention—Susan always seemed impressed with his knowledge and listened intently to his every word. When George Trepal and his wife abruptly decided to move from their home in Alturas, Susan made arrangements to rent his house.

When detectives started searching George's garage, little did they know what "treasures" they would find—manuals on how to poison people and a jar containing thallium(I) nitrate. The police determined that George had gained access to the Carrs' home easily, as they often left their doors unlocked during the day, even if nobody was at home. He took a six-pack of glass Coca-Cola bottles, removed the metal caps, and added thallium(I) nitrate. He was known to have a home bottle capper for brewing and could thus easily reseal the bottles. He then returned the tampered drinks to the Carrs' house. Peggy and her sons all drank the Coke, but Peggy was the only one to become critically ill. She was petite in comparison to her sons, who were better able to recover from the poison.

George Trepal was convicted of murder, as well as charges for tampering with food products. He was sentenced to death and awaits his punishment on death row in Florida.

The case was not without some controversy, however. The evidence had been sent to the main FBI laboratory for analysis. During investigation of alleged misconduct at the FBI laboratory, it was determined that the chemist who analyzed the evidence neglected to do some basic tests, mislabeled his laboratory notebook, and failed to write some data into his notebook. Furthermore, while testifying at trial, he overstated the results of his analysis.

These allegations are troublesome, to say the least, especially in a death-penalty case. However, the attorneys for George Trepal had arranged for an independent analysis of all the evidence at Georgia Tech and did not challenge at trial the accuracy of the analysis. Only after the allegation of misconduct by the FBI chemist did George Trepal use this information during his appeals. The prosecutor even noted that while Trepal was attempting to impeach the accuracy of the state's evidence, he was simultaneously trying to prevent the state from gaining access to the results from the independent analysis, apparently playing even more mind games.

During the appeals process, the courts determined that the analysis was flawed. However, it was not flawed to the extent that it would overturn the original conviction. If George Trepal had been granted a retrial, the properly analyzed evidence would still have pointed to his guilt.

CHAPTER SUMMARY

• The columns of the periodic table are called groups or families. Elements within a group undergo similar chemical reactions. The groups are numbered 1–18 from left to right, but some groups also have common names such as alkali metals, alkaline earth metals, halogens, and noble gases.

• The rows of the periodic table are called periods and are numbered 1–7. The only periods with common names are the lanthanide and actinide periods, which make up the inner transition metals located beneath the main body of the periodic table.

• Covalent compounds are created when two nonmetal atoms form a bond by sharing electrons. The formulas of covalent compounds cannot be predicted. Therefore, the number of atoms of each element in the compound is indicated by prefixes in the name.

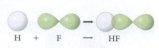

H   +   F   →   HF

• Ionic compounds consist of two oppositely charged ions. The cation has a positive charge that results when an atom loses at least one electron. The anion has a negative charge that results when an atom gains at least one electron. Metals tend to form cations; nonmetals tend to form anions.

• Ionic compounds tend to have high melting points and form rigid structures. The strength of the attraction between ions is directly related to the magnitude of their charges.

• Ionic compounds are named by listing the cation first, followed by the anion. If the cation does not have a predictable ionic charge, Roman numerals are used to indicate the charge.

• Chemical equations are a shorthand method for quickly, efficiently, and clearly communicating information about a reaction. The coefficient in front of each compound in a chemical equation is used to balance the equation.

• One mole of any compound is equal to the molecular mass of the substance expressed in grams and contains 6.022×10^{23} molecules.

• Reactions can be classified into groups based on the types of substances in the reaction and how they react. Common reaction types are synthesis, decomposition, metathesis-precipitation, metathesis-neutralization, redox-single displacement, and redox-combustion.

• A limiting reagent is one that is completely consumed in a reaction, leaving an excess amount of one or more of the other reactants. Determination of the limiting reagent is important because it controls the maximum amount of a product that can form.

• The concentration of a compound in a solution can be measured by determining how much light passes through a sample and how much of the light is absorbed by the solution.

KEY TERMS

acid (p. 117)
alkali metal (p. 91)
alkaline earth metal (p. 91)
anion (p. 96)
Avogadro's number (p. 110)
base (p. 117)
cation (p. 96)
coefficient (p. 108)
combustion reaction (p. 117)
covalent bond (p. 93)

crystal lattice (p. 103)
decomposition reaction (p. 116)
denatured alcohol (p. 106)
double displacement
   reaction (p. 117)
family (p. 90)
group (p. 90)
halogen (p. 91)
inner transition metal (p. 92)
ion (p. 96)

ionic bond (p. 96)
ionic compound (p. 96)
limiting reactant (p. 119)
melting point (p. 104)
metathesis (p. 117)
molar mass (M) (p. 110)
mole (p. 110)
monatomic ion (p. 102)
neutralization reaction (p. 117)
noble gas (p. 91)

oxidation (p. 117)
period (p. 92)
polyatomic ion (p. 102)
precipitation reaction (p. 117)
product (p. 107)
reactant (p. 107)

redox reaction (p. 117)
reduction (p. 117)
salt (p. 106)
single displacement
    reaction (p. 117)
spectrophotometry (p. 120)

standard solution (p. 121)
stoichiometry (p. 114)
synthesis reaction (p. 116)
theoretical yield (p. 119)
transition metal (p. 91)

MAKING MORE CONNECTIONS: Additional Readings, Resources, and References

Gerber, S. M., and Saferstein, R. *More Chemistry and Crime*, Washington D.C.: American Chemical Society, 1997.

For information about another famous poisoning, the Tylenol murders by cyanide poisoning, see numerous articles in the *Chicago Tribune* from 10/01/82 through 10/06/82. Also, go to: www.trutv.com/library/crime/terrorists_spies/terrorists/tylenol_murders/index.html

For more information on the Trepal case study, see these sources:

An official recap of the case can be found at www.floridacapitalcases.state.fl.us/case_updates/121965.doc

For information on the Department of Justice review of the FBI analysis of evidence, go to www.usdoj.gov/oig/special/9704a/21trepal.htm

Goreck, S., and Good, J. *Poison Mind*, St. Martin's Paperbacks, New York, 1996.

REVIEW QUESTIONS AND PROBLEMS

Questions

1. Make a sketch of the periodic table and label each of the following areas: alkali metals, alkaline earth metals, transition metals, halogens, and noble gases. (4.1)
2. What do elements within the same group have in common? (4.1)
3. Which elements tend to form cations? Which elements tend to form anions? Where is each of these groups of elements located on the periodic table? (4.3)
4. How do two atoms share electrons in a covalent bond? (4.2)
5. What type of elements form covalent compounds? (4.2)
6. Are the melting points of covalent compounds typically higher or lower than those of ionic compounds? Why? (4.3)
7. What happens to an atom when it becomes a cation? An anion? (4.3)
8. What is a monatomic ion? Provide several examples. (4.3)
9. What is a polyatomic ion? Provide several examples. (4.3)
10. How does the nature of the crystal lattice affect the melting point of an ionic solid? (4.3)
11. Why does MgS have a much higher melting point than NaF? (4.3)
12. Many compounds have common names that are regularly used. Choose two examples and provide the common name, scientific name, and formula. (4.4)

13. What is the importance of a balanced chemical equation? (4.5)
14. When balancing a chemical equation, why is it important to change the coefficients in front of compounds but not the subscripts in the formulas of compounds? (4.6)
15. Why is the concept of the mole necessary? (4.7)
16. What are the main features of the synthesis, decomposition, metathesis-precipitation, metathesis-neutralization, redox-single displacement, and redox-combustion reactions? (4.9)
17. Why is it necessary to determine the limiting reagent in chemical reactions? (4.10)
18. The calculated theoretical yield of a reaction is seldom obtained in a laboratory experiment. Can you think of anything that might lower the actual yield to a value less than the theoretical yield? (4.10)
19. Why does a solution containing a greater amount of FD&C Red Dye No. 40 appear darker red than a solution that contains a smaller amount of the dye? (EA)
20. Sodium cyanide, when dissolved in water, forms a colorless solution that does not absorb visible light. What can be done experimentally to the solution so that visible light can be used to measure the amount of cyanide with a spectrophotometer? (ITL)

Problems

21. Identify all halogens from the following four lists of elements. Write *none* if there are no halogens listed. (4.1)
    (a) Li, Sr, U, Fe, I, Sb
    (b) Pt, Ca, Br, Na, Mg, Cu
    (c) F, Co, Be, Rb, La, Th
    (d) Pb, Eu, Y, Cs, Ca, Cl

22. Identify all alkaline earth metals from the following four lists of elements. Write *none* if there are no alkaline earth metals listed. (4.1)
    (a) Li, Sr, U, Fe, I, Sb
    (b) Pt, Ca, Br, Na, Mg, Cu
    (c) F, Co, Be, Rb, La, Th
    (d) Pb, Eu, Y, Cs, Ca, Cl

23. Identify the group, group name (if any), and period for the following elements. (4.1)
    (a) K
    (b) Ru
    (c) Ne
    (d) As

24. Identify the group, group name (if any), and period for the following elements. (4.1)
    (a) Mg   (c) Si
    (b) Se   (d) Al

25. Identify the following elements. (4.1)
    (a) Period 6, Group 10
    (b) Halogen, Period 2
    (c) Noble gas, Period 4
    (d) Period 4, Group 14

26. Identify the following elements. (4.1)
    (a) Period 5, Group 5
    (b) Alkali metal, Period 6
    (c) Period 5, Group 11
    (d) Period 3, Group 15

27. Name the following covalent compounds. (4.2)
    (a) BF_3 (c) ClF_3
    (b) SF_4 (d) OF_2

28. Name the following covalent compounds. (4.2)
    (a) NF_3 (c) CF_4
    (b) SCl_2 (d) CS_2

29. Write the proper formula for the following covalent compounds. (4.2)
    (a) Disulfur dichloride
    (b) Dinitrogen pentasulfide
    (c) Sulfur tetrafluoride
    (d) Sulfur trioxide

30. Write the proper formula for the following covalent compounds. (4.2)
    (a) Phosphorus pentachloride
    (b) Diphosphorus triiodide
    (c) Bromine monofluoride
    (d) Tribromine octoxide

31. Using the periodic table as a guide, determine the charge that each of the following atoms will have in an ionic compound. Indicate *multiple* for any atom that can commonly have more than one charge. (4.3)
    (a) K
    (b) N
    (c) Pb
    (d) Br

32. Using the periodic table as a guide, determine the charge that each of the following atoms will have in an ionic compound. Indicate *multiple* for any atom that can commonly have more than one charge. (4.3)
    (a) Sn   (c) Ca
    (b) F   (d) Al

33. Write the formulas for the ionic compounds that will form between the following ions. (4.3)
    (a) Aluminum ion and chloride ion
    (b) Cesium ion and fluoride ion
    (c) Cadmium ion and oxide ion
    (d) Strontium ion and iodide ion

34. Write the formulas for the ionic compounds that will form between the following ions. (4.3)
    (a) Calcium ion and oxide ion
    (b) Barium ion and nitride ion
    (c) Aluminum ion and sulfide ion
    (d) Cadmium ion and bromide ion

35. What is the formula and charge for each of the following polyatomic ions? (4.3)
    (a) Nitrate ion
    (b) Hydroxide ion
    (c) Cyanide ion
    (d) Phosphate ion

36. What is the formula and charge for each of the following polyatomic ions? (4.3)
    (a) Ammonium ion
    (b) Acetate ion
    (c) Carbonate ion
    (d) Sulfate ion

37. What is the formula for the compounds formed between the following ions? (4.3)
    (a) Cesium ion and acetate ion
    (b) Potassium ion and carbonate ion
    (c) Barium ion and phosphate ion
    (d) Aluminum ion and permanganate ion

38. What is the formula for the compounds formed between the following ions? (4.3)
    (a) Zinc ion and carbonate ion
    (b) Lithium ion and hydroxide ion
    (c) Magnesium ion and cyanide ion
    (d) Silver ion and nitrate ion

39. Determine the charge of the cation in the following ionic compounds. (4.3)
    (a) $CuCl_2$
    (b) $Co(NO_3)_2$
    (c) $CuCl$
    (d) CrO

40. Determine the charge of the cation in the following ionic compounds. (4.3)
    (a) $FeSO_4$ (c) $Cu(OH)_2$
    (b) $CoCO_3$ (d) $CrPO_4$

41. Name the following ionic compounds. (4.3)
    (a) Al_2S_3 (c) KOH
    (b) MgO (d) $AlPO_4$

42. Name the following ionic compounds. (4.3)
    (a) CaS (c) $Mg(OH)_2$
    (b) $BaSO_4$ (d) NaF

43. Name the following compounds. (4.3)
    (a) $CuCl_2$
    (b) $Co(NO_3)_2$
    (c) $CuCl$
    (d) CrO

44. Name the following compounds. (4.3)
    (a) $FeSO_4$
    (b) $CoCO_3$
    (c) $Cu(OH)_2$
    (d) $CrPO_4$

45. Write the chemical equation that corresponds to the following description: Three moles of sodium cyanide react with one mole of iron(III) nitrate to form one mole of iron(III) cyanide and three moles of sodium nitrate. (4.5)

46. Write the chemical equation that corresponds to the following description: One mole of calcium metal reacts with two moles of water to form one mole of calcium hydroxide and one mole of hydrogen gas. (4.5)

47. Balance the following equations. (4.6)
    (a) $BaCl_2 + Na_3PO_4 \rightarrow Ba_3(PO_4)_2 + NaCl$
    (b) $Na_2S + Fe(NO_3)_2 \rightarrow NaNO_3 + FeS$
    (c) $C_3H_8 + O_2 \rightarrow CO_2 + H_2O$
    (d) $Ca(C_2H_3O_2)_2 + KOH \rightarrow KC_2H_3O_2 + Ca(OH)_2$

48. Balance the following equations. (4.6)
    (a) $AgNO_3 + LiBr \rightarrow AgBr + LiNO_3$
    (b) $(NH_4)_2CO_3 + Fe(NO_3)_2 \rightarrow FeCO_3 + NH_4NO_3$
    (c) $H_2O_2 \rightarrow H_2O + O_2$
    (d) $H_2SO_4 + NaOH \rightarrow$ water $+ Na_2SO_4$

49. Predict the products of the following metathesis (double replacement) reactions and write a balanced chemical equation. (4.6, 4.9)
    (a) Ammonium bromide + silver acetate →
    (b) Potassium sulfate + lead(II) nitrate →
    (c) Sodium phosphate + magnesium chloride →
    (d) Lithium carbonate + copper(I) sulfate →

50. Predict the products of the following metathesis (double replacement) reactions and write a balanced chemical equation. (4.6, 4.9)
    (a) Cobalt(II) nitrate + sodium hydroxide →
    (b) Barium acetate + ammonium sulfate →
    (c) Strontium iodide + calcium sulfate →
    (d) Barium sulfide + iron(II) sulfate →

51. Calculate the molar mass of the compounds you determined in Problem 37. (4.7)

52. Calculate the molar mass of the compounds you determined in Problem 38. (4.7)

53. How many moles of each substance are present in the following quantities? (4.7)
    (a) 21.0 g of $AgNO_3$
    (b) 0.295 g of C_3H_8
    (c) 25.0 g of iron(II) nitrate
    (d) 4.10 g of CO_2

54. How many moles of each substance are present in the following quantities? (4.7)
    (a) 66.0 g of $CuNO_3$ (c) 21.0 g of O_2
    (b) 115 g of Ca   (d) 2.30 g of fluorine gas

55. What is the mass in grams of the following quantities of substances? (4.7)
    (a) 1.50 mol of N_2O_3
    (b) 0.375 mol of chlorine gas
    (c) 2.25 mol of iron(II) oxide
    (d) 1.33 mol of $NH_4C_2H_3O_2$

56. What is the mass in grams of the following quantities of substances? (4.7)
    (a) 0.423 mol of KBr
    (b) 0.0335 mol of NH_3
    (c) 2.50 mol of sulfur trioxide
    (d) 0.683 mol of sulfur hexafluoride

57. Calculate the number of grams of barium sulfate produced if 27.8 g of barium chloride reacts completely according to the reaction: (4.8)

$$BaCl_2(aq) + Na_2SO_4(aq) \rightarrow 2NaCl(aq) + BaSO_4(s)$$

58. Calculate the number of grams of sodium sulfate consumed if 52.6 g of barium sulfate is produced according to the reaction: (4.8)

$$BaCl_2(aq) + Na_2SO_4(aq) \rightarrow 2NaCl(aq) + BaSO_4(s)$$

59. Calculate the number of grams of sodium chloride produced if 40.5 g of barium chloride reacts completely according to the reaction: (4.8)

$$BaCl_2(aq) + Na_2SO_4(aq) \rightarrow 2NaCl(aq) + BaSO_4(s)$$

60. Calculate the number of grams of propane (C_3H_8) consumed if 36.0 g of water is produced according to the reaction: (4.8)

$$C_3H_8(g) + 5O_2(g) \rightarrow 3CO_2(g) + 4H_2O(g)$$

61. Calculate the number of grams of carbon dioxide produced if 88.0 g of propane (C_3H_8) is consumed according to the reaction: (4.8)

$$C_3H_8(g) + 5O_2(g) \rightarrow 3CO_2(g) + 4H_2O(g)$$

62. Identify the type of chemical reaction based on the description given below: (4.9)
    (a) The explosion last night at the railroad terminal was linked to a small leak in the tanker carrying diesel fuel.
    (b) If the swimming pool pH is greater than 8.0, add muriatic acid to lower the pH.
    (c) Carbon dioxide from the atmosphere will react with concentrated sodium hydroxide to form an insoluble carbonate.
    (d) Warning: The whitewall tire cleaner should not be used on metal because it contains hydrofluoric acid.

63. Identify the type of chemical reaction based on the description given below: (4.9)
    (a) Hydrogen gas is generated by the action of hydrochloric acid on zinc metal.
    (b) Guncotton is a nitrated cellulose fiber that leaves almost no ash after ignition.
    (c) Latent fingerprints can be visualized by spraying a dilute silver nitrate solution onto a fingerprint containing the chloride ion. The resulting reaction forms solid silver chloride.
    (d) Take two teaspoons of milk of magnesia for occasional acid reflux episodes.

64. Determine the limiting reagent and theoretical yield of the product for the reaction represented below. (4.10)

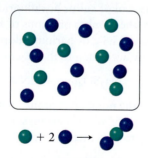

65. Determine the limiting reagent and theoretical yield of the product for the reaction represented below. (4.10)

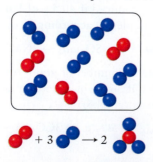

66. Determine the limiting reagent and the theoretical yield of $Fe(OH)_3$ for the following reaction, given a solution containing 15.0 g of $Fe(NO_3)_3$ and 15.0 g of KOH. (4.10)

$$Fe(NO_3)_3(aq) + 3KOH(aq) \rightarrow Fe(OH)_3(s) + 3KNO_3(aq)$$

67. Determine the limiting reagent and the theoretical yield of $Fe(OH)_3$ for the following reaction, given a solution containing 75.0 g of $Fe(NO_3)_3$ and 25.0 g of KOH. (4.10)

$$Fe(NO_3)_3(aq) + 3KOH(aq) \rightarrow Fe(OH)_3(s) + 3KNO_3(aq)$$

68. The following ions can all be found in the cells of our body: Na^+, K^+, Ca^{2+}, Mg^{2+}, Cl^-, Br^-, NO_3^-. (4.2)
    (a) Name each ion.
    (b) How many different ionic compounds could be formed from the ions above? Write the name and formula for each possible compound.

69. Iodine, phosphorus, and lithium metal are very reactive in their elemental states. What group number, group name (if any), and period correspond to each element? (4.1)

70. Calculate the number of grams of carbon monoxide produced from the detonation of 200.0 g of TNT ($C_7H_5N_3O_6$), given the following reaction: (4.8)

$$2C_7H_5N_3O_6(s) \rightarrow 12CO(g) + 5H_2(g) + 3N_2(g) + 2C(s)$$

71. Calculate the number of grams of nitrogen gas produced from the detonation of 125.0 g of TNT ($C_7H_5N_3O_6$), given the following reaction: (4.8)

$$2C_7H_5N_3O_6(s) \rightarrow 12CO(g) + 5H_2(g) + 3N_2(g) + 2C(s)$$

72. In the Marsh test for arsenic, why would it be imperative for hydrogen gas to be the excess reagent and As_2O_3 to be the limiting reagent? The initial reaction is: (4.10)

$$6H_2 + As_2O_3 \rightarrow AsH_3 + H_2O$$

73. Cyanide analysis is done using a spectrophotometer. However, the human eye is very sensitive to changes in color. Estimate the concentration of cyanide in the following solution, based on the solutions of known concentration in Figure 4.12. (ITL)

50 mg/L   100 mg/L   200 mg/L   400 mg/L   Unknown

Case Study Problems

74. A suspected case of food tampering involving a soft drink containing FD&C Yellow Dye No. 5 has occurred. It is suspected that part of the contents was dumped and a household cleaner was added to it. Explain how one might use spectrophotometry to prove tampering by analyzing the evidence and an unadulterated sample. (ITL)

75. For the case explained in Problem 76, determine the wavelength of light you would use to measure the concentration of FD&C Yellow Dye No. 5 by examining the spectrum provided in Figure 4.10. (ITL)

76. If the concentration of FD&C Yellow Dye No. 5 differs from the evidence to that of an unadulterated sample, does it prove the allegations true? If not, what further analysis or tests might be necessary? (ITL)

77. Another famous product-tampering case occurred in the early 1980s, when someone replaced the active ingredient in Tylenol capsules with potassium cyanide, a very deadly poison. How might the detectives have been able to link fingerprints to the killer who laced the Tylenol with cyanide? *Hint:* A large number of people from the manufacturing plant, the store where the Tylenol was sold, potential customers, etc., could have handled the Tylenol. Consider where their fingerprints would be found and how the killer could be identified. (The killer's fingerprints were never actually found.) (CP)

78. George Trepal was convicted and sentenced to death for the poisoning of a neighbor with the poisonous compound thallium(I) nitrate laced into bottles of soda. A bottle of thallium(I) nitrate was found in his garage. If another thallium(I) salt (chloride or sulfate) had been used, Trepal could claim innocence. Samples of the tampered soda were sent to the laboratory for analysis. A preliminary screening method for the presence of the chloride ion is the addition of silver nitrate, because a precipitate forms with the chloride ion. Write a balanced chemical equation for the reaction of thallium(I) chloride with silver nitrate to produce thallium(I) nitrate and silver chloride. If a sample contained 0.54 g of thallium(I) chloride and 1.18 g of silver nitrate, determine the limiting reactant and the theoretical yield of silver chloride. (CS)

79. The Trepal case continued with the screening for the sulfate ion by the addition of barium chloride, as a precipitate forms with the sulfate ion. Write a balanced chemical equation for the reaction of thallium(I) sulfate with barium chloride to produce thallium(I) chloride and barium sulfate. If a sample contained 4.06 g of thallium(I) sulfate and 4.41 g of barium chloride, determine the limiting reactant and the theoretical yield of barium sulfate. (CS)

80. One of the criticisms of the FBI analyst was that he never tested an unadulterated bottle of Coca-Cola for the presence of chloride or sulfate ions, as was done with the evidence. Why would this test be important? (CS)

81. Another criticism of the FBI chemist was that during trial he said that thallium(I) nitrate *was* added to the bottle of Coca-Cola when he should have said that the evidence was *consistent* with thallium(I) nitrate having been added. Explain the difference between the two statements, from a scientific perspective. (CS)

CHAPTER 5

Chemistry of Bonding: Structure and Function of Drug Molecules

CASE STUDY: Exploring Chemotherapy Drugs

"In the fields of observation, chance only favors the mind which is prepared."

—Louis Pasteur

Many great discoveries in science have been attributed to serendipity, which is to discover something of great significance accidentally. However, as Louis Pasteur, the creator of the pasteurization process and the vaccine for rabies tells us, only the mind that is prepared will have the insight to recognize those discoveries. The story of one of the most powerful chemotherapy drugs fits Pasteur's explanation. It was only through the rigorous application of the scientific method that the chemotherapy drug cisplatin was discovered.

Cisplatin was first discovered in the year 1845 by the Italian chemist Michele Peyrone. The biological significance of cisplatin, however, would not be realized for well over 100 years, and even then only in an unlikely manner. In 1961, Barnett Rosenberg, a physicist, had joined the biophysics department at Michigan State University. Rosenberg had a hypothesis that an electric field might have an effect on *mitosis*, the process by which plant and animal cells reproduce to create two cells identical to the original parent cell.

Rosenberg designed an experiment that would expose cells undergoing mitosis to electric fields. To prove his hypothesis, though, he had to make sure that the electric field didn't have any other effects on cells. Therefore, before he started his experiment on animal and plant cells, he decided to test his system using the bacteria *Escherichia coli* (*E. coli*). Bacteria reproduce through a process called *binary fission,* not mitosis. If Rosenberg's hypothesis was correct, the reproduction of *E. coli* should be unaffected by the presence of electric fields and his experimental design should have no effect on the bacteria. Rosenberg then planned to apply electric fields to cells undergoing mitosis and would be able to credit any changes in the cells to electric fields.

The experimental results surprised, confounded, and shocked Rosenberg. The application of an electric field caused the *E. coli* to halt reproduction but not cell growth. The typical rod structure of *E. coli* turned into long filaments as the bacteria stretched to more than 300 times their normal length! With his original hypothesis proven false, Rosenberg designed new

---

As you read through the chapter, consider the following questions:

• How do drug molecules "know" where they are needed?

• Why is it that people who undergo chemotherapy commonly lose their hair during treatment? Is there some connection to how the drugs attack cancer cells?

• How important is the three-dimensional shape of a molecule to the function of the molecule in the body?

---

experiments to better understand how the electric field was affecting the bacteria, though it became apparent that the electric field was not affecting the bacteria but rather the electrodes themselves. The electrodes were creating a compound during the application of the voltage that was causing the new and interesting phenomenon.

After many experiments, Rosenberg began to close in on the source of the mysterious compound that halted the reproduction of *E. coli* but not its growth. The electrodes used to apply the voltage to the solution were made out of platinum. Platinum is commonly used in electrochemistry because it is considered inert and usually will not react with other components in the solution. When Rosenberg created his experiment, however, all the conditions were present for the platinum to react to form a compound called *cisplatin*.

MAKE THE CONNECTION

To understand how cisplatin attacks cancer cells, we must first understand how molecules form three-dimensional shapes.

5.1 Nature of Covalent Bonds

LEARNING OBJECTIVE

Explain the valence bond theory of covalent bonding.

When an unconscious person is brought to an emergency room, the doctor will have a more difficult time diagnosing the patient, as the emergency room staff will not be able to learn of the patient's medical history and recent activities. Is the patient a diabetic who hasn't eaten properly, or perhaps suffering alcohol poisoning, or a drug overdose, or carbon monoxide poisoning? The list of possible causes could go on. Once the person's vital statistics are obtained such as pulse, temperature, and blood pressure, the doctor will order blood samples to be sent to the laboratory for analysis.

The analysis of human blood can be quite complex and time-consuming because of the many natural components in blood and the great variety of drugs, both legal and illegal, that could be present in the sample. Drug molecules may pass through the body unchanged or break down into new compounds, adding to the complexity of the analysis. Some drug molecules also tend to bind tightly with particular protein molecules present in blood. This binding can artificially lower the apparent concentration of the drug if only the unbound form of the drug is measured.

To narrow down the vast number of possible drugs that could be present in a sample, screening tests are used to determine, in general, what classes or types of drugs the sample may contain. Screening tests, which are fairly quick and inexpensive, give a positive result for a broad class of compounds.

One common method of drug screening is called an **immunoassay**. The basis of immunoassay is that all molecules of a particular compound have a unique three-dimensional shape. The immunoassay brings together the target molecule (the drug or substance being tested) with another molecule specifically chosen because its structure has an opening into which only the target molecule will fit. This customized fit is very similar to the way illegal drugs or, for that matter, legal prescription drugs, function in the body. The three-dimensional shape of the drug molecule is such that it can bind to a receptor molecule in the body, triggering the drug's desired effect.

Figure 5.1 shows a large circular molecule of cyclodextrin with a benzene molecule within the center space. This example is a simplified illustration of how molecules can interact based on their size and shape. The opening in the center of a molecule of cyclodextrin is of sufficient size and polarity that the benzene molecule fits inside the cyclodextrin molecule. The three-dimensional shape and bonding angles make it possible for this interaction to occur. Understanding the three-dimensional shapes of molecules is vitally important in understanding how those molecules function in our body. We will examine both of these topics in more depth as the chapter progresses.

FIGURE 5.1 Molecules interact based on their size and three-dimensional shape. For example, cyclodextrin is a large circular molecule with an open center cavity into which the benzene molecule can easily enter.

The nature of the bonding between the individual atoms of a molecule determines the three-dimensional shape of the molecule. In our earlier discussions of chemical compounds, we saw that atoms form two types of bonds—ionic or covalent. Ionic bonds involve the gain or loss of electrons from neutral atoms to form positive and negative ions that are attracted to one another. Metal atoms form cations while nonmetal atoms form anions.

Covalent bonds form when electrons are shared between two atoms. The sharing of electrons occurs because the relative difference in electronegativities of the two atoms is not sufficiently large for either atom to remove an electron from the other atom. Covalent bonds form when two nonmetal atoms bond together and share electrons. But how exactly does a pair of atoms share electrons?

Electrons are located in atomic orbitals, the regions around the nucleus in which there is the greatest probability of finding an electron. For a covalent bond to form between two atoms, orbitals must overlap. The overlapped orbitals create a region of higher electron density in which two electrons are simultaneously located in the orbitals of both atoms.

This model of covalent bonding is called the **valence bond theory**. Figure 5.2 illustrates the simplest case of a covalent bond forming between two hydrogen atoms. Each hydrogen atom has an electron in a spherically shaped *s* orbital. When *p* orbitals are involved in the formation of a single bond between atoms, the overlap occurs at the end of one lobe of the orbital.

FIGURE 5.2 Overlap of atomic orbitals. The formation of a covalent bond is based on the sharing of electrons. Electrons are shared between two atoms by overlapping a region of three-dimensional space in which the electrons have a high probability of being found.

■ WORKED EXAMPLE 1

Draw the covalent bond between the *s* orbital of a hydrogen atom and the *p* orbital of a fluorine atom found in hydrofluoric acid (HF).

SOLUTION

The H atom has one electron in an *s* orbital. The F atom has one unpaired electron in a *p* orbital (dumbbell shape). The overlap of the *s* and *p* orbitals to form a covalent bond is represented as follows:

Practice 5.1

Draw the covalent bond between the F atoms in fluorine gas (F_2).

Answer

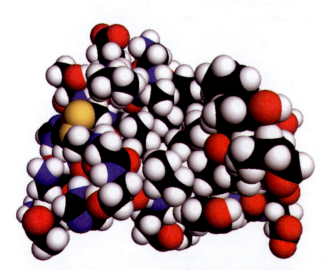

The complex shape of insulin ($C_{257}H_{383}N_{65}O_{77}S_6$) is due to the local geometry of each atom and the restriction placed on the location of each neighboring atom.

The valence bond theory works well for explaining how bonding occurs in a covalent compound. However, for determining the shapes of molecules, the theory works only in the simplest of cases, such as those illustrated above. For even slightly more complex molecules, another method for determining the geometry is necessary. For larger molecules such as insulin ($C_{257}H_{383}N_{65}O_{77}S_6$), the overall three-dimensional shape of the molecule is a result of the local geometry at each atom influencing the possible location of the neighboring atoms.

There are two types of information needed to determine the three-dimensional shape of a molecule: the locations of all of the valence electrons and how those electrons interact with one another. The *Lewis theory of bonding*, covered in the next section, addresses where valence electrons are located. The *valence shell electron pair repulsion theory* addresses how those valence electrons interact with one another, as it determines the actual geometry of the molecules from the information obtained from the Lewis theory.

5.2 Lewis Structures of Ionic Compounds

In Chapter 4, we saw the predictive power of the periodic table in determining the formula for ionic compounds. For instance, all alkali metals (Group 1) form cations with a charge of $+1$, all alkaline earth metals (Group 2) form cations with a charge of $+2$, and the halogens (Group 7) form anions with a charge of -1.

The **Lewis theory** explains this trend, stating that the atoms are attempting to achieve the same electron configuration as the closest noble gas by gaining or losing electrons as needed. The noble gases are chemically unreactive and do not bond to other atoms under normal conditions. The reason for this unusual stability is that the valence shell of the noble gas atoms has been completely filled. When the other elements in the periodic table react, the individual atoms strive to fill their outer valence shells and attain the stable electronic structure of the noble gas elements. Put another way, the elements react in such a way as to become **isoelectronic** with the noble gas elements. For example, sodium has one more electron than neon, the closest noble gas. Sodium metal is extremely reactive, but the sodium ion, Na^+, is very stable. By losing one electron and forming the Na^+ ion, sodium becomes isoelectronic with neon and has a filled valence shell. The sodium ion will not react further to gain or lose more electrons under standard conditions.

This pattern of elements forming bonds to become isoelectronic with the noble gases serves as the basis for the **octet rule**. Sometimes called the *rule of eight*, it states that elements react to attain a total of eight valence electrons, a configuration that corresponds to a filled *s*-orbital and *p*-orbital set. The exceptions to this rule are those elements that achieve an electron configuration identical to helium, which needs only two electrons.

The Lewis symbol for an element uses dots to represent the valence electrons for an atom. The electrons are placed around the elemental symbol, as shown at the top of Figure 5.3. The number of valence electrons is determined by counting the number of *s* and *p* electrons in the outermost level of the atom, as illustrated beneath the Lewis symbols in the figure.

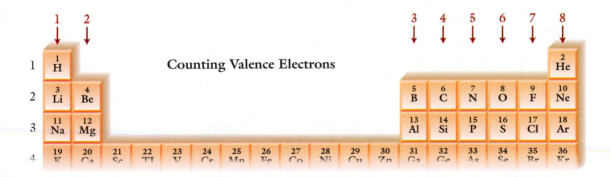

FIGURE 5.3 The Lewis dot structure of an element shows the number of valence electrons in the outer shell of the atom. The number of valence electrons is determined by counting the number of *s* and *p* electrons in the valence shell of the atom.

■ WORKED EXAMPLE 2

Draw the Lewis dot structures for the following elements:

(a) Li

(b) Mg

(c) N

SOLUTION

(a) Referring to the periodic table, Li has only one s electron: Li·

(b) Likewise, Mg has two s electrons: $\dot{M}g\cdot$

(c) As has two s electrons and three p electrons (do not count d electrons!): $\cdot\ddot{N}:$

Practice 5.2

Draw the Lewis dot structures for the following elements:

(a) O   (b) Xe   (c) Al

Answer

(a) $\cdot\ddot{O}:$ (b) $:\ddot{X}e:$ (c) $\cdot\dot{A}l\cdot$

Lewis dot structures can be used to show the gain and loss of electrons in the formation of an ionic bond between a metal and a nonmetal, as illustrated in Figure 5.4. The reaction shows the formation of calcium oxide (CaO), in which the calcium atom loses two valence electrons to an oxygen atom. In losing two electrons, the calcium 2+ ion has become isoelectronic with argon, completing its octet. The oxygen atom, with six original valence electrons, gains the two electrons from calcium and becomes isoelectronic with neon. The reaction produces the stable ionic compound calcium oxide. The formulas and charges of simple ionic compounds can be predicted and understood by writing the Lewis dot structures.

$$Ca\cdot + \cdot\ddot{O}: \longrightarrow [Ca]^{2+}[:\ddot{O}:]^{2-}$$

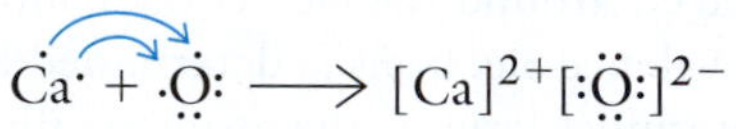

FIGURE 5.4 Formation of an ionic compound.

■ WORKED EXAMPLE 3

Draw the Lewis dot structures for the ionic bonding in MgF_2.

SOLUTION

The Mg atom will lose its two valence electrons, one to each F atom forming the ionic compound MgF_2.

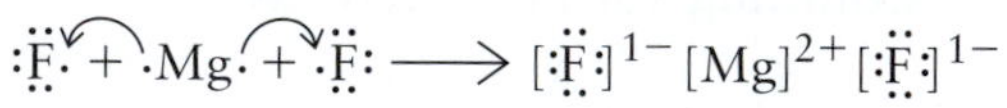

$$:\ddot{F}\cdot + \cdot Mg\cdot + \cdot\ddot{F}: \longrightarrow [:\ddot{F}:]^{1-}[Mg]^{2+}[:\ddot{F}:]^{1-}$$

Practice 5.3

Draw the Lewis dot structures for the ionic bonding in Na_2S.

Answer

$$[Na]^{1+}[:\ddot{S}:]^{2-}[Na]^{1+}$$

5.3 Lewis Structures of Covalent Compounds

Lewis dot structures can also illustrate covalent bonds between atoms in simple molecules. Figure 5.5 shows the valence electrons and covalent bond in the molecules Cl_2 and O_2. In the chlorine reaction, each chlorine atom has seven valence electrons and is striving to have eight. Each chlorine atom gains one electron, becoming isoelectronic with argon. Unlike atoms that react to form ionic compounds, each chlorine atom in Cl_2 has an equal tendency to obtain an electron. Therefore, it is impossible for one chlorine atom to form an anion and the other to form a cation. The only way a chlorine atom can achieve an octet in forming Cl_2 is for each atom to share its one unpaired electron with the other atom. The **single bond** that forms between two chlorine atoms consists of the two shared electrons. In the oxygen reaction, each oxygen atom has six valence electrons and is striving for eight. To achieve the octet, each oxygen atom must share two electrons with the other, thus forming a **double bond**. It is also possible to share three sets of electrons, forming a **triple bond** to achieve an octet.

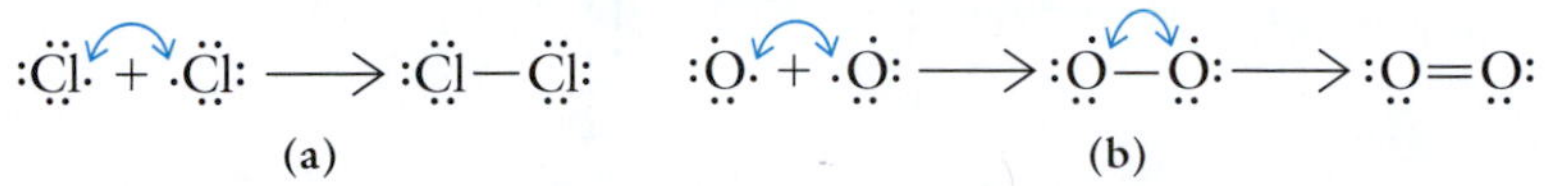

$$:\ddot{\underset{..}{Cl}}\cdot + \cdot\ddot{\underset{..}{Cl}}: \longrightarrow :\ddot{\underset{..}{Cl}}-\ddot{\underset{..}{Cl}}: \qquad :\ddot{\underset{.}{O}}\cdot + \cdot\ddot{\underset{.}{O}}: \longrightarrow :\ddot{\underset{.}{O}}-\ddot{\underset{.}{O}}: \longrightarrow :\ddot{O}=\ddot{O}:$$

(a)   (b)

FIGURE 5.5 Lewis symbols for simple covalent compounds.

■ WORKED EXAMPLE 4

Draw the Lewis structure for the covalent bonding in N_2.

SOLUTION

Each N atom has five valence electrons and needs three to complete an octet. By sharing three electrons each and forming a triple bond, both N atoms obtain an octet of valence electrons.

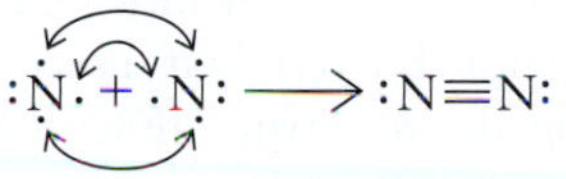

$$:\overset{..}{N}\cdot + \cdot\overset{..}{N}: \longrightarrow :N\equiv N:$$

Practice 5.4

Draw the covalent bonding in ammonia, NH_3.

Answer

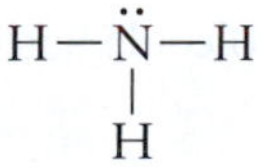

$$H-\overset{..}{N}-H$$
$$|$$
$$H$$

The Lewis structures of complex molecules cannot be determined by simply attempting to fill the vacant spots on the Lewis dot structure of individual atoms. A set of rules, listed below, must be followed in order to obtain the proper structure. Sulfur dioxide (SO_2), the compound responsible for the characteristic odor of a burning match, is used to illustrate each step. It is not necessary to write the Lewis dot structure of each atom, as is done for simpler compounds.

Step 1. Determine the total number of valence electrons in the compound. For neutral molecules, add the valence electrons contributed by each atom. For polyatomic cations, *subtract* the charge magnitude of the ion from the total number of valence electrons, because electrons are *lost* in the formation of the cation. For polyatomic anions, *add* the charge magnitude of the ion to the total number of valence electrons, because electrons are *gained* in the formation of anions. In the example of SO_2, there are six valence electrons from sulfur and six valence electrons from each of the two oxygen atoms, for a total of $6 + 2 \times 6 = 18$.

Step 2. Draw a skeletal structure that consists of single bonds from the central atom to each of the outer atoms. Subtract two electrons for each single bond used from the total number of valence electrons. The first atom in a formula is generally the central atom. Hydrogen is *never* a central atom.

$$\text{O}-\text{S}-\text{O} \quad \begin{array}{r} 18 \\ -4 \\ \hline 14 \end{array}$$

Step 3. Place pairs of electrons around the outer atoms to give each atom an octet of electrons. Subtract the electrons used from the total available. Hydrogen needs only a duet (two electrons total). Because they are not part of a bond, each of these pairs of electrons is called a **lone pair** or a **nonbonding pair**.

$$:\ddot{\text{O}}-\text{S}-\ddot{\text{O}}: \quad \begin{array}{r} 18 \\ -4 \\ \hline 14 \\ -12 \\ \hline 2 \end{array}$$

Step 4. If there are surplus electrons, place those electrons as lone pairs on the central atom.

$$:\ddot{\text{O}}-\ddot{\text{S}}-\ddot{\text{O}}: \quad \begin{array}{r} 18 \\ -4 \\ \hline 14 \\ -12 \\ \hline 2 \\ -2 \\ \hline 0 \end{array}$$

Step 5. If and only if the central atom does not have an octet of electrons resulting from steps 1 through 4, form a double bond by moving a lone pair of electrons from one of the outer atoms. (Use a triple bond only if a second double bond to another outer atom does not produce the desired structure with octets around all atoms.)

$$:\ddot{\text{O}}-\ddot{\text{S}}-\ddot{\text{O}}: \longrightarrow :\ddot{\text{O}}-\ddot{\text{S}}=\text{O}:$$

The structure obtained from this procedure could not have been determined by merely examining the simple Lewis structures of the individual atoms. In general, it is best to use these rules whenever the molecule has three or more atoms.

WORKED EXAMPLE 5

Draw the Lewis structure for CO_3^{2-}.

SOLUTION

Step 1. The total number of valence electrons is $4 + (3 \times 6) + 2 = 24$.

Step 2. The skeletal structure has three single bonds, so subtract six electrons from the total number of valence electrons: $24 - 6 = 18$.

$$\begin{array}{c} \text{O}-\text{C}-\text{O} \\ | \\ \text{O} \end{array}$$

Step 3. Complete the octet of outer atoms and subtract the number of electrons used from the total number of valence electrons remaining: $18 - 18 = 0$.

$$\begin{array}{c} :\ddot{\text{O}}-\text{C}-\ddot{\text{O}}: \\ | \\ :\ddot{\text{O}}: \end{array}$$

Step 4. Since there are no electrons left to place on the central atom, the diagram remains unchanged.

Step 5. Complete the octet of the central atom by forming multiple bonds:

$$:\ddot{O}-C-\ddot{O}: \implies \left[:\ddot{O}-C=\ddot{O}:\right]^{2-}$$
$$\quad\quad\quad |\quad\quad\quad\quad\quad\quad |$$
$$\quad\quad\quad :\ddot{O}:\quad\quad\quad\quad\quad :\ddot{O}:$$

Practice 5.5

Draw the Lewis structure for ammonia, NH_3.

Answer

$$H-\ddot{N}-H$$
$$\quad\quad |$$
$$\quad\quad H$$

WORKED EXAMPLE 6

Examine these Lewis structures for any errors.

$$H=O=H \quad :\ddot{O}=C=\ddot{O}: \quad H-\underset{\underset{H}{|}}{\overset{\overset{H}{|}}{C}}-H$$

(a)   (b)   (c)

SOLUTION

Structure (a) is incorrect because hydrogen atoms cannot have double bonds.
Structure (b) is incorrect because oxygen atoms cannot have 10 valence electrons.
Structure (c) is correct.

Practice 5.6

Draw the correct Lewis structures for Worked Example 6.

Answer

$$H-\ddot{O}-H \quad :\ddot{O}=C=\ddot{O}:$$

(a)   (b)

5.4 Resonance Structures

The Lewis structure for the carbonate ion is shown in Worked Example 5. The formation of a double bond between one of the oxygen atoms and the central carbon atom was needed to complete the octet of the central carbon atom. Given that all oxygen atoms are chemically identical, what driving force would cause one oxygen atom to share a set of lone pair electrons over the other oxygen atoms in the compound? It is therefore impossible for one oxygen atom to form a double bond to carbon while the other two identical oxygen atoms remain as single bonds.

LEARNING OBJECTIVE

Draw resonance structures for compounds that have multiple equivalent Lewis structures to accurately depict the bonding in a compound.

One way to study the bonding in a molecule is to examine the distance between the atoms within the bond, as illustrated in Figure 5.6 (on page 140). When a carbon atom forms a single bond to an oxygen atom, the bond length is 144 pm. If a double bond exists between the oxygen and carbon atoms, the atoms are pulled closer together and the length decreases to 126 pm. The shortest bond length between oxygen and carbon atoms is found in a triple bond between the two, with the distance being 116 pm.

To determine the nature of the bond between carbon and oxygen atoms in the carbonate ion, the bond length of each bond is measured. As stated earlier, there can be no difference

Carbon–Oxygen Bond Type	Bond Length (pm)	Model
Single	144	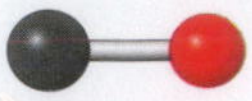
Double	126	
Triple	116	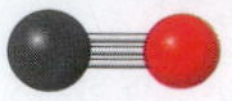

FIGURE 5.6 Carbon–oxygen bond types and lengths. The length of the bond between two atoms differs, depending on the nature of the bonding between the two atoms. The carbon–oxygen single bond is the longest bond at 144 pm, and the carbon–oxygen triple bond is the shortest at 116 pm.

in the bonding from one oxygen atom to another. This has been verified experimentally, as the bond length for all carbon–oxygen bonds in the carbonate ion is 138 pm. It should also be noted that the experimental bond length does not correlate to either the single, double, or triple bond. Rather, it is more than a single bond but less than a double bond.

When drawing a Lewis structure in which the formation of multiple bonds can be located between multiple atoms of the same element, it is proper to write each possible structure. In fact, the true structure is not shown by any of the Lewis structures. These multiple equivalent structures are called *resonance structures* (see Figure 5.7).

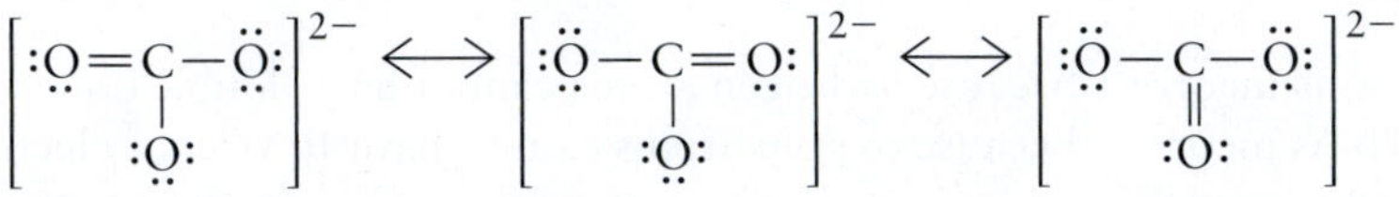

$$\left[:\overset{..}{O}=C-\overset{..}{O}: \atop \quad :\overset{..}{O}: \right]^{2-} \longleftrightarrow \left[:\overset{..}{O}-C=\overset{..}{O}: \atop \quad :\overset{..}{O}: \right]^{2-} \longleftrightarrow \left[:\overset{..}{O}-C-\overset{..}{O}: \atop \quad :O: \right]^{2-}$$

FIGURE 5.7 Resonance structures for the carbonate ion. The carbonate ion has three equivalent Lewis structures.

One common misconception is that the compound alternates between the resonance structures, when in actuality the true structure is a hybrid of the resonance structures. A good analogy is a dog that is a cross between two different breeds, such as a bulldog and a beagle. Your mind works to envision the appearance of this dog by using knowledge of the parental breeds. This is also how you should create a mental image of the actual structure of the compound from resonance structures.

WORKED EXAMPLE 7

Draw the three Lewis structures representing the resonance structures of sulfur trioxide (SO_3).

SOLUTION

The location of the double bond can occur at any of the oxygen atoms because they are all equivalent.

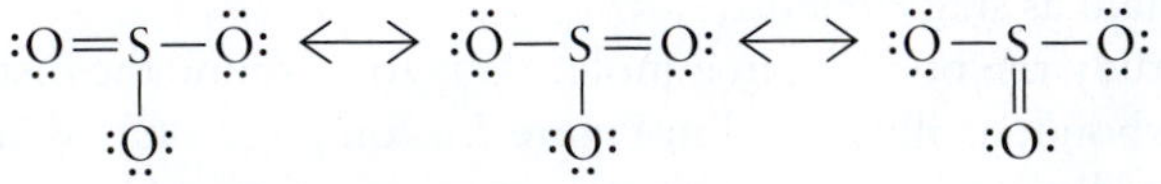

$$:O=S-\overset{..}{O}: \longleftrightarrow :\overset{..}{O}-S=O: \longleftrightarrow :\overset{..}{O}-S-\overset{..}{O}:$$

Practice 5.7

Draw the two Lewis structures representing the resonance structures for sulfur dioxide (SO_2).

Answer

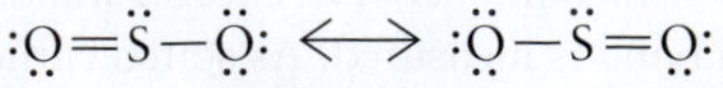

$$:O=\overset{..}{S}-\overset{..}{O}: \longleftrightarrow :\overset{..}{O}-\overset{..}{S}=O:$$

5.5 VSEPR Theory

The shape of molecules, such as cisplatin from the case study (Exploring Chemotherapy Drugs), is dictated by the three-dimensional location of each component atom. The shape is important because the interaction of the drug molecule within the human body is controlled by the shape of the molecule. It is possible to predict the shape by using the **valence shell electron pair repulsion theory (VSEPR theory)**. This theory is based on the principle that electrons in bonds and lone pairs repel one another and, in doing so, move as far apart from one another as possible. To use the VSEPR theory, it is necessary to start with the proper Lewis structure. The molecule's shape can then be determined by applying the principles of VSEPR theory. The shapes of complex molecules so derived are most often visualized through computer modeling software.

One common mistake students initially make is to look at the Lewis structure they have drawn and assume that the molecule's geometry is determined by the appearance of the Lewis structure. For example, because the Lewis structure for water is H—Ö—H, some students incorrectly state that it is a linear molecule. However, it is known from experimental evidence that the molecular shape of a water molecule is bent, not linear. This geometry can be predicted accurately by following the rules of VSEPR theory.

Another point of importance is that VSEPR theory first provides information on **electron geometry**—the location of valence *electrons*—and, based on this information, then determines **molecular geometry**—the location of *atoms* within the molecule. The electron geometry and the molecular geometry can be different. As we shall see, the difference depends on the presence or absence of lone pair electrons on the central atom of the molecule. The steps for determining electron geometries are provided below. The rules for determining molecular geometries will be given after the steps for determining electron geometries.

Water (H_2O) has a bent molecular geometry.

Step 1. Draw the Lewis structure of the molecule. (Starting with the correct Lewis structure is essential!)

Step 2. Determine how many regions of electron density surround the central atom. An **electron region** is one set of lone pair electrons, a single bond, a double bond, or a triple bond. The following examples show how to determine the number of electron regions:

Ö=C=Ö   :Ö=S̈—Ö: :Ö=S—Ö: (with :Ö: below)   H—C—H (with H above and below)   H—Ö—H   H—N̈—H (with H below)

2 regions   3 regions   3 regions   4 regions   4 regions   4 regions

Step 3. Determine the electron geometry from the descriptions shown in Figure 5.8. (Note that the lines now represent electron regions, *not* single bonds.)

When two electron regions surround a central atom, they are farthest apart if there is a 180° angle between them. This is called a **linear** arrangement because the two electron regions lie on a line. When a third electron region is added to a central atom, the angles between the regions change to 120° in order to maximize their distance apart. This arrangement is called **trigonal planar** because the three electron regions are located within the same geometrical plane and the tips of each region form a triangle. When four electron regions are located around a central atom, the optimal angle between electron regions is 109.5° with the shape of a tetrahedron. The resulting shape is called **tetrahedral** from the prefix *tetra-* meaning "four." The green dashed lines connecting the bottom three regions of the tetrahedron have been added to help you visualize the three-dimensional shape.

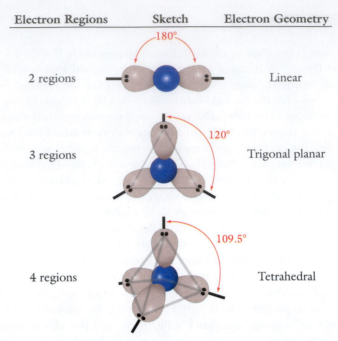

Electron Regions	Sketch	Electron Geometry
2 regions	180°	Linear
3 regions	120°	Trigonal planar
4 regions	109.5°	Tetrahedral

FIGURE 5.8 Determining electron geometry from the Lewis structure and the number of electron regions.

◼ WORKED EXAMPLE 8

Determine the electron geometry of both ions in the potentially explosive ammonium nitrate (NH_4NO_3).

SOLUTION

Step 1. The Lewis structures for the two ions are

$$
\left[\begin{array}{c} H \\ | \\ H-N-H \\ | \\ H \end{array}\right]^{+} \qquad \left[\begin{array}{c} :\ddot{O}-N=\ddot{O}: \\ | \\ :\ddot{O}: \end{array}\right]^{-}
$$

Step 2. In NH_4^+ there are four electron regions and in NO_3^- there are three electron regions.

Step 3. From Figure 5.8, we see that the electron geometry of NH_4^+ is tetrahedral and the electron geometry of NO_3^- is trigonal planar.

Practice 5.8

Determine the electron geometry of nitrogen triiodide, NI_3, a potentially explosive molecule.

Answer

☐ There are four electron regions, and the electron geometry is tetrahedral.

Space-filling models of the ammonium and nitrate ions.

Once the electron geometry is known, the next step is to determine the molecular shape by examining the location of the outer atoms in relation to the central atom. The key concept to remember during this part of the process is that lone pair electrons, if present, help to force the atoms into the positions they occupy but are not considered part of the molecular geometry. If no lone pair electrons are present, the molecular geometry is identical to the electron geometry.

Figure 5.9 shows the relationship between the number of electron regions in a molecule and the resulting geometry of the molecule. The geometry of a molecule that has

Electron Regions	Number of Lone Pairs	Sketch	Molecular Geometry
3	0		Trigonal planar
3	1		Bent
4	0		Tetrahedral
4	1		Trigonal pyramidal
4	2		Bent

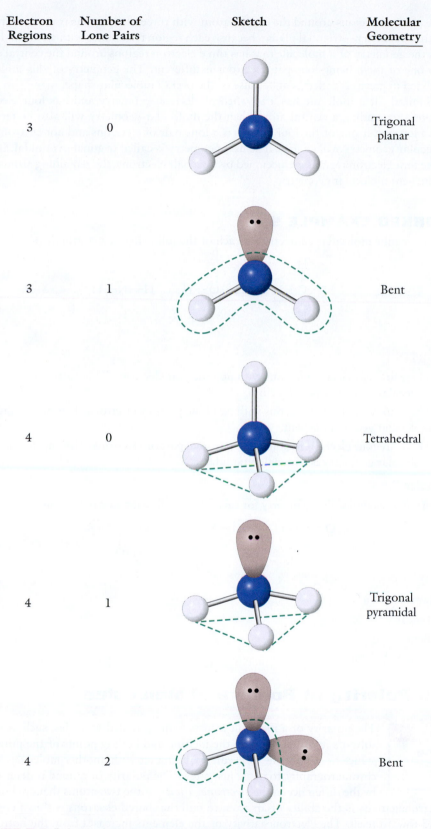

FIGURE 5.9 Relationship between the number of electron regions and molecular geometry.

three electron regions around the central atom with three outer atoms is identical to the electron geometry—trigonal planar, because each region terminates with an atom. However, the geometry of a molecule that has three electron regions around the central atom with one of them being lone pair electrons is different. The geometry of this molecule (outlined in green) is called **bent** because of the overall molecular shape.

Similarly, if a molecule has a tetrahedral electron geometry and has four electron regions surrounding a central atom, then the molecular geometry will also be tetrahedral. However, if one of the four regions is a lone pair of electrons and not an atom, the molecular geometry will change. The new geometry is called **trigonal pyramidal**. If two of the four electron regions are occupied by lone pair electrons, the remaining atoms take on the bent molecular geometry.

■ WORKED EXAMPLE 9

Determine the molecular geometry for each of the following Lewis structures.

$$\begin{array}{ccc} & \overset{\displaystyle H}{\underset{\displaystyle H}{H-\overset{|}{\underset{|}{C}}-H}} & H-\overset{..}{\underset{..}{O}}-H & H-\overset{..}{\underset{|}{N}}-H \\ & & & H \\ & (a) & (b) & (c) \end{array}$$

SOLUTION

(a) There are four electron regions with no lone pair electrons. This means the molecular geometry is tetrahedral.

(b) There are four electron regions with two lone pairs of electrons. This means the molecular geometry is bent.

(c) There are four electron regions with one lone pair of electrons. This means the molecular geometry is trigonal pyramidal.

Practice 5.9

Determine the molecular geometry for each of the following Lewis structures.

$$\ddot{O}=C=\ddot{O} \qquad :\ddot{O}=\ddot{S}-\ddot{O}: \qquad :\ddot{O}=S-\ddot{O}:$$
$$\qquad\qquad\qquad\qquad\qquad\qquad\qquad :\ddot{O}:$$
$$(a) \qquad\qquad (b) \qquad\qquad (c)$$

Answer

(a) Linear
(b) Bent
(c) Trigonal planar

5.6 Polarity of Bonds and Molecules

The polarity of a molecule affects many physical variables, such as which solvents it will dissolve in, the melting and boiling points of the pure substance, and how one molecule will interact with another molecule during chromatography, to name just a few. The polarity of a bond is determined by the difference in the *electronegativity* of the two atoms that are bonded. **Electronegativity** is the ability of an atom to pull the shared electrons within a covalent bond toward itself. The electronegativity of the elements increases from the bottom to the top of the columns on the periodic table and from the left to the right of the groups. Fluorine has the greatest electronegativity and the noble gases have zero electronegativity.

TABLE 5.1	Electronegativity of the Nonmetal Elements
Element	Relative Electronegativity
Fluorine	4.0
Oxygen	3.5
Chlorine	3.0
Nitrogen	3.0
Bromine	2.8
Carbon	2.5
Sulfur	2.5
Iodine	2.5
Selenium	2.4
Hydrogen	2.1
Phosphorus	2.1

The noble gases do not form covalent bonds because they have a full shell of valence electrons. Table 5.1 lists the relative electronegativity of the nonmetal elements, with fluorine having the maximum value of 4.0.

A **nonpolar covalent bond** occurs between two atoms when the electronegativity of both atoms is equal. The shared electrons are attracted equally to both atoms. Nonpolar covalent bonds occur between diatomic molecules such as nitrogen gas (N_2) and oxygen gas (O_2) and are thought to occur between two atoms of different electronegativity, as long as the difference in electronegativities is between 0.0 and 0.3.

A **polar covalent bond** occurs when the difference in electronegativities is between 0.4 and 1.9. The greater the difference in electronegativities, the more strongly the shared electrons are pulled toward the atom with the highest electronegativity. If the difference in electronegativities is greater than 2.0, the bond is considered to be ionic.

■ WORKED EXAMPLE 10

Are the following bonds nonpolar covalent or polar covalent?

(a) H—P

(b) C—H

(c) Cl—Br

(d) O—H

SOLUTION

(a) The difference in electronegativity is $2.1 - 2.1 = 0$, so it is a nonpolar covalent bond.

(b) The difference in electronegativity is $2.5 - 2.1 = 0.4$, so it is a polar covalent bond.

(c) The difference in electronegativity is $3.0 - 2.8 = 0.2$, so it is a nonpolar covalent bond.

(d) The difference in electronegativity is $3.5 - 2.1 = 1.4$, so it is a polar covalent bond.

Practice 5.10

Which element in the bonds from Worked Example 10 will have the partially negative charge?

Answer

(a) Neither   (c) Neither

(b) Carbon   (d) Oxygen

One misconception that occurs with polar and nonpolar covalent bonds is the assumption that nonpolar compounds contain nonpolar covalent bonds and polar compounds contain polar covalent bonds. This is *not* the case. Figure 5.10a shows a polar covalent bond between carbon and chlorine. The electron density is being pulled toward the chlorine atom, as the arrow indicates. When determining the polarity of a molecule such as carbon tetrachloride (CCl_4), it is imperative to examine the molecule's geometry. Figure 5.10b shows the electron density being pulled equally in all directions about the tetrahedral geometry. When electron density is pulled equally in all directions, the net result is a nonpolar molecule. If the electron density of the molecule is not symmetrically distributed, the molecule is polar. Any molecule in which the central atom contains lone pair electrons tends to be polar, too.

(a)   **(b)**

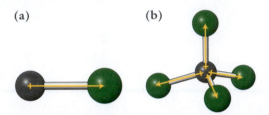

FIGURE 5.10 (a) The carbon (black) to chlorine (green) bond is polar in nature due to the higher electronegativity of the chlorine atom. (b) The carbon tetrachloride molecule is nonpolar because the polarities of the four carbon–chlorine bonds effectively negate one another.

■ WORKED EXAMPLE 11

Draw the Lewis structure for ammonia (NH_3), and determine whether or not this is a polar molecule.

SOLUTION

The Lewis structure for ammonia: 8 valence electrons, of which 6 are used in forming bonds to hydrogen and the remaining 2 valence electrons are placed on the nitrogen atom as a lone pair of electrons:

$$H-\ddot{N}-H$$
$$|$$
$$H$$

The molecular geometry for a molecule with four electron regions and one lone pair of electrons is trigonal pyramidal. The electronegativity of nitrogen is greater than that of hydrogen, drawing the partial negative charge toward the lone pair electrons, producing a polar molecule as shown at left.

Practice 5.11

Which molecular geometries will always produce a polar molecule?

Answer

Bent and trigonal pyramidal.

Ammonia has tetrahedral electron geometry with trigonal pyramidal molecular geometry. There is a lone pair of electrons on the nitrogen atom, and the electron density is all being directed to one portion of the molecule. Ammonia is polar.

Assigning a term such as *polar* or *nonpolar* to a molecule is a dramatic oversimplification of the physical phenomena involved. Some molecules such as water are very polar, while others such as hexane are very nonpolar. There are numerous variables, such as size and the functional groups attached, that determine the polarity of a molecule. Table 5.2 lists many common compounds and gives their polarity on a relative scale, with 0 being nonpolar and 9.0 being polar. The solubility is the maximum amount of the compound that can be dissolved in water.

TABLE 5.2	Polarity of Common Organic Solvents	
Compound	Polarity Index	Solubility in Water (%)
Hexane	0.0	0.001
Pentane	0.0	0.004
Toluene	2.4	0.051
Benzene	2.7	0.18
1-butanol	4.0	0.43
1-propanol	4.4	100
Butanone	4.7	24
Propanone	5.1	100
Ethanol	5.2	100
Water	9.0	100

5.7 Molecular Geometry of Fats and Oils

When you go to the supermarket to buy food, you will more than likely find food labels advertising "No Trans Fats." A typical Nutrition Facts food label on product packaging now breaks down the types of fats present in food to include saturated fats, trans fats, polyunsaturated fats, and mono-unsaturated fats, as shown in the side margin. The labels group together both oils and fats under the Total Fat category. Each type of fat and oil has different chemical properties, which give rise to different physiological effects when consumed.

The main physical difference between a fat and an oil is that an oil is a liquid at room temperature whereas a fat is a semi-solid. For example, butter is a semisolid at room temperature, yet vegetable oil is a liquid.

The main difference between their molecules is that fats contain only single bonds between the carbon atoms, whereas oils contain double bonds. It is chemically possible to convert an oil to a fat by a process called **hydrogenation** in which the double-bonded carbon (C=C) in oil molecules reacts, breaking the double bond, and two hydrogen atoms take the bond's place. Margarine is created by partially hydrogenating vegetable oils.

If the molecule is completely saturated with hydrogen atoms (contains no C=C bonds), then the molecule is classified as a *saturated fat*. Examples of saturated fats would include butter and lard (pork fat). If there is one C=C bond present in the molecule, it is a *monounsaturated fat,* and if there are two or more C=C bonds, the molecule is *polyunsaturated*. Monounsaturated and polyunsaturated molecules are liquids at room temperature and, therefore, are classified as oils.

Trans fats are a subclass of unsaturated fats that have been identified as particularly dangerous, as they can increase the risk of coronary heart disease. Trans fats are seldom found in natural fats and oils, although they are commonly created in the process of hydrogenation of the carbon double bond. We must look at the

LEARNING OBJECTIVE

Describe how the arrangements of atoms in three-dimensional space can lead to new compounds.

Nutrition Facts
Serving Size 1 Bar (85g)
Servings Per Container 4

Amount Per Serving

Calories 170	Calories from Fat 50

	% Daily Value *
Total Fat 6g	9%
Saturated Fat 4g	19%
Trans Fat 0g	
Polyunsaturated Fat 0.5g	
Monounsaturated Fat 1g	
Cholesterol 13mg	4%
Sodium 83mg	3%
Total Carbohydrate 33g	11%
Dietary Fiber 4g	16%
Sugar 25g	
Protein 3g	

(© Eugene Feygin/Alamy)

FIGURE 5.11 Trans and cis isomers of dichloroethene. When identical atoms are located across the double bond, the molecule is in the trans isomer arrangement (a). If the atoms are located on the same side of the double bond, the molecule is in the cis isomer arrangement (b).

geometry of the carbon atoms and the arrangement of atoms around the double bond to further understand the difference in molecules.

Fat and oil molecules are very large molecules, so a simpler example will be used to illustrate the differences in the geometry. Consider the molecules shown in Figure 5.11(a) and (b), where both molecules have the same formula, $C_2H_2Cl_2$. The only difference between the two molecules is the location of the hydrogen and chlorine atoms. In Figure 5.11(a), the hydrogen atoms are located across the double bond from one another in the trans position. (*trans* means "across from.") The second molecule, in Figure 5.11(b), has the hydrogen atoms on the same side in what is called the *cis* position. Molecules that share the same exact chemical formula but have different structures are called **isomers**. Isomers might share the same formula but they will have different melting points, boiling points, densities, and chemical reactivity.

Natural vegetable oils are unsaturated fats that contain carbon double bonds in the cis geometry. When food processors make semisolid fats, such as margarine, out of vegetable oil, they partially hydrogenate the oils. In this process, some of the double bonds in the oils are converted to the trans geometry. When an oil molecule switches to the trans geometry, the molecule still contains a C=C bond but behaves as if it were a fully saturated fat, and the oil will solidify.

Figure 5.12 illustrates how the cis and trans isomers affect fat and oil molecules. The first molecule (a) is stearic acid, a saturated fat that contains only single bonds between carbon atoms and is a semisolid at room temperature. The second molecule (b) is oleic acid, a monounsaturated fat, found in olive oil, that contains a single C=C bond in the cis conformation. Notice how the double bond puts a "kink" in the chain. This kink forces a greater distance between adjacent molecules, which weakens the attractions between them. As a result, oleic acid is a liquid at room temperature. The final molecule (c) is elaidic acid, which is an unsaturated fat with the C=C in the trans arrangement. This transformation allows the molecule to lie flat, much like stearic acid. Elaidic acid is a common component in partially hydrogenated vegetable oils and is a semisolid at room temperature.

In general, the amount of fats should be minimized, and when fats are consumed, our diets should lean toward the mono- and polyunsaturated fats. Up until the 1990s it was incorrectly believed that the trans fats created in this process were a healthier alternative

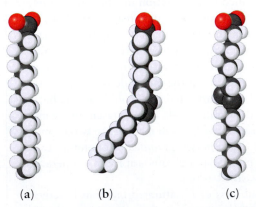

FIGURE 5.12 Saturated and unsaturated fats and oils. Stearic acid (a) is a saturated fat commonly found in animal fats. Oleic acid (b) is a monounsaturated oil found in vegetable oils that contains a C=C in the cis conformation. Elaidic acid (c) is a trans fat molecule that is created in the partial hydrogenation of vegetable oils.

to animal fats, which are fully saturated. We now know that trans fats are just as dangerous, if not more so than saturated fats as a risk factor for coronary heart disease and should be eliminated from diets.

■ WORKED EXAMPLE 5.12

2-Butene is a common industrial compound created by breaking up larger molecules obtained from the refining of petroleum in a process called cracking. Determine if the molecule below is in the cis or the trans conformation.

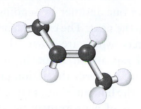

SOLUTION

Examining the C=C shows that the hydrogen atoms are located across the bond from one another. Therefore, this molecule is in trans conformation.

Practice 5.12

Draw the cis isomer of 2-butene, formula $CH_3CH=CHCH_3$. Use the structures shown in Figure 5.11 as models.

Answer

$$\underset{H_3C}{\overset{H}{\diagdown}}C=C\underset{CH_3}{\overset{H}{\diagup}}$$

Visualizing large molecules can be a challenge for even the most experienced chemists, who usually employ computer-aided software to determine the structure of molecules. Some researchers are now using computer-generated images in virtual reality chambers that allow scientists to explore the molecular structures from the vantage point of standing *inside* the molecules, as shown in Figure 5.13!

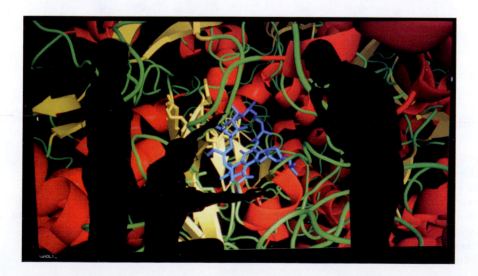

FIGURE 5.13 Scientists can now test potential drug candidates in the computer laboratory before ever creating the molecules in the chemistry laboratory. Students at Arkansas University view complex interactions of drug molecules with protein molecules using PyMOL in immersive 3D provided by Virtalis. (Courtesy of Arkansas University and Virtalis Limited.)

5.8 Drug Receptors and Brain Chemistry

Many illegal drugs alter the biochemical processes that take place in the brain. One of the dangers of abusing these drugs is that they can cause permanent changes to the normal operation of the brain cells. The three-dimensional shape of drug molecules influences their interaction with the human brain. Before we explore this interaction, we must first discuss how the central nervous system functions.

An estimated 100 billion nerve cells called **neurons** make up the human nervous system. Neurons communicate with one another by sending an electrical signal, called an **action potential**, from one cell to the next. The end of a neuron contains small packets of chemicals called **neurotransmitters**, which are chemical compounds that can travel outside of the neuron, cross a small gap, and arrive at the next neuron. When a neurotransmitter binds with another neuron, it causes the action potential to continue on its path toward a specific location within the human brain. Figure 5.14 illustrates the process. The action potential triggers the release of neurotransmitters from the top neuron. The neurotransmitters travel by diffusion to receptor sites on the second neuron, completing transmission of the action potential. The neurotransmitter molecules then diffuse back to the original neuron through uptake channels and are recycled for use with the next action potential. This entire process occurs on the microsecond time scale.

Illegal drugs create their mind-altering effects by several different mechanisms. Cocaine functions by interfering with the uptake of neurotransmitters, as shown in the right-hand side of Figure 5.14. Cocaine molecules block the uptake of the neurotransmitter, a process that floods the gap between the cells with an excess of neurotransmitter molecules and causes an ampli-fication of the signal. Because the affected nerve cells are in the pleasure-sensing region of the brain, the person who has taken the cocaine experiences a high.

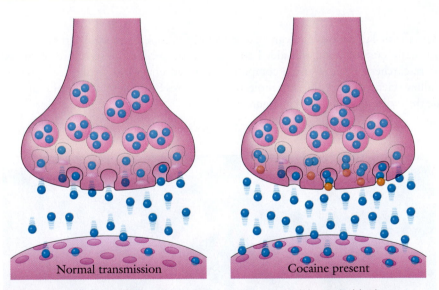

Normal transmission

Cocaine present

FIGURE 5.14 Neuron communication. Neurotransmitters (shown in blue) are chemicals released from one neuron that diffuse across a small gap to receptor sites on another neuron. This process allows communication between neurons. Once a neurotransmitter has sent the signal to the receptor, it diffuses back to the original neuron, enters through uptake channels, and is stored for future use. Cocaine molecules (shown in yellow) block the uptake channels, preventing released neurotransmitters from diffusing back into the neuron. This floods the gap with neurotransmitters, thus magnifying the signal being sent.

Although the cocaine molecules are eventually removed, the system does not return completely to normal. People who use illegal drugs build up a tolerance that requires higher doses to continue producing the same pleasurable effect. The user must take in more cocaine to block more uptake channels and produce a large enough neurotransmitter concentration to achieve the same level of sensation. These higher doses have a damaging effect on neurons. When people stop using drugs, the damage to the neurons is not repaired. Such damage can lead to depression because the mood-sensing neurons no longer function properly under the normal release of neurotransmitters.

How exactly does cocaine interfere with the uptake of the neurotransmitters? The uptake channel of the neuron is a portion of the cell membrane that functions as a one-way tunnel for the neurotransmitter. The tunnel walls are strands of protein molecules arranged so that neurotransmitters of a particular shape and size can pass through. Cocaine is able to lodge itself into this tunnel, blocking the neurotransmitters from reentering the cell. To block this tunnel, cocaine molecules must interact strongly with portions of the protein molecules through intermolecular forces. The distinct three-dimensional structure of cocaine makes this interaction possible. Figure 5.15 illustrates this interaction. The red coil ribbons represent protein molecules forming the neurotransmitter uptake channel. Key portions of the molecules are drawn in as molecular models. The cocaine molecule is highlighted to help illustrate its location.

A scanning electron microscope image of a neuron (shown in red). (Quest/Photo Researchers, Inc.)

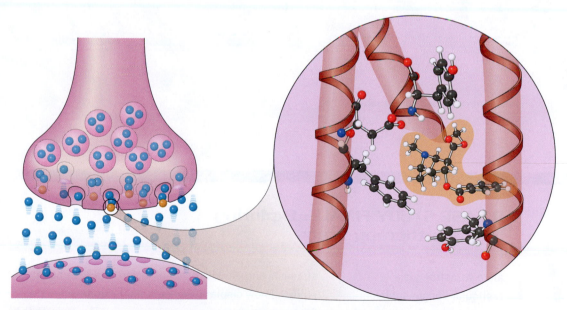

FIGURE 5.15 Cocaine blocking a neurotransmitter uptake transporter. The cocaine molecule is the correct size, shape, and polarity to wedge itself into the uptake channels of neurotransmitters. Once the uptake of neurotransmitters is blocked, the concentration of the neurotransmitters in the gap between neurons is greatly magnified, which results in the magnification of the signal being sent between neurons. The increased signal produces a pleasurable sensation, or "high." (Adapted from Dahl)

Immunoassay Methods

A lock-and-key analogy is often used to explain how an immunoassay test works. The shapes of a lock and key are designed in such a way that only a key with one particular shape can fit into the lock properly. Certain types of molecules have three-dimensional structures that enable them to behave as though they are locks or keys. The molecule that functions as a key is called an **antigen**, and the molecule that functions as the lock is called an **antibody**.

The human body produces antibodies to fight off infections. When a foreign molecule is introduced into the body, antibodies are created to attack the foreign molecules (antigens). The antibody is able to attack the antigen because it has a three-dimensional structure that is specific for binding to the antigen. A bound antigen is unable to disrupt normal cellular functions. Immunoassay techniques depend on similar antigen-antibody reactions that are specific for particular drugs.

There are several methods in which the basic principles of immunoassay are used to detect molecules. Figure 5.16 illustrates the method called **radioimmunoassay** (**RIA**), a technique in which radioactive isotopes play a part in the analysis. The first step in the detection of an antigen such as insulin is to mix it with a known amount of insulin that is labeled with a radioactive iodine atom, as shown in step 1 of Figure 5.16.

RIA Simulation

Recall from Section 3.5 that isotopes are forms of an element that have the same number of protons but different numbers of neutrons. Iodine-131 is an isotope that emits radiation that can be detected by an electronic instrument. When radioactive iodine is

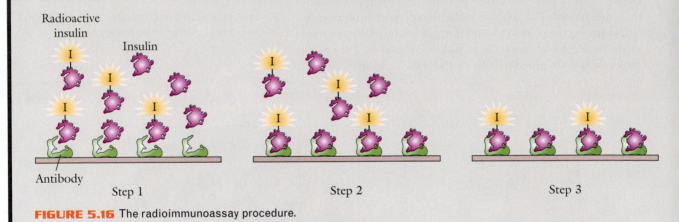

Radioactive insulin

Insulin

Antibody

Step 1

Step 2

Step 3

FIGURE 5.16 The radioimmunoassay procedure.

5.9 CASE STUDY FINALE: Exploring Chemotherapy Drugs

Lance Armstrong is a testicular cancer survivor, who has gone on to set a record by winning the Tour de France seven consecutive times. His battle against cancer was aided by the use of cisplatin, one of the most powerful chemotherapy drugs for testicular, ovarian, lung, stomach, and bladder cancers. Scientists have been studying how cisplatin works to fight cancer since the FDA approved its use over 30 years ago. By studying the molecular structure of cisplatin and how it interacts with DNA inside cancer cells, scientists have started to develop a new generation of chemotherapy drugs using knowledge derived from cisplatin.

The large protein molecule forms the "lock" with an opening that is specific for the "key" molecule in yellow. The ribbon is superimposed on the large protein molecule to better illustrate the overall shape of the protein molecule. (Science Photo Library/Photo Researchers, Inc.)

incorporated into a molecule such as insulin, the drug molecule is radioactively labeled.

For an RIA analysis, the amount of radioactively labeled insulin is known exactly, but the amount of insulin in the sample, if any, is unknown. In step 2, the labeled and unlabeled insulin are allowed to react with an antibody that has been physically immobilized on a plate. The labeling of the insulin molecules does not change their ability to bond to the antibody. Therefore, both the labeled and unlabeled insulin have an equal chance of binding to the antibody. If the

two types of antigens are present in equal amounts, an equal number of each type of molecule will bond to the antibody. However, if the sample being analyzed contains no insulin, the plate will be filled with only radioactively labeled insulin. After the two antigens are allowed to react with the antibodies, the excess antigens are washed away. The final step is to determine the ratio of labeled to unlabeled antibodies. The greater the radioactivity of the immunoassay test when finished, the less insulin in a sample and vice versa.

WORKED EXAMPLE 13

Is it necessary to separate out the various compounds of blood or urine using chromatography before using the RIA method?

SOLUTION

Only those molecules that fit into the antibody will react. Therefore, if other molecules are present, they will simply be washed away between steps 2 and 3, making chromatography unnecessary.

Practice 5.13

Will a radioimmunoassay test that shows positive results for insulin have a larger or smaller signal from the radioactive antigen in step 3 than a radio-immunoassay test that comes back negative for insulin?

Answer

The greater the insulin concentration in the sample, the smaller the observed signal from the radioactive isotopes will be in the final step.

Barnett Rosenberg chose experimental conditions for testing *E. coli* that would react with the normally inert platinum electrodes to produce cisplatin. Through his diligent efforts to understand the cause of his unexpected results, he determined that the compound cisplatin, $Pt(NH_3)_2Cl_2$ was responsible. The molecular geometry of cisplatin has Pt in the center position with NH_3 and Cl surrounding the Pt ion. In a twist on the geometry studied so far in this chapter, the cisplatin molecule adopts a square planar geometry with the platinum ion in the center and the ammonia and chloride ions occupying the corner positions as shown here.

The position of the ammonia molecules and chloride ions is critical to the biological reactivity of cisplatin. The cis isomer has the ammonia molecules on adjacent corners and the chloride ions on the opposite side, as shown in Figure 5.17 (a). The trans isomer has the ammonia molecules on opposite corners, as well as the chloride ions as shown in Figure 5.17(b). The transplatin molecule shows almost no ability to attack cancer cells, a major biological difference in spite of such a minor geometric difference.

Cl⟍   ⟋NH₃   H₃N⟍   ⟋Cl

      Pt   Pt

Cl⟋   ⟍NH₃   Cl⟋   ⟍NH₃

       (a)   (b)

FIGURE 5.17 The isomers of $Pt(NH_3)_2Cl_2$ have different biological reactivity. The cis isomer (a) is a powerful chemotherapy drug, whereas the trans isomer (b) has virtually no value as a chemotherapy drug.

Why does cisplatin have the ability to kill cancer cells, but not transplatin? For that information, we have to take a closer look at DNA. The DNA in a cell is replicated during the process of mitosis to form an identical copy, which can then be transferred to the new cell as it splits from the original cell. For DNA to make a copy of itself, it must unravel the two strands of DNA, and this is where cisplatin comes in to do its work.

You may recall from a biology class that DNA strands are held together when the nucleic acid adenine bonds to thymine and guanine bonds to cytosine, which can come apart during the replication of DNA. This bonding is depicted in Figure 5.18. When cisplatin is present in a cell, it binds to the nitrogen atom on two adjacent guanine bases. This new bonding causes a literal kink to form in the DNA and triggers the cell to attempt to repair the DNA. However, cancer cells do not contain the necessary proteins that can remove cisplatin, so the cell undergoes the programmed cell death called *apoptosis*.

Cisplatin does have some adverse side effects, such as nausea, hearing loss, and temporary hair loss. Hair loss is common among chemotherapy drugs, as they are targeting fast-growing cancer cells and preventing their reproduction. Hair cells are affected for the same reason, because they are fast-growing cells compared to the rest of the cells in our body. The newer drugs developed based on our knowledge of cisplatin have been able to reduce nausea and can now be taken orally rather than administered intravenously. Today, people from all over the world have had a second chance on life thanks to the astute observations and application of the scientific method to an experiment that did not work the way the researchers thought it would.

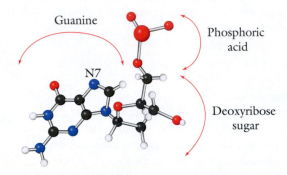

FIGURE 5.18 Interaction of cisplatin with DNA. Cisplatin is believed to function by forming a bond between two nitrogen atoms (N7) on consecutive guanine bases on a strand of DNA, which effectively kinks the DNA. The kink prevents replication and triggers cell death.

CHAPTER SUMMARY

• The sharing of electrons to produce a covalent bond arises through an overlap of the orbitals in which the bonding electrons have the highest probability of being found.

• The Lewis theory of bonding states that valence electrons are the electrons responsible for the formation of both ionic and covalent bonds. Ionic bonding is achieved by the transfer of electrons between two atoms to produce positively and negatively charged ions. The newly formed ions achieve a completed octet of electrons, a stable state in which each ion is isoelectronic with one of the noble gases. Covalent bonds share electrons to achieve a stable octet of electrons.

• Lewis structures for complex molecules can be determined by verifying the total available valence electrons, then distributing them according to the rules for drawing Lewis structures.

• Writing a proper Lewis structure is a critical step for examining both the electron geometry and molecular geometry of molecules. Several common mistakes to avoid include failure to subtract or add electrons to the total valence electron number for cations and anions, placement of more than a single bond on hydrogen, and use of double or triple bonds when the central atom already has a completed octet.

• Resonance structures exist whenever multiple equivalent Lewis structures can represent a compound in which a double bond occurs between equivalent atoms within the structure.

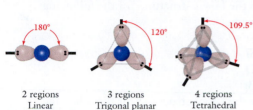

| 2 regions | 3 regions | 4 regions |
| Linear | Trigonal planar | Tetrahedral |

• The VSEPR theory states that electrons around a central atom in bonds or lone pairs will repel one another and will assume positions as far apart from one another as possible. Electron geometry determines molecular geometry, the structural arrangement of the atoms in the molecule.

• Complex molecules have geometries that are dictated by the local geometry of each atom within the molecule. The complex three-dimensional shape of molecules influences the mechanisms by which drug molecules interact with the human body. The molecular geometry of drugs can be exploited for the detection of illegal drugs by employing immunoassay methods.

• Immunoassays are based on the principle that a drug molecule (antigen) of interest has a unique three-dimensional structure that allows it to bind specifically to another molecule called an antibody.

• In radioimmunoassay tests, the molecules from an evidence sample compete for the binding sites with a known amount of radioactively labeled molecules that have identical bonding properties. If labeled and unlabeled molecules are present in equal numbers, equal amounts will bond to the antibody molecules. The stronger the signal from the radioactive isotope, the lower the amount of the target molecule in the evidence sample.

KEY TERMS

action potential (p. 150)
antibody (p. 152)
antigen (p. 152)
bent geometry (p. 144)
cis isomer (p. 148)
double bond (p. 137)
electronegativity (p. 144)
electron geometry (p. 141)
electron region (p. 141)
hydrogenation (p. 147)
immunoassay (p. 133)

isoelectronic (p. 135)
Lewis theory (p. 135)
linear geometry (p. 141)
lone pair (p. 138)
molecular geometry (p. 141)
neurons (p. 150)
neurotransmitter (p. 150)
nonbonding pair (p. 138)
nonpolar covalent bond (p. 145)
octet rule (p. 135)
polar covalent bond (p. 145)

radioimmunoassay (RIA) (p. 152)
single bond (p. 137)
tetrahedral geometry (p. 141)
trans isomer (p. 148)
trigonal planar geometry (p. 141)
trigonal pyramidal geometry (p. 144)
triple bond (p. 137)
valence bond theory (p. 134)
valence shell electron pair repulsion
(VSEPR) theory (p. 141)

MAKING MORE CONNECTIONS: Additional Readings, Resources, and References

The following journal articles give specific details relating to several applications provided in the chapter:

Dahl, Svein G. *The Journal of Pharmacology and Experimental Therapeutics*, vol. 307, no. 1, pp. 34–41.

Hirota, O. S., Suzuki, A., Ogawa, T., and Ohtsu, Y. "Application of Semi-Microcolumn Liquid Chromatography to Forensic Analysis," *Analysis Magazine*, 1998, vol. 26, no. 5.

The following Web site contains the opinion of a Massachusetts court in *Commonwealth v. Christina Martin*: http://caselaw.findlaw.com/ma-supreme-judicial-court/1010762.html

A large number of newspaper articles regarding Christina Martin's appeal can be found in the *Standard-Times* newspaper of New Bedford, MA, during the height of her case in 1998 and 1999.

REVIEW QUESTIONS AND PROBLEMS

Questions

1. Discuss the concept of electron sharing between atoms and how this takes place. To illustrate your explanation, sketch a diagram of two atoms sharing a pair of electrons. (5.1)

2. Which electrons are responsible for the formation of ionic and covalent bonds? (5.1–5.2)

3. What determines whether an atom will form a covalent bond or an ionic bond? (5.1)

4. What determines whether an atom will form a cation or an anion? (5.2)

5. What benefit do atoms receive by the exchange or sharing of electrons in reactions? (5.2)

6. What determines the charge an ion will have? (5.2)

7. Under what conditions will a compound form a double bond between two atoms? (5.3)

8. List all of the possible examples of an electron region. (5.5)

9. Explain why resonance structures do not actually represent accurate bonding within a molecule. (5.4)

10. What is the basic principle of the valence shell electron pair repulsion (VSEPR) theory? (5.5)

11. How does the number of electron regions around a central atom influence the electron geometry of a molecule? (5.5)

12. What are the possible electron geometries? Draw a sketch of each. (5.5)

13. Does the nature of the electron region (single bond, double bond, triple bond, lone pair) affect the electron geometry of the molecule? Explain. (5.5)

14. What are the possible molecular geometries? Draw a sketch of each. (5.5)

15. Does the nature of the electron regions affect molecular geometry? Explain. (5.5)

16. Discuss how the local geometry of individual atoms contributes to the overall geometry of a larger molecule. (5.7)

17. What are stereoisomers? Discuss their significance in drug chemistry. (5.7)

18. Explain how neurotransmitters enable communication between neurons. (5.8)

19. Describe the role of molecular shape of neurotransmitters in transmission of action potentials from one neuron to another. (5.8)

20. Discuss the mechanism by which illegal drugs such as cocaine interfere with the process indicated in Question 19. (5.8)

21. How can illegal drugs such as cocaine cause a permanent change in brain function? (5.8)

22. Discuss how molecular shape is crucial to immunoassay methods of analysis. (5.8)

Problems

23. Draw the Lewis structures for the following elements: (5.2)
    (a) Sr
    (b) Si
    (c) O
    (d) Br

24. Draw the Lewis structures for the following elements: (5.2)
    (a) Ca
    (b) Rb
    (c) F
    (d) B

25. Draw the Lewis structures for the following ionic compounds: (5.2)
    (a) Na_2S
    (b) $MgCl_2$
    (c) SrO
    (d) Li_3P

26. Draw the Lewis structures for the following ionic compounds: (5.2)
    (a) NaBr
    (b) CaS
    (c) KF
    (d) K_2O

27. Draw the Lewis structure showing the bonding between the nitrogen and iodine atoms in NI_3. (5.3)

28. Draw the Lewis structure showing the bonding between the two bromine atoms in Br_2. (5.3)

29. Draw the Lewis structures for the following polyatomic ions. Draw resonance structures if applicable. (5.3–5.4)
    (a) CN^-
    (b) PO_4^{3-}
    (c) CO_3^{2-}
    (d) OH^-

30. Draw the Lewis structures for the following polyatomic ions. Draw resonance structures if applicable. (5.3–5.4)
    (a) NO_3^-
    (b) NH_4^+
    (c) SO_4^{2-}
    (d) ClO_3^-

31. Draw the Lewis structures for the following covalent compounds. Draw resonance structures if applicable. (5.3–5.4)
    (a) PF_3
    (b) NO_2
    (c) CCl_4
    (d) AsH_3

32. Draw the Lewis structures for the following covalent compounds. Draw resonance structures if applicable. (5.3–5.4)
    (a) H_2S
    (b) O_3
    (c) SO_3
    (d) IF_3

33. Determine the electron geometry for the following ions from Problem 29: (5.5)
    (a) CN^-
    (b) PO_4^{3-}
    (c) CO_3^{2-}
    (d) OH^-

34. Determine the electron geometry for the following ions from Problem 30: (5.5)
    (a) NO_3^-
    (b) NH_4^+
    (c) SO_4^{2-}
    (d) ClO_3^-

35. Determine the electron geometry for the following compounds from Problem 31: (5.5)
    (a) PF_3
    (b) NO_2
    (c) CCl_4
    (d) AsH_3

36. Determine the electron geometry for the following compounds from Problem 32: (5.5)
    (a) H_2S
    (b) O_3
    (c) SO_3
    (d) IF_3

37. Determine the molecular geometry for the following ions from Problem 29: (5.5)
    (a) CN^-
    (b) PO_4^{3-}
    (c) CO_3^{2-}
    (d) OH^-

38. Determine the molecular geometry for the following ions from Problem 30: (5.5)
    (a) NO_3^-
    (b) NH_4^+
    (c) SO_4^{2-}
    (d) ClO_3^-

39. Determine the molecular geometry for the following compounds from Problem 31: (5.5)
    (a) PF_3
    (b) NO_2
    (c) CCl_4
    (d) AsH_3

40. Determine the molecular geometry for the following compounds from Problem 32: (5.5)
    (a) H_2S
    (b) O_3
    (c) SO_3
    (d) IF_3

41. Fill in the table. (5.5)

Electron Regions	Number of Lone Pairs	Electron Geometry	Molecular Geometry
2	0		
3	1		
4	0		

42. Fill in the table. (5.5)

Electron Regions	Number of Lone Pairs	Electron Geometry	Molecular Geometry
3	0		
4	2		
4	1		

43. Provide the Lewis structure, electron geometry, and molecular geometry for each of the following substances. (5.5)
    (a) NO_2^-
    (b) ClO_4^-
    (c) ClO_2^-
    (d) NH_3

44. Provide the Lewis structure, electron geometry, and molecular geometry for each of the following substances. (5.5)
    (a) SO_2
    (b) SO_3^{2-}
    (c) PO_2^{3-}
    (d) BrO_4^-

45. Determine whether the following bonds are polar covalent or nonpolar covalent. If the bond is polar covalent, indicate the polarity by drawing an arrow in the direction of electron density. (5.6)
    (a) F—P
    (b) H—Br
    (c) N—Cl
    (d) S—C

46. Determine whether the following bonds are polar covalent or nonpolar covalent. If the bond is polar covalent, indicate the polarity by drawing an arrow in the direction of electron density. (5.6)
    (a) H—Se
    (b) P—Br
    (c) S—S
    (d) N—I

47. Determine whether the following bonds are polar covalent or nonpolar covalent. If the bond is polar covalent, indicate the polarity by drawing an arrow in the direction of electron density. (5.6)
    (a) O—P
    (b) I—Se
    (c) Br—N
    (d) C—Se

48. Determine whether the following bonds are polar covalent or nonpolar covalent. If the bond is polar covalent, indicate the polarity by drawing an arrow in the direction of electron density. (5.6)
    (a) C—C   (c) I—F
    (b) O—H   (d) P—S

49. Determine whether the following compounds from Problem 31 are polar or nonpolar: (5.6)
    (a) PF_3
    (b) NO_2
    (c) CCl_4
    (d) AsH_3

50. Determine whether the following compounds from Problem 32 are polar or nonpolar: (5.6)
    (a) H_2S
    (b) O_3
    (c) SO_3
    (d) IF_3

51. Discuss why immunoassays are considered a screening method. (CS)

52. Identify the molecular geometry of each carbon atom (black) in the following structure. (5.7)

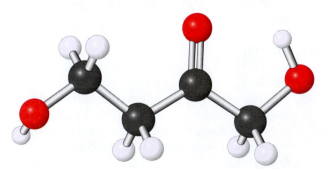

53. Identify the electron geometry of each oxygen atom (red) in the following structure. Each oxygen atom contains two lone pairs of electrons. (5.7)

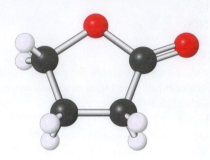

54. Identify the electron geometry and molecular geometry of each atom in the structure of the prescription medicine Ritalin. Each oxygen atom (red) has two lone pairs of electrons, and the nitrogen (blue) has one lone pair of electrons. (5.7)

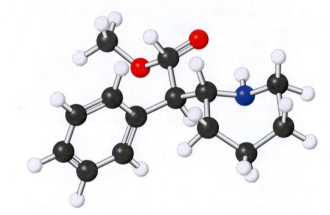

Case Study Problems

55. Christina Martin was convicted of murdering her long-time boyfriend by spiking his food with LSD because her teenage daughter told her that he had been sexually assaulting her. Christina did not fall under suspicion by the police until after her boyfriend had been buried, when the police learned Christina had reportedly told friends and family of her plans. Thirty days after his burial, Christina's boyfriend was exhumed, and it was discovered that his grave had filled with groundwater. A sample of tissue recovered during the exhumation tested positive for LSD by radioactive immunoassay analysis (RIA). However, the sample came back negative for LSD when tested with a more advanced method that can detect the molecular mass of any compound present in the sample. The prosecution relied heavily on the RIA analysis of a questionable sample while downplaying the gas chromatography-mass spectroscopy (GC-MS) data results as insignificant to the jury. The defense failed to challenge the evidence in the first trial, which resulted in a conviction. In your opinion, and relying on the limited amount of information provided in this case, could a conviction have been secured if the evidence had been properly presented? Remember that the standard for a jury to convict a defendant is that the person must be guilty "beyond a reasonable doubt."

56. Discuss how a typical blood sample that tests positive for LSD by radioimmunoassay could be considered stronger evidence than the sample taken in the Martin case, even if the same analysis procedure were used. Also take into consideration that ergot fungus is commonly found in soil and is known to be cross-reactive with LSD in immunoassay tests.

57. Christina Martin appealed the court decision on the basis of ineffective counsel, and the courts agreed that a competent defense lawyer should have challenged the validity of the scientific evidence presented in court. The original sentence was dismissed, and the prosecutor offered her a plea deal, which released her from prison with time already served. The circumstantial evidence of her guilt included attempts to purchase drugs from local teenagers, witnesses who testified that she had indicated she was going to get revenge, and her daughter claiming at trial that she was the one to poison the victim and not her mother. In your opinion, was the circumstantial evidence stronger than the scientific evidence, and would it be enough to secure a conviction?

CHAPTER 6

Aqueous Solutions: Part I

CASE STUDY: Exploring Chemistry in the Kitchen

"Marie [Curie] read and re-read her cookery book and annotated it conscientiously in the margins, reporting her trials, failures and successes in brief phrases of scientific accuracy . . . But cooking was as difficult and mysterious as chemistry."

—Madame Curie, *by Eve Curie*

Madame Curie, the first female recipient of a Nobel Prize and the only person to ever receive two Nobel Prizes in science, approached cooking the same way she approached an experiment: changing variables, making observations, creating a hypothesis on how to improve the experiment, and testing that hypothesis. For those who have ever struggled to replicate a recipe, a better understanding of the chemistry inside the kitchen can make that process much easier.

Consider Table 6.1, which compares the recipe for a brownie with that for a muffin. You will quickly notice that the ingredients are nearly the same for the two different

TABLE 6.1	
Brownies	**Muffins**
$1\frac{2}{3}$ cups white sugar	1/2 cup white sugar
4 eggs	1 egg
1 cup vegetable oil	1/3 cup unsalted butter
$1\frac{1}{3}$ cups all-purpose flour	$1\frac{2}{3}$ cups all-purpose flour
1 teaspoon salt	1/2 teaspoon salt
1/2 cup unsweetened cocoa powder	1 ounce semisweet chocolate squares
1 cup semisweet chocolate chips	
1 teaspoon vanilla extract	1 teaspoons vanilla extract
	3/4 cup buttermilk
	1 teaspoon baking soda

desserts: Sugar, eggs, oil-butter fat source, flour, salt, chocolate source, and vanilla. Of course, the proportions differ between the two recipes, but that alone does not explain why a brownie comes out chewy and dense while a muffin comes out firm and cakelike. The key is in the last two ingredients in the muffin recipe, which have no counterpart in the brownie recipe.

Muffins contain buttermilk and baking soda. What exactly are these ingredients, and how does chemistry explain the difference in the two tasty

As you read through the chapter, consider the following questions:

• What role do chemical reactions have in cooking?

• If a substance is sour tasting, does that have a chemical significance?

• What does the baking soda do in a recipe? Is it the same thing as baking powder?

experiments? Buttermilk is traditionally the milk left over from the process of making butter. Buttermilk has less fat content than milk and has a characteristic sour taste. Baking soda is better known in the chemistry laboratory as sodium bicarbonate, $NaHCO_3$.

MAKE THE CONNECTION

But to understand how buttermilk and sodium carbonate can create such a large difference, we will first study the different types of reactions in solutions.

6.1 Aqueous Solutions

LEARNING OBJECTIVE

Distinguish between a solvent and a solute.

From everyday observations, you are undoubtedly familiar with the process of dissolving one substance into another. Take a teaspoon of ordinary sugar (sucrose), stir it into a cup of water, and what happens? The white crystals of sugar are no longer visible in the mixture of sugar and water, and the resulting solution now has a sweet taste that the water alone did not have.

But try to dissolve a teaspoon of sand (SiO_2) or oil in water, and you will notice something different. The sand settles to the bottom of the container after stirring, and the oil floats to the top. Neither sand nor oil is incorporated into the water in the same way as sugar. Why is there a difference between the behavior of the sugar, sand, and oil? To understand why some substances dissolve in water and others do not, it is necessary to go beyond what can be seen with the naked eye.

In this chapter, we discuss properties of solutions, focusing on **aqueous** (water-based) solutions because of their importance in chemistry and forensic science (and in cooking). In an aqueous solution, water is the **solvent** (the liquid that dissolves a chemical compound) and is the major component of the solution. The substance that dissolves into the water is called the **solute** and is the minor component of the solution. In the example given above, sugar was the solute (see Figure 6.1). We do not refer to the sand or oil as solutes because they do not form solutions with water.

The typical kitchen is full of solutions such as vinegar, which contains acetic acid as one of the main solutes. If it is balsamic, apple cider, or wine vinegar, there are many additional compounds dissolved in the vinegar solution that produce the flavor and color of the vinegar. Some solutions have compounds added to help preserve food and prevent it from spoiling. Many recipes call for all the liquid ingredients to be mixed together before any of the dry ingredients are added. How each ingredient influences the flavors, textures, and outcome of a recipe is best understood by knowing the chemistry. A few examples of solutions found in the kitchen are provided below.

FIGURE 6.1 An aqueous solution forms as the solute (sugar) dissolves into the solvent (water). (Martyn F. Chillmaid/Science Photo Library)

• Wine: A complex mixture of sugars, alcohol, water, and tannins (if red wine).
• Soft Drinks: A mixture of water, sugar (or sugar substitute), dissolved carbon dioxide gas (the carbonation), and flavoring compounds.
• Broth: A mixture of water, sodium chloride, and the flavorful compounds created by the reaction of amino acids (from the protein) with sugar.
• Vanilla extract: A mixture of alcohol and vanillin.

6.2 Solution Properties

LEARNING OBJECTIVE

Distinguish between electrolytic and nonelectrolytic solutions.

Aqueous solutions exhibit different kinds of properties, depending on the nature of the compound that is dissolved in the water. One of these properties is the ability or inability of the solution to conduct an electric current. A compound that can conduct electricity when dissolved in water is called an **electrolyte.** All compounds that dissociate into ions in water are electrolytes—examples are sodium chloride and hydrochloric acid. A compound that cannot conduct electricity when dissolved in water is a **nonelectrolyte**—examples are nonionic compounds such as sugar and ethanol. Sports drinks advertise their high electrolyte content and often contain sugar for energy.

Solutions in which ionic substances are dissolved in water conduct electricity by completing an electrical circuit. This occurs because the dissolved ions are no longer held together—they are free to move throughout the solution. When two electrodes are placed into the solution, the positively charged cations are attracted to the negative electrode and move toward it. The negatively charged anions are attracted to the positive electrode and move toward it. The movement of ions between the two electrodes constitutes a flow of charge through the solution.

Even though all electrolytes conduct electricity in solution, they do not all do so to the same extent. Figure 6.2 shows an experimental apparatus in which a light bulb is connected to a power supply and to two electrodes in an open circuit. When the electrodes are immersed in solution, the circuit will be closed and electricity will flow, provided that the solution contains an electrolyte.

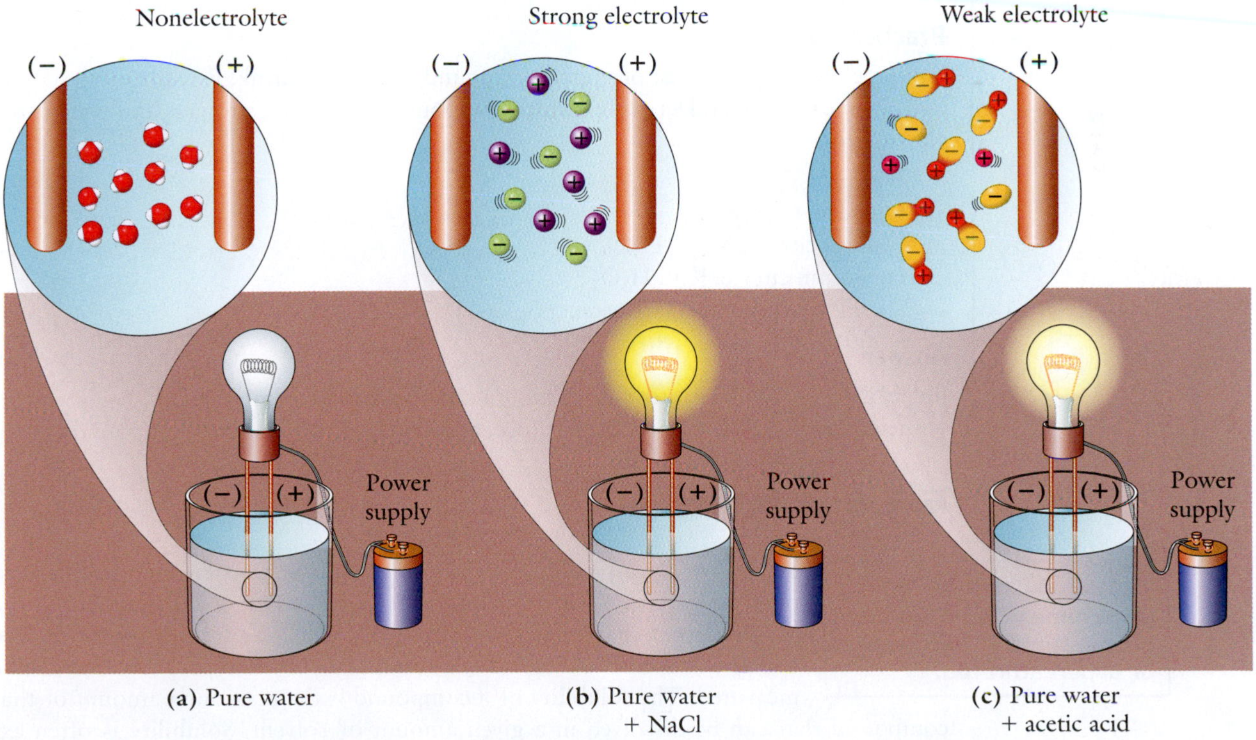

FIGURE 6.2 The ability of a solution to conduct electricity is directly related to the ability of the solute to dissolve and separate into component ions during the dissolution process. Nonelectrolytes will not conduct electricity, but strong electrolytes are excellent conductors of electricity. Weak electrolytes are poor conductors, as they exist in both a dissociated ionic form and an undissociated form.

Figure 6.2a shows the electrodes immersed in pure water: Electricity cannot flow between the electrodes, so the light bulb remains dark. Figure 6.2b shows an electrode immersed in a **strong electrolyte**: The light bulb is bright because the ionic compound is fully dissociated into free ions and a large amount of electricity is flowing. Figure 6.2c shows an electrode immersed in a **weak electrolyte**: The light bulb is dim because the ionic compound is only partially dissociated into free ions. Therefore, only a small amount of electricity can pass through the solution. Examples of weak electrolytes are acetic acid and ammonium hydroxide. If a nonelectrolyte is dissolved in water, ions do not form and electricity will not pass through the light bulb.

Chapter 4 provided many examples of ions that are important in forensic science investigations. The ability of ions to conduct electricity is one method used to detect the presence of ions after they have been separated from a mixture by chromatography. DNA analysis is also based on the movement of ions between electrodes.

■ WORKED EXAMPLE 1

Which of the following compounds when dissolved in water would be an electrolyte? Nonelectrolyte?

(**a**) Sucrose (sugar), $C_{12}H_{22}O_{11}$

(**b**) Baking soda, Na_2CO_3

(**c**) Potassium nitrite (a preservative), KNO_2

SOLUTION

(**a**) $C_{12}H_{22}O_{11}$ is a covalent compound that doesn't form ions: nonelectrolyte.

(**b**) Na_2CO_3 is an ionic compound: electrolyte.

(**c**) KNO_2 is an ionic compound: electrolyte.

Practice 6.1

Given below is a partial list of ingredients found in a sports drink, listed by both name and chemical formula. Determine which compounds are nonelectrolytes and which are electrolytes.

Water, H_2O
Fructose, $C_6H_{12}O_6$
Sodium citrate, $Na_3C_6H_5O_7$
Potassium citrate, $K_3C_6H_5O_7$
Sucralose (the artificial sweetener Splenda®), $C_{12}H_{19}Cl_3O_8$

Answer

Electrolytes: Sodium citrate and potassium citrate are ionic compounds.
Nonelectrolytes: Water (solvent), fructose, and sucralose are covalent compounds.

6.3 Solubility

When you eat a pickle, you get a salty and sour taste. The juice used to preserve the cucumber contains a very large amount of salt. Salt works as a preservative, as bacteria cannot survive in such concentrated solutions. The ability of a solute to dissolve in a solvent is a physical property that can be measured. The **solubility** of a compound is the maximum amount of that compound that can be dissolved in a given amount of solvent. Solubility is often expressed in units of grams solute per 100 mL solvent. Table 6.2 shows the solubility in water of several compounds that are commonly found in kitchens. Notice that 200 grams of sucrose will dissolve in 100 mL of water, but only 35.9 grams of sodium chloride will dissolve in the same volume.

TABLE 6.2	Solubility of Several Cooking Ingredients in Water at 25°C
Compound	Solubility (g/100 mL)
Potassium nitrate (preservative)	310
Sucrose (sugar)	200
Sodium chloride (salt)	35.9
Potassium bitartrate (cream of tartar)	0.6

A solution that contains the maximum amount of solute that can be dissolved is considered **saturated** (the values listed in Table 6.2 correspond to saturated solutions). A solution that contains less solute than the solubility value is referred to as an **unsaturated** solution. A more mathematical approach to expressing concentration is discussed in Section 6.4.

If a saturated solution is cooled, then the solubility of the compound decreases, and under normal circumstances the excess amount of solute will come out of solution as a solid. But an interesting situation can arise when a saturated solution is made at an elevated temperature and is then carefully and slowly cooled down. The excess solute may remain in solution, resulting in a **supersaturated** solution. There is more solute dissolved in a supersaturated solution than the solvent should be able to hold at the now-lowered temperature. Supersaturated solutions are not stable and will often return to the saturated level by crystallizing out the excess dissolved solute, as shown in Figure 6.3.

Supersaturated

FIGURE 6.3 Rock candy is made by dissolving large amounts of sugar in hot water, which creates an unsaturated solution. As the solution cools, it will become saturated. As the water evaporates from the solution, the sugar will crystallize out onto the stick rather than become a supersaturated solution. (a, b, c, d: Richard Meghna, Fundamental Photographs)

■ **WORKED EXAMPLE 2**

A 500.0-mL solution contains 175 g of sodium chloride. Is the solution unsaturated, saturated, or supersaturated?

SOLUTION

Using Table 6.2, a saturated solution contains 35.9 g of NaCl per 100 mL. Using this as a conversion factor,

$$\frac{35.9 \text{ g NaCl}}{100 \text{ mL}} \times 500.0 \text{ mL} = 179.5 \text{ g NaCl}$$

The solution is unsaturated because it contains only 175 g of sodium chloride.

Practice 6.2

How many grams of sugar are needed to make 25.0 of the saturated solution?

Answer

❏ 50.0 g of sugar

6.4 Mathematics of Solutions: Concentration Calculations

LEARNING OBJECTIVE

Determine the concentration of a solution.

For most purposes in the laboratory, it is important to know how much solute is dissolved in the solvent. An unsaturated solution of sodium chloride that contains 25 g per 100 mL is quite different from an unsaturated solution containing 0.25 g per 100 mL.

The solubility values in Table 6.2 indicate concentration by giving the number of grams of solute present in 100 mL of solution, and they represent the mass of solute required to produce a saturated solution. However, most solutions used in the laboratory are unsaturated solutions. The concentration unit most commonly used in chemistry is **molarity** (M), which is defined as the number of moles of solute per liter of solution.

$$\text{Molarity} = \frac{\text{moles of solute}}{\text{liters of solution}}$$

You have already learned to calculate the number of moles of a compound from the atomic mass of the compound. For example, the molecular mass of sodium chloride is 53.35 amu. Therefore, one mole of NaCl has a mass of 53.35 g. If 53.35 g are dissolved into 1.00 liter, the concentration of the resulting solution would be 1.00 M. Expressing the concentration in units of molarity accurately communicates the concentration of a solution. Even if the mass of solute given is not equal to the molar mass or the volume is not expressed in liters, it is possible to calculate the molarity of a solution using conversion factors, as shown below.

■ **WORKED EXAMPLE 3**

What is the molarity of a solution prepared by dissolving 73.5 g of sucrose ($C_{12}H_{22}O_{11}$) in water to a final volume of 250.0 mL?

SOLUTION

It may be easier to do this calculation in two steps. First, calculate the moles of NaCN:

$$73.5 \text{ g of sucrose} \times \frac{1 \text{ mol sucrose}}{342.3 \text{ g sucrose}} = 0.215 \text{ mol } C_{12}H_{22}O_{11}$$

Then calculate the volume of solution in liters:

$$250.0 \text{ mL} \times \frac{1 \text{ L}}{1000 \text{ mL}} = 0.2500 \text{ L}$$

The final step is to combine these two values in the formula for molarity:

$$M = \frac{0.215 \text{ mol } C_{12}H_{22}O_{11}}{0.2500 \text{ L}} = 0.860 \text{ M } C_{12}H_{22}O_{11}$$

Practice 6.3

What is the molarity of a sucrose ($C_{12}H_{22}O_{11}$) solution prepared by dissolving 6.85 g in water to a final volume of 500.0 mL?

Answer

❏ 0.200 M ($C_{12}H_{22}O_{11}$)

■ WORKED EXAMPLE 4

The salt shaker in your kitchen most likely contains iodized salt, meaning that potassium iodide (KI) has been added in small amounts to the sodium chloride. Adding KI prevents goiters from forming, a medical condition in which the thyroid gland swells. Calculate the molarity of a potassium iodide solution made by dissolving 20.0 g KI in 0.250 L of water.

SOLUTION

$$M = \frac{\text{moles of KI}}{\text{liters of solution}}$$

$$20.0 \text{ g KI} \times \frac{1 \text{ mol KI}}{166.0 \text{ g KI}} = 0.120 \text{ mol KI}$$

$$M = \frac{0.120 \text{ mol KI}}{0.250 \text{ L}} = 0.480 \text{ M KI}$$

Practice 6.4

Citric acid is a common compound in citrus fruits and is also used to flavor soft drinks and sour candy. A solution is prepared by dissolving 38.0 g of citric acid ($C_6H_8O_7$) in water, to a final volume of 2.00 L. What is the concentration of the citric acid solution?

Answer

❏ 0.0989 M $C_6H_8O_7$

In addition to calculating the molarity of a solution, it is necessary to calculate how many grams of a solute are needed to make a solution of a given concentration. For example, how many grams of citric acid ($C_6H_8O_7$) are needed to make 1.50 L of 0.0900 M solution?

$$\underbrace{\frac{0.0900 \text{ mol } C_6H_8O_7}{1 \text{ L}}}_{\substack{\text{Molarity} = \frac{\text{moles solute}}{\text{liters of solution}}}} \times 1.50 \text{ L} \times \underbrace{\frac{192.1 \text{ g } C_6H_8O_7}{1 \text{ mol}}}_{\text{Molar mass}} = 25.9 \text{ g } C_6H_8O_7$$

WORKED EXAMPLE 5

Calculate the mass of citric acid needed to prepare 0.750 L of a 0.200 M solution.

SOLUTION

$$\frac{0.200 \text{ mol } C_6H_8O_7}{1 \text{ L}} \times 0.750 \text{ L} \times \frac{192.1 \text{ g } C_6H_8O_7}{1 \text{ mol}} = 38.4 \text{ g } C_6H_8O_7$$

Practice 6.5

Calculate the mass of sodium chloride required to make 300.0 mL of a 0.250 M solution.

Answer

4.38 g

Dilution Calculations

The forensic scientist has to work with solution dilutions regularly. Recall that the spectrophotometry experiment discussed in Chapter 4 required solutions of known concentrations. These standards are made from diluting a **stock solution**, an accurately prepared solution of known concentration that can be diluted to make solutions of lower concentration. Dilution calculations are all determined using the formula

$$\underbrace{M_1V_1}_{\text{Initial}} = \underbrace{M_2V_2}_{\text{Final}}$$

The initial solution concentration (M_1) is multiplied by the initial volume of solution (V_1) and set equal to the final solution concentration (M_2) multiplied by the final volume of solution (V_2). It should be noted that as long as the units for the concentrations and volumes are the same on both sides of the equation, any units may be used—be they liters, milliliters, molarity, or percentages.

WORKED EXAMPLE 6

Phosphoric acid is used as a flavoring agent in several soft drinks. Urban legends circulating on the Internet would have you believe the acid content is enough to dissolve nails overnight. Concentrated phosphoric acid is purchased as 14.7 M. Calculate how many mL of concentrated phosphoric acid would be diluted to create a 2.0 L bottle of soda at a concentration of 0.0058 M.

SOLUTION

$$M_1V_1 = M_2V_2$$

$$\left.\begin{array}{l} M_1 = 14.7 \text{ M} \\ V_1 = ? \end{array}\right\} \text{Initial solution}$$

$$\left.\begin{array}{l} M_2 = 0.0058 \text{ M} \\ V_2 = 2.0 \text{ L} \end{array}\right\} \text{Final solution}$$

$$(14.7 \text{ M})(V_1) = (0.0058 \text{ M})(2.0 \text{ L}) \Rightarrow$$

$$V_1 = \frac{(0.0058 \text{ M})(2.0 \text{ L})}{14.7 \text{ M}} \times 1000 \text{ mL}/1 \text{ L} = 0.79 \text{ mL}$$

Practice 6.6

Concentrated phosphoric acid is 85.5% H_3PO_4. Calculate the final % H_3PO_4 of the soft drink from Worked Example 6.

Answer

0.034% H_3PO_4

WORKED EXAMPLE 7

The directions for preparing a potassium nitrite solution call for 13.9 mL of a 1.00 M stock solution to be diluted to 75.0 mL. What is the final molarity of the solution?

SOLUTION

$$M_1V_1 = M_2V_2$$
$$\left.\begin{array}{l} M_1 = 1.00 \text{ M} \\ V_1 = 13.9 \text{ mL} \end{array}\right\} \text{ Initial solution}$$
$$\left.\begin{array}{l} M_2 = ? \\ V_2 = 75.0 \text{ mL} \end{array}\right\} \text{ Final solution}$$

$$(1.00 \text{ M})(13.9 \text{ mL}) = (M_2)(75.0 \text{ mL}) \Rightarrow$$

$$M_2 = \frac{(1.00 \text{ M})(13.9 \text{ mL})}{(75.0 \text{ mL})} = 0.185 \text{ M HgCl}_2$$

Practice 6.7

What concentration (M) should a potassium iodide stock solution be if the procedure calls for 50.0 mL of stock solution to be diluted to 200.0 mL for a 0.600 M KI solution?

Answer

2.40 M KI

6.5 Acid Chemistry

In previous sections of this chapter, we have considered some examples of aqueous solutions, ways of expressing their concentrations, and factors that affect the solubility of compounds. In this section and the next, we turn our attention to two common types of aqueous solutions—acids and bases—that are important in the chemistry laboratory, and especially in the kitchen.

> **LEARNING OBJECTIVE**
> Distinguish between strong and weak acids.

One of the most common acids found in the kitchen is vinegar, which is approximately 5% acetic acid ($HC_2H_3O_2$). It is used in many recipes to add a sour taste, but that is not the only role it can play. For example, some recipes for chocolate cake call for vinegar, yet you do not end up with a sour-tasting cake; so we need to continue exploring acids. Vinegar is also commonly used to clean the hard water deposits left inside coffee makers and is an ingredient in many homemade cleaning products.

A simple definition of an **acid** is any compound that can release the hydrogen ion, H^+, into solution. This definition was first proposed by Svante Arrhenius in the late 1800s, and it is common to refer to the H^+ ion as an *Arrhenius acid*. Most acids are easily identified by the fact that the formula starts with the hydrogen atom. Acids have characteristic properties that distinguish them from other compounds. For example, acids:

- Taste sour
- Dissolve many metals (e.g., iron)
- Conduct electricity
- React with bases to form water and a salt
- Turn blue litmus paper red

Acids can be classified as **weak** or **strong**, indicating whether the acid is a weak or strong electrolyte. For example, a weak acid such as acetic acid only partially dissociates into its ions (H^+ and $C_2H_3O_2^-$) in solution. Therefore, a solution of acetic acid is a weak conductor of electricity. A strong acid, such as hydrochloric acid (HCl), completely dissociates into ions in solution and is a strong conductor. Table 6.3 lists some commonly used strong and weak acids and the products that contain them.

TABLE 6.3	Names, Formulas, and Sources of Various Acids		
Name	Formula	Strength	Common Use
Hydrochloric acid	HCl	Strong	Stomach acid
Nitric acid	HNO_3	Strong	Acid rain
Sulfuric acid	H_2SO_4	Strong	Car batteries
Acetic acid	$HC_2H_3O_2$	Weak	Vinegar
Carbonic acid	H_2CO_3	Weak	Soft drink
Hydrofluoric acid	HF	Weak	Wheel cleaner
Phosphoric acid	H_3PO_4	Weak	Soft drink

Acids will turn blue litmus paper red whereas bases will turn red litmus paper blue. (Top: Leonard Lessin/Peter Arnold Inc.; bottom: Science Photo Library)

Hydrofluoric acid is an example of a weak acid, but lest you think weak acids are not dangerous, hydrofluoric acid can easily dissolve glass! It is important to understand that the *weak* and *strong* qualifiers do not refer either to concentration or danger level.

Acids undergo a reduction-oxidation reaction with certain metals. For example, the hydrogen ion (H^+) will be reduced to hydrogen gas (H_2) when iron metal is added to an acid. The iron metal is in turn oxidized to form iron(II) ions.

Acids can also react with compounds that contain the carbonate ion (CO_3^{2-}) or the bicarbonate ion (HCO_3^-) to produce carbon dioxide gas.

$$HCl(aq) + NaHCO_3(s) \rightarrow CO_2(g) + H_2O(l) + NaCl(aq) \qquad \textbf{(Reaction 6.1)}$$

This is the classic vinegar–baking soda reaction many children do with their parents or teachers. Figure 6.4 shows the classic science fair volcano powered by Reaction 6.1.

Acid compounds can be formed by the reaction of nonmetal oxides with water. For example:

$$CO_2(g) + H_2O(l) \rightarrow H_2CO_3(aq) \qquad \textbf{(Reaction 6.2)}$$

Carbon dioxide in our atmosphere reacts with the water to produce carbonic acid, a weak acid. A small amount of acid is natural in rain. However, when other nonmetal oxides, such as sulfur trioxide (SO_3), enter the atmosphere from burning fossil fuels, the amount of acid in rain increases.

FIGURE 6.4 The baking soda-vinegar reaction powers the model volcano used in countless science fair projects and illustrates the release of carbon dioxide when acids meet carbonates. The same reaction is used by geologists to test if a mineral contains carbonates. (Mathew Johll)

6.6 Base Chemistry

LEARNING OBJECTIVE
Distinguish between strong and weak bases.

The Arrhenius definition of a **base** is any compound that produces the hydroxide ion, OH^-, in solution. In neutralization reactions, a base reacts with an acid to form a salt and water. Table 6.4 lists common bases that can be found in everyday consumer products. The terms *caustic*, *lye*, *alkaline*, and *alkali* are all associated with compounds that are bases.

TABLE 6.4 Names, Formulas, and Sources of Various Bases

Name	Formula	Strength	Common Use
Ammonium hydroxide	NH_4OH	Weak	All purpose cleaners
Sodium hydroxide	NaOH	Strong	Drain cleaners
Potassium hydroxide	KOH	Strong	All purpose cleaners
Calcium hydroxide	$Ca(OH)_2$	Strong	Toilet bowl cleaners

Bases also have characteristic properties that set them apart from other compounds. For example, they:

- Taste bitter
- Feel slippery to the touch
- React with oils and fats to form soap
- Conduct electricity
- React with acids to form water plus a salt
- Turn red litmus paper blue

A quick examination of Table 6.4 makes it clear that bases are used to clean up the kitchen after you are finished there. Bases have a bitter, unappealing taste, in contrast with most foods, which are slightly acidic. In fact, several toxic plants contain alkaline components that make animals averse to ingesting them. Antacids are the rare exception, as they are basic compounds ingested specifically to neutralize excess stomach acid.

The alkali metals (Group 1) and the alkaline earth metals (Group 2) form bases when they react with water. Hydrogen gas is also produced during this reaction. In the case of the alkali metals, sufficient heat is also released that the hydrogen gas ignites. The formation of potassium hydroxide is shown in Reaction 6.3. To prevent the alkali metals from reacting with moisture in the air, they are typically stored in mineral oil. The alkaline earth metals undergo a similar reaction, as illustrated by the formation of calcium hydroxide in Reaction 6.4, but they are much slower reactions than the alkali metal reactions.

$$2K(s) + 2H_2O(l) \rightarrow 2KOH(aq) + H_2(g) \qquad \textbf{(Reaction 6.3)}$$

$$Ca(s) + 2H_2O(l) \rightarrow Ca(OH)_2(aq) + H_2(g) \qquad \textbf{(Reaction 6.4)}$$

The alkali and alkaline earth metals will also form bases when the metal oxides react with water. Reactions 6.5 and 6.6 illustrate the formation of potassium hydroxide and calcium hydroxide by these reactions.

$$K_2O(s) + H_2O(l) \rightarrow 2KOH(aq) \qquad \textbf{(Reaction 6.5)}$$

$$CaO(s) + H_2O(l) \rightarrow Ca(OH)_2(aq) \qquad \textbf{(Reaction 6.6)}$$

Ammonium hydroxide is also called *aqueous ammonia* because it is formed by the reaction of ammonia and water, as illustrated in the following reaction:

$$NH_3(g) + H_2O(l) \rightarrow NH_4OH(aq) \qquad \textbf{(Reaction 6.7)}$$

The reason most basic compounds are used as cleaning agents is the ability of bases to react with grease and fats to produce water-soluble soap. In fact, the traditional recipe for making soap is to combine animal fat with lye (NaOH). This type of reaction is called a **saponification** reaction. When a kitchen sink is clogged, it is usually due to an accumulation of grease on the inside of the pipes. Pouring concentrated sodium hydroxide down the drain converts the grease into soap, which then dissolves in the water, releasing the clog.

The action of soap depends on the "like dissolves like" principle. The soap molecule has a very long, nonpolar carbon chain with an ionized group at one end, as shown in Figure 6.5. The ionized region of the soap molecule associates with the positive polar region of water molecules. This interaction is strong enough to keep the nonpolar region of the molecule in solution. When a long nonpolar grease molecule comes in contact with the soap molecule, the nonpolar carbon chains associate with one another and the grease dissolves in the water. The nature of the interactions between molecules will be explored in greater detail in Chapter 7.

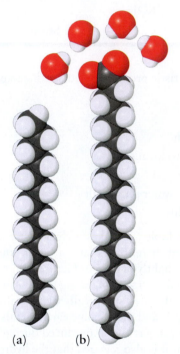

(a)   (b)

FIGURE 6.5 How soap dissolves oil and grease. (a) Oil and grease molecules consist of long chains of carbon and hydrogen atoms. (b) The soap molecule has a similar long-chain structure but also has an ionized polar end. The polar end interacts strongly with water, serving to bridge the solubility gap between water and oil.

6.7 Neutralization Reactions

Acids and bases will undergo a **neutralization reaction** upon mixing to produce water and a salt, as shown in Reaction 6.10. In daily use, the term *salt* means sodium chloride (NaCl). However, in chemistry the term **salt** is used for any ionic compound. For example, the reaction of hydrochloric acid with potassium hydroxide produces water and potassium chloride (a salt) according to Reaction 6.8.

$$\underbrace{HCl(aq)}_{\text{Acid}} + \underbrace{KOH(aq)}_{\text{Base}} \rightarrow \underbrace{H_2O(l)}_{\text{Water}} + \underbrace{KCl(aq)}_{\text{A salt}}$$

(**Reaction 6.8**)

Hydrochloric acid and potassium hydroxide are both strong electrolytes and separate into their respective ions in aqueous solution. The hydrogen ions will react with hydroxide ions to form water, as shown in Reaction 6.8. The potassium ion and chloride ion remain in solution unchanged and do not react.

$$H^+(aq) + OH^-(aq) \rightarrow H_2O(l) \qquad \text{(Reaction 6.9)}$$

Note that the product of a neutralization reaction is water. Neutralization reactions can be monitored by the addition of an **indicator**, which is a compound that will change colors depending on the amount of acid or base present in solution. In the case of the indicator phenolphthalein, if the solution is acidic, as shown in Figure 6.6a, the solution is colorless. When there is a slight excess amount of hydroxide present in solution, phenolphthalein will turn to a faint pink color, as shown in Figure 6.6b. Upon further addition of the base, the color of the indicator becomes more intense, indicating that too much base has been added, as shown in Figure 6.6c.

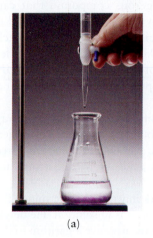
(a)

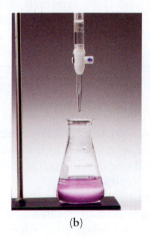
(b)

(c)

FIGURE 6.6 (a) A solution of acid and phenolphthalein is colorless. (b) When the base is added, it neutralizes the acid and a faint pink color appears. (c) When excess base is added, the pink color of the solution becomes more evident. (All: Charles D. Winters/Science Photo Library)

6.8 The pH Scale and Buffers

This chapter so far has focused on the chemistry of aqueous solutions in the chemistry lab and the kitchen. However, this section looks at a more dangerous incident. We will explore a case on how pool chemicals were used to terrorize protestors at a peaceful demonstration against the arrest of Dr. Martin Luther King, Jr., during the civil-rights movement.

LEARNING OBJECTIVE
Describe what the pH scale measures and what a buffer does.

Dr. Martin Luther King, Jr., serving a 10-day jail sentence for attempting to eat at a segregated restaurant in St. Augustine, Florida, in June 1964. (Bettmann/Corbis) Left. James Brock, manager of the Monson Motor Lodge, pours concentrated hydrochloric acid into the hotel pool in an attempt to drive civil rights protestors out. (Bettmann/Corbis) Right.

In June 1964, Dr. Martin Luther King, Jr., came to St. Augustine, Florida, and was promptly arrested and given a 10-day jail sentence for attempting to eat at the Monson Motor Lodge Restaurant, a segregated restaurant. Dr. King's arrest initiated a string of protests at the motor lodge, resulting in more arrests and larger protests. By June 18, a large number of civil rights activists had joined the sit-in at the Monson Motor Lodge.

At the same time, protestors were having wade-ins at the segregated beaches of St. Augustine. Perhaps inspired by the idea of a wade-in, a group of protestors at the motor lodge decided to go swimming in the segregated motor lodge pool. The hotel manager, James Brock, reacted to this new development by retrieving from the swimming pool chemical supply area two jugs of muriatic acid, more commonly known in the chemistry laboratory as hydrochloric acid. Mr. Brock proceeded to pour two jugs of the concentrated acid into the pool to drive the protestors out, an incident captured on film as shown on page 173.

Concentrated hydrochloric acid is an extremely dangerous compound that can easily cause severe chemical burns, which was exactly why Mr. Brock chose to add it to the swimming pool. What effect did the hydrochloric acid have when added to such a large volume of water? Were the swimmers in danger? How could the effect be determined? The answer lies in understanding a measure of acidity known as the *pH scale*. What happened to the protesters in the swimming pool is described at the end of Section 6.9.

But what does **pH** mean and why is it used? The concept of pH was developed as a way of expressing the acidity of a solution. An acid is any compound that produces hydrogen ions in solution. However, a strong acid, such as hydrochloric acid, will produce more free hydrogen ions in solution than will a weak acid, such as acetic acid—even if the strong and weak acids are at equal molar concentrations.

In 1909 Søren Sørenson, a Danish scientist, published work in which he determined the acidity of a sample by measuring how much free hydrogen ion was in solution. He developed the concept of pH to express the acidity: The *p* refers to the negative logarithm to the power of 10, and the *H* refers to the hydrogen ion. For every unit change in pH, the concentration of hydrogen ions changes by a factor of 10, as pH is a logarithmic scale. Figure 6.7 illustrates the logarithmic properties of the pH scale. A solution that has pH = 1 has an acid concentration 10 times more concentrated than pH = 2, 100 times more concentrated than pH = 3, and 1000 times more concentrated than pH = 4.

The pH scale goes from 0 to 14; a pH of 7 is neutral. This is the pH of pure water, which has a very low concentration of hydrogen ion (0.0000001 M) naturally present. Pure water also has the same concentration of hydroxide ions (0.0000001 M). If the pH is less than 7, the solution is acidic and the hydrogen ion concentration is greater than the hydroxide concentration. If the pH is greater than 7, the solution is basic and the concentration of hydroxide ions is greater than the concentration of hydrogen ions. Figure 6.8 illustrates the pH scale and the pH of common solutions.

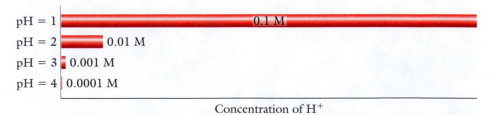

FIGURE 6.7 The pH scale is a logarithmic scale, meaning that the difference between two adjacent pH units is a tenfold difference in the amount of acid. The difference between solutions that have a pH unit difference of 2 is a hundredfold difference in hydrogen-ion concentration.

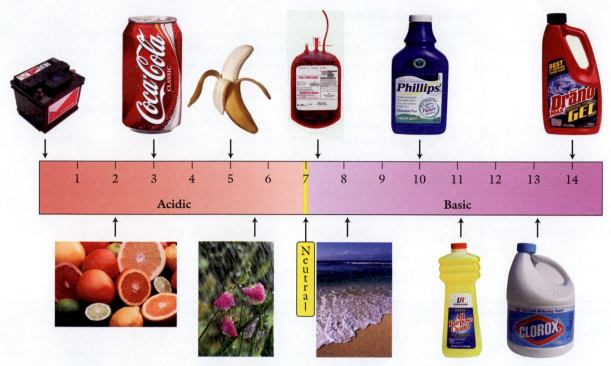

FIGURE 6.8 The pH scale. The pH of common solutions, including rain and sea water, can range over the entire pH scale. (a: Ingram Publishing/Alamy; b: the Coca-Cola Company; c: Brand X Pictures/Getty; d: Davies & Starr/The Image Bank/Getty; e, f, j, k: The Photo Works; g: Photodisc Green/Getty Images; h: Tony Cortazzi/Alamy; i: Corbis)

6.9 Mathematics of Solutions: Calculating pH

To calculate the pH of a strong acid solution, the molarity of free hydrogen ions, [H^+], must be calculated first. For strong acids, the concentration of the hydrogen ion is the same as the concentration of the acid because of the full dissociation of the acid into its component ions. The pH is calculated according to the following equation:

$$pH = \underbrace{-\log}_{p} \underbrace{[H^+]}_{H} \longleftarrow \text{The brackets mean the molarity of } H^+ \text{ in the solution.}$$

■ WORKED EXAMPLE 8

What is the pH of a 0.010 M HCl solution?

SOLUTION

HCl is a strong electrolyte that fully dissociates into H^+ and Cl^-. Therefore, [H^+] = 0.10 M. The logarithm of 0.01 to the base 10 is

$$\log [H^+] = \log (0.010) = -2$$

The pH of the HCl is

$$pH = -\log [H^+] = -[-2] = 2$$

Practice 6.8

What is the pH of a 0.0180 M HNO_3 solution?

Answer

pH = 1.74

WORKED EXAMPLE 9

It is suspected that a can of soda was tampered with by addition of sodium hydroxide. The concentration of H^+ is determined to be 0.00094 M. Does this support the suspicion of tampering by the addition of sodium hydroxide?

SOLUTION

Determine the pH of the soda: pH = $-\log$ (0.00094) = 3.03. This is an acidic pH, reasonable for the pH of soda according to Figure 6.8. If sodium hydroxide had been added, the pH would have been higher. Therefore, the pH does not support the suspicion of tampering with sodium hydroxide.

Practice 6.9

If $[H^+]$ of the soda from Worked Example 9 were determined to be 0.000000014 M, would this support the suspicion of tampering by sodium hydroxide?

Answer

The pH of the soda is 7.85, so yes, tampering has occurred. Further tests would be needed to confirm whether NaOH was the substance added.

Returning to the swimming pool incident at the Monson Motor Lodge, we now estimate the effect of the acid on the pH of the water. Recall that hydrochloric acid is a strong acid; therefore, it fully dissociates in water. The final concentration of the hydrogen ion in the swimming pool would correspond to the final concentration of the hydrochloric acid. To calculate the pH of the water in the swimming pool, a few assumptions about the size of the swimming pool will have to be made, because the motor lodge no longer exists. Capacities of typical in-ground swimming pools range from 20,000 to 40,000 gallons. A capacity of 20,000 gallons would represent a worst-case scenario, as this provides the least dilution of the acid. The concentration of hydrochloric acid used for swimming pools is typically 30 to 32%. However, the concentration in the worst-case scenario is 37.2% (12.1 M), the most concentrated form of hydrochloric acid. Assuming 2 gallons of HCl were poured into the pool, the resulting pH is calculated below.

$$\left.\begin{aligned} M_1 &= 12.1 \text{ M} \\ V_1 &= 2 \text{ gal} \end{aligned}\right\} \text{Initial solution}$$

$$\left.\begin{aligned} M_2 &= ? \\ V_2 &= 20{,}000 \text{ gal} \end{aligned}\right\} \text{Final solution}$$

$$M_1 V_1 = M_2 V_2$$

$$(12.1 \text{ M})(2 \text{ gal}) = M_2 (20{,}000 \text{ gal}) \Rightarrow$$

$$M_2 = \frac{(12.1 \text{ M})(2 \text{ gal})}{20{,}000 \text{ gal}} = 0.00121 \text{ M} \Rightarrow$$

$$\text{pH} = -\log 0.00121 = 2.92$$

The pH of the swimming pool at the motor lodge would probably not have been this low because most swimming pools are maintained at a pH of 7.2 to 7.8 by a buffer. A **buffer** is a combination of a weak acid and a soluble salt that contains the same anion as the weak acid. A buffer solution has the ability to consume both hydrogen ions and hydroxide ions, which prevents, to an extent, a dramatic change in pH.

If a strong acid is added to a solution buffered with a weak acid and its salt, the hydrogen ions will combine with the anion of the salt to form a molecule of the weak acid. This has the effect of removing the hydrogen ion from the solution. If a strong base is added to a buffered solution, it is neutralized by the weak acid. The products of this reaction are water and the salt of the weak acid. If excessive amounts of an acid or base are added, the capacity of the buffer is overwhelmed and the pH will change. The ability of a buffer to resist a change in pH is illustrated below.

$$\text{Swimming pool buffer:} \quad \underbrace{HCO_3^-}_{\substack{\text{Reacts with} \\ OH^-}} + \underbrace{CO_3^{2-}}_{\substack{\text{Reacts with} \\ H^+}}$$

$$CO_3^{2-} + H^+ \rightarrow HCO_3^- \qquad \text{(Reaction 6.10)}$$
$$HCO_3^- + OH^- \rightarrow H_2O + CO_3^{2-} \qquad \text{Reaction 6.11)}$$

Reaction 6.10 illustrates the neutralization of a strong acid by reacting with the carbonate ion, forming the bicarbonate ion (HCO_3^-), a weak acid. A strong base is neutralized in Reaction 6.11 by reacting with the bicarbonate ion to produce water and the carbonate ion. In both situations, the pH of the solution is stabilized because neither the hydroxide ion nor the hydrogen ions remain in solution.

There are many situations in which it is desirable to hold the pH close to a particular value. Blood, for example, has to have a pH value of between 7.35 and 7.45. The medical terms *acidosis* and *alkalosis* refer to conditions in which the buffered pH of blood is overwhelmed, producing serious medical consequences.

The final pH of the swimming pool after the muriatic acid was added would not have been overly dangerous. But the civil rights activists who were near James Brock as he poured the acid into the pool faced a very real danger because the concentrated acid was not diluted instantly.

6.10 Net Ionic Reactions

"I've found it! I've found it," he shouted to my companion, running towards us with a test-tube in his hand. "I have found a re-agent which is precipitated by hemoglobin, and by nothing else." Had he discovered a gold mine, greater delight could not have shone upon his features. . . . As he spoke, he threw into the vessel a few white crystals, and then added some drops of a transparent fluid. In an instant the contents assumed a dull mahogany colour, and a brownish dust was precipitated to the bottom of the glass jar.

—Sir Arthur Conan Doyle, *A Study in Scarlet*

In one of the earliest dramatizations of a forensic laboratory, Sherlock Holmes uses his knowledge of chemistry to discover a reagent that selectively precipitates hemoglobin. Chemists to this day still use a variety of precipitation reactions, as we saw in Chapter 4. For a precipitation reaction to occur, both reactants must dissolve in water. When the ions from the reactants mix, a solid forms.

For example, if we mix aqueous solutions of lead(II) nitrate and potassium iodide, a bright yellow substance, lead(II) iodide, forms and eventually settles out of solution as a precipitate. The equation for this reaction is

$$Pb(NO_3)_2(aq) + 2KI(aq) \rightarrow PbI_2(s) + 2KNO_3(aq) \qquad \text{(Reaction 6.12)}$$

Recall that when ionic compounds dissolve in water, the ions are solvated (surrounded by solvent molecules) and become independent of one another. The Pb^{2+} ions move through solution independently of the two NO_3^- ions. The same is true for the K^+ and I^- ions found in the reactant side of the reaction. The products of a precipitation reaction will typically consist of one soluble compound and one solid compound (the precipitate). Knowing which product forms the precipitate is a key step for this section and comes from applying a set of rules for solubility, given below. Precipitation reactions are

The reaction of lead(II) nitrate with potassium iodide to form a precipitate of lead(II) iodide. (David Taylor/Photo Researchers)

also commonly used to treat industrial wastewater before discharging it. When applying the solubility rules to solve questions, note that many of the rules also have exceptions.

Rules for Soluble Compounds

1. All compounds containing ammonium ions (NH_4^+)
2. All compounds containing alkali metal ions (Li^+, Na^+, and K^+)
3. All compounds containing acetate ions ($C_2H_3O_2^-$)
4. All compounds containing nitrate ions (NO_3^-)
5. Most compounds containing chloride (Cl^-), bromide (Br^-), or iodide (I^-) ions, with the exceptions of compounds with these cations: Ag^+, Hg_2^{2+}, or Pb^{2+}
6. Most compounds containing sulfate ions (SO_4^{2-}), with the exception of compounds with these cations: Ca^{2+}, Ba^{2+}, Sr^{2+}, or Pb^{2+}

Rules for Insoluble Compounds

7. All carbonate (CO_3^{2-}), phosphate (PO_4^{3-}), and chromate (CrO_4^{2-}) ions, except when the cation comes from those listed in rules 1 and 2
8. All sulfide ions (S^{2-}), except when the cation is from those listed in rules 1 and 2 or is Ca^{2+}, Sr^{2+}, or Ba^{2+}

WORKED EXAMPLE 10

Consulting the solubility rules, identify each of the following compounds as soluble or insoluble in water.

(a) $Fe(NO_3)_2$ (b) $AlPO_4$
(c) CuS (d) KCl

SOLUTION

(a) Soluble (rule 4)
(b) Insoluble (rule 7)
(c) Insoluble (rule 8)
(d) Soluble (rules 1 and 5)

Practice 6.10

What ion would you use to form an insoluble compound with the following ions?
(a) I^- (b) Li^+ (c) PO_4^{3-} (d) SO_4^{2-}

Answer

(a) Ag^+, Pb^{2+}, Hg_2^{2+}
(b) Cannot make an insoluble lithium compound.
(c) Any cation except those listed in rules 1 and 2
(d) Ca^{2+}, Ba^{2+}, Sr^{2+}, or Pb^{2+}

WORKED EXAMPLE 11

It is suspected that a beverage has been tampered with by the addition of ice remover. Ice removers are typically made up of sodium chloride, calcium chloride, magnesium chloride, or a combination of these three salts. Using the solubility rules, create a strategy for adding reagents that would cause precipitation to determine which salt(s), if any, were in the beverage. Indicate whether any ion cannot be determined by precipitation.

SOLUTION

There is more than one correct solution to this problem.

1. All three salts share the chloride anion. The presence of the chloride ion can be verified by adding Ag^+.

2. The sodium ion does not form a precipitate with any of the reagents listed in the solubility rules, so the presence of sodium cannot be determined with this method.

3. The calcium ion would form a precipitate with the addition of SO_4^{2-}.

4. The magnesium ion would form a precipitate with the addition of S^{2-}.

5. As the presence of the calcium, magnesium, and chloride ion is not proof of tampering, a control sample of the beverage that is known not to have been tampered with should also be tested.

Practice 6.11

Why would you not want to use Na_2SO_4 or Na_2S to precipitate calcium and magnesium in Worked Example 3?

Answer

Sodium is one of the ions that are suspected to have been added, so future tests on the sample would be compromised by the addition of sodium during the preliminary testing of the sample.

The equation for Reaction 6.12 can be rewritten so that all of the aqueous ionic compounds are written as free ions, as shown below in a **total ionic equation**, which is a more accurate depiction of what is occurring in solution. Notice that the precipitate, PbI_2, is not written as free ions because it is an insoluble compound according to solubility rule 5.

$$Pb^{2+}(aq) + 2NO_3^-(aq) + 2K^+(aq) + 2I^-(aq) \rightarrow PbI_2(s) + 2K^+(aq) + 2NO_3^-(aq)$$

A close examination of this equation shows that the potassium ion exists on both the reactant and product sides of the equation without undergoing a change. The nitrate ion is also found on both sides of the equation and therefore does not undergo any reaction or change. Ions that do not actively form a new compound in a chemical reaction are called **spectator ions**. The spectator ions in this equation can be cancelled out to produce the **net ionic equation**, as shown below.

$$Pb^{2+}(aq) + 2NO_3^-(aq) + 2K^+(aq) + 2I^-(aq) \rightarrow PbI_2(s) + 2K^+(aq) + 2NO_3^-(aq)$$
$$Pb^{2+}(aq) + 2I^-(aq) \rightarrow PbI_2(s) \qquad \textbf{(Reaction 6.13)}$$

Reaction 6.13 shows the net effect of Reaction 6.12, which is the combination of lead(II) ions with iodide ions to form the solid precipitate lead(II) iodide. The spectator ions, while they do not react, play a vital role in the reaction: They provide charge balance for the reacting ions because it is impossible to add just Pb^{2+} or I^- ions to a reaction.

WORKED EXAMPLE 12

A scientist wishes to detect the presence of lead ions by precipitation with potassium iodide. However, the laboratory does not have any potassium iodide in stock. Suggest an alternative compound that could be used to produce lead(II) iodide.

SOLUTION

For this reaction, the important part of potassium iodide is the iodide ion. The potassium serves as a spectator ion. If the stock room has another soluble iodide salt such as NaI or NH_4I, the identical net ionic reaction will occur.

Practice 6.12

The precipitation of $AgCl(s)$ calls for the addition of $AgNO_3(aq)$ to $CaCl_2(aq)$. Write the overall balanced equation, the total ionic equation, and the net ionic equation. If $CaCl_2(aq)$ is not available, suggest another alternative chloride salt that could be used.

Answer

The overall balanced equation is $2AgNO_3(aq) + CaCl_2(aq) \rightarrow Ca(NO_3)_2(aq) + 2AgCl(s)$

The total ionic equation is

$$2Ag^+(aq) + 2NO_3^-(aq) + Ca^{2+}(aq) + 2Cl^-(aq) \rightarrow Ca^{2+}(aq) + 2NO_3^-(aq) + 2AgCl(s)$$

The net ionic equation is $2Ag^+(aq) + 2Cl^-(aq) \rightarrow 2AgCl(s)$

which must be simplified to $2Ag^+(aq) + Cl^-(aq) \rightarrow AgCl(s)$

❏ NaCl could be used as an alternative chloride salt that is soluble in water.

■ WORKED EXAMPLE 13

What is the net ionic equation for the precipitation of magnesium phosphate from the addition of a magnesium nitrate solution to a sodium phosphate solution?

SOLUTION

The overall balanced equation is

$$3Mg(NO_3)_2(aq) + 2Na_3PO_4(aq) \rightarrow Mg_3(PO_4)_2(s) + 6NaNO_3(aq)$$

The total ionic equation is

$$3Mg^{2+}(aq) + 6NO_3^-(aq) + 6Na^+(aq) + 2PO_4^{3-}(aq) \rightarrow$$
$$Mg_3(PO_4)_2(s) + 6Na^+(aq) + 6NO_3^-(aq)$$

The net ionic equation is

$$3Mg^{2+}(aq) + 2PO_4^{3-}(aq) \rightarrow Mg_3(PO_4)_2(s)$$

Practice 6.13

What are the spectator ions in the reaction of an ammonium chromate solution with a copper(II) bromide solution?

Answer

❏ NH_4^+ and Br^-

6.11 CASE STUDY FINALE: Exploring Chemistry in the Kitchen

This chapter opened by comparing the ingredients found in a recipe for brownies to those for muffins. The physical difference between a brownie and a muffin is the tiny air pockets inside the muffin that cause the dough to rise and produce its shape. This process is called *leavening*. Both recipes shared common ingredients, except that the muffin recipe called for buttermilk and baking soda. These two ingredients must be responsible for the leavening of the muffins.

Buttermilk has a distinctive sour taste indicative of acidic compounds, and it does contain lactic acid. While acids are often added to a recipe to impart a sour taste, that is not the case for muffins, as the buttermilk reacts with the baking soda. Recall that baking soda is sodium bicarbonate, which will react with acids to produce carbon dioxide, creating the leavening effect.

Leavening can occur by other means, such as the use of yeast or baking powder. Yeast is a microorganism that consumes sugar and produces carbon dioxide and ethanol. It is most widely used to make breads and, of course, to brew beer. Baking powder is very different from baking soda. The two are not interchangeable, as many careless bakers can verify. Baking powder is a mixture of sodium bicarbonate, starch, and a dry acidic compound such as cream of tartar. The baking powder has all of the necessary ingredients for leavening and is activated upon the addition of liquid ingredients that dissolve the compounds and allow them to react in solution. You will notice that recipes that call for baking powder do not contain highly acidic ingredients. Recipes that call for the use of baking soda, however, usually include acidic ingredients such as buttermilk, vinegar, or fruit juice.

Some recipes call for both baking soda and baking powder. In these recipes, the baking soda is added to neutralize a basic ingredient such as egg whites, which allows the baking powder to work more effectively.

The final ingredient that is critical to the leavening process is the type of flour used in the recipe. There are three main types of flour: bread flour, all-purpose flour, and cake flour. The difference between the three types of flour is the amount of protein they contain.

Bread flour contains 12–15% protein, all-purpose flour contains 8–12% protein, and cake flour contains 8–10% protein.

The amount of protein is critical, because the protein molecules, specifically gluten, form a giant molecular network that traps gas bubbles as they form from the leavening agent. As the dough bakes, the heat from the oven causes the bubbles to swell, creating the lifting effect. The dough then sets, and the carbon dioxide can dissipate.

Higher-protein flours will produce a more extensive network, which gives strength to the dough. Bread has a sturdy shape that doesn't crumble upon eating, in contrast to cake. With a cake, you want the final product to crumble in your mouth upon eating.

All-purpose flour lies between the two extremes and is actually a source of frustration among novice bakers. Because the amount of protein in the flour can vary from one bag to the next, your baking can vary from one batch to another. Many experienced bakers use the recipe as a guideline, trusting their own experience of handling dough to decide whether more or less flour is needed.

Eve Curie was correct in her assessment that cooking can be as difficult and mysterious as chemistry at times. However, with continued experimentation and study, success can be had in both fields.

CHAPTER SUMMARY

• A solution is made up of a solvent (the major constituent) and a solute (the minor component). An aqueous solution specifically refers to the use of water as the solvent.

• Electrolytes are compounds that, when dissolved in water, can conduct electricity since they dissociate into component ions that are free to move between oppositely charged electrodes. Nonelectrolytes are compounds that will not conduct electricity because they do not dissociate into ions upon dissolving.

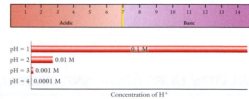

• Compounds that fully dissociate into ions in solution are called strong electrolytes; those that partially dissociate are called weak electrolytes.

• Saturated solutions contain the maximum concentration of dissolved solute that the solution can hold at a given temperature. Solutions with less than this concentration are termed unsaturated, and those unstable solutions containing more dissolved solute are termed supersaturated.

• Concentration of solute in solutions is commonly expressed as molarity (M), which is mathematically equal to the moles of solute divided by the liters of solution. Stock solutions are carefully prepared solutions of known concentration from which dilute solutions can be made.

• Net ionic equations show the ionic species in solution that come together to form an insoluble precipitate. The other ions in solution shown in the total ionic equation provide charge balance but are interchangeable with other ions.

• An acid is a compound that will produce hydrogen ions when dissolved in water. Acids can be either strong (fully dissociated) or weak (partially dissociated). A base is a compound that will produce hydroxide ions when dissolved in water. Bases can be either strong (fully dissociated) or weak (partially dissociated).

• Neutralization reactions occur when an acid and a base are mixed together. The products of a neutralization reaction are water and a salt (an ionic compound).

• The pH of a solution is a logarithmic measurement of the free hydrogen ion concentration. The pH scale goes from 0 (acidic) to 14 (basic), with a pH of 7 representing a neutral pH value.

• A buffer is a mixture of a weak acid and a soluble salt of the weak acid. This mixture is capable of reacting with both strong acids and strong bases to prevent a substantial change in the pH of the original solution.

KEY TERMS

acid (p. 169)
aqueous (p. 162)
base (p. 171)
buffer (p. 176)
electrolyte (p. 173)
indicator (p. 173)
molarity (M) (p. 166)
net ionic equation (p. 179)
neutralization reaction (p. 172)

nonelectrolyte (p. 163)
pH (p. 182)
salt (p. 181)
saponification (p. 180)
saturated (p. 172)
solubility (p. 164)
solute (p. 162)
solvent (p. 162)
spectator ion (p. 179)

stock solution (p. 168)
strong acid (p. 178)
strong electrolyte (p. 164)
supersaturated (p. 165)
total ionic equation (p. 179)
unsaturated (p. 165)
weak acid (p. 178)
weak electrolyte (p. 164)

MAKING MORE CONNECTIONS: Additional Readings, Resources, and References

Doyle, A. C. *Sherlock Holmes: The Complete Novels and Stories*, vol. 1, New York: Random House, 2003.
U.S. Department of Justice, *Processing Guide for Developing Latent Prints*, Washington, D.C.: U.S. Department of Justice, 2000.

For more information about the swimming pool incident in St. Augustine, Florida: www.drbronsontours.com/bronsonhistorypageamericancivilrights.html

REVIEW QUESTIONS AND PROBLEMS

Questions

1. What is a solvent? What is the solvent used in most solutions? (6.1)
2. What is a solute? How can you tell the difference between the solvent and solute? (6.1)

3. How does a strong electrolyte conduct electricity in solution? To illustrate your answer, draw a sketch that includes a cathode, anode, and arrows to indicate the direction the ions are traveling. (6.2)

4. Explain how a weak electrolyte differs from a strong electrolyte in its ability to conduct electricity. Modify your sketch from Question 3 to reflect this difference. (6.2)

5. Explain the difference between electrolytes and nonelectrolytes. Give examples of both. (6.2)

6. How could you determine whether a solution is saturated or not? (6.3)

7. Explain how a supersaturated solution is prepared. (6.3)

8. What is the definition of an acid? Of a base? (6.5, 6.6)

9. What are the properties of an acid? Of a base? (6.5, 6.6)

10. What is the difference between a weak acid and a strong acid? A weak base and a strong base? (6.5, 6.6)

11. Are weak acids inherently safer than strong acids? (6.5)

12. Give examples of two types of reactions that acids can undergo. (6.5)

13. Give examples of two types of reactions that bases can undergo. (6.6)

14. What are the products of a neutralization reaction? (6.7)

15. How does a chemist's use of the term *salt* differ from the everyday use of the same term? (6.7)

16. What is the net ionic equation for the reaction of a strong acid with a strong base? (6.7)

17. What is the purpose of the pH scale? Why was it developed? (6.8)

18. If the pH of a solution increases from pH = 5.0 to pH = 8.0, what happens to the concentration of hydrogen ions in the solution? (6.8)

19. What is the purpose of a buffer? How does it work? (6.8)

20. What could be done experimentally to determine whether a solution contains a buffer? (6.9)

21. What is the role of spectator ions in a precipitation reaction? Can a precipitation reaction occur without spectator ions? (6.10)

22. How can you determine whether a chemical reaction can be classified as a precipitation reaction? (6.10)

23. What is the total ionic equation attempting to show? Why is it useful? (6.10)

24. What is the net ionic equation attempting to show? Why is it useful? (6.10)

25. Explain the difference between a saturated solution and an unsaturated solution. (6.3)

26. You may have heard of the old saying, "If you're not part of the solution, you're part of the problem." Explain the old chemistry joke, "If you're not part of the solution, you're part of the precipitate." (6.10)

Problems

27. Identify each of the following compounds as either an electrolyte or a nonelectrolyte. (6.2)
    (a) $Fe(NO_3)_3$
    (b) C_2H_5OH
    (c) HNO_3
    (d) SO_2

28. Identify each of the following compounds as either an electrolyte or a nonelectrolyte. (6.2)
    (a) Na_3PO_4
    (b) N_2O
    (c) PF_5
    (d) $Mg(C_2H_3O_2)_2$

29. Identify each of the following as either a strong electrolyte or a weak electrolyte. (6.2)
    (a) HF
    (b) $NaCl$
    (c) H_2SO_4
    (d) H_2CO_3

30. Identify each of the following as either a strong electrolyte or a weak electrolyte. (6.2)
    (a) H_3PO_4
    (b) HCl
    (c) NaF
    (d) NH_4OH

31. The solubility of sodium cyanide is 48.0 g/100 mL of water at 25°C. Determine whether the following solutions are saturated, unsaturated, or supersaturated. (6.3)
    (a) 118 g of NaCN in 250.0 mL
    (b) 12.0 g of NaCN in 25.0 mL
    (c) 3.95 g of NaCN in 8.33 mL
    (d) 214 mg of NaCN in 0.400 mL

32. The solubility of cadmium cyanide is 1.70 g/100 mL of water at 25°C. Determine whether the following solutions are saturated, unsaturated, or supersaturated. (6.3)
    (a) 5.750 g of $Cd(CN)_2$ in 330.0 mL
    (b) 325.5 g of $Cd(CN)_2$ in 20.5 L
    (c) 4.25 g of $Cd(CN)_2$ in 250.0 mL
    (d) 97.0 mg of $Cd(CN)_2$ in 5.00 mL

33. How many grams of potassium cyanide would be needed to make a saturated solution for the volumes of water indicated below? The solubility of potassium cyanide is 50.0 g/100 mL. (6.3)
    (a) 0.500 L   (c) 2.25 L
    (b) 5.15 mL   (d) 250.0 mL

34. How many grams of potassium cyanide would be needed to make a saturated solution for the volumes indicated below? The solubility of potassium cyanide is 50.0 g/100 mL. (6.3)
    (a) 2.10 L   (c) 12.0 mL
    (b) 75.0 mL   (d) 200.0 μL

35. What is the molarity of each solution listed below, given the indicated mass of solute and final volume of water? (6.4)
(a) 31.45 g of NaCl in 2.00 L
(b) 14.41 g of MgS in 0.500 L
(c) 0.4567 g of $CuSO_4$ in 630.0 mL
(d) 25.5 mg of NaCN in 10.0 mL

36. What is the molarity of each solution listed below, given the indicated mass of solute and final volume of water? (6.4)
(a) 121.45 g of KOH in 75.0 mL
(b) 23.49 g of NH_4OH in 125.0 mL
(c) 217.5 g of $LiNO_3$ in 2.00 L
(d) 15.25 g of $Pb(C_2H_3O_2)_2$ in 65.0 mL

37. How many moles of calcium carbonate are in the following solutions? (6.4)
(a) 75.0 mL of 0.997 M $CaCO_3$
(b) 25.0 mL of 2.50 M $CaCO_3$
(c) 175 mL of 0.501 M $CaCO_3$
(d) 0.85 L of 3.42 M $CaCO_3$

38. How many moles of barium nitrate are in the following solutions? (6.4)
(a) 26.20 mL of 2.21 M $Ba(NO_3)_2$
(b) 19.23 mL of 8.5×10^{-2} M $Ba(NO_3)_2$
(c) 2.58 L of 0.0250 M $Ba(NO_3)_2$
(d) 1.77 L of 1.55 M $Ba(NO_3)_2$

39. Determine the number of grams of calcium carbonate to make each solution described in Problem 43. (6.4)

40. Determine the number of grams of barium nitrate to make each solution described in Problem 44. (6.4)

41. What would be the final concentration of a hydrochloric acid solution prepared by diluting 100.0 mL of concentrated HCl (12.1 M) to each of the following volumes? (6.4)
(a) 125.0 mL
(b) 2.00 L
(c) 1.50 L
(d) 625.0 mL

42. What would be the final concentration of a sodium hydroxide solution prepared by diluting 50.00 mL of concentrated NaOH (19.3 M) to each of the following volumes? (6.4)
(a) 0.80 L   (c) 1.33 L
(b) 350.0 mL   (d) 1.15 L

43. How many milliliters of concentrated HCl (12.1 M) are needed to make the following amounts of acid? (6.4)
(a) 5.00 L of 2.15 M
(b) 1.00 L of 3.50 M
(c) 2.00 L of 0.650 M
(d) 100 mL of 0.250 M

44. How many milliliters of concentrated NaOH (19.3 M) are needed to make the following amounts of base? (6.4)
(a) 8.00 L of 0.330 M
(b) 300 mL of 1.25 M
(c) 0.250 L of 0.310 M
(d) 10.0 mL of 0.025 M

45. Name each of the following acids. Indicate whether it is a strong or weak acid. (6.5)
(a) H_2CO_3
(b) HNO_3
(c) H_3PO_4
(d) H_2SO_4

46. Name each of the following bases. Indicate whether it is a strong or weak base. (6.6)
(a) NH_4OH
(b) NaOH
(c) $Ca(OH)_2$
(d) KOH

47. The burning of sulfur-containing fossil fuels leads to the production of sulfur trioxide, an acidic oxide. Write the balanced chemical equation for the reaction between sulfur trioxide and water. (6.5)

48. Nitric acid can be produced from its acidic oxide, dinitrogen pentoxide, reacting with water. Write the balanced chemical equation for the reaction. (6.5)

49. Balance the equation for the reaction of each substance below with water. (6.6)
(a) K
(b) Rb
(c) Sr
(d) Mg

50. Balance the equation for the reaction of each substance below with water. (6.6)
(a) Be
(b) Cs
(c) Ca
(d) Na

51. Balance the equation for the reaction of each compound below with water. (6.6)
(a) K_2O
(b) Rb_2O
(c) SrO
(d) MgO

52. Balance the equation for the reaction of each compound below with water. (6.6)
(a) BeO
(b) Cs_2O
(c) CaO
(d) Na_2O

Figure for Problem 55. (a, b: The Photo Works; c: Ingram Publishing/Alamy; d: The Coca-Cola Company)

53. Write the balanced chemical equation for the neutralization of acetic acid with ammonium hydroxide. (6.7)

54. Write the balanced chemical equation for the neutralization of phosphoric acid with sodium hydroxide. (6.7)

55. For each of the consumer products pictured at the top of the page, indicate whether the product contains an acid or base, provide the compound name and formula, and determine whether it is a strong or weak electrolyte. (CP)

56. For each of the consumer products pictured at the bottom of the page, indicate whether the product contains an acid or base, provide the compound name and formula, and determine whether it is a strong or weak electrolyte. (CP)

57. Calculate the pH of the following solutions of HCl. (6.9)
    (a) 0.0027 M   (c) 0.075 M
    (b) 0.091 M   (d) 3.08×10^{-5} M

58. Calculate the pH of the following solutions of HNO_3. (6.9)
    (a) 0.00035 M   (c) 0.000015 M
    (b) 0.50 M   (d) 0.0033 M

59. Determine whether each of the following compounds is soluble or insoluble. (6.10)
    (a) Iron(II) carbonate
    (b) Strontium carbonate
    (c) Lead(II) sulfate
    (d) Lithium sulfate

60. Determine whether each of the following compounds is soluble or insoluble. (6.10)
    (a) Silver sulfide
    (b) Ammonium sulfide
    (c) Lead (II) bromide
    (d) Copper (II) bromide

61. Write the balanced equation, total ionic equation, and net ionic equation for the following unbalanced chemical reaction. (6.10)

$$Na_2SO_4(aq) + CaCl_2(aq) \rightarrow NaCl\,(aq) + CaSO_4(s)$$

62. Write the balanced equation, total ionic equation, and net ionic equation for the following unbalanced chemical reaction. (6.10)

$$Cu(NO_3)_2(aq) + K_3PO_4(aq) \rightarrow$$

Figure for Problem 56. (All: The Photo Works)

63. Write the balanced equation, total ionic equation, and net ionic equation for the following reaction. (6.10)

Aluminum nitrate + sodium sulfide →
Aluminum sulfide + sodium nitrate

64. Write the balanced equation, total ionic equation, and net ionic equation for the following unbalanced chemical reaction. (6.10)

Silver nitrate + potassium iodide →

65. Based on the saturated solubility values given in Table 6.2, calculate the molar concentrations of the four cyanide solutions shown in the table, assuming that the final volume of solution is 100.0 mL. (6.4)

66. Physical developer (PD) is a set of solutions used to develop latent fingerprints on different porous and nonporous surfaces. It is often used to develop fingerprints on currency. Below is the FBI recipe for PD. Solution 1 serves as a cleaning solution before using a mixture of solutions 2 through 4 in which the fingerprint is developed due to a redox reaction causing the deposition of metallic silver within the ridges of the fingerprint. Determine the concentration of each ingredient in the four solutions. The final volume for each solution is 1.00 L. (6.4)
Solution 1: 25.0 g of maleic acid ($C_4H_4O_4$)
Solution 2:
   30.0 g of iron(III) nitrate
   80.0 g of iron(II) ammonium sulfate
     ($Fe(NH_4)_2(SO_4)_2$)
   20.0 g of citric acid ($C_6H_8O_7$)
Solution 3:
   3.00 g of N-dodecylamine acetate ($C_{14}H_{31}NO_2$)
   4.00 g of synperonic-N (skip—this is a mixture)
Solution 4: 200.0 g of silver nitrate

67. Europium chloride hexahydrate, $EuCl_3 \cdot 6H_2O$, is used to prepare a fluorescent dye used to visualize fingerprints that have been detected using superglue fuming (cyanoacrylate fuming). The solution is prepared by mixing 1.00 g of europium chloride hexahydrate in a volume of 800.0 mL of distilled water. Calculate the molarity of this solution. (Hint: Water is part of the structure and must be accounted for in the molar mass.) (6.4)

68. Ninhydrin forms a pink-colored compound when it reacts with amino acids left in a fingerprint. The solution is prepared by mixing 6.00 g of ninhydrin with 1000.0 mL of acetone. What is the molarity of ninhydrin, given its chemical formula is $C_9H_6O_4$? (6.4)

69. Amido black is a dye used to visualize fingerprints left in blood. A water-based solution of amido black uses 0.10 M citric acid. How many grams of citric acid are required to prepare 500.0 mL of solution, given the formula of citric acid is $C_6H_8O_7$? (6.4)

70. Recalculate the pH of the swimming pool from the case study in Section 6.9, assuming that the pool was 40,000 gallons and that the 2 gallons of hydrochloric acid had a concentration of 10.2 M. (6.9)

71. For determining the toxicity of different compounds, we use the concept of LD_{50}, which is the lethal dose for 50% of test subjects. The LD_{50} in rats for arsenic trioxide is 15 mg per 1 kg of body mass. Assuming the LD_{50} in humans is similar, determine the molarity of arsenic trioxide in a 355-mL (12-oz) drink that would correlate to the LD_{50}. Assume the average person has a mass of 70 kg. (CP)

72. The LD_{50} for potassium cyanide in rats is 5.0 mg per 1 kg of body mass. Assuming the LD_{50} in humans is similar, calculate the concentration of cyanide in the stomach of a person corresponding to the LD_{50}. Assume a human stomach has a volume of 1.0 L and the average mass of a person is 70.0 kg. Compare your answer with the concentration from ingesting a tampered capsule containing 0.50 g of potassium cyanide. (CP)

73. Quicklime is an older term used for calcium oxide, which will react with water to produce calcium hydroxide. Write the balanced equation for this reaction. After reading the following 1898 poem by Oscar Wilde, comment on the poetic license taken with the chemistry involved and what would have actually happened. (CP)

The Ballad of Reading Gaol
The Warders strutted up and down,
   And kept their herd of brutes,
Their uniforms were spick and span,
   And they wore their Sunday suits,
But we knew the work they had been at
   By the quicklime on their boots.
For where a grave had opened wide,
   There was no grave at all:
Only a stretch of mud and sand
   By the hideous prison-wall,
And a little heap of burning lime,
   That the man should have his pall.
For he has a pall, this wretched man,
   Such as few men can claim:
Deep down below a prison-yard,
   Naked for greater shame,

He lies, with fetters on each foot,
    Wrapt in a sheet of flame!
And all the while the burning lime
    Eats flesh and bone away,
It eats the brittle bone by night,
    And the soft flesh by the day,
It eats the flesh and bones by turns,
    But it eats the heart alway.
For three long years they will not sow
    Or root or seedling there:
For three long years the unblessed spot
    Will sterile be and bare,
And look upon the wondering sky
    With unreproachful stare.

Case Study Problems

74. The case study in Chapter 4 centered on the conviction of George Trepal in 1991 for murdering his neighbor Peggy Carr by poisoning her with thallium nitrate. His motive was a longstanding feud over loud music and uncontrolled pet dogs at the Carr home. The FBI agent handling the analysis did an extremely poor job of testifying in regard to the strength of the evidence. He had also failed to do some basic tests, although the courts have since ruled in appeals that these errors did not greatly affect the decision because independent laboratories have verified much of what should have been done the first time. The results are summarized below.

Sample	Tl Detected	Tl Quantified	NO_3^- Detected	NO_3^- Quantified
Q1	Yes	Yes	Yes	No
Q2	Yes	Yes	Yes	No
Q3	Yes	Yes	No test	No

Once the case was determined to be a thallium poisoning, the FBI agent needed to know whether it was thallium nitrate, thallium chloride, or thallium sulfate. Trepal had access only to thallium nitrate. To determine the anion, the FBI scientist added diphenylamine (DPA), which detects nitrate. He also checked for chloride and sulfate by adding silver nitrate and barium nitrate, respectively, to trigger a precipitation reaction. Write the precipitation reactions. (6.3)

75. The FBI scientist would testify that thallium nitrate had been added to bottles of cola found at the victim's home. His superiors would later state that he should have said the samples "are consistent with thallium nitrate being added." What is the difference between the two statements? (CP)

76. The presence of thallium is a clear indication of product tampering due to the compound's toxicity. However, the presence of the nitrate ion in food or beverages would in itself not be unusual. How could it be proven whether the nitrate in the analyzed sample came from tampering or was in fact naturally present? If the FBI scientist had quantified the nitrate ion concentration, would this have strengthened the data supporting the theory that thallium nitrate had been added to the cola? (CP)

77. The directions for preparing Kool-Aid drink mix call for 1 package of drink mix, 1 cup of sugar (sucrose), and enough water to make 2 quarts. What is the concentration (M) of the sucrose? Some useful conversions: 1 quart = 946 mL and 1 cup of sugar = 200 g. (6.4)

78. The directions for a sour-tasting frozen treat made from Kool-Aid drink mix call for 2 packages of drink mix, 1/3 cup of sugar, and 3/4 cup water. Calculate the molarity of the sugar solution as prepared using the conversions provided in problem 77. The ingredient label lists citric acid as the main ingredient. Explain why the drink mix would taste sour as compared to the normal preparation, given in problem 77.

CHAPTER 7

Aqueous Solutions: Part II

CASE STUDY: Exploring Antibiotics and Drug-Resistant Infections

There is a killer stalking the hallways of the hospital. They know exactly how he kills his victims, how he got there—and that there is almost nothing they can do to stop him. The killer is *Pseudomonas aeruginosa,* a bacterial infection that has evolved to become resistant to most antibiotics, one of the so-called *superbugs.* *Pseudomonas aeruginosa* is commonly found in the

environment (soil and water) and is most likely coating your hands and body even as you read this information.

Your immune system is normally strong enough to fight off infections. When hospitalized, however, a person's weakened immune system lets the body become a target for infection. *Pseudomonas aeruginosa* commonly targets the lungs and the urinary system and also causes infections in burn patients.

Nearly 2 million infections are acquired each year by patients in a hospital setting. These infections develop in patients who already have compromised health, which hinders recovery and lengthens recovery time. Nearly 100,000 of those infected will succumb, resulting in death. Drug-resistant strains of bacteria are spreading faster than scientists can develop new antibiotics to combat these deadly bacteria.

The drug-resistant bacteria strains evolve in response to being exposed to an antibiotic that does not kill off all of the infectious bacteria during that exposure. This evolutionary response happens when patients do not follow the directions properly while taking the antibiotic—for instance, by not completing the full prescription or by not taking the antibiotics at required intervals. Drug-resistant bacteria can also develop when physicians overprescribe antibiotics in situations where they are not warranted, such as a viral infection. The more times bacteria are exposed to an antibiotic, the greater the likelihood of the strain of

As you read through the chapter, consider the following questions:

• How do antibiotic drug molecules recognize and attack bacteria cells and not human cells?

• What forces exist between molecules in solution and how could these be used by drug molecules?

• What happens to the physical properties of a liquid when a solute is added?

bacteria developing a resistance. For this reason, most physicians rotate through a list of antibiotics when prescribing them.

The rise of the drug-resistant superbug is also due in part to the failure of our system of drug discovery and development. When a new antibiotic is developed, there is a good chance that the drug-resistant bacteria will be vulnerable, as they have not yet been exposed to the new drug. The problem is that pharmaceutical companies are not actively engaged in developing new antibiotics because these drugs have very low profit potential and quite large research and development costs.

MAKE THE CONNECTION
Antibiotics have several different unique methods for attacking and killing bacteria, which we can explore further after you first learn more about how molecules interact with one another.

7.1 Intermolecular Forces and Surface Tension

LEARNING OBJECTIVE
Describe intermolecular forces and how they create the property called surface tension.

It is a common misconception that a drop of liquid will assume the shape of a teardrop as it falls. If no external forces are applied, a drop of liquid will actually form the shape of a perfect sphere. Why does it do this? To answer this question, we must explore the forces that exist between the molecules of a substance in the liquid state.

In the liquid, solid, or aqueous phases, particles (molecules, atoms, or ions) are very close to one another, and several types of attractive forces, called **intermolecular forces**, can develop between adjacent particles. Intermolecular forces are not as strong as the covalent or ionic bonds that exist within the particles. However, intermolecular forces significantly affect a substance's physical properties such as its boiling point and melting point.

Intermolecular forces also control the shape of a drop of liquid. Any molecule within the center of a droplet will be attracted to neighboring molecules through intermolecular forces in all possible directions, as shown by the blue arrows in Figure 7.1. However, if a molecule is at the surface of the droplet, it is attracted only to molecules located adjacent to it or beneath it, as shown by the blue arrows. The result is that the surface molecule is being pulled with a net force toward the center of the droplet, as shown by the brown arrow. Each molecule on the surface is being pulled inward in the same way. This effect minimizes the surface area of the liquid, and the shape that minimizes surface area for a given volume is the sphere.

Surface tension is a property of liquids that is a measure of how much force is needed to overcome the pull of the intermolecular forces on molecules at the surface of the liquid: The higher surface tension of a liquid, the stronger the intermolecular forces.

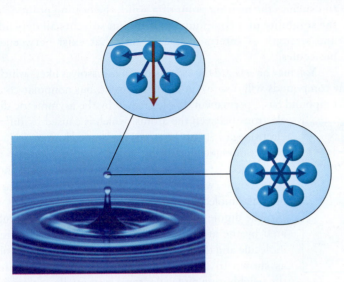

FIGURE 7.1 Intermolecular forces within a liquid drop. The attractive forces surrounding a molecule on the surface of a drop pull the molecule inward. Molecules within the liquid drop are pulled equally in all directions. (Photo by John Gillmoure/Corbis)

WORKED EXAMPLE 1

Calculate the surface area of a sphere with a volume of 1 cm^3, given the formulas for a sphere: volume $= \dfrac{4}{3}\pi r^3$ and surface area $= 4\pi r^2$.

SOLUTION

The formulas for the volume and surface area of a sphere are

$$V = \frac{4}{3}\pi r^3 \qquad \text{and} \qquad A = 4\pi r^2$$

First, you need to solve for r given that $V = 1$ cm^3. We have

$$V = \frac{4}{3}\pi r^3 \Rightarrow 1 \text{ cm}^3 = \frac{4}{3}\pi r^3$$

$$r^3 = 1 \text{ cm}^3 \times \frac{3}{4\pi} \Rightarrow r = \sqrt[3]{1 \text{ cm}^3 \times \frac{3}{4\pi}} = 0.620 \text{ cm}$$

Now, the formula for the surface area can be used:

$$A = 4\pi r^2 = 4\pi(0.620 \text{ cm})^2 = 4.84 \text{ cm}^2$$

Practice 7.1

Calculate the surface area of a cube with a volume of 1 cm^3, given formulas for the cube: volume $=$ length^3 and surface area $= 6 \times$ length^2. What percent increase in surface area does a cube have compared with a sphere of the same volume?

Answer

❏ Surface area $= 6$ cm^2, 124% larger than a sphere of the same volume.

7.2 Types of Intermolecular Forces

LEARNING OBJECTIVE

Identify the different intermolecular forces and how they affect the physical properties of a substance.

Intermolecular forces play a key role in many physical properties besides surface tension, covered in the previous section. Other physical properties such as the melting point of a solid, the boiling point of a liquid, and the solubility of a compound in different solvents all depend on the type and strength of intermolecular forces that exist between neighboring molecules.

You may have heard the expression "like dissolves like," which refers to the fact that polar compounds will dissolve in a polar solvent, but nonpolar compounds will not. A **polar** compound has a permanent dipole created by the asymmetric distribution of electrons between the atoms, which is caused by differences in the electronegativity of the atoms (as covered in Section 5.6). The ability of a compound to dissolve in a solvent depends on the attractive intermolecular forces that exist between the solute and the solvent. For example, water is a polar molecule, whereas oil is a mixture of nonpolar molecules; therefore, oil will not dissolve in water.

The **dipole–dipole force** is a type of intermolecular force created by the attractive force between the positive region of one molecule and the negative region of a neighboring polar molecule, as shown in Figure 7.2. The intermolecular forces between the neighboring molecules form continually as the molecules move through solution.

Nonpolar oil molecules do not mix with the polar water molecules. (© Aaron Graubart/Getty Images)

Gum–Chocolate Demo

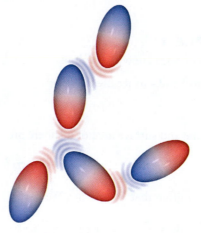

FIGURE 7.2 Dipole–dipole intermolecular force. The oppositely charged regions of polar molecules are attracted to one another through electrostatic attraction.

A particularly strong type of dipole–dipole interaction is the **hydrogen bond**. Hydrogen bonding is not to be confused with the covalent bonding of hydrogen atoms with other atoms *within* a molecule. Intermolecular forces always refer to forces *between* molecules. Hydrogen bonding between molecules occurs for compounds in which a hydrogen atom is bonded directly to a nitrogen, oxygen, or fluorine atom, as in NH_3, H_2O, or HF.

The hydrogen bond results from the highly electronegative atoms (N, O, or F) pulling the shared electrons toward themselves. Recall that hydrogen has only one proton and one electron. When the shared electrons are pulled toward the electronegative atom, the remaining proton from hydrogen is exposed. This highly positive region is then attracted very strongly to the electron-rich electronegative element (N, O, or F) on adjacent molecules.

The hydrogen bond is an especially important bond between the strands of DNA. This bond is strong enough to keep the DNA molecule in its double helix form, yet it can be overcome when the DNA strands unravel in the processes required for protein synthesis or cell reproduction. More information about the role of hydrogen bonding in DNA is presented in Chapter 14.

The electrons, however, are in continual motion within the orbitals of the atoms, creating an **electron cloud**. Because the motion of electrons is random, a majority of the electrons at some instant in time will be on one side of the molecule. This bunching of electrons on one side of the molecule sets up a temporary dipole in which one region is more negative and the other region is more positive, creating the conditions for the **induced dipole-induced dipole** intermolecular force.

The induced dipole causes a change in the molecule next to it. The temporarily negative end of the molecule will repel the electrons in a neighboring molecule, distorting its electron cloud. The neighboring molecule then has a temporary induced dipole, and an attractive force is created between the oppositely charged ends of the induced dipoles. The induced dipole forces are short-lived because the electrons keep moving, as shown in Figure 7.3.

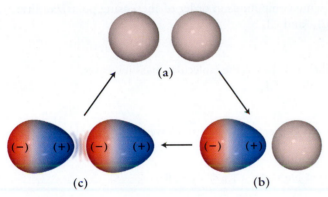

FIGURE 7.3 Induced dipole–induced dipole intermolecular force. (a) In a nonpolar molecule, the electron density is evenly dispersed. (b) However, due to the random motion of the electrons, a temporary dipole may form. (c) The temporary dipole induces a dipole in a neighboring molecule, creating a weak attractive force.

The attractive forces created in this manner are called *dispersion forces*, also known as *London forces*. Dispersion forces are present in all molecules, yet they also are the weakest of the intermolecular forces. Collectively, dispersion forces and dipole–dipole forces are referred to as **van der Waals forces**.

Acetone–Water–Ethanol Rates of Evaporation

For nonpolar substances (such as the halogens shown in Table 7.1), dispersion forces are the only attractive forces present and are vital in determining the physical properties of the compounds. For example, the melting point of a nonpolar compound is affected

TABLE 7.1	Physical Properties of Nonpolar Halogens		
Halogen	Molecular Mass (g/mol)	Melting Point (°C)	Physical State at Room Temperature
Fluorine, F_2	38.0	−219.6	Gas
Chlorine, Cl_2	70.9	−101.5	Gas
Bromine, Br_2	179.8	−7.3	Liquid
Iodine, I_2	253.8	−113.7	Solid

by the intermolecular forces present. The molecules of a nonpolar solid are held together in a rigid structure with attractive dispersion forces between neighboring molecules. To melt a solid compound, enough energy must be provided for each molecule to overcome the attractive intermolecular forces holding it in place. The weaker the intermolecular forces, the less energy the molecules need to break free of the solid structure and move about in the liquid state.

The strength of the temporary dipole depends on how easily the electron cloud is distorted; this is called the *polarizability* of a molecule. Table 7.1 lists the physical properties of the nonpolar halogens. Notice that fluorine gas, the lightest of the halogens, has the lowest melting point. As the molecular mass of the diatomic elements increases, so does the melting point of the compound. Iodine, the diatomic element with the highest mass, has the highest melting point. The trend illustrates that the greater the molecular mass of the compound, the higher the melting point. This implies that a molecule's polarizability increases as its size increases.

■ WORKED EXAMPLE 2

Place the following compounds in order of increasing polarizability: CBr_4, CF_4, CCl_4, and CI_4.

SOLUTION

The polarizability increases as the molecular mass increases:
$CF_4 < CCl_4 < CBr_4 < CI_4$.

Practice 7.2

Place the noble gas elements in order from lowest melting point to highest melting point.

Answer

❏ The order is He < Ne < Ar < Kr < Xe < Rn.

■ WORKED EXAMPLE 3

Which of the bonds will have the largest dipole?
(a) H—F
(b) H—H
(c) H—O
(d) H—S

SOLUTION

The magnitude of the dipole in a bond is a function of the difference in electronegativity of the two atoms. The trend for electronegativity on the periodic table increases from bottom to top in groups (columns) and increases from left to right in periods (rows). The largest difference in electronegativity is between the atoms in bond (a).

Practice 7.3

Two compounds often identified in cases of accidental poisoning are ethanol (CH_3CH_2OH), from ingestion of alcoholic beverages, or ethylene glycol ($HOCH_2CH_2OH$), the main ingredient in antifreeze. Which compound has the highest melting point?

Answer

❏ Ethylene glycol

WORKED EXAMPLE 4

Which of the following molecules will form hydrogen bonds?

(a)
```
    H
    |
H — C — O — H
    |
    H
```

(b)
```
    H       H
    |       |
H — C — O — C — H
    |       |
    H       H
```

(c) H — O — H

(d)
```
    H   O   H
    |   ||  |
H — C — C — C — H
    |       |
    H       H
```

SOLUTION

Compounds (a) and (c) have hydrogen atoms bonded to a highly electronegative atom (oxygen) as required to form hydrogen bonds.

Practice 7.4

Methanol has the formula CH_3OH and propanol has the formula C_2H_5OH. Which compound would have the highest boiling point? Why?

Answer

Propanol would have the highest boiling point, as both compounds can form hydrogen bonds. However, propanol is a bigger molecule and therefore more polarizable.

7.3 Mixed Intermolecular Forces

Thus far, intermolecular forces have been explained by comparing how one molecule will interact with an identical molecule in a pure substance. It is important to realize that intermolecular forces occur between all molecules that are in proximity to one another—whether they constitute a pure compound or a complex mixture.

LEARNING OBJECTIVE

List and describe other intermolecular forces that form in mixtures.

Pharmaceutical scientists must fabricate drug molecules that can reach the intended target while remaining unreactive when present in a complex mixture of substances such as blood. Blood has solid components (red blood cells, white blood cells, and platelets) suspended in plasma that has polar, nonpolar, and ionic substances dissolved in it. The polar components include water, glucose, and urea. Nonpolar compounds include oxygen and carbon dioxide gas. The proteins albumin, hemoglobin, and immunoglobulin are large polar molecules that are suspended in the solution. Electrolyte ions such as sodium, potassium, and chloride ions are also present in blood.

Oxygen gas exists in two forms in blood: oxygen bound to hemoglobin within the red blood cells, which accounts for 99% of the oxygen, and 1% in the form of O_2 dissolved in the plasma. Interactions occur among all of the molecules and ions present in blood. Do water molecules in blood interact differently with dissolved oxygen gas than with sugar? The intuitive answer is that water interacts with a neutral, nonpolar molecule in different ways than it interacts with a polar molecule. To clarify these differences, we must examine some additional types of attractive forces that occur in mixtures.

Water and glucose are both polar. Therefore, their interaction is a type of mixed dipole–dipole force, as covered in the previous section. When an ionic compound dissolves in water (discussed in more detail in the next section), each ion is attracted to the oppositely charged end of the polar water molecule and creates the **ion–dipole force**. The ion–dipole intermolecular force is a critical factor in dissolving ionic compounds.

Salting Out Ethanol with K_2CO_3

■ WORKED EXAMPLE 5

Which components in blood exhibit ion–dipole interactions?

SOLUTION

Water and urea, both polar compounds, exhibit ion–dipole interactions with the electrolyte ions (Na^+, K^+, Cl^-).

Practice 7.5

What intermolecular force plays a dominant role in an aqueous solution of alcohol (CH_3OH)?

Answer

❏ Hydrogen bonding (a particularly strong dipole–dipole interaction)

Another intermolecular force present in a solution is the **dipole–induced dipole force**, which originates when a polar molecule causes the electrons of an adjacent nonpolar molecule to be distorted. This occurs because the negative region of a polar molecule will repel the evenly distributed electrons of a nonpolar molecule. Likewise, the positive region of the polar molecule will attract the evenly distributed electrons of a nonpolar molecule. Either source of distortion induces a temporary dipole in the nonpolar molecule. The induced dipole of the nonpolar molecule is weakly attracted to the permanent dipole of the polar molecule. However, the distorted electron cloud will return to normal and the attraction ends.

The dipole–induced dipole intermolecular force is weaker than the dipole–dipole intermolecular force. Despite being a weak force, the dipole–induced dipole force allows a nonpolar gas such as oxygen to dissolve in a polar solvent such as water—a phenomenon that is necessary for life to exist. The solubilities of several gases in water are listed in Table 7.2. The solubility of each gas in water is directly related to the pressure of each gas above the water.

TABLE 7.2	Solubility of Nonpolar Gases in Water
Gas	Solubility (g/100 mL water)
H_2	0.000160
N_2	0.000190
O_2	0.000434

John C. Kotz and Paul Treichel, *Chemistry & Chemical Reactivity*, 5th ed., Brooks/Cole Publishing, 2002.

■ WORKED EXAMPLE 6

Which components in blood exhibit dipole–induced dipole intermolecular forces?

SOLUTION

The polar molecules such as water and urea interact with nonpolar solutes such as the dissolved gases CO_2 and O_2.

Practice 7.6

Is nitrogen gas more or less soluble in blood than oxygen gas? Explain why.

Answer

The dipole–induced dipole intermolecular force is the main interaction between the gases and the water in blood. The oxygen gas is more soluble because it is a larger molecule and therefore is more polarizable.

7.4 The Process of Dissolution

LEARNING OBJECTIVE

Explain how water dissolves certain ionic and molecular compounds.

The key factor in the formation of an aqueous solution is the interaction of water molecules with the solute such that the solute molecules or ions are pulled away from each other and dispersed throughout the water. To understand the process of dissolution, let's start by examining the forces at work in a water molecule. Figure 7.4 is a three-dimensional map of water that shows the effect of the large difference in electronegativity between the hydrogen and oxygen atoms within a bond. The region where electrons are most often found is colored red and the region where electrons are rarely found is colored blue. It is important to remember that water is neutral—it has no net negative or positive charge. However, the red area of the molecule represents a negative region near the oxygen atom and the blue area represents a positive region near the hydrogen atoms.

Because opposite charges attract, the negative region of water will be attracted to any substance with a positive charge or a positive polar region. The positive region of water is attracted to negative charges or negative polar regions. If the solute is an ionic compound such as KCl, ion–dipole forces form between the ions and the oppositely charged region of water. However, when sugar, a polar molecule itself, dissolves in water, it is due to dipole–dipole forces.

To illustrate how water interacts with ions, let's examine how potassium chloride dissolves when placed into water. Potassium chloride is an ionic compound with the formula KCl. The potassium ion has a +1 charge and the chloride ion (Cl^-) has a −1 charge. As the KCl crystalline powder is placed into water, water molecules surround the surface of the crystal. Figure 7.5 shows that the K^+ and Cl^- ions separate as each ion becomes surrounded by water molecules, or *solvated*. The water molecules orient in such a way that negative regions (at the oxygen end) cluster around the K^+ ions while positive regions (at the hydrogen end) cluster around the Cl^- ions. The attractive forces between water molecules and the ions are sufficient to cause a breaking of the ionic bond between K^+ and Cl^- ions that held the solid crystal together.

Not all ionic compounds dissolve in water. For example, lead(II) chloride is not soluble in water. The attractive force that water has for the ions is not strong enough to overcome the electrostatic attraction between the ions in the ionic bond.

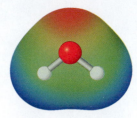

FIGURE 7.4 Electron density of water. The electrons within the hydrogen–oxygen bond are pulled toward the oxygen atom due to the difference in the electronegativity of the two atoms. The result is a polar molecule with the negative region (red) on the oxygen atom and the positive region (blue) located at the hydrogen atoms.

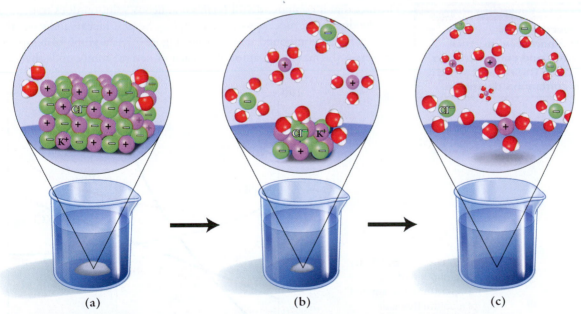

(a)   (b)   (c)

FIGURE 7.5 Dissolution of potassium chloride in water. (a) When solid potassium chloride begins to dissolve, the water molecules arrange themselves on the surface of the ionic solid in a way that the polar regions of water are lined up with the oppositely charged ion. (b) When enough water molecules surround an ion, it will be completely removed from the crystal lattice. (c) The process continues as new water molecules approach the crystal until the crystal has been fully dissolved.

Molecular compounds do not form ions when they dissolve in water. However, water molecules still have to interact with the solute molecules if dissolution is to occur. For example, glucose (blood sugar) is a molecular compound that is capable of dissolving in water. When you examine the model of glucose in Figure 7.6, it is apparent that, like water, it is a polar molecule. The positive regions of glucose molecules are attracted to the negative regions of water molecules. Glucose is soluble in water because both are polar compounds.

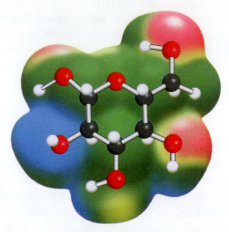

FIGURE 7.6 Electron density of glucose. Glucose ($C_6H_{12}O_6$) is a polar compound that dissolves in water.

7.5 Rate of Dissolving Soluble Compounds

LEARNING OBJECTIVE

Describe how temperature, surface area, and concentration affect the rate at which a compound dissolves.

The fact that a compound will form an aqueous solution does not mean that the dissolving process will occur rapidly. Several factors influence the rate at which a compound will dissolve. Figure 7.7 is a graph of temperature versus the maximum amount of each of several compounds that can

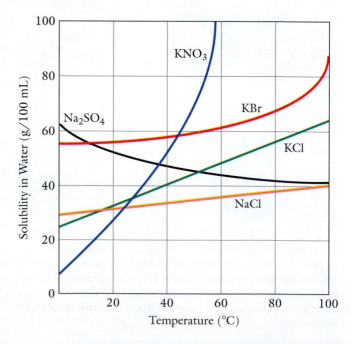

FIGURE 7.7 Solubility as a function of temperature. The amount of a solute that can dissolve in a solvent increases as a function of temperature. It is a common procedure to heat a solution that is being prepared in the laboratory.

dissolve in solution. For most compounds, as the temperature increases, the solubility of a substance increases. Because the rate at which compounds dissolve also increases with increasing temperature, it is common to speed up the dissolution of a compound in the laboratory by heating the sample. Figure 7.7 also shows that not all compounds will increase their solubility as temperature increases. However, sodium sulfate is one of the exceptions to this rule—it actually decreases its solubility as temperature increases.

If a sugar cube is dropped into a cup of water, it will slowly dissolve. But if the same amount of powdered sugar is added to a cup of water, it will dissolve almost immediately. This occurs because the surface area of the solid compound affects the rate of dissolution. By increasing the surface area of the solid, more solvent molecules interact with the solid and the speed of the dissolution process increases. A solid cube that measures 1 cm × 1 cm × 1 cm has a surface area of 6 cm^2 (1 cm^2 for each of the six sides of the cube). By simply halving the cube, the surface area increases to 8 cm^2 and, as shown in Figure 7.8, the surface area increases to 12 cm^2 by splitting the original cube into eight smaller cubes.

The rate at which a solute is dissolved by a solvent is also influenced by the amount of solute that has already been dissolved. When the solute is first added, there is only pure solvent surrounding the solid particles at the interface of the solute and solvent, and the solid interacts with the solvent quickly. However, as the solute continues to dissolve, the solvent molecules surround those particles of the solute that are already dissolved. For the solid solute to dissolve, fresh solvent is needed to interact with the solid. Stirring the solution will bring fresh solvent to the surface. When solutions are prepared, it is common to both heat and stir solutions in the laboratory to speed up the dissolving process.

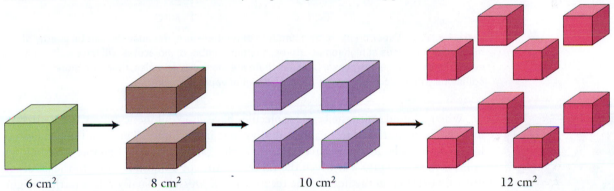

6 cm^2 8 cm^2 10 cm^2 12 cm^2

FIGURE 7.8 Total surface area as a function of particle size. The surface area of a solute increases dramatically with decreasing particle size. Each division of the original cube results in an increase in the total surface area of the original solid. A large surface area allows a larger amount of solvent to reach the solute, thus increasing the rate of dissolution.

7.6 Colligative Properties: Boiling Point of Solutions

Properties that depend only on the number of particles dissolved in solution and not on the identity of the particles are called **colligative properties**. Colligative properties include vapor pressure lowering, boiling point elevation, freezing point depression, and osmotic pressure—all of which are covered in this chapter.

What happens to the boiling point of water when a solution is created? We know that pure water boils at 100°C under standard conditions, yet a saturated sodium chloride solution would boil at 108°C. To understand how the boiling point of a solution is altered, it is first necessary to explore further the physical process that occurs when water boils.

FIGURE 7.9 Vapor pressure of a liquid. A small percentage of molecules in a liquid will have sufficient energy to go into the gas phase.

The individual molecules that make up a liquid move through it at various speeds. Some of the molecules have sufficient kinetic energy (the energy of motion) to overcome the intermolecular forces of the surrounding molecules at the surface of the liquid and escape into the gas phase. The term **vapor** is used to describe the gas phase of a compound that is normally a liquid at room temperature. The **pressure** of a gas is due to the particles colliding with walls of the container that holds the gas and the liquid. The **vapor pressure** of a liquid is the pressure exerted by the vapor molecules, as illustrated in Figure 7.9.

The vapor pressure of a liquid results from molecules that have sufficient energy to escape the surface of the liquid and collide with the container walls. As a liquid is heated, the molecules gain kinetic energy. The increased kinetic energy results in more molecules that have the ability to escape the surface of the liquid and change into the vapor state, as illustrated in Figure 7.10.

T=20°C   T=30°C

FIGURE 7.10 Vapor pressure as a function of temperature. Because the kinetic energy of molecules increases at high temperatures, a larger number of molecules will have sufficient energy to exist in the gas phase. Because there are more gas particles, there are more collisions with the container wall, creating a higher vapor pressure.

The boiling point of a pure solvent changes with the addition of a solute to form a solution. (A. Pasieka/ Photo Researchers, Inc.)

When a liquid boils, the liquid molecules have enough kinetic energy to go directly into the vapor state and form a bubble anywhere in the liquid. The liquid molecules no longer have to be at the surface to go into the gaseous state. Therefore, the **boiling point** is the temperature at which the vapor pressure of a liquid is equal to the external pressure exerted on the liquid. The term *external pressure* usually refers to the weight of the atmosphere pushing down on the liquid in an open container, and boiling points are typically determined with 1 atmosphere (atm) of pressure on the liquid. The formation of a vapor bubble within the liquid can occur only when the vapor pressure is equal to the external pressure. If the vapor pressure is less than the external pressure, bubbles of vapor do not form, as the atmospheric pressure would crush them.

The boiling point for 2-propanol, used in rubbing alcohol, is 83°C, whereas water boils at 100°C. The difference in the boiling points can be explained by the relative strengths of the intermolecular forces that exist between the different molecules. Water forms hydrogen bonds to neighboring water molecules through both hydrogen atoms. The 2-propanol molecule can form only one hydrogen bond to neighboring molecules. The amount of energy required to vaporize rubbing alcohol is therefore less than that required for water, as the intermolecular forces formed between alcohol molecules are not as strong as those of water molecules. The vapor pressure of rubbing alcohol therefore is equal to the external pressure at a lower temperature compared to water, and the alcohol boils.

The vapor pressure of a liquid is altered by the presence of a solute, which, in turn, alters the boiling point of the solution. When a substance is dissolved in water, it forms a homogeneous solution, so there is an even distribution of solute particles throughout the solution, including the surface. Figure 7.11 illustrates that if the solute is occupying some sites at the surface, those sites are unavailable for the liquid to use in forming vapor molecules.

FIGURE 7.11 Vapor pressure of pure liquids versus solutions. Solute molecules replace solvent molecules at the surface of a solution, as shown in the beaker to the right. When fewer solvent molecules can escape into the vapor phase, vapor pressure is lowered.

By adding a solute, the vapor pressure is lowered, but for the liquid to boil, the vapor pressure must be equal to the external pressure. The boiling point of the solution is then elevated beyond that of the pure solvent because additional heat must be added to compensate for the vapor pressure decrease.

■ WORKED EXAMPLE 7

A sports drink has been sent to a crime lab as part of an investigation into suspected tampering. If the boiling point of the sample matches the boiling point of an unaltered sample of the sports drink, does this result mean that the sample was not tampered with?

SOLUTION

If the boiling point of the suspected solution matches an unaltered solution, this would indicate that it is unlikely the sample was tampered with. However, several possibilities could still exist. A toxic substance might be present at such extremely low levels that the boiling point is not significantly altered, or the toxic substance may have evaporated from solution or decomposed.

Practice 7.7

If the suspect sports drink sample does not match the boiling point of an unaltered sample, does it mean that it has been tampered with?

Answer

If the physical properties of the two solutions do not match, there is definitely a difference between the two samples. It does not mean the sample was maliciously tampered with. Further testing of the sample would be needed to determine the nature of the difference, and further investigation would be needed to account for the difference in the solutions.

The boiling point of an aqueous solution containing 1 mole of sugar (a molecular compound) dissolved in 1 kilogram of water is 100.512°C, an elevation of 0.512°C above the boiling point of pure water. The boiling point of a solution containing 1 mole of NaCl (an ionic compound) in the same quantity of water is 101.024°C, an elevation twice as great as the increase observed for the sugar solution. Why is this the case?

The extent to which **boiling point elevation** of a solution occurs (in comparison to the boiling point of the pure solvent) depends strictly on the number of particles of solute in solution, *not* their identity. When potassium chloride dissolves in water, the salt dissociates into its component ions—two particles for each unit of potassium chloride. One mole of potassium chloride has twice the effect on the boiling point elevation as dissolving the same number of moles of sugar because sugar is a molecular compound that does not dissociate into multiple ions.

A key factor in boiling point elevation measurements is that the solute must be non-volatile, which in this case means that the solute does not boil at a temperature lower than that of the solvent. If the solute has a low boiling point, our model of boiling point elevation is no longer completely accurate. Ethanol, for example, is volatile because it will boil at 78°C. In a mixture of ethanol and water, alcohol will be lost during the heating of the solution before the boiling point of water is reached. Because the compound in the solution might be volatile, boiling point measurements are not typically used to determine possible adulterants in liquids. An alternative to measuring boiling point is to measure the freezing point of a solution, as will be discussed in Section 7.8.

7.7 Mathematics of Boiling Point Elevation

The change in the boiling point of a solution compared with the boiling point of the pure solvent can be calculated using the following equation:

$$\Delta T_{bp} = K_{bp} m_{particles} \tag{1}$$

ΔT_{bp} is the change in the boiling point temperature, K_{bp} is the boiling point constant that is unique to each solvent, and $m_{particles}$ is the molality of the solute particles. **Molality** (m) is a method of expressing a solute concentration that is similar to, but not the same as, molarity. The molality of a solution is the number of moles of solute per kilogram (1000 grams) of solvent:

$$\text{Molality} = \frac{\text{moles of solute}}{\text{kilograms of solution}}$$

The reason molarity (moles/liter) is not suitable is that the volume of liquids changes as a function of temperature. Therefore, the molarity of a solution at room temperature is different from the molarity of the same solution near the boiling point. Molality, however, is based strictly on the mass of the solvent, and the mass does not change. The following equation shows how the molality of the solute particles is calculated:

$$m_{particles} = \frac{\text{moles of solute} \times \text{number of particles per solute}}{\text{kilograms of solvent}}$$

■ WORKED EXAMPLE 8

What is the boiling point of a solution made from 0.450 g of KCN dissolved in 10.0 mL of water? The K_{bp} for water is 0.512°C/m.

SOLUTION

The first step is to calculate the moles of KCN:

$$0.450 \text{ g KCN} \times \frac{1 \text{ mol KCN}}{65.12 \text{ g}} = 0.00691 \text{ mol}$$

The second step is to determine the number of particles: KCN is an ionic compound made up of K^+ and CN^- particles, so there are two particles. The third step is to calculate the mass of water in kilograms:

$$10.0 \text{ mL H}_2\text{O} \times \underbrace{\frac{1 \text{ g H}_2\text{O}}{1 \text{ mL H}_2\text{O}}}_{\text{Density of water}} \times \frac{1 \text{ kg}}{1000 \text{ g}} = 0.0100 \text{ kg}$$

The fourth step is to calculate the molality of the solute particles:

$$m_{particles} = \frac{0.00691 \text{ mol KCN} \times 2 \text{ particles/mol KCN}}{0.0100 \text{ kg}} = 1.38 \text{ } m$$

The fifth step is to calculate the increase in the boiling point:

$$\Delta T_{bp} = K_{bp}m_{particles} = 0.512°C/m \times 1.38\ m = 0.708°C$$

Finally, we calculate the elevated boiling point by adding the increase to the normal boiling point:

$$T_{bp} = 100°C + 0.314°C = 100.708°C$$

Practice 7.8

Determine the molality of a 100.0-mL water sample to which antifreeze (ethylene glycol, $C_2H_6O_2$) has been added, if the boiling point of the solution rises to 102.56°C. From the molality, determine the number of grams of ethylene glycol added to the sample.

Answer

❑ 5.00 m ethylene glycol and 31.0 g of ethylene glycol

■ WORKED EXAMPLE 9

Which of the following aqueous solutions has the highest boiling point?

    Solution A: 0.10 mol of K_2S dissolved in 100.0 mL of water
    Solution B: 0.25 mol of $C_6H_{12}O_6$ (glucose) dissolved in 100.0 mL of water
    Solution C: 0.20 mol of $NaNO_3$ dissolved in 200.0 mL of water

SOLUTION

The boiling point change for each solution could be calculated by using equation 1 (on page 202). However, this question can be answered in a simpler way by noting that the solution that has the most particles per unit volume will have the highest boiling point.

    Solution A: K_2S has 3 particles, therefore 0.30 mol of particles per 100 mL.
    Solution B: $C_6H_{12}O_6$ has 1 particle, therefore 0.25 mol of particles per 100 mL.
    Solution C: $NaNO_3$ has 2 particles, therefore 0.40 mol of particles per 200 mL, equivalent to 0.20 mol of particles for 100 mL.

    Solution A has the highest boiling point.

Practice 7.9

Which of the following aqueous solutions has the lowest boiling point?

    Solution A: 60.0 g of NH_4NO_3 dissolved in 100.0 mL of water
    Solution B: 60.0 g of $C_6H_{12}O_6$ (glucose) dissolved in 100.0 mL of water
    Solution C: 60.0 g of Na_2S dissolved in 100.0 mL of water

Answer

❑ Solution B

7.8 Colligative Properties: Freezing Point of Solutions

In 1867, Alfred Nobel developed a method for stabilizing nitroglycerin, a highly explosive liquid compound, into a more stable and commercially viable product that he named *dynamite*. Dynamite helped spur an industrial and construction boom by enabling engineers to excavate large areas of land with explosives.

> **LEARNING OBJECTIVE**
> **Explain why solutions have lower freezing points than the pure solvent.**

Alfred Nobel (1833–1896), inventor of dynamite and founder of the Nobel Prize. (Bettmann/Corbis)

The problem with using pure nitroglycerin as an explosive is that it is sensitive to shock. Just jarring or dropping a container of nitroglycerin can detonate the material. However, nitroglycerin becomes even more dangerous as it freezes and starts to form a solid—the friction of the crystals rubbing against each other can trigger an explosion. Because the freezing point of nitroglycerin is only 13.5°C (56.3°F), nitroglycerin becomes particularly dangerous to handle in cold weather.

In **freezing point depression**, the temperature at which a liquid freezes is lowered by adding a nonvolatile solute to form a solution. The stabilization of nitroglycerin to make commercial dynamite takes advantage of freezing point depression by creating a solution of nitroglycerin with ethylene glycol dinitrate, which is another explosive compound, as shown in Figure 7.12. The freezing point of ethylene glycol dinitrate is −22.8°C (−9.04°F). The final product is an equal mixture of the two explosive compounds and results in a freezing point of approximately −20°C (−4°F), which greatly reduces the risk of shock detonation.

The making of dynamite is a good example of how the principle of freezing point depression, a colligative property, was used for the production of a more stable explosive. This same principle is used in winter to melt ice from roads and to prevent radiator fluid from freezing. But what is occurring on a molecular scale to lower the freezing point in these circumstances?

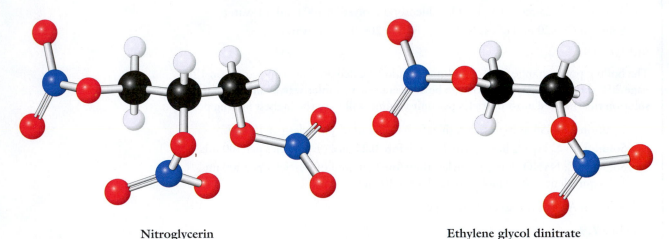

Nitroglycerin   Ethylene glycol dinitrate

FIGURE 7.12 Nitroglycerin and ethylene glycol dinitrate are two explosive molecules that contain both the fuel (carbon atoms, in black) and the oxygen (red atoms) necessary for combustion. The other atoms in the compound are nitrogen (blue) and hydrogen (white).

Two processes occur when the temperature of a liquid is at its freezing point. Some of the molecules in the liquid slow down and are captured by the solid. However, some of the solid molecules have enough energy to enter the liquid. As shown in Figure 7.13a, the two processes are in **equilibrium**. There is no net change in the number of molecules in either phase: For every molecule that joins the solid, another joins the liquid.

When a solute such as NaCl is dissolved in a solvent, it blocks part of the liquid solvent from interacting with the molecules in the solid phase. However, it does not prevent the solid particles from entering the liquid phase, as shown in Figure 7.13b. The result is that the rate at which molecules bind to the solid decreases, but the rate at which molecules enter the liquid remains unchanged. As a result, the solid begins to melt. To reestablish the freezing process, fewer solid molecules must have enough energy to enter the liquid phase. This occurs by decreasing the temperature. Equilibrium is reestablished at the new freezing point.

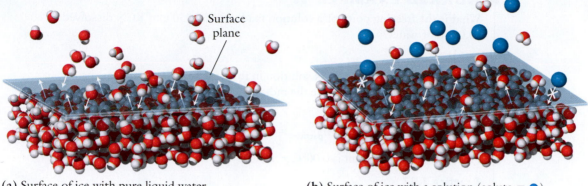

(a) Surface of ice with pure liquid water   **(b)** Surface of ice with a solution (solute = ●)

FIGURE 7.13 Molecular view of freezing point depression. (a) In the pure substance at the freezing point, molecules in the solid phase can escape the surface at the same rate that molecules in the liquid phase return to the surface. (b) In the solution, solute molecules replace solvent molecules interacting with the solid surface, thereby decreasing the rate at which solvent molecules enter the solid phase. The presence of solute molecules does not affect the rate of water molecules leaving the solid phase and entering the liquid phase. Therefore, more water molecules can leave the surface than can return to the surface, resulting in the net melting of the solid.

■ WORKED EXAMPLE 10

What is the advantage of measuring the freezing point instead of the boiling point in determining whether a liquid sample contains a solute?

SOLUTION

If volatile compounds are present in the solution, they will not be lost during the freezing process, whereas they are preferentially lost in measuring the boiling point of water.

Practice 7.10

Place the following compounds in order from the greatest change to the least change in the freezing point of a liquid, assuming that an equal number of molecules are added to each solution: $NaCN$, $MgCl_2$, and $Al(NO_3)_3$.

Answer

❑ The order is $Al(NO_3)_3 > MgCl_2 > NaCN$.

7.9 Mathematics of Freezing Point Depression

The change in the freezing point temperature of a solution can be calculated using equation 2. You will notice that this equation is strikingly similar to equation 1 for boiling point elevation. However, it is important to note that the freezing point constant K_{fp}, which is unique to each solvent, is *not* the same as the K_{bp} value for the same solvent. The K_{fp} for water is $-1.86°C/m$, whereas the K_{bp} for water is $0.512°C/m$.

$$\Delta T_{fp} = K_{fp}m_{particles} \qquad (2)$$

> **LEARNING OBJECTIVE**
>
> Illustrate how the freezing point of a solution will change because of the concentration of the solute.

■ WORKED EXAMPLE 11

What is the freezing point of a solution made from 0.450 g of KCN dissolved in 10.0 mL of water?

SOLUTION

The concentration of the KCN solution in this example is the same as the one in Worked Example 8. We can use the molality calculated there to get the decrease in freezing temperature.

$$\Delta T_{fp} = K_{fp}m_{particles} = -1.86°C/m \times 1.38 \ m = -2.57°C$$

Hence, the freezing point is $0.00°C - 2.57°C = -2.57°C$.

Practice 7.11

In Practice 7.8, it was determined that 31.0 g of ethylene glycol ($C_2H_6O_2$) added to 100 mL of water produced a 5.00 m solution. Determine the freezing point of that same solution.

Answer

−9.3°C

7.10 Colligative Properties: Osmosis

To understand osmosis on a molecular level, it is helpful to first consider the concept of diffusion. **Diffusion** is the process in which solute particles move by purely random motion from a region of high concentration to a region of lower concentration within a solution. As a solid solute dissolves, the solute particles (molecules or ions) will move randomly, colliding with one another, solvent molecules, and the walls of the container. This process continues until the solute is evenly distributed throughout the solution.

Osmosis differs from diffusion in that the movement in solution is of the water molecules across a semipermeable membrane, such as a cell wall. A **semipermeable membrane** is a thin membrane with extremely small holes that allow small molecules such as water to move back and forth through the membrane but prevent larger molecules or ions from passing through. Water will move from a region of low solute concentration to a region of high solute concentration, as illustrated in Figure 7.14.

The molecular view of osmosis is that water molecules from both sides of the membrane can pass through it. However, the solute particles on the concentrated side cannot pass through. If pure water were on both sides, the same number of water molecules would strike and pass through the membrane in each direction. When a solution is on one side, there are fewer water molecules per volume of solution, so fewer water molecules will strike the membrane and a net gain of water occurs on the concentrated side of the membrane.

The resulting difference in volumes on either side of the membrane introduces a new force—the additional pressure pushing down the higher column of solution. This pressure starts to counteract the process of osmosis. Osmosis continues until the pressure pushing down on the higher column of water is enough to create an equal transfer of water molecules across the membrane. The **osmotic pressure** of a solution is the pressure needed to prevent a net change in water volume across the membrane.

Water purification can be accomplished by **reverse osmosis**. In this process, a pressure greater than the osmotic pressure is applied to impure water, forcing water molecules to move across the membrane preferentially toward the pure water side.

When an intravenous (IV) fluid is administered to a patient, it is critical that the concentration of solutes in the IV fluid match the concentration of solutes in the blood plasma. An IV fluid that meets this condition is called an **isotonic solution**. If pure water were used in an IV fluid, the concentration of solutes within the body cells would become

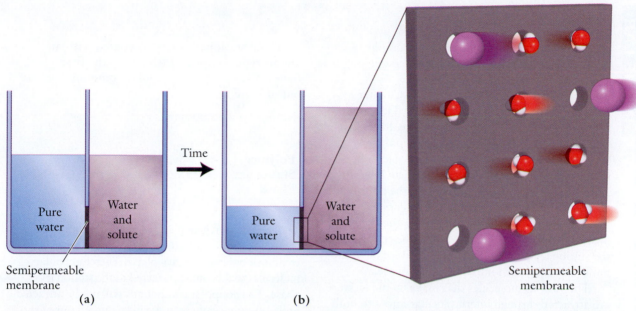

FIGURE 7.14 Molecular view of osmosis. Water molecules move across a semipermeable membrane in either direction. (a) However, solute molecules replace water molecules on the solution side of the membrane, thereby decreasing the rate at which water molecules cross the membrane toward the pure solvent side. This results in a net gain in the volume of solution on the solute side, as shown in (b).

greater than the concentration of solutes in the blood plasma. Water would then move by osmosis into the body cells, causing them to swell and possibly rupture. The same problem can result if a **hypotonic solution**—one that has a lower solute concentration than blood—is used as an IV fluid. A different problem results if the IV fluid is too concentrated. The water within the body cells would move by osmosis to the outside of the cell, causing the cell to shrivel to the point of being destroyed. A solution that is more concentrated than blood is called a **hypertonic solution**. The effect of three solution conditions on red blood cells is illustrated in Figure 7.15.

Many times suspicious situations are the result of unusual, but not criminal, circumstances. Medical doctors will often run a battery of tests while attempting to ascertain the cause of an individual's condition. When these situations are investigated, it is critical to determine whether anything unusual is present in a person's blood and, if so, what it is. The **osmolality** of the blood is a measure of how many particles are dissolved in blood.

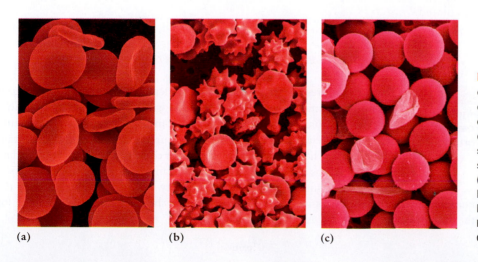

FIGURE 7.15 Red blood cells and osmosis. Red blood cells can undergo osmosis if exposed to a solution with a different concentration of solute molecules. The images show blood cells exposed to (a) an isotonic solution, (b) a hypertonic solution, and (c) a hypotonic solution. (a: Dennis Kunkel/Phototake; b, c: Kalab/Custom Medical Stock)

HPLC

High performance liquid chromatography (HPLC) is a more sophisticated version of chromatography than the thin-layer chromatography (TLC) introduced in Chapter 1 but is based on the same principles. In TLC, the stationary phase is a coated glass plate. In HPLC, the stationary phase consists of small polymer or SiO_2 particles contained in a stainless steel tube that is about 1 cm in diameter and 20 cm long. In TLC, the mobile phase was added to the bottom of a beaker and allowed to move by capillary action across the stationary phase. In HPLC, the solvent is mechanically pumped through the tube containing the stationary phase. The sample mixture of solute molecules is injected into the mobile phase.

The various molecules that make up the mixture are attracted through intermolecular forces to both the stationary phase and the mobile phase, but to different degrees. Those molecules that are most attracted to the mobile phase pass through the system first. Those molecules that are attracted to the stationary phase pass through the system last. The stationary material used in HPLC can be coated with different compounds. The coating is usually chosen to optimize the differences in intermolecular forces of the compounds in the mixture.

If the stationary phase is coated with an amino-type compound (shown in the first figure), will polar or nonpolar compounds come out of the column first?

Amino-type compound

$$\text{Polymer Stationary Phase} \Big) -O-\underset{\underset{CH_3}{|}}{\overset{\overset{CH_3}{|}}{Si}}-CH_2CH_2CH_2CH_2NH_2$$

The nitrogen group at the end of the amino-type stationary phase molecule is very electronegative, creating a polar stationary phase. Therefore, polar molecules will be most attracted to the stationary phase. The nonpolar compounds will not be attracted to the stationary phase and will be pushed out of the column very quickly with the mobile phase.

If the stationary phase is coated with an octyl-type compound, will polar or nonpolar compounds come out first? (See the next figure.)

Octyl-type compound

$$\text{Polymer Stationary Phase} \Big) -O-\underset{\underset{CH_3}{|}}{\overset{\overset{CH_3}{|}}{Si}}-CH_2CH_2CH_2CH_2CH_2CH_2CH_2CH_3$$

The *osmol gap* is a measure of the difference between the expected and actual osmotic pressures of a person's blood. A higher than normal value for the blood's osmotic pressure indicates the presence of an unusual compound or unusually high concentration of a blood component. Further blood work must be done to identify the unknown substance. Most modern medical laboratories will use devices designed to measure the freezing point depression of samples, from which the osmol gap can be calculated.

7.11 Mathematics of Osmotic Pressure

The osmotic pressure (π) of a solution is calculated by using equation 3 and has units of atmospheres (atm). Where M is the molarity of the solution, R is the gas constant (0.0821 L · atm/K · mol) and T is the temperature expressed in Kelvin. Osmotic pressure measurements are made at a constant temperature. Therefore, molarity is used for concentration units rather than molality, as was needed for boiling point and freezing point calculations. Osmotic pressure is a colligative property, so the concentration of the number of solute particles dissolved in solution is used in the calculation, as was done for the molality calculations discussed earlier in the chapter.

$$\pi = M_{particles} RT \tag{3}$$

The octyl-type stationary phase molecule contains mostly C—H bonds, which produce a nonpolar stationary phase. Therefore, polar compounds will come out first because the nonpolar compounds will be more attracted to the nonpolar stationary phase.

The detectors work in HPLC by measuring how much light each compound absorbs as it leaves the

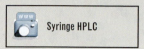

column and enters the detector. A schematic of an HPLC system is shown in Figure 7.16. The advantage of HPLC is that it can be used to analyze and quantify extremely small samples and detect trace amounts of impurities, whereas TLC is useful only for identifying the main components of a mixture.

The DEA uses HPLC to determine which trace impurities are present in cocaine or heroin. DEA scientists can then match the profiles of impurities to a specific method used by one of the drug cartels to make the drug. In this manner, agents can trace illegal drugs seized in a raid directly to the cartel that manufactured it. When a cartel tries a new method of smuggling, the DEA can determine from seized samples which cartel is responsible. If a new distributor of illegal drugs is coming into a town, the DEA can determine who it is. The intelligence gained from trace impurity analysis provides vast amounts of information that would be impossible to gather without informants at the highest levels of each cartel.

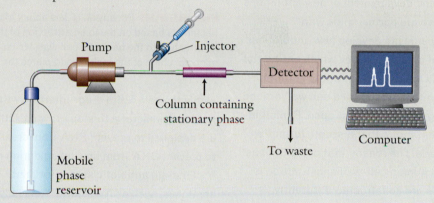

FIGURE 7.16 High performance liquid chromatography.

WORKED EXAMPLE 12

An isotonic intravenous saline solution contains 9.00 g of NaCl per 1.00 L of solution. Calculate the osmotic pressure of this solution at 25°C.

SOLUTION

Calculate the molarity of NaCl: $9.00 \text{ g NaCl} \times \dfrac{1 \text{ mol NaCl}}{58.44 \text{ g NaCl}} \times \dfrac{1}{1.00 \text{ L}} = 0.154 \text{ M}$

However, NaCl will dissociate into 2 particles, Na^+ and Cl^-:

$$M_{particles} = 2 \times 0.154 \text{ M} = 0.308 \text{ M}$$

Using equation 3:

$$\pi = \underbrace{\frac{0.308 \text{ mol}}{\text{L}}}_{0.308 \; M_{particles}} \times \underbrace{\frac{0.0821 \text{ L} \cdot \text{atm}}{\text{mol} \cdot \text{K}}}_{R} \times \underbrace{298 \text{ K}}_{25°C+273} = 7.54 \text{ atm}$$

Practice 7.12

A solution is prepared by dissolving 18.0 g of KCl into 500.0 mL of water. What is the osmotic pressure of this solution at 20°C?

Answer

23.2 atm

7.12 CASE STUDY FINALE: Exploring Antibiotics and Drug-Resistant Infections

A supply of new antibiotics is one of the critical tools that physicians can use to combat drug-resistant strains of bacteria. However, as Figure 7.17 shows, there has been a dramatic decline in the number of new antibiotics being developed over the last 25 years. The Infectious Disease Society of America is attempting to draw attention to this critical weakness and encourage the U.S. Congress to support research and development of new antibiotics and develop public-private partnerships with pharmaceutical companies to develop new antibiotics.

There are several mechanisms by which antibiotics work to kill bacteria. One method is for the antibiotic drug to interfere with the creation of a cell wall around the bacteria, causing the cell to burst open. Penicillin is an example of an antibiotic that disrupts formation of cell walls in bacteria. The reason that penicillin has no adverse effect on human cell-wall construction is that human cell walls are constructed differently.

Folic acid is an essential vitamin that is needed in both human and bacteria cells. Human cells can obtain folic acid from dietary sources, and it diffuses into the cells. Bacteria cells, however, must synthesize folic acid

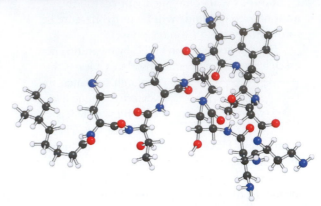

FIGURE 7.18 Polymyxin B. The ability of polymyxin B to kill bacteria is based on the ionic attraction of the ring portion of its structure to the bacterial cell wall, which then allows the nonpolar chain of the molecule to pierce the nonpolar cell wall.

within the cell. Sulfa drugs interfere with this synthesis, forcing the bacteria to stop growing. Other antibiotics are designed to prevent RNA from creating proteins or to prevent DNA from unwinding during replication.

First-aid antibiotic sprays for minor cuts and scratches, such as Neosporin™, usually contain several different antibiotics to cover a variety of possible bacterial infections. One of the main ingredients is the antibiotic polymyxin B, shown in Figure 7.18. The large circular region will have a net positive charge at physiological pH, which serves to bind the antibiotic to a negatively charged site in the outer membrane. The long nonpolar carbon-chain portion of polymyxin B is then able to pierce the nonpolar cell membrane of the bacteria. The bacteria cell is semipermeable and subject to osmosis, leading to cell death.

Polymyxin B has a disadvantage in that it can also affect human cells and is classified as a *neurotoxin* (attacking neurons) and *nephrotoxic* (attacking kidney cells). The use of polymyxin B as topical first aid spray is safe, so long as the drugs do not enter your circulatory system. The shortage of new antibiotics to combat the multiple drug-resistant *Pseudomonas aeruginosa* has led physicians to the use of polymyxin B, despite its possible toxicity, as infection presents a greater risk to the patient than the side effects of the antibiotic.

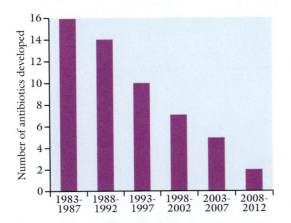

FIGURE 7.17 The declining discovery of antibiotics. The number of antibiotic drugs being developed has dramatically declined over the last 25 years despite the need for new antibiotic drugs to fight drug-resistant bacteria. (Adapted from the Infectious Disease Society of America's Public Policy Statement, April 2011.)

CHAPTER SUMMARY

• Intermolecular forces are electrostatic attractions between positive and negative regions of molecules. Intermolecular forces affect the physical properties of solutions.

• The spherical shape of a liquid droplet is due to the net inward pull of surface molecules by other molecules in proximity to the surface molecules. A sphere is the most efficient shape a droplet can form.

• The dipole–dipole intermolecular force is established between polar molecules and increases with the strength of the dipole within a molecule. A particularly strong form of dipole–dipole interaction is the hydrogen bond, which occurs only when a sufficiently electronegative element (N, O, or F) is bonded directly to a hydrogen atom.

• Dispersion forces result from a distortion of the electron cloud surrounding the atoms. This distortion sets up a temporary dipole within the molecule, which induces a dipole on molecules surrounding it.

• The ease of distorting an electron cloud is called the polarizability of the molecule. The polarizability of atoms and molecules increases with size.

• In solutions, other intermolecular attractions can develop, such as ion–dipole and dipole–induced dipole interactions.

• Ionic compounds and polar molecular compounds dissolve in water because of the polar nature of water. The difference in the electronegativity of hydrogen and oxygen is such that shared electrons reside closer to the oxygen atom, providing a negative polar region. The relative absence of electrons near the hydrogen atoms results in a positive polar region.

• The water molecules orient their polar regions to attract the oppositely charged region of the solute. By surrounding the solute particle, the water is able to overcome the attractive forces holding the solute particle in the solid.

• The temperature of the solvent, the surface area of the solute, and the concentration of the solute within a solution all affect the rate at which a substance dissolves.

• Diffusion is the passive mixing of solute and solvent particles by random molecular motion from a region of high solute concentration to a region of low solute concentration.

• The melting point, boiling point, vapor pressure, and osmotic pressure of solutions are called colligative properties—those properties that simply depend on how many solute particles are present in solution, not their identity.

• HPLC relies on the intermolecular forces of attraction between the compounds of a mixture to the stationary phase and mobile phase to separate the components.

KEY TERMS

boiling point (p. 200)
boiling point elevation (p. 201)
colligative property (p. 199)
diffusion (p. 206)
dipole–dipole force (p. 192)
dipole–induced dipole force (p. 196)
dispersion force (p. 193)
electron cloud (p. 193)
equilibrium (p. 204)
freezing point depression (p. 204)
high performance liquid
    chromatography (HPLC) (p. 208)

hydrogen bonding (p. 192)
hypertonic solution (p. 207)
hypotonic solution (p. 207)
induced dipole-induced
    dipole (p. 193)
intermolecular force (p. 190)
ion-dipole force (p. 195)
isotonic solution (p. 206)
London forces (p. 193)
molality (p. 202)
osmolality (p. 207)
osmosis (p. 206)

osmotic pressure (p. 206)
polar (p. 192)
polarizability (p. 194)
pressure (p. 200)
reverse osmosis (p. 206)
semipermeable membrane (p. 206)
surface tension (p. 191)
van der Waals force (p. 193)
vapor (p. 200)
vapor pressure (p. 200)

MAKING MORE CONNECTIONS: Additional Readings, Resources, and References

For more information on antibiotics and antibiotic-resistant bacteria: http://cid.oxfordjournals.org/content/52/suppl_5/S397.full.pdf+html, http://www.idsociety.org/10x20.htm, http://www.neosporin.com/firstaid/neosporin.asp?sec=0&page=16

For more information about osmotic pressure, see the entry for J. H. van't Hoff: nobelprize.org/nobel_prizes/chemistry/laureates/1901/

For van't Hoff's interesting lecture: nobelprize.org/nobel_prizes/chemistry/laureates/1901/hoff-lecture.pdf

For a summary of osmotic pressure and related concepts: www.bbc.co.uk/dna/h2g2/A686766

For more information about Alfred Nobel: nobelprize.org/nobel/alfred-nobel/index.html

REVIEW QUESTIONS AND PROBLEMS

Questions

1. List the types of intermolecular forces that may exist in a polar solvent. (7.2)

2. List the type of intermolecular force that exists in a nonpolar solvent. (7.2)

3. What are the conditions required for the formation of the hydrogen bond? (7.2)

4. Why is the hydrogen bond the strongest intermolecular force? (7.2)

5. Why is the induced dipole–induced dipole (dispersion) intermolecular force the weakest? (7.2)

6. What type of intermolecular forces will develop when polar molecules are dissolved in a polar solvent? (7.2)

7. What type of intermolecular forces will develop when ionic compounds are dissolved in a polar solvent? (7.3)

8. Will nitrogen gas be more or less soluble in water as compared with oxygen gas, based on the polarizability of the molecules? (7.3)

9. Although oxygen gas is nonpolar, it will dissolve to a very small extent in the polar solvent water. Which intermolecular force(s) is/are responsible for the interaction between the two compounds? (7.3)

10. It has been shown that alcohol movement into the bloodstream is slowed by large amounts of starch-containing foods. Explain how this occurs, given the fact that starch molecules contain many —OH groups. (7.3)

11. Sketch a beaker containing an aqueous K_2S solution. Include water molecules interacting properly with the solute particles. Label all molecules and ions. (7.4)

12. How is the dissolution process different for ionic compounds such as table salt (NaCl) compared with polar molecular compounds such as sugar? (7.4)

13. It used to be common for mechanics to use gasoline to clean grease from their hands and arms (a very dangerous practice that should not be done!). Explain why they did this. (7.4)

14. Water, a polar molecule, is labeled 1 in the following figure. Hexane, a nonpolar organic solvent, is labeled 2. Which compounds will mix together? Molecule (a) is ethanol, (b) is cyclohexane, and (c) is decane. (7.4)

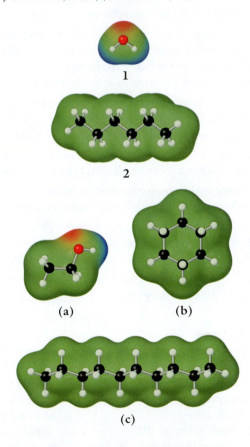

(a)   (b)

(c)

15. What factors affect the rate of dissolving a soluble compound? (7.5)

16. What three things can be done to increase the rate at which a substance is dissolved? (7.5)

17. Why would it not be a good idea to calibrate thermometers by using boiling water? (7.6)

18. Explain what the vapor pressure of a liquid is and what happens to the vapor pressure as the temperature of the liquid increases. Sketch a molecular view of the same liquid at two different temperatures that illustrates the differences in vapor pressure. (7.6)

19. Explain what is meant by the boiling point of a solution and how vapor pressure plays a role. (7.6)

20. Explain why two different liquids will have different boiling points in terms of the types of intermolecular forces present. (7.6)

21. Explain how the vapor pressure of a solution is lowered by the presence of a solute, and sketch a molecular view of this process. (7.6)

22. Explain how the freezing point of a solution is lowered by the presence of a solute, and sketch a molecular view of this process. (7.8)

23. If a saltwater sample has a freezing point of $-4°C$, draw a molecular view of what happens when the temperature is lowered to $-5°C$. (7.8)

24. In winter it is common to spread sodium chloride on highways to melt ice. However, salt is corrosive and not suitable for use on airplanes. Using the Internet as a resource, determine what is contained in the solutions used to deice airplanes. (7.8)

25. Calcium chloride is a more expensive alternative to sodium chloride that can be used to melt ice during winter and is usually sold as "Super Deicer!" Is calcium chloride more efficient than sodium chloride at melting ice? (7.8)

26. Explain why freezing point depression is the preferred method for the detection of tampered solutions. (7.8)

27. Until the twentieth century, a common method for preserving food was to use salt in high concentration brines (pickling) or to simply store the meat in barrels lined with salt. Explain how these methods used the principle of osmosis to prevent bacteria or mold growth. (7.10)

28. Explain how drinking water can be obtained from seawater based on the process of osmosis. (7.10)

29. Explain the role of intermolecular forces in HPLC analysis used to trace the point of origin of a heroin sample seized by the DEA. (ITL)

30. Will carbon monoxide or carbon dioxide diffuse faster through blood, assuming they are present at equal concentrations? (7.10)

Problems

31. Calculate the surface area of a regular octahedron with a volume of 1 cm^3. For the regular octahedron, the volume and surface area are given by:

$$V = \frac{\sqrt{2}}{3} a^3 \quad \text{and} \quad A = 2a^2 \sqrt{3}$$

where a is the length of an edge. What percent increase in surface area does a regular tetrahedron have as compared with a sphere of the same volume? (7.1)

32. Calculate the surface area of a regular tetrahedron with a volume of 1 cm^3. For the regular tetrahedron, the volume and surface area are given by:

$$V = \frac{\sqrt{2}}{12} (L)^2 \quad \text{and} \quad A = (L)^2 \sqrt{3}$$

where L is the length of an edge. What percent increase in surface area does a regular tetrahedron have as compared with a sphere of the same volume? (7.1)

33. Place the following list of atoms in order of increasing polarizability: Mg, Be, Sr, Ca. (7.2)

34. Place the following list of atoms in order of increasing polarizability: Rb, Na, Al, Cl. (7.2)

35. List all of the intermolecular forces that will be present in the following liquids, and underline the most dominant force for each liquid. The type of compound is indicated in parentheses. (7.2)
    (a) H_2O (polar)
    (b) CH_3OH (polar)
    (c) Br_2 (nonpolar)

36. List all of the intermolecular forces that will be present in the following liquids, and underline the most dominant force for each liquid. The type of compound is indicated in parentheses. (7.2)
    (a) NH_3 (polar)
    (b) C_3H_8 (nonpolar)
    (c) N_2 (nonpolar)

37. List all of the intermolecular forces that will be present in the following aqueous solutions, and underline the most dominant force for each liquid. The type of compound is indicated in parentheses. (7.2–7.3)
    (a) O_2 (nonpolar)
    (b) CH_3OH (polar)
    (c) $NH_4C_2H_3O_2$ (ionic)

38. List all of the intermolecular forces that will be present in the following aqueous solutions, and underline the most dominant force for each liquid. The type of compound is indicated in parentheses. (7.2–7.3)
    (a) NaCl (ionic)
    (b) CO_2 (nonpolar)
    (c) $C_6H_{12}O_6$ (polar)

39. Which compound will have a higher boiling point, HI or HBr? What is the most prominent intermolecular force present? (7.2)

40. Which compound will have a higher boiling point, HI or HF? What is the most prominent intermolecular force present? (7.2)

41. Based on the graph shown in Figure 7.7, determine whether a 40.0-g sample of each salt in 100 mL of solution at a temperature of 20°C is unsaturated, saturated, or supersaturated. (7.5)

42. Based on the graph shown in Figure 7.7, determine whether a 50.0-g sample of each salt in 100 mL of solution at a temperature of 20°C is unsaturated, saturated, or supersaturated. (7.5)

43. Based on the graph shown in Figure 7.7, estimate the solubility of KCl at the following temperatures. (7.5)
    (a) 10°C   (b) 30°C   (c) 60°C   (d) 90°C

44. Based on the graph shown in Figure 7.7, estimate the solubility of KNO_3 at the following temperatures. (7.5)
    (a) 10°C   (b) 30°C   (c) 60°C   (d) 90°C

45. Based on the graph shown in Figure 7.7, estimate the temperature at which the following amounts of KNO_3 dissolved in 100 mL of solution constitute a saturated solution. (7.5)
    (a) 80.0 g   (b) 20.0 g   (c) 65.0 g   (d) 30.0 g

46. Based on the graph shown in Figure 7.7, estimate the temperature at which the following amounts of KCl dissolved in 100 mL of solution constitute a saturated solution. (7.5)
    (a) 25.0 g   (b) 30.0 g   (c) 40.0 g   (d) 60.0 g

47. Place the following solutions in order of their decreasing vapor pressure of water. (7.6)

    Solution A: 0.30 M $CaCl_2$
    Solution B: 0.25 M Na_3PO_4
    Solution C: 0.45 M LiI

48. Place the following solutions in order of their decreasing vapor pressure of water. (7.6)

    Solution A: 0.20 M $(NH_4)_2SO_4$
    Solution B: 0.30 M $CuSO_4$
    Solution C: 0.45 M KI

49. Place the following solutions in order of highest boiling point to lowest boiling point: 0.66 m NaCl, 0.50 m Na_2S, and 0.15 m glucose ($C_6H_{12}O_6$). (7.7)

50. Place the following solutions in order of highest boiling point to lowest boiling point: 0.15 m KI, 0.15 m Na_3PO_4, and 0.15 m $NH_4C_2H_3O_2$. (7.7)

51. Calculate the boiling point for each solution in Problem 49, given that the $K_{bp} = 0.512°C/m$ for water. (7.7)

52. Calculate the boiling point for each solution in Problem 50, given that the $K_{bp} = 0.512°C/m$ for water. (7.7)

53. Calculate the change in the boiling point of water ($K_{bp} = 0.512°C/m$) under the following conditions. (7.7)
    (a) 47.0 g of KI dissolved in 500.0 g of water
    (b) 28.0 g of NH_4NO_3 dissolved in 250.0 g of water
    (c) 33.0 g of $CO(NH_2)_2$ (urea) dissolved in 100.0 g of water

54. Calculate the change in the boiling point of water ($K_{bp} = 0.512°C/m$) under the following conditions. (7.9)
    (a) 154.8 g of $C_2H_6O_2$ (ethylene glycol) dissolved in 500.0 g of water
    (b) 97.2 g of $NH_4C_2H_3O_2$ dissolved in 250.0 g of water
    (c) 30.4 g of NaOH dissolved in 100.0 g of water

55. Calculate the freezing point for each solution listed in Problem 53, given that the $K_{fp} = -1.86°C/m$ for water. (7.9)

56. Calculate the freezing point for each solution listed in Problem 54, given that the $K_{fp} = -1.86°C/m$ for water. (7.9)

57. Calculate the freezing point for each solution listed in Problem 49, given that the $K_{fp} = -1.86°C/m$ for water. (7.9)

58. Calculate the freezing point for each solution listed in Problem 50, given that the $K_{fp} = -1.86°C/m$ for water. (7.9)

59. Calculate the change in the freezing point of water ($K_{fp} = -1.86°C/m$) under the following conditions. (7.9)
    (a) 13.2 g of $FeCl_3$ dissolved in 50.00 g of water
    (b) 149.0 g of $NaNO_3$ dissolved in 350.0 g of water
    (c) 327.4 g of $(NH_4)_2SO_4$ dissolved in 2.20 kg of water

60. Calculate the change in the freezing point of water ($K_{fp} = -1.86°C/m$) under the following conditions. (7.9)
    (a) 7.76 g of KCl, dissolved in 50.00 g of water
    (b) 6.56 g of of caffeine ($C_8H_{10}N_4O_2$) dissolved in 75.0 g of water
    (c) 56.2 g of KCN dissolved in 1000.0 g of water

61. Determine the molality of an aqueous glucose ($C_6H_{12}O_6$) solution under the following conditions. (7.7, 7.9)
    (a) Freezing point = $-2.8°C$
    (b) Boiling point = 102.7°C
    (c) Boiling point = 101.3°C

62. Determine the molality of an aqueous ethylene glycol solution ($C_2H_6O_2$) under the following conditions. (7.7, 7.9)
    (a) Boiling point = 117.0°C
    (b) Freezing point = $-6.90°C$
    (c) Freezing point = $-11.5°C$

63. Assuming each solution from Problem 61 contained 1.00 kg of solvent, determine the moles of glucose and the mass of glucose dissolved in each solution. (7.7, 7.9)

64. Assuming each solution from Problem 62 contained 1.00 kg of solvent, determine the moles of ethylene glycol and the mass of ethylene glycol dissolved in each solution. (7.7, 7.9)

65. Draw a system showing the natural osmotic process of seawater being placed on one side of the membrane and pure water on the opposite side of the membrane. Be sure to indicate the movement of all solute and solvent particles. (7.10)

66. Draw a membrane system showing the reverse osmosis of seawater to pure water, with the seawater placed on one side of the membrane and pure water on the opposite side of the membrane. Be sure to indicate the movement of all solute and solvent particles. (7.10)

67. Calculate the osmotic pressure of the following solutions. (7.11)
    (a) 16.3 g NaF in 1.00 L of H_2O at 20°C
    (b) 5.61 g $C_6H_{12}O_6$ in 1.00 L of H_2O at 20°C
    (c) 2.90 g $MgCl_2$ in 500.0 mL of H_2O at 25°C

68. Calculate the osmotic pressure of the following solutions. (7.11)
    (a) 17.0 g Li_2S in 1.00 L of H_2O at 22°C
    (b) 41.2 g $C_6H_{12}O_6$ in 1.00 L of H_2O at 7°C
    (c) 18.3 g $Ca(NO_3)_2$ in 200.0 mL of H_2O at 15°C

Case Study Problems

69. In the case of a fatal drunk-driving accident, the alcohol consumed throughout the evening can be determined by sampling the vitreous humor. However, moments before the fatal crash, the deceased consumed an extremely large portion of alcohol. Will this be reflected in the vitreous humor or not? (CP)

70. A liquid sample is sent to a laboratory for analysis. The contents are suspected to be either propanol (boiling point = 97.2°C) or heptane (boiling point = 98.4°C). Explain whether simply measuring the boiling point with a traditional thermometer would be sufficient for identifying the compound. (CP)

71. For Problem 70, will the additional information that the density of propanol is 0.80 g/cm³ and the density of heptane is 0.68 g/cm³ make a difference in your answer? Are there other physical properties that could be used to help distinguish the two liquids? Explain. (CP)

72. Thin-layer chromatography was discussed in the In the Lab box of Chapter 1. Read this again and explain how the separation of different compounds occurs using the information on intermolecular forces from Chapter 7. (CP)

73. Immunoassay methods were discussed in the In the Lab box of Chapter 5. Read this again and explain how intermolecular forces play a role in this method. (CP)

74. A jogger is found collapsed and unresponsive along a beach in southern Wales. The police are called and an ambulance takes the woman to the local hospital. The jogger does not appear to be dehydrated, and her medical history provides no clues as to her present condition. Toxicology screening is ordered, and the contents of a half-full water bottle found with the jogger are sent to the laboratory for analysis.

    The toxicology screen comes back negative for alcohol and the commonly abused classes of drugs. However, the preliminary freezing point measurement of the water bottle contents reveals it is a solution, not pure water. Does this imply the jogger was poisoned? List several compounds that could be present in water that the jogger herself may have added. How would a forensic scientist determine whether more than one compound had been added to the water? (CP)

75. A young man in his early twenties was found dead with several cans of petroleum-based cleaners and a plastic bag by his side. Investigators made the initial hypothesis that the young man had died of asphyxiation from inhaling the volatile fumes from the cleaners. The estimated time of death was 16 hours before discovery. When the autopsy was conducted several days later, only trace amounts of volatile petroleum-based compounds could be detected in his lungs, well beneath the lethal level. Is the original hypothesis of death by asphyxiation from inhalants still valid? Explain why only trace levels would be found. (CP)

76. Arson investigators collect samples of debris from the suspected point of origin of the blaze to test for the presence of an accelerant, such as gasoline. Investigators also collect a sample of debris from a region believed not to contain any accelerant. The evidence is analyzed by exposing a small fiber coated with the polymer (see figure) to the air that is trapped in the vial above the debris. If gasoline is present, the compounds are adsorbed onto the fiber and subsequently analyzed. Explain in terms of intermolecular forces why the compounds that constitute gasoline would be present in the airspace and why they would collect in the fiber. Why is it important to collect and analyze a sample that is believed to contain *no* accelerant? (CP)

Organic Chemistry and Polymers

CASE STUDY: Exploring Biodegradable Polymers

Being an educated consumer is easier said than done. Advertising agencies will exaggerate and make claims that—upon closer examination—hold little merit. For example, you may have watched a commercial for over-the-counter pain medications that will make the claim, "No other pain reliever proven more effective at providing relief." This is a very clever way of saying that all products of this kind are regulated by law and must contain the same exact amount of active ingredient. No other product has been proven more effective, as they all are equal in dose, if not price.

Questionable advertising is not the only peril facing consumers. Bacteria-contaminated food causes many massive food recalls each year in which millions of pounds of produce and raw meats are ordered off the shelves. In the past few years, attention has been drawn to the dangers of a diet high in trans fats, as discussed in Chapter 5. As a result, consumers are making an impact on the types of products being offered, in the form of increased consumer demand for products that are manufactured in an environmentally sustainable fashion, which has made industry reexamine supply chains.

Our consumer world is packaged and wrapped in plastic polymers—from our clothing, to the shopping bags we use, to the throw-away packaging, to the products themselves. Industrially produced polymer products are one area of great environmental concern. Polymers are fabricated from petroleum-based chemical compounds obtained in the refining of crude oil, leading to increased demand on a nonrenewable resource. Polymers make up 12.3% of garbage (30 million tons) headed to landfills in the United States. Only 7% of the plastic (2.1 million tons) is recovered and recycled into new products. One reason why polymers are used in such a large array of products is their durability and strength, although these same properties prevent polymers from decomposing or degrading in landfills.

Some power plants can burn household waste to generate energy. These facilities reduce by 90% the

amount of landfill space that might be taken by garbage and convert the garbage to energy. While they do generate carbon dioxide, they also prevent the formation of methane that is associated with landfills. Furthermore, they reduce the amount of fossil fuels that would be consumed in the generation of power. Polymers are an excellent source of energy, too. Yet if they are not burned under optimal conditions, they can emit a variety of toxic substances.

MAKE THE CONNECTION

Is there any way in which polymers can be manufactured and used that is more environmentally friendly? To better answer this question you first need to understand basics of organic chemistry . . .

> **As you read through the chapter, consider the following questions:**
> - **How are polymers made?**
> - **Why don't polymers break down in the environment?**
> - **How does the chemical structure of a polymer affect its physical properties?**
> - **What do the recycling numbers represent?**

8.1 Introduction to Organic Chemistry

LEARNING OBJECTIVE

Differentiate between organic and inorganic compounds.

In the simplest terms, an **organic compound** is composed primarily of carbon and hydrogen atoms. Conversely, an **inorganic compound** is composed of elements from the rest of the periodic table. You are probably familiar with several of the simplest of the organic compounds because of their use as fuels: methane (CH_4), propane (C_3H_8), and butane (C_4H_{10}). The name of the compounds provides information about the number of carbon atoms in each compound, as shown in Table 8.1. More information about how to determine the formulas and structures of organic compounds will be provided in later sections.

Polymers fit the definition of an organic compound. For example, high density polyethylene (HDPE) is a long chain of carbon atoms, each one bonded to two hydrogen atoms. The large variety of polymers is due to the enormous variety of organic compounds that can be used to form polymers. Other elements such as oxygen, nitrogen, phosphorus, and sulfur can be found as minor components of organic compounds.

Until the early nineteenth century, organic compounds had to be isolated from living or once-living sources (organisms) such as plants and animals, because scientists were unable to synthesize organic compounds in the laboratory. However, in 1828 Friedrich Wöhler was able to successfully synthesize urea, an organic compound, in the laboratory. Modern organic chemistry is no longer limited to compounds found in living systems. Organic compounds now include a seemingly endless array of substances resulting from the discovery of new natural molecules and from the synthesis of new molecules in the laboratory.

It might seem that not too many compounds would result from combinations with just carbon, hydrogen, and a few other elements as building blocks. But as you will learn in this chapter, the variety of ways to combine

TABLE 8.1	Carbon Prefixes
Prefix	Carbon Atoms
meth-	1
eth-	2
prop-	3
but-	4
pent-	5
hex-	6
hept-	7
oct-	8
non-	9
dec-	10

these elements into organic compounds is almost limitless. Nearly 11,500 new substances are added each day to the Chemical Abstracts Service (CAS) Registry System, an internationally used database of chemical compounds. Most of the new substances are organic compounds. Of the more than 37 million known, the vast majority of them contain carbon.

In addition to polymer chemistry, one of the major areas of research based on organic chemistry is the creation of new pharmaceutical compounds. For some pharmaceuticals, such as narcotic painkillers, the desired medical effect lends itself to abuse. The chemical structure of illegal narcotics and prescription drugs can be very similar. Notice, for example, the similarities between heroin, an illegal narcotic, and Vicodin, a drug prescribed for pain, shown in Figure 8.1.

As you progress through the chapter, you will learn how to interpret these structures, but the similarities between heroin and Vicodin are strikingly apparent just by casual observation. In fact, it is common for pharmaceutical companies to study slight modifications of a basic drug structure to determine whether the new molecule will be more effective and have fewer side effects than the original drug. The example of Vicodin and heroin shows that simply adding or removing a few atoms of carbon or oxygen changes how that molecule interacts with the human body.

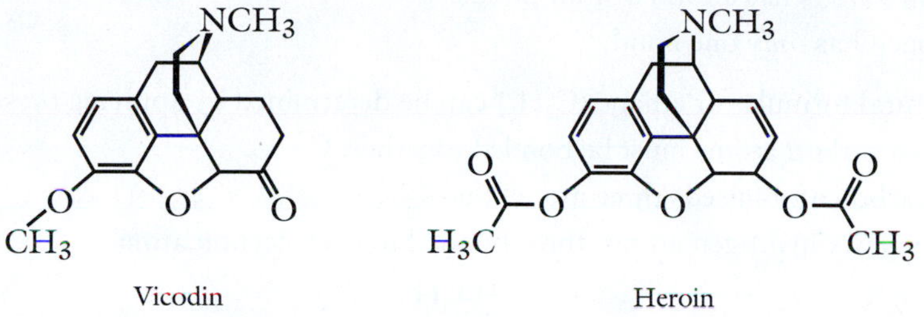

Vicodin

Heroin

FIGURE 8.1 Chemical structures. The central portions of Vicodin ($C_{18}H_{21}NO_3$) and heroin ($C_{21}H_{23}NO_5$) drug molecules are very similar. The molecules differ only by three carbon, two hydrogen, and two oxygen atoms.

8.2 Alkanes

The simplest class of organic compounds is the **alkanes**. The simplest of all alkanes is the compound methane (CH_4), which consists of a carbon atom bonded to four hydrogen atoms (see Figure 8.2a). The next simplest compound is ethane (C_2H_6). The alkane class of organic compounds has chemical formulas that follow the pattern C_nH_{2n+2}, where n is the number of carbon atoms as shown in Figure 8.2.

Methane is the major component in natural gas. Propane is commonly used as a fuel to heat homes and to heat barbeque grills. Butane is the fuel found in common

> **LEARNING OBJECTIVE**
> Name alkanes and draw their structures.

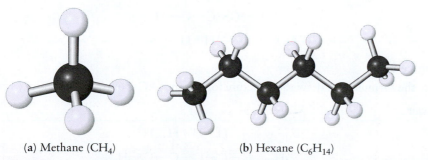

(a) Methane (CH_4)   (b) Hexane (C_6H_{14})

FIGURE 8.2 Molecular structures of simple alkanes. The structures of methane (a) and hexane (b) follow the same pattern, C_nH_{2n+2}, where n is the number of carbon atoms.

cigarette lighters, and gasoline is commonly ranked by its octane rating. These examples illustrate that one of the major uses of alkane compounds is as a fuel source.

Alkanes are commonly used as solvents in the laboratory because they are fairly nonreactive; that is, they dissolve nonpolar compounds without undergoing chemical reactions. In Chapters 6 and 7, the process of dissolution and the role of intermolecular forces were discussed.

Methane (CH_4) consists of one carbon atom bonded to four hydrogen atoms. The next simplest alkane, ethane (C_2H_6), consists of two carbon atoms bonded together, with each carbon having three bonds to hydrogen atoms. This information is not readily apparent from just looking at the chemical formula. For this reason, structural formulas are often written for organic compounds. A **structural formula** shows not only the type and number of atoms present in a molecule, but also the way these atoms are arranged. Writing out the structural formula of alkanes is fairly simple if you follow these basic rules:

Rules for Organic Structures

1. Carbon atoms bond together to form a chain.
2. Carbon *always* has a total of four bonds.
3. Hydrogen has *only* one bond.

The structural formula for ethane (C_2H_6) can be determined by applying these rules:

1. The two carbon atoms must be bonded together: C—C.
2. Each carbon atom needs three more bonds.
3. There are six hydrogen atoms, three bonded to each carbon atom:

$$\begin{array}{c} \text{H} \quad \text{H} \\ | \quad\; | \\ \text{H}-\text{C}-\text{C}-\text{H} \\ | \quad\; | \\ \text{H} \quad \text{H} \end{array}$$

■ WORKED EXAMPLE 1

Given the rules for organic compounds, what is the formula and structure for the compound propane that contains three carbon atoms?

SOLUTION

Using the formula, $n = 3$ and $C_3H_{2(3)+2} = C_3H_8$. If the carbon atoms are bonded together, the basic structure must be: C—C—C. Because a carbon atom forms a total of four bonds, each end carbon has three hydrogen atoms and the center carbon has two hydrogen atoms, for a total of eight hydrogen atoms.

$$\begin{array}{c} \text{H} \quad \text{H} \quad \text{H} \\ | \quad\; | \quad\; | \\ \text{H}-\text{C}-\text{C}-\text{C}-\text{H} \\ | \quad\; | \quad\; | \\ \text{H} \quad \text{H} \quad \text{H} \end{array}$$

Practice 8.1

Write the formula and draw the structure for butane (4 C atoms).

Answer

$$C_4H_{10} \quad \begin{array}{c} \text{H} \quad \text{H} \quad \text{H} \quad \text{H} \\ | \quad\; | \quad\; | \quad\; | \\ \text{H}-\text{C}-\text{C}-\text{C}-\text{C}-\text{H} \\ | \quad\; | \quad\; | \quad\; | \\ \text{H} \quad \text{H} \quad \text{H} \quad \text{H} \end{array}$$

As the number of carbon atoms increases, writing the structural formula showing all the bonds can become cumbersome. The **condensed structural formula** of a compound shows the atoms in the same order as the structural formula but leaves out the lines representing bonds. For example, the condensed structural formula for butane is $CH_3CH_2CH_2CH_3$.

The formulas and condensed structural formulas for the first ten alkane compounds are given in Figure 8.3. As stated earlier, the names also indicate the number of carbon atoms present: *meth-* represents 1, *eth-* represents 2, *prop-* represents 3, and so on. You should commit these prefixes to memory.

Formula	$\longrightarrow$		Condensed Structural Formula
Methane	CH_4		CH_4
Ethane	C_2H_6		CH_3CH_3
Propane	C_3H_8		$CH_3CH_2CH_3$
Butane	C_4H_{10}		$CH_3CH_2CH_2CH_3$
Pentane	C_5H_{12}		$CH_3CH_2CH_2CH_2CH_3$
Hexane	C_6H_{14}		$CH_3CH_2CH_2CH_2CH_2CH_3$
Heptane	C_7H_{16}		$CH_3CH_2CH_2CH_2CH_2CH_2CH_3$
Octane	C_8H_{18}		$CH_3CH_2CH_2CH_2CH_2CH_2CH_2CH_3$
Nonane	C_9H_{20}		$CH_3CH_2CH_2CH_2CH_2CH_2CH_2CH_2CH_3$
Decane	$C_{10}H_{22}$		$CH_3CH_2CH_2CH_2CH_2CH_2CH_2CH_2CH_2CH_3$

FIGURE 8.3 Alkane formulas.

The structures of Vicodin and heroin shown in Figure 8.1 illustrate another method of drawing the structures of organic compounds. In this method, lines represent the bonds between carbon atoms that are understood to be at the beginning and end of the lines and at the places where the lines change direction. To simplify the **line structure**, hydrogen atoms are not typically shown when bonded to carbon. For example, butane has four carbon atoms bonded together in a chain, as represented by the line structure shown below.

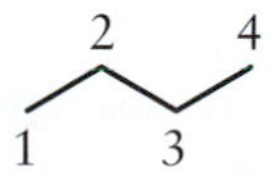

WORKED EXAMPLE 2

Write the line structure for hexane, the solvent commonly used to extract nonpolar drugs in the laboratory.

SOLUTION

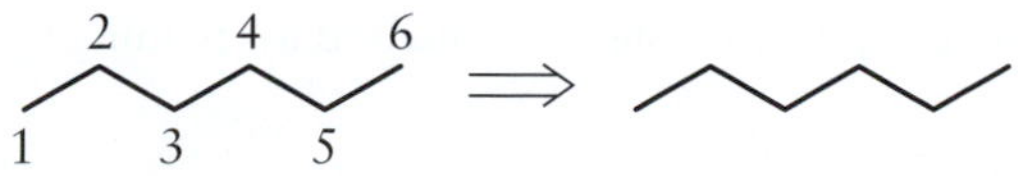

Practice 8.2

Write the line structure for heptane.

Answer

WORKED EXAMPLE 3

What is the name and formula for the following compound?

SOLUTION

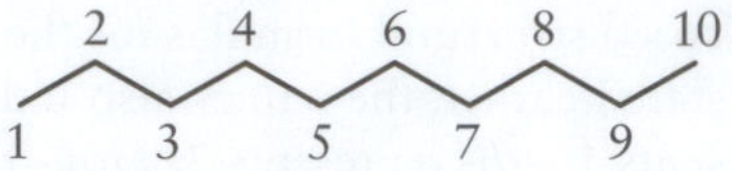

```
  2     4     6     8      10
 1     3     5     7     9
```

The number of carbon atoms is 10, so the compound is decane and the formula is $C_nH_{2n+2} = C_{10}H_{22}$.

Practice 8.3

What is the name and formula for the following compound?

Answer

☐ Octane (C_8H_{18})

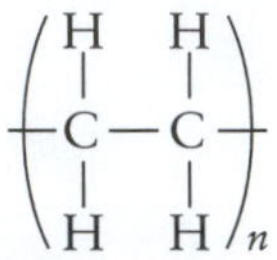

The molecular structure of HDPE contains chains of carbon atoms bonded to hydrogen atoms as shown in this structure. Several thousand carbon atoms are bonded together in HDPE chains.

$$\left(\begin{array}{cc} H & H \\ | & | \\ -C & -C- \\ | & | \\ H & H \end{array} \right)_n$$

Products made out of high density polyethylene (HDPE) are marked with the recycling number 2. Number 2 plastics are comprised of long alkane chains of carbon atoms consisting of several thousand or more carbon atoms linked together. The extremely long polymer chains create strong induced intermolecular forces between the chains, producing strong and durable products. The process of making such extremely large molecules is discussed in the following section.

8.3 Alkenes and Alkynes

Alkenes

LEARNING OBJECTIVE

Name alkenes and alkynes and draw their structures.

The carbon–carbon bonds in alkanes are single bonds in which two carbon atoms share one pair of electrons. Compounds that contain a carbon–carbon double bond (a sharing of two pairs of electrons) belong to the **alkene** class of organic compounds. Alkenes have the general formula C_nH_{2n} and the *-ene* ending to their names. The simplest alkene is $CH_2=CH_2$, which is called *ethene*. The rules for drawing the structural formula, condensed structural formula, and line structure still apply.

As the length of the carbon chain increases, there are different possible locations for the double bond, as shown in Figure 8.4 of butene. Does the placement of the double bond within the molecule change the nature of the compound? If the answer is yes, then the physical properties of the three compounds shown in Figures 8.4a, 8.4b, and 8.4c will differ. In fact they do: The boiling point of the compounds in Figures 8.4a and 8.4c is −6.3°C, but that of the compound in Figure 8.4b is 4°C.

The compounds shown in Figures 8.4a and 8.4c are, in fact, the same compound, just flipped end for end. The compound in Figure 8.4b is different. Two compounds that have the identical chemical formula but different structural formulas are called **isomers**. Distinguishing one isomer from another requires a systematic method for naming the compounds.

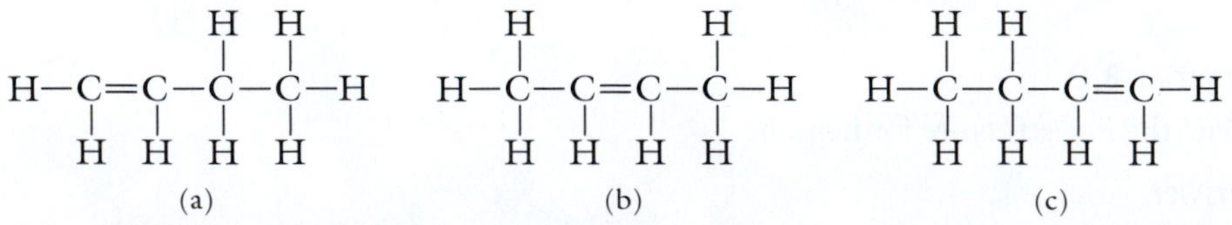

FIGURE 8.4 Double bond placement in butene.

Chapter 5 first introduced isomers as you investigated the health effects of cis and trans fats in Section 5.7 and again in the Case Study Finale, which discussed the mechanism of the chemotherapy drug cisplatin. The isomers covered in Chapter 5 are *stereoisomers*, which are isomers that have the same bonding format but differ in the placement of the atoms in space between isomers. The isomers covered in this section differ in the placement of the double or triple bond and are classified as *constitutional isomers*.

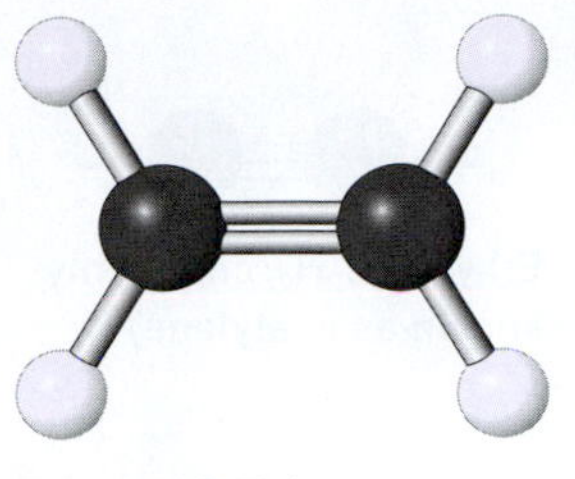

Ethene (C_2H_4)

In the past, there was no agreed-upon method for naming organic compounds. Today the naming of compounds follows suggested rules from the International Union of Pure and Applied Chemistry (IUPAC), the same organization that regulates changes or additions to the periodic table. Many compounds are commonly referred to by their old names, such as ethylene glycol, a toxic compound used as antifreeze in automobiles. The IUPAC name for this poison is 1,2-ethanediol.

The IUPAC rules for naming alkenes are listed as follows, with an example to illustrate their application.

1. Number the carbon atoms from left to right and also from right to left:

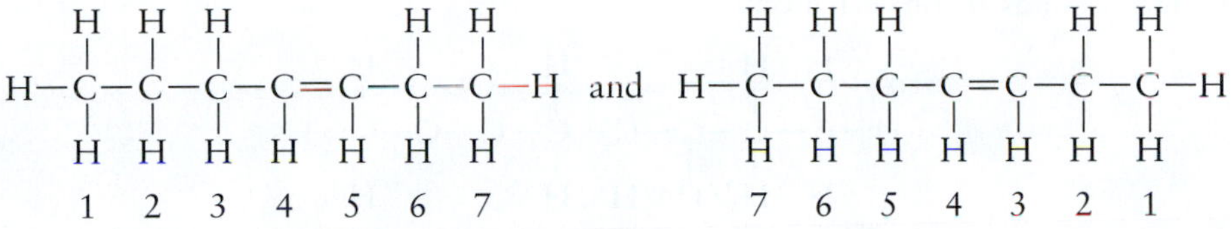

2. Indicate carbon chain length by using the prefixes in Table 8.1: *hept-*.
3. Add the *-ene* ending to the name: heptene.
4. Choose the structure that has the double bond between the lower carbon atom values. The lower of these two values is used to name the compound: 3-heptene.

The correct name for this compound is 3-heptene, not 4-heptene, which illustrates why it is important to number the alkene chain in both directions. Applying these naming rules to the structures drawn in Figure 8.4, the compound in Figures 8.4a and 8.4c is 1-butene, and the compound in Figure 8.4b is 2-butene.

WORKED EXAMPLE 4

Write the condensed structural formula and the line structure for 2-pentene.

SOLUTION

1. Recall that the prefix *pent-* indicates a five-carbon chain. Start by drawing the carbon chain and place the double bond after the second carbon: C—C═C—C—C.
2. Apply the rule that all carbon atoms need four bonds, with hydrogen atoms making up the remaining atoms: $CH_3CH{=}CHCH_2CH_3$.
3. Confirm that the chemical formula follows the C_nH_{2n} pattern for alkenes from the condensed structural formula in step 2. Because there are five carbon atoms and ten hydrogen atoms, the chemical formula is C_5H_{10}.
4. Draw the line structure based on step 1 with a double line to show the double bond:

Practice 8.4

Write the condensed structural formula and the line structure for 3-octene.

Answer

$CH_3CH_2CH{=}CHCH_2CH_2CH_2CH_3$ 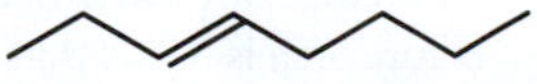

Ethyne (C_2H_2, commonly known as acetylene)

Alkynes

Two carbon atoms can form not only single and double bonds but also triple bonds in which three pairs of electrons are shared. Compounds containing a carbon–carbon triple bond are known as **alkynes** and have the general formula C_nH_{2n-2}. The simplest of the alkynes is H—C≡C—H, which is called *ethyne*.

With the exception of using the *-yne* ending to name alkynes, the same considerations and rules apply to naming alkynes and drawing their structures as applied to alkenes. Just as alkenes form isomers in which the double bond is located in different places along the carbon chain, alkyne isomers result from different placements of the triple bond.

■ WORKED EXAMPLE 5

Write the structural formula for the compound $CH_3CH_2CHCHCH_2CCCH_3$.

SOLUTION

1. Draw the expanded structure to clarify where the hydrogen atoms belong and how many bonds are on each atom.

$$\begin{array}{c} \quad\;\; H \;\; H \qquad\qquad H \qquad\qquad H \\ \quad\;\; | \quad\; | \qquad\qquad | \qquad\qquad | \\ H-C-C-C-C-C-C-C-C-H \\ \quad\;\; | \quad\; | \quad\; | \quad\; | \quad\; | \qquad\qquad | \\ \quad\;\; H \;\; H \;\; H \;\; H \;\; H \qquad\qquad H \end{array}$$

2. Identify carbon atoms with fewer than four bonds. Add double and triple bonds as needed:

$$\begin{array}{c} \quad\;\; H \;\; H \qquad\qquad H \qquad\qquad H \\ \quad\;\; | \quad\; | \qquad\qquad | \qquad\qquad | \\ H-C-C-C=C-C-C\equiv C-C-H \\ \quad\;\; | \quad\; | \quad\; | \quad\; | \quad\; | \qquad\qquad | \\ \quad\;\; H \;\; H \;\; H \;\; H \;\; H \qquad\qquad H \end{array}$$

Practice 8.5

Write the condensed structural formula for 3-hexyne.

Answer

$CH_3CH_2CCCH_2CH_3$

A **polymer** is a very large molecule created by reacting together many smaller molecules called monomers. The molar mass of a polymer can be in the tens of thousands or more, but the reaction to form it starts with a single monomer such as ethylene. The reaction sequence shown below illustrates the formation of HDPE plastic from the ethylene monomer.

$$\begin{array}{c} H \;\; H \quad\; H \;\; H \quad\; H \;\; H \qquad\qquad\qquad \left(H \;\; H \quad H \;\; H \quad H \;\; H \right) \\ | \quad\; | \qquad | \quad\; | \qquad | \quad\; | \qquad\qquad\qquad\;\; | \quad\; | \quad\;\; | \quad\; | \quad\;\; | \quad\; | \\ C=C \;+\; C=C \;+\; C=C \longrightarrow \; -C-C-C-C-C-C- \\ | \quad\; | \qquad | \quad\; | \qquad | \quad\; | \qquad\qquad\qquad\;\; | \quad\; | \quad\;\; | \quad\; | \quad\;\; | \quad\; | \\ H \;\; H \quad\; H \;\; H \quad\; H \;\; H \qquad\qquad\qquad \left(H \;\; H \quad H \;\; H \quad H \;\; H \right) \end{array}$$

The formation of high density polyethylene is an example of an addition reaction in which all reactant atoms are conserved and used in the polymer product.

8.4 Branched Isomers

Another type of constitutional isomer is formed when the chain of carbon atoms splits off into multiple branches, as shown in Figure 8.5. All three structures share the same chemical formula, C_5H_{12}. However, each compound has properties different from the others. For example, the boiling point of the compound in Figure 8.5a is 36°C, that of Figure 8.5b is 28°C, and that of Figure 8.5c is 9.5°C.

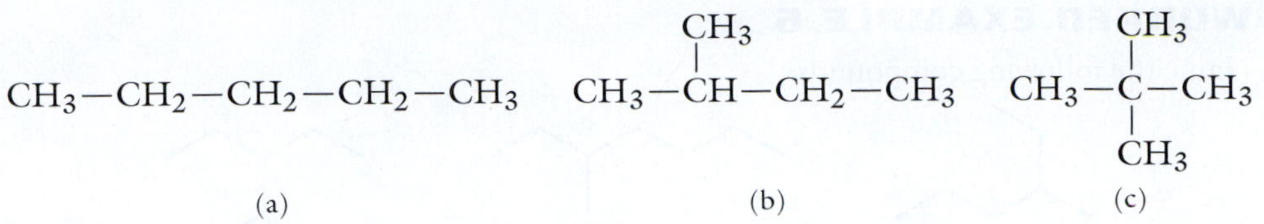

$$CH_3-CH_2-CH_2-CH_2-CH_3$$

$$CH_3-\overset{\overset{\displaystyle CH_3}{|}}{CH}-CH_2-CH_3$$

$$CH_3-\overset{\overset{\displaystyle CH_3}{|}}{\underset{\underset{\displaystyle CH_3}{|}}{C}}-CH_3$$

(a)   (b)   (c)

FIGURE 8.5 Isomers of pentane.

The name of each isomer must accurately communicate its structure. The IUPAC rules used to indicate where double or triple bonds are located along the carbon chain are also used with minor modifications for branched-chain isomers:

Naming Branched-Chain Isomers

1. Find the longest unbranched carbon chain.

2. Number the carbon atoms so that the carbon atom with the branch attached has the lowest value.

3. Indicate the branch length by using the prefixes in the table below (based on the Table 8.1 prefixes, modified with a -*yl* ending).

4. Indicate the name of the compound with the prefixes in Table 8.1 and the -*ane* ending applied to the number of carbon atoms in the longest branch.

5. If a branch appears more than once, use the prefixes *di-*, *tri-*, *tetra-*, etc.

6. In front of the name, indicate the location of the branches by writing the numbers of the carbon atoms at which the branches are located, separated by commas.

Prefix	Branch Length
methyl-	1
ethyl-	2
propyl-	3
butyl-	4
pentyl-	5
hexyl-	6
heptyl-	7
octyl-	8
nonyl-	9
decyl-	10

The compound in Figure 8.5a has no branches and is named *pentane*, or sometimes *n*-pentane for "*normal* (unbranched) pentane." The compound in Figure 8.5b has a four-carbon (4-C) chain with a 1-C branch on carbon atom number 2. Therefore, the name is 2-methylbutane. The compound in Figure 8.5c has a 3-C chain with two 1-C branches on carbon atom number 2. Therefore, the name is 2,2-dimethylpropane. The complexity and nuances of naming organic compounds can increase dramatically for compounds that have multiple branches and multiple double or triple bonds. We will focus on naming simple branched compounds in this textbook.

WORKED EXAMPLE 6

Name the following compounds:

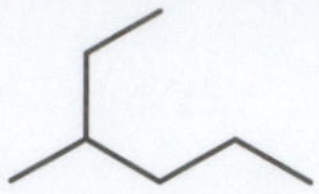

(a) 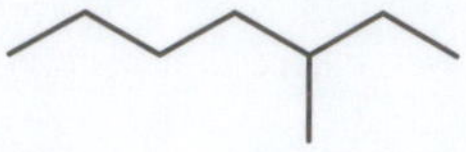   (b) 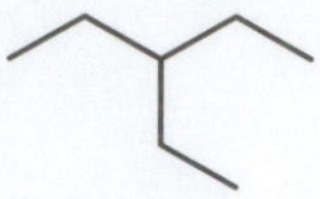   (c)

SOLUTION

(a) The longest chain is a 6-C chain and the branch is a 1-C chain. The branch is located at carbon atom number 3, so the name is 3-methylhexane.

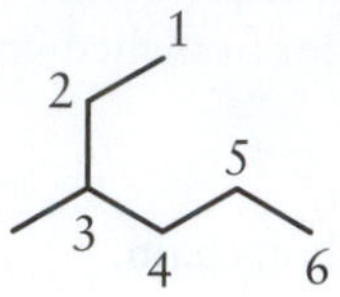

(b) The longest chain is a 7-C chain and the branch is a 1-C chain. The branch is located at carbon atom number 3, so the name is 3-methylheptane.

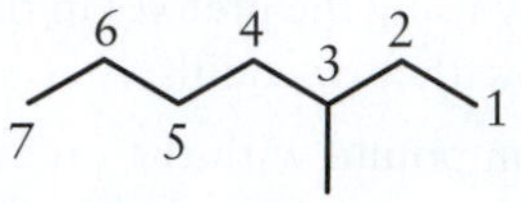

(c) The longest chain is a 5-C chain and the branch is a 2-C chain. The branch is located at carbon atom number 3, so the name is 3-ethylpentane.

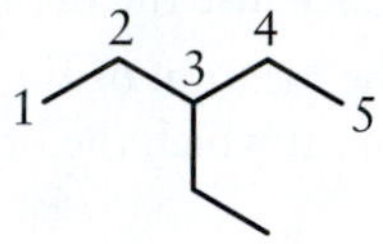

Practice 8.6

Draw the following molecules:
(a) 4-propylheptane
(b) 2-methylhexane

Answer

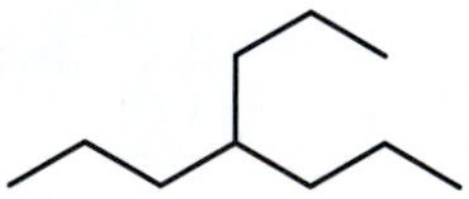

(a) 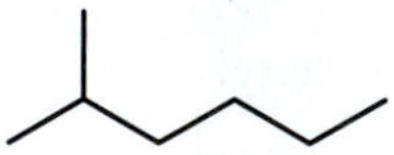   (b)

Branched compounds play an important role in polymer chemistry. The plastic shopping bags common at grocery stores are an example of a branched polymer called low density polyethylene (LDPE). Low density polyethylene is a *copolymer,* which means that a second type of monomer, such as 1-butene, is added during the polymerization process as shown below.

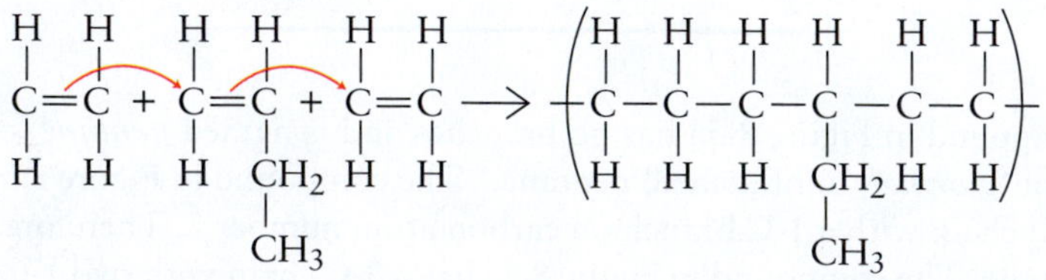

The side branches on the polymer chain prevent the molecules from having contact as close in LDPE as they do in HDPE. This creates a lower density, hence the name; but this also decreases the strength of the intermolecular forces between the polymer chains, which results in a more flexible plastic.

8.5 Cyclic Compounds

In the preceding sections, you have seen carbon compounds in the forms of long-chain compounds and branched-chain compounds. There is another form possible, called the **cycloalkanes**.

As the name implies, the cyclic rings consist of carbon atoms connected with single bonds. Figure 8.6 shows the structures for several cycloalkane compounds. Cyclohexane is a common solvent used in laboratories.

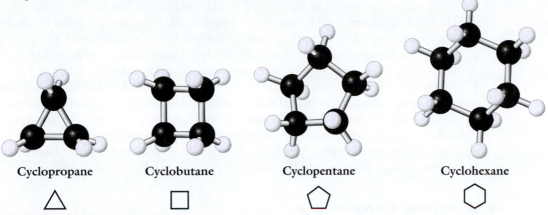

Cyclopropane　　Cyclobutane　　Cyclopentane　　Cyclohexane

FIGURE 8.6 Cycloalkanes.

WORKED EXAMPLE 7

What is the chemical formula for each of the cycloalkanes in Figure 8.6?

SOLUTION

In alkane ring structures, each carbon forms single bonds to two other carbons. Because carbon always forms four bonds in total, the other two bonds are to hydrogen atoms. Therefore, the formula for cyclopropane is C_3H_6, for cyclobutane is C_4H_8, for cyclopentane is C_5H_{10}, and for cyclohexane is C_6H_{12}.

Practice 8.7

The rule for alkane formulas is C_nH_{2n+2}. What is the general formula for the cycloalkanes?

Answer

C_nH_{2n}

Just as the addition of double bonds increased the variety of linear carbon-chain compounds, double bonds can also be present in cyclic carbon compounds. For example, in Figure 8.7 the cyclic compound has alternating double bonds and single bonds around a six-carbon ring, called a *benzene ring*. **Benzene** is a very commonly used industrial and

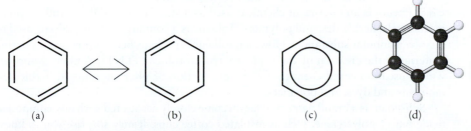

　(a)　　　　　　(b)　　　　　　(c)　　　　　(d)

FIGURE 8.7 Benzene ring representations.

laboratory solvent belonging to a class called **aromatic compounds**. In the early days of discovery of organic compounds, those that were fragrant were grouped together and classified as aromatic. Most of these aromatic compounds contain a benzene ring.

It is appropriate to take a closer look at the bonding of benzene, as illustrated in Figure 8.7. The difference between the structure in Figure 8.7a and the structure in Figure 8.7b is that the double bonds have been shifted by one carbon. Does this mean that the two structures represent different compounds? Shifting the bonds in these two structures does not really change the molecule, but the bonding in benzene turns out to be more complex than alternating double and single bonds. The length of a double bond is, on average, shorter than a single bond. However, when the bonds of benzene are measured, the length of carbon–carbon bonds are the same! The length of the bond in benzene is longer than a double bond but shorter than a single bond.

The actual structure of benzene is not accurately represented by either Figure 8.7a or Figure 8.7b—it is somewhere between the two. The structures in Figure 8.7a and Figure 8.7b are called **resonance structures**, a term used when more than one equivalent structure can be drawn for a compound. (Resonance structures were described fully in Section 5.4.) The structure in Figure 8.7c is actually a shorthand version of benzene that chemists often use to represent the resonance structures of benzene. Figure 8.7d is the molecular model of benzene. Notice that each carbon is trigonal planar, which creates a flat molecule that lies in a single plane.

■ WORKED EXAMPLE 8

Xanax is a prescription drug used to treat anxiety disorders and panic attacks. How many benzene rings are in Xanax? How many other ring structures are present?

SOLUTION

In the structure below, the benzene rings are colored blue. The yellow ring has three nitrogen atoms and two carbon atoms. The green ring has four carbon atoms and two nitrogen atoms.

Practice 8.8

What is the formula for benzene?

Answer

C_6H_6

Several other common compounds that are simple derivatives of benzene are shown in Figure 8.8. Phenol is an important chemical used as a starting material to produce pharmaceuticals, herbicides, and polymers. Toluene is a common organic solvent used to dissolve nonpolar substances. However, toluene is perhaps better known as a starting material in the creation of the explosive trinitrotoluene (TNT). Aniline is another widely used industrial chemical for the production of pharmaceuticals, herbicides, pigments and dyes, and polymers.

Styrofoam is a brand name that has become closely associated with any compound made out of polystyrene, such as insulated coffee cups. It may surprise you to know that plastic silverware is made from the same exact polymer; the only difference is how

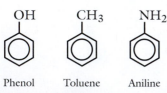

FIGURE 8.8 Compounds based on benzene.

the polymer is processed. The structure of polystyrene is shown in Figure 8.9a. While the fashion merits of Spandex may be debatable, the basic structure of the polyurethane shown in Figure 8.9b illustrates the role that aniline plays as a starting material.

Polycarbonate, shown in Figure 8.9c, is used for many applications, as it produces a clear, durable, and shatterproof polymer. However, one of the main compounds used to create polycarbonate is known as bisphenol A, also shown in Figure 8.9c. Bisphenol A simulates the estrogen hormone, and safety concerns have arisen about infants' and children's toys made from a substance that could leach bisphenol A. Many companies manufacturing such items as baby bottles have altered their products so that bisphenol A is no longer a concern.

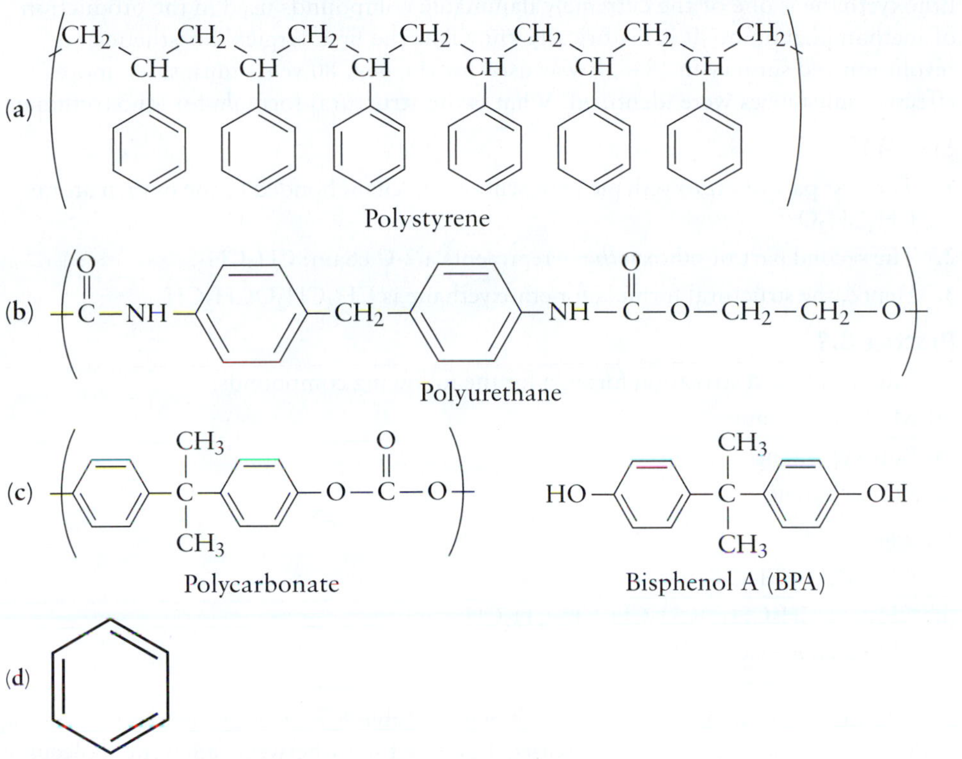

(a) Polystyrene

(b) Polyurethane

(c) Polycarbonate   Bisphenol A (BPA)

(d)

FIGURE 8.9 Cyclic ring structures in polymers. The structures of polystyrene, polyurethane, and polycarbonate all contain cyclic benzene rings.

8.6 Ethers, Ketones, and Esters

Ethers

In both the pharmaceutical and polymer examples discussed so far in the chapter, it is common to see oxygen atoms in the structures. A closer examination reveals that the oxygen atoms can have a variety of bonding arrangements, as shown in Figure 8.10, the chemical structure of the powerful pain-relief drug oxycodone. It is also apparent that oxygen is found in several different bonding arrangements. In some cases, an oxygen atom is simply bonded between two carbon atoms. This arrangement is called the **ether** functional group and is often indicated as R—O—R, where R represents any carbon functional group. A **functional group** is an atom or group of atoms in an organic compound that imparts a distinct set of physical and chemical properties to the molecule.

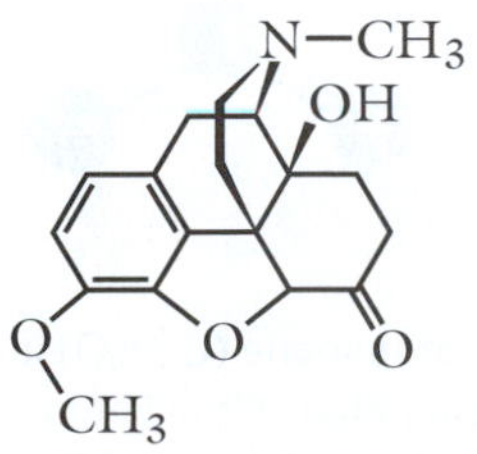

> **LEARNING OBJECTIVE**
> Recognize, draw, and name compounds containing the ether, ketone, and ester functional groups.

Oxycodone

FIGURE 8.10 Chemical structure of oxycodone.

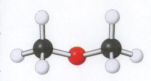

Methoxymethane (C_2H_6O) is a member of the ether functional group. All ethers share the R—O—R structure.

The rules for naming an ether compound are summarized below using the compound $CH_3OCH_2CH_3$ as an example:

Naming Ethers

1. The name of the ether starts with the name of the shortest R group, modified with the ending -*oxy*: methoxy.
2. The name of the longest R group makes up the remainder of the name: methoxyethane.

WORKED EXAMPLE 9

Ethoxyethane is one of the extremely flammable compounds used in the production of methamphetamine. It is historically known as the first surgical anesthetic and revolutionized surgery in 1846. It was used for the next 80 years until safer, more effective substances were identified. What is the structural formula for ethoxyethane?

SOLUTION

1. The first part of *ethox*yethane represents a 2-C chain bonded to the oxygen atom: CH_3CH_2O.
2. The second part of ethox*yethane* represents a 2-C chain: CH_2CH_3.
3. Hence, the structural formula for ethoxyethane is $CH_3CH_2OCH_2CH_3$.

Practice 8.9

Give the condensed structural formula for the following compounds:

(a) Methoxypropane

(b) Butoxypentane

(c) Ethoxybutane

Answer

(a) $CH_3OCH_2CH_2CH_3$

(b) $CH_3CH_2CH_2CH_2OCH_2CH_2CH_2CH_2CH_3$

(c) $CH_3CH_2OCH_2CH_2CH_2CH_3$

The boiling points of the ethers, shown in Table 8.2, increase as the size of the molecule increases because of increased London forces between adjacent molecules. Methoxypropane and ethoxyethane are isomers of each other. Notice that the placement of the oxygen atom changes the boiling point of the two compounds by 5°C.

TABLE 8.2	Boiling Point of Selected Ethers		
Compound	Formula	Molecular Weight	Boiling Point
Methoxymethane	C_2H_6O	46.1	−22°C
Methoxyethane	C_3H_8O	60.1	7°C
Methoxypropane	$C_4H_{10}O$	74.1	39°C
Ethoxyethane	$C_4H_{10}O$	74.1	34°C

Ketones

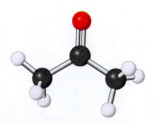

2-propanone (C_3H_6O) is a member of the ketone functional group. All ketones share the R−C−R structure.

A second functional group observed in the drug molecules in Figure 8.10 is the **ketone** group, in which an oxygen atom is double-bonded to a central carbon atom. The general formula for a ketone group is R−C−R, where R again represents any carbon group. Note that the C=O cannot occur on the last carbon of the chain. Therefore, the simplest ketone is 2-propanone, which can be simplified to propanone, as the second carbon is the only possible location for the C=O. This compound is also known by its common name, *acetone*.

Glucose undergoes a series of oxidation–reduction reactions within living cells to release the stored chemical energy. Insulin is a hormone that helps transport glucose from the blood into the cell to undergo this metabolic process. Diabetic patients require supplemental insulin because their bodies no longer produce sufficient amounts. One indicator that a diabetic patient hasn't been taking his or her insulin shots is the characteristic smell of acetone on the breath, a by-product of the incomplete oxidation of glucose. The rules for naming ketones are listed below, using the following compound as an example:

$$\underset{CH_3CH_2\overset{\displaystyle O}{\overset{\displaystyle \|}{C}}CH_2CH_2CH_3}{}$$

Naming Ketones

1. Count the number of carbon atoms in the longest continuous chain: 6.
2. Modify the name of the alkane chain from Figure 8.3 to end with -*one*: hexane to hexanone.
3. Identify the number of the carbon atom with the C=O group, always choosing the lowest possible value: 3-hexanone.

The ketone functional group is more polar than the ether functional group. Therefore, a ketone has stronger intermolecular forces than an ether of comparable size. This results in higher boiling points for ketones, as shown in Table 8.3. The position of C=O on isomers 2-pentanone and 3-pentanone causes a difference, albeit a slight one, in the boiling point of each compound.

TABLE 8.3	Boiling Point of Selected Ketones		
Compound	Formula	Molecular Weight	Boiling Point
2-propanone	C_3H_6O	58.1	56°C
2-butanone	C_4H_8O	72.1	80°C
2-pentanone	$C_5H_{10}O$	86.1	100°C
3-pentanone	$C_5H_{10}O$	86.1	102°C

WORKED EXAMPLE 10

The compound 2-butanone is used in clandestine drug laboratories. Draw the condensed formula structure of this compound.

SOLUTION

1. The prefix (2-*but*anone) indicates a 4-C chain: C—C—C—C.
2. The ending (2-buta*none*) indicates that one of the carbon atoms is double bonded to an oxygen atom: C=O.
3. The number in front (2-butanone) indicates that the second carbon has the C=O.
4. Hydrogen atoms make up the remainder of the bonds:

$$\underset{CH_3\overset{\displaystyle O}{\overset{\displaystyle \|}{C}}CH_2CH_3}{}$$

Practice 8.10

Draw and name all straight-chain isomers of 3-heptanone.

Answer

$$CH_3CH_2\overset{\displaystyle O}{\overset{\|}{C}}CH_2CH_2CH_2CH_3$$

3-heptanone

$$CH_3\overset{\displaystyle O}{\overset{\|}{C}}CH_2CH_2CH_2CH_2CH_3$$

2-heptanone

$$CH_3CH_2CH_2\overset{\displaystyle O}{\overset{\|}{C}}CH_2CH_2CH_3$$

4-heptanone

Esters

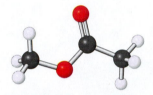

Methylethanoate ($C_3H_6O_2$) is a member of the ester functional group. All esters share the

$$R-O-\overset{\displaystyle O}{\overset{\|}{C}}-R \text{ structure.}$$

Another oxygen-containing functional group found in the drug molecule shown in Figure 8.10 is the **ester** functional group, represented by $R-O-\overset{\displaystyle O}{\overset{\|}{C}}-R$. Esters are commonly found in nature—they are the molecules that give us fragrance and flavor. Esters are also used in developing fingerprints using superglue fumes, which polymerize in the presence of moisture found in an invisible fingerprint, leaving a visible gray fingerprint.

The name of an ester compound has two parts, which are derived from the two separate R groups. The rules for naming esters are listed below, using the following compound as an example:

$$CH_3CH_2CH_2O\overset{\displaystyle O}{\overset{\|}{C}}CH_3$$

Naming Esters

1. The first R group is the one not directly attached to the C=O. Count the number of carbon atoms in it (using the prefixes in Table 8.1) and apply the *-yl* ending: propyl.
2. Count the number of carbon atoms in the second R group, including the carbon atom in the C=O group, and modify the name of the alkane chain to end with *-oate*: propyl ethanoate.

■ WORKED EXAMPLE 11

Draw the condensed structure formula for ethyl butanoate.

SOLUTION

1. The ethyl R group is a 2-C chain: C—C.
2. The butanoate R group is a 4-C chain: C—C—C—C.
3. The ending *-oate* indicates an ester functional group connecting the two R groups:

$$C-C-O-\overset{\displaystyle O}{\overset{\|}{C}}-C-C-C$$

4. Include hydrogen atoms as needed to give each carbon atom a total of four bonds:

$$CH_3CH_2O\overset{\displaystyle O}{\overset{\|}{C}}CH_2CH_2CH_3$$

Practice 8.11

Name all possible straight carbon chain isomers of ethyl butanoate.

Answer

The isomers are methyl pentanoate, propyl propanoate, butyl ethanoate, and pentyl methanoate.

As would be expected, the boiling points of the ester compounds increase with molecular weight, as shown in Table 8.4. When comparing esters with other organic compounds of similar molecular mass, the boiling point of esters is greater than that of ethers, but less than that of ketones. The polarity of the functional groups follows this same pattern: ketones > esters > ethers.

TABLE 8.4	Boiling Point of Selected Esters		
Compound	Structure	Molecular Weight	Boiling Point
Methyl methanoate	$C_2H_4O_2$	60.1	32°C
Methyl ethanoate	$C_3H_6O_2$	74.1	56°C
Ethyl methanoate	$C_3H_6O_2$	74.1	54°C
Ethyl ethanoate	$C_4H_8O_2$	88.1	77°C

You may have noticed that the term *plastic* has not been used so far in our discussion of polymers. The term *plastic* has become somewhat interchangeable with polymer in everyday use, which is not accurate. Any material that is capable of being shaped or formed is **plastic,** which is a physical property of a material. For example, modeling clay is plastic. In making plastic explosives, explosive compounds are mixed with clay and lubricants to make them moldable. So it is true that most polymers are plastic, yet not all plastic substances are polymers.

Some polymers in their pure state do not have the desired physical properties for a given application. **Plasticizers** are compounds that can be added into the reaction mixture to increase the spacing between polymer chains and thereby impart a new set of physical properties. Polyvinyl chloride—better known as PVC and marked with recycling number 3—is a good example. PVC is very rigid and strong. Manufacturers will often add an ester-based plasticizer (see margin image) to soften the PVC and make it more workable. One caution is that plasticizers can leach out of the plastic and are not permanently bonded to the polymer. That "new car smell" comes from the leaching of plasticizers out of the plastics and foams used to manufacture the car.

The diester phthalate molecule is used as a plasticizer in many polymers. The R groups can vary from short methyl groups, to long alkane chains, to complex branched groups, depending on the specific physical properties required for the polymer.

■ WORKED EXAMPLE 12

Draw the condensed structural formulas for the following compounds and place them in order from lowest boiling point to highest.

(a) 2-butanone (b) Methoxypropane (c) Methyl ethanoate (d) Pentane

SOLUTION

The boiling points of the compounds are based on the total intermolecular forces present in each molecule. The polarity of the functional groups, which directly affects the strength of the intermolecular forces present, follows the pattern: ketones > esters > ethers. Because pentane does not have an oxygen atom as the other compounds do, it would have the weakest intermolecular forces present.

(d) $CH_3CH_2CH_2CH_2CH_3$ < (b) $CH_3OCH_2CH_2CH_3$ < (c) $CH_3O\overset{\displaystyle O}{\overset{\displaystyle \|}{C}}CH_3$ < (a) $CH_3\overset{\displaystyle O}{\overset{\displaystyle \|}{C}}CH_2CH_3$

Practice 8.12

Sketch oxycodone from Figure 8.10 and polyurethane from Figure 8.9b. Circle each ketone functional group, draw a box around each ether functional group, and draw a dashed circle around the ester functional groups.

Answer

Oxycodone

Polyurethane

8.7 Amines

Organic molecules containing nitrogen belong to a category of compounds called **amines.** The amine functional group is common to many drugs, as this functional group is able to interfere with many biological processes. For example, amphetamines increase the release of neurotransmitters in the brain, and some antidepressants function by blocking the uptake of neurotransmitters in the brain. Other antidepressants work by blocking the degradation of the neurotransmitters, while caffeine and nicotine alter the firing of neurons. In all of these examples, the chemical structure of the compound contains an amine functional group.

Nitrogen differs from carbon in that it forms three bonds to other atoms. In **primary amines,** the nitrogen atom is bonded to one carbon atom and two hydrogen atoms. A primary amine has the general formula R—NH_2. Figure 8.11a shows the structure of the primary amine phenylpropanolamine (PPA), which is a stimulant and decongestant. PPA has been banned in the United States and Canada but is still widely available in Europe despite several studies that have connected PPA to increased risk of strokes in younger women.

The compound in Figure 8.11b is ephedrine and is an example of a **secondary amine.** In secondary amines, the nitrogen atom is bonded to two carbon groups and one hydrogen atom. The general formula for secondary amines is R_2NH. Ephedrine is a common decongestant and stimulant used in over-the-counter cold medicines. There is now a limit imposed by law on how much ephedrine one may buy, as it is used in the illegal manufacturing of methamphetamine.

The compound in Figure 8.11c is bromopheniramine, an example of a **tertiary amine.** Tertiary amines have three carbon atoms bonded directly to the nitrogen atom, and their general formula is R_3N. Bromopheniramine is an antihistamine used in over-the-counter cold medicines.

Methylamine (CH_5N), a primary amine.

Phenylpropanolamine

(a)

Ephedrine

(b)

Bromopheniramine

(c)

FIGURE 8.11 Examples of (a) primary, (b) secondary, and (c) tertiary amines.

Nylon was first developed in the 1930s by DuPont as a replacement for silk, which was becoming scarce as trade relations with Japan declined in the build-up to World War II. Nylon was used in the war to manufacture parachutes, flak jackets, and, of course, women's nylons. The polymerization reaction used to create nylon, shown below, is a **condensation reaction** that yields water as a product. This differs from an addition reaction in which all atoms of the reactants are part of the polymer product. The number of carbon atoms in each of the starting compounds can vary, and the resulting polymer will have differing properties. The main commercial form of nylon is called 6,6-Nylon, as both reactants have six carbon atoms, producing a polymer repeating unit with a total of 12 carbon atoms.

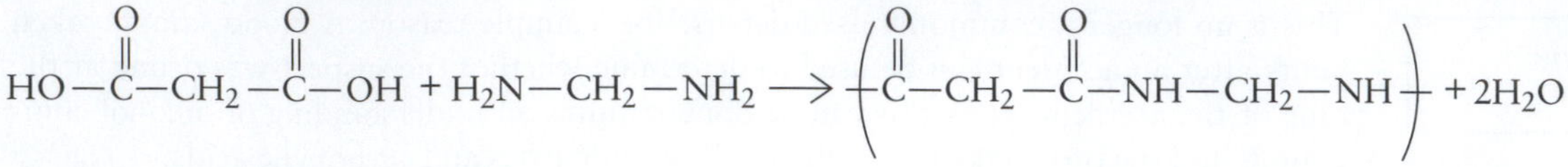

$$HO-\overset{\overset{\textstyle O}{\|}}{C}-CH_2-\overset{\overset{\textstyle O}{\|}}{C}-OH + H_2N-CH_2-NH_2 \longrightarrow \left(\overset{\overset{\textstyle O}{\|}}{C}-CH_2-\overset{\overset{\textstyle O}{\|}}{C}-NH-CH_2-NH\right) + 2H_2O$$

Amines are named by listing in alphabetical order the carbon groups attached to the nitrogen atom. The simplest primary amine is methylamine, CH_3NH_2. The simplest secondary amine is dimethylamine, $(CH_3)_2NH$, and the simplest tertiary amine is trimethylamine, $(CH_3)_3N$.

WORKED EXAMPLE 13

Draw the structures for the following compounds:

(a) Propylamine (b) Butylethylamine (c) Ethylmethylpropylamine

SOLUTION

(a) Propylamine has a 3-C chain on a nitrogen atom with two hydrogen atoms:

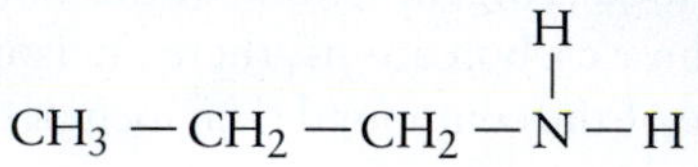

$$CH_3-CH_2-CH_2-\overset{\overset{\textstyle H}{|}}{N}-H$$

(b) Butylethylamine has a 4-C chain and a 2-C chain on a nitrogen atom with one hydrogen atom:

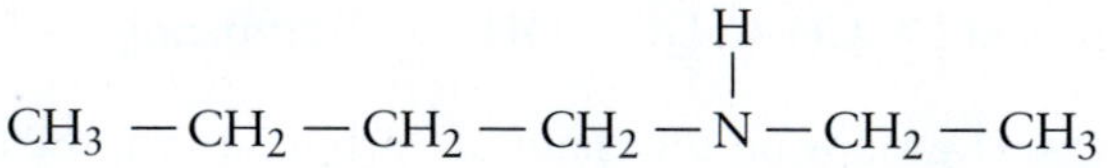

$$CH_3-CH_2-CH_2-CH_2-\overset{\overset{\textstyle H}{|}}{N}-CH_2-CH_3$$

(c) Ethylmethylpropylamine has a 2-C chain, a 1-C chain, and a 3-C chain on a nitrogen atom: $CH_3-CH_2-\overset{\overset{\textstyle CH_3}{|}}{N}-CH_2-CH_2-CH_3$

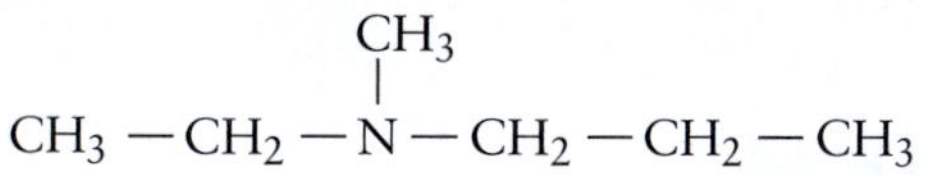

Practice 8.13

Name the following molecules:

(a) $CH_3-CH_2-CH_2-\overset{\overset{\textstyle H}{|}}{N}-CH_2-CH_2-CH_2-CH_2-CH_3$

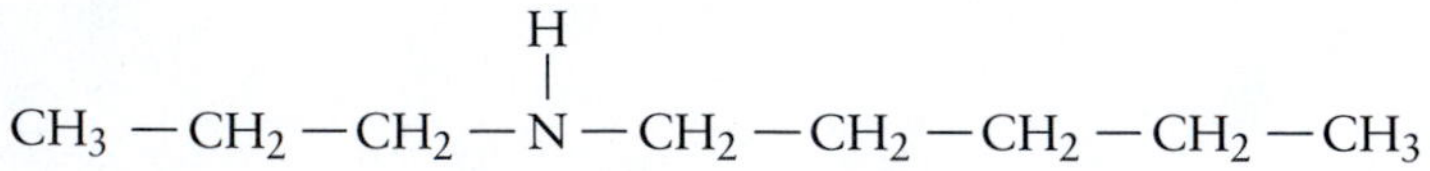

(b) $CH_3-CH_2-\overset{\overset{\textstyle CH_3}{|}}{N}-CH_2-CH_2-CH_2-CH_2-CH_2-CH_3$

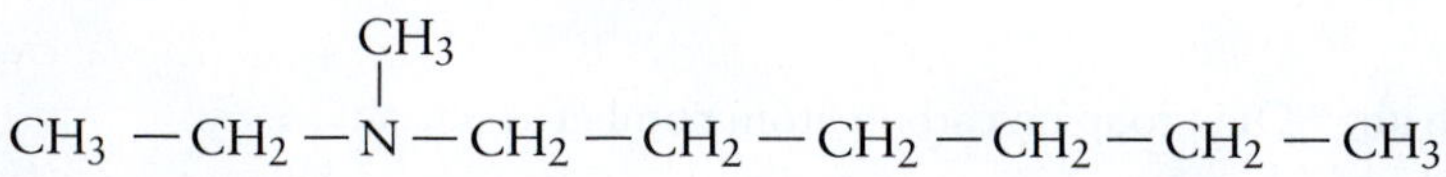

(c) $CH_3-CH_2-\overset{\overset{\textstyle H}{|}}{N}-CH_2-CH_3$

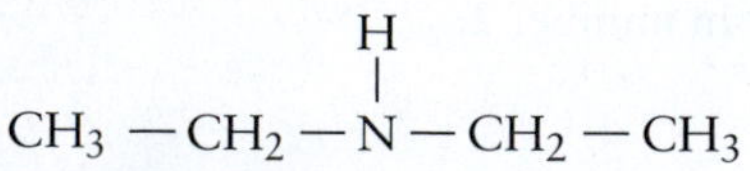

Answer

(a) Pentylpropylamine   (b) Ethylhexylmethylamine   (c) Diethylamine

8.8 Alcohols, Aldehydes, and Carboxylic Acids

Alcohol

LEARNING OBJECTIVE

Recognize, draw, and name alcohols, aldehydes, and carboxylic acids.

Breathalyzer Chemistry

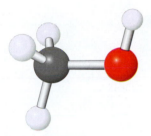

Methanol (CH_4O) is a member of the alcohol functional group. All alcohols share the R—OH structure.

In Germany, a first offender for drunk driving is heavily fined and may lose his or her driver's license permanently. Initially, the severe penalties encouraged drivers to flee the accident scene if alcohol was involved. Hours later, when police apprehended a suspect with alcohol on the breath, the suspect might claim the alcohol was consumed after the accident to calm nerves. This is no longer a commonly used defense for a simple reason: A blood sample taken hours after an accident can be used to determine whether the suspect was drunk at the time of the accident. How this can be done requires an understanding of alcohol compounds and the products of their metabolism: aldehydes and carboxylic acids.

The **alcohol** functional group is —OH. The simplest alcohol is methanol (CH_3OH), in which a methyl group (—CH_3) is bonded to the oxygen of the —OH group. Ethanol (CH_3CH_2OH) is the main compound consumed in alcoholic beverages. It should be pointed out that almost all alcohol compounds are toxic to varying degrees and can easily lead to death if consumed in quantities that exceed the toxic threshold.

While ethanol is the major component in alcohol beverages, there can be other compounds that contain the alcohol functional group in the beverage. More than 800 different compounds have been identified in various alcoholic beverages, which contribute to the color and taste of the beverage. The term **congener** refers to any compound that shares the same functional group with another compound. The metabolism of ethanol and alcohol congeners produces compounds with greater toxicity, called *aldehydes*.

Alcohols are named by modifying the ending of the alkane name with *-ol*. For alcohol compounds with at least three carbon atoms, there are isomers based on the location of the —OH group. The name of the compound then includes the lowest carbon number for the —OH group. For example:

$$1\text{-propanol} = CH_3CH_2CH_2OH \qquad 2\text{-propanol} = \overset{\displaystyle OH}{\underset{\displaystyle |}{CH_3CHCH_3}}$$

If there are two —OH groups on the chain of carbon atoms, the ending is *-diol*, as in 1,2-ethanediol, the toxic compound present in antifreeze.

■ WORKED EXAMPLE 14

Write the structural formulas for the following compounds that are all alcohol congeners found in alcoholic beverages:

(a) 1-propanol

(b) 2-butanol

(c) Methanol

SOLUTION

(a) A 3-C chain with an —OH group on carbon atom number 1:
   HO—CH_2—CH_2—CH_3

(b) A 4-C chain with an —OH group on carbon atom number 2:
   $$CH_3-\underset{\displaystyle \underset{\textstyle OH}{|}}{CH}-CH_2-CH_3$$

(c) One carbon atom with an —OH group: CH_3—OH

Practice 8.14

Name the following compounds:

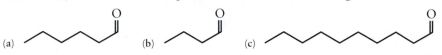

(a)   (b)   (c)

Answer

☐ (a) 4-heptanol   (b) 1-propanol   (c) 1-hexanol

Aldehydes

The initial step in the metabolism of alcohol is to produce a compound with an **aldehyde** functional group. The aldehyde functional group consists of a terminal carbon atom having a double bond to an oxygen atom and a single bond to hydrogen. The aldehyde

$$\text{O} \atop \| $$

general formula is R—CH. The simplest aldehyde is methanal (HCHO), which consists of one carbon atom with a double bond to oxygen and a single bond to each hydrogen. The common name for methanal is *formaldehyde*, the compound used in embalming fluid. Aldehydes are named by modifying the ending of the carbon chain name with -*al*.

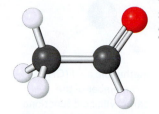

Ethanal (C_2H_4O) is a member of the aldehyde functional group. All aldehydes share the R$-$CH structure.

■ WORKED EXAMPLE 15

Write the structures of the following aldehydes, all of which are produced during the metabolism of alcoholic beverages.

(a) Propanal   (b) Butanal   (c) Methanal

SOLUTION

(a) A 3-C chain ending with C=O: CH_3CH_2CH with O double-bonded

(b) A 4-C chain ending with C=O: $CH_3CH_2CH_2CH$ with O double-bonded

(c) A 1-C chain ending with C=O: HCH with O double-bonded

Practice 8.15

Name the following compounds, which are produced during decomposition of the human body and are studied as possible markers for electronic grave site detection.

(a)   (b)   (c)

Answer

☐ (a) Hexanal   (b) Butanal   (c) Decanal

Ethanal is the first compound produced by the metabolism of ethanol and is extremely poisonous. Ethanal is thought to be responsible for the flushed complexion, headaches, dizziness, and nausea that accompany alcohol consumption and hangovers. Hangovers are more severe when alcoholic beverages with a high level of alcohol congeners are consumed because the congeners remain in the body longer. Many of the alcohol congeners are soluble in fat due to the nonpolar carbon chains that interact with the nonpolar fat molecules. Therefore, the congeners accumulate in fat during periods of high alcohol consumption and are slowly released after the alcohol consumption ceases. As the congeners are released back into the bloodstream, they are metabolized to their corresponding toxic aldehyde compounds.

Fortunately, the aldehyde compounds are further metabolized in the body to less troublesome compounds called *carboxylic acids*. The prescription compound Antabuse is used as a deterrent to drinking because it prevents the metabolism of aldehydes to carboxylic acids, permitting a buildup of aldehydes that very quickly cause an individual to become nauseated and have severe headaches.

Carboxylic Acids

The **carboxylic acid** functional group consists of a terminal carbon atom that contains both a double-bonded oxygen atom and an —OH group. The general formula is

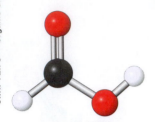

Methanoic acid (CH_2O_2) is a member of the carboxylic acid functional group. All carboxylic acids share the R—COOH

$$R-\overset{\displaystyle O}{\overset{\displaystyle \|}{C}}-OH$$

or R−C−OH structure.

represented as $R-\overset{O}{\overset{\|}{C}}-OH$ or R—COOH. The terminal —OH group is *not* an alcohol group, nor is it a hydroxide ion. Carboxylic acids are weak acids that partially dissociate to form H^+ and $R-COO^-$ ions in solution. The simplest carboxylic acid is methanoic acid and contains one carbon atom. The metabolism of ethanol (CH_3CH_2OH) first involves its conversion to ethanal (CH_3CHO), which is an aldehyde. The ethanal in turn is converted to ethanoic acid (CH_3COOH), which is a carboxylic acid. You previously learned the common name of ethanoic acid—*acetic acid*, the main ingredient in vinegar.

The name of the carboxylic acid compounds is created by modifying the ending of the name of the carbon chain with *-oic* and adding the word *acid*, as in *methanoic acid*.

■ WORKED EXAMPLE 16

Write the line structures for the following compounds.
(a) 1,2-ethanediol
(b) Propanoic acid
(c) 2-methylpropanedioic acid

SOLUTION

(a) A 2-C chain with an —OH on each carbon atom:

(b) A 3-C chain with a carboxylic acid group:

(c) A 3-C chain with —CH$_3$ on carbon number 2 and —COOH on

carbon numbers 1 and 3:

Practice 8.16

Why is it impossible to have the compound 3-hexanoic acid?

Answer

The carboxylic acid functional group must occur on a terminal carbon atom.

The final step in the metabolism of alcohol is to form carbon dioxide and water. The human body will metabolize ethanol, which is present at much higher concentrations than the alcohol congeners, before it will metabolize the congeners. Therefore, the amount of alcohol consumed hours earlier can be determined by measuring the relative amounts of the alcohol congeners remaining in the blood.

Table 8.5 provides a summary of organic compound functional groups and their nomenclature.

TABLE 8.5	Summary of Organic Compounds			
Class	Formula	Example	Naming	Example Name
Alkanes	C_nH_{2n+2}	$CH_3CH_2CH_3$	C chain length + -ane	Propane
Alkenes	C_nH_{2n}	$CH_2\!=\!CH_2CH_3$	C=C location + C chain length + -ene	1-propene (or propene)*
Alkynes	C_nH_{2n-2}	$CH\!\equiv\!CCH_3$	C≡C location + C chain length + -yne	1-propyne (or propyne)*
Ethers	R—O—R	$CH_3OCH_2CH_3$	Short C chain + -oxy + long C chain + -ane	Methoxyethane
Ketones	R—C(=O)—R	CH_3CCH_3 (C=O)	C=O location + C chain length + -one	2-propanone (or propanone)*
Esters	R—O—C(=O)—R	CH_3OCCH_3 (C=O)	C chain not touching C=O, C chain length + -oate	Methyl ethanoate
Amines	R_3N	$CH_3NHCH_2CH_3$	Name each C chain (alphabetical order) + -amine	Ethylmethylamine
Alcohols	R—OH	CH_3OH	C chain length + -ol	Methanol
Aldehydes	R—CH(=O)	CH_3CH (C=O)	C chain length + -al	Ethanal
Carboxylic acids	R—C(=O)—OH	CH_3COH (C=O)	C chain length + -oic + acid	Ethanoic acid

*When there is only one possible location for a functional group, providing the number is unnecessary.

8.9 How to Extract Organic Compounds: Solubility and Acid–Base Properties

We have discussed carboxylic acids as examples of organic acids, but there are also organic compounds, such as organic amines, that function as bases. According to the definitions previously given, acids are compounds that produce free H^+ ions in an aqueous solution, and bases are compounds that produce free OH^- ions in an aqueous solution. This is often referred to as the **Arrhenius definition** of acids and bases.

LEARNING OBJECTIVE

Identify and describe how to separate acidic, basic, and neutral organic compounds dissolved in organic solvents.

However, we can describe acids and bases in a broader way by using the **Brønsted-Lowry definition:** An acid is any compound that can donate a hydrogen ion; a base is any compound that can accept a hydrogen ion. The Brønsted-Lowry acid–base definition expands the number of compounds that we consider to be acids or bases. All compounds that were identified as acids and bases from the Arrhenius definition still fit the new Brønsted-Lowry definition.

Why do amines fit the definition of a base? If we examine the valence electron arrangement of a nitrogen atom, we see that nitrogen has five valence electrons available to form bonds. However, in all of the amine structures, only three covalent bonds are found on nitrogen. Because each atom in a bond shares one electron, nitrogen is sharing only three of its five valence electrons. The other two are called **lone pair electrons**—electrons that are paired up in an orbital and are not part of a covalent bond. The lone pair electrons impart a highly negative region around the nitrogen atom that can attract a positive ion such as the H$^+$ ion. Figure 8.12 shows the ability of an amine to act as a Brønsted-Lowry base, and a carboxylic acid to act as a Brønsted-Lowry acid.

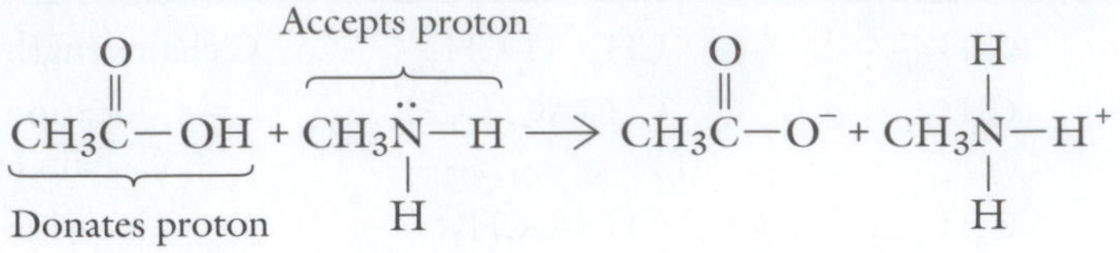

$$\underbrace{CH_3\overset{\overset{\displaystyle O}{\|}}{C}-OH}_{\text{Donates proton}} + CH_3\overset{\overset{\displaystyle \cdot\cdot}{N}-H}{\underset{H}{|}} \longrightarrow CH_3\overset{\overset{\displaystyle O}{\|}}{C}-O^- + CH_3\overset{\overset{\displaystyle H}{|}}{\underset{H}{N}}-H^+$$

$$\overbrace{}^{\text{Accepts proton}}$$

FIGURE 8.12 Brønsted-Lowry acid–base reaction.

Carboxylic acids, especially those with long carbon chains, tend to dissolve in non-polar organic solvents such as vegetable oil. However, if you deprotonate (remove a proton from) the carboxylic acid and convert it into an anion, it becomes soluble in water and not in the organic solvent. The amine compounds also dissolve in organic solvents when in the neutral state, but when protonated (given a proton) to form a cation, amines become insoluble in organic solvents and soluble in water. Knowledge of these principles of solubility and acid–base chemistry allows a scientist to use the pH of a solvent like a switch to control the solubility of compounds in either organic solvents or water.

A procedure for the extraction of acidic, basic, and neutral organic molecules from an organic solvent is shown in Figure 8.13. The pear-shaped container is called a *separatory funnel* and is specifically designed for this type of experimental procedure. All the molecules in the organic solvent are initially neutral.

• The first step is to add to the organic solvent either an acid such as HCl or a base such as NaOH. An acid will protonate an amine and a base will deprotonate a carboxylic acid. Because the HCl and NaOH are aqueous reagents, they will not mix with the nonpolar organic solvent and two layers will form, as shown in Figure 8.13a.

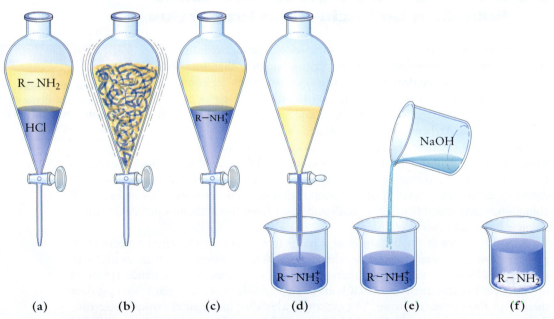

(a) (b) (c) (d) (e) (f)

FIGURE 8.13 Extraction of acidic or basic organic compounds.

- For a weak acid or a weak base dissolved in the organic solvent to react with either the HCl or the NaOH in the separate aqueous layer, the solution must be vigorously shaken, as shown in Figure 8.13b.

- As the organic molecules are converted to their ionic form, they become insoluble in the organic solvent and soluble in the aqueous solvent. The two layers are then allowed to settle and separate, as depicted in Figure 8.13c.

- In the next step, the more dense solution (typically the aqueous solution) is drained from the bottom of the separatory funnel, as shown in Figure 8.13d. Recall that the aqueous solution now contains the ionic form of the organic compound.

- Figure 8.13e shows the neutralization of the solution (with acid or base as needed). This causes the ionic organic compound to become neutral.

- When the ionic form of the organic compound is neutralized, it is insoluble in aqueous solution. The neutral molecule will crystallize out of solution and can easily be filtered, as shown in Figure 8.13f.

- The neutral organic molecules in the organic solvent always stay there and can typically be recovered by evaporating the organic solvent and leaving the neutral organic molecules behind.

■ WORKED EXAMPLE 17

Many over-the-counter pain medicines contain aspirin and caffeine. How do you isolate the two compounds from a tablet? The structures of aspirin and caffeine are shown in the figure below.

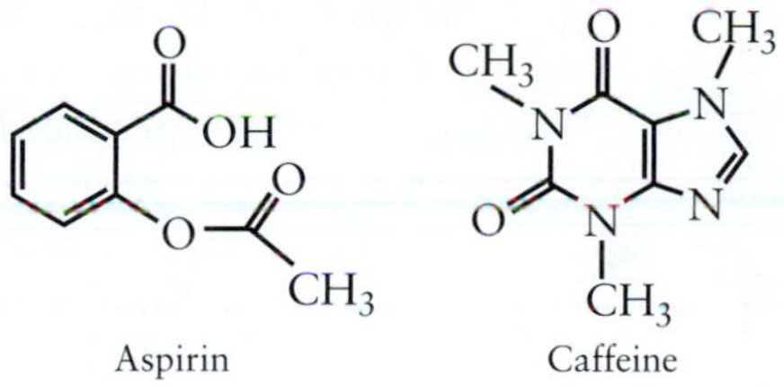

Aspirin   Caffeine

SOLUTION

Dissolve the tablet in an organic solvent. The aspirin molecule can be extracted into an aqueous layer by the addition of NaOH, which deprotonates the carboxylic acid functional group. Caffeine can be extracted into an aqueous layer by the addition of HCl, which protonates the tertiary nitrogen atoms. The amine cations and the carboxylic acid anions can be removed by neutralizing the solutions.

Practice 8.17

Oil of Jasmine is an extract from the jasmine flower. One of the main aromatic chemical compounds is called indole. Given the structure of indole below, what type of an amine is it? Would you add an acid or a base to indole to ionize it?

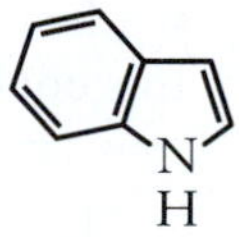

Answer

The indole molecule is a secondary amine, which could be protonated by the addition of an acid.

Infrared Spectroscopy

Infrared radiation can be used to positively identify a chemical compound and is commonly used for analyzing organic compounds such as polymers and prescription drugs. To understand how infrared radiation is used to identify compounds, it is necessary to look into what happens when infrared light strikes an object.

The infrared region of the spectrum begins where the red region of visible light ends. Infrared radiation is invisible to the human eye, but we sense it as heat when it strikes our skin. Obviously, the molecules that make up our skin are affected by infrared radiation, but how? When infrared radiation interacts with our skin, the bonds between the molecules of our skin absorb the energy, and the temperature of the skin increases.

Likewise, the bonds that form between atoms in a molecule are capable of absorbing infrared radiation. The bonds are not rigid, inflexible connections that force the atoms into a single fixed position. Rather, the bonds act more like springs that keep the two atoms together. The bonds can therefore be compressed, stretched, or bent just like a spring. Figure 8.14 shows the three main vibrations that can occur in bonds between atoms when infrared radiation strikes a molecule. A molecule will absorb the infrared radiation if the energy of the radiation matches the energy of one of the vibrations.

There are two main regions of interest when measuring the absorption of infrared radiation: the **functional group region** (4000 cm^{-1} to 1300 cm^{-1}) and the **fingerprint region** (1300 cm^{-1} to 900 cm^{-1}). Note that for infrared radiation, the traditional unit used to specify the frequency or wavelength of the radiation is the *wave number*, which is the number of waves per centimeter.

The functional group region shows characteristic absorption patterns or bands for each of the functional groups covered in this chapter. For example, a carbon atom with a single bond to a hydrogen atom (C—H) absorbs infrared radiation in the range of 2850 to 2980 cm^{-1}, whereas an oxygen atom single bonded to a hydrogen atom (O—H) absorbs radiation from 3300 to 3550 cm^{-1}.

The fingerprint region provides a pattern of absorption bands for each molecule and is created from all of the atoms within the molecule undergoing vibrations. This creates a unique pattern not reproduced by any other molecule.

The energy associated with each type of vibration from Figure 8.14 depends on what type of bond exists between the atoms (single, double, or triple) and on what types of atoms are involved in the bonding. Figure 8.15 shows the infrared spectrum of butane and 1-butanol. Notice the change in both the functional group region and the fingerprint region, caused by the addition of a simple alcohol functional group.

In the laboratory, the infrared spectrum is obtained using an instrument called a **Fourier Transform Infrared Spectrometer (FTIR)**. The Fourier transform is a complex mathematical process (beyond the scope of this book) used to obtain the signal from the raw data. For analysis of a compound, the sample is placed inside the instrument and the spectrum is obtained within minutes. The experimentally obtained spectrum is compared with a library of standard spectra

Symmetrical stretch

Unsymmetrical stretch

Bend

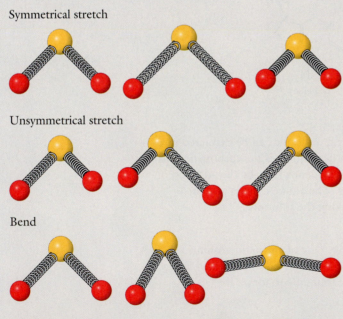

FIGURE 8.14 The infrared active vibrations of a molecule. The symmetrical stretch involves the simultaneous stretching and then compression of bonds. The unsymmetrical stretching involves the simultaneous compression of one bond and stretching of another. The bending motion involves two atoms flexing in toward each other and then outward away from each other.

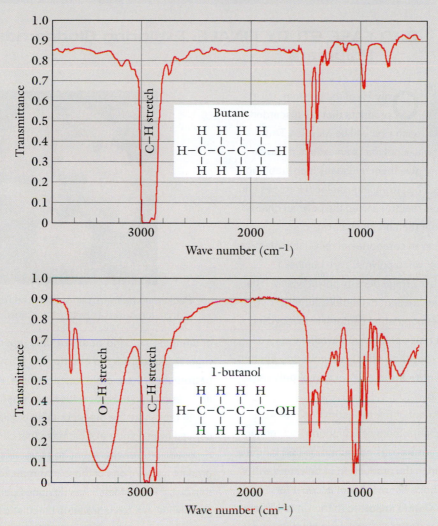

FIGURE 8.15 Infrared absorption spectrum of butane and 1-butanol. As the infrared radiation striking the sample is scanned through the functional group region (4000 cm^{-1} to 1300 cm^{-1}), any functional groups within the compound will absorb energy. This creates a distinctive pattern, as can be seen for both the C—H stretch and the O—H stretch. Each compound creates a unique pattern in the fingerprint region (1300 cm^{-1} to 900 cm^{-1}).

prepared by the manufacturer of the instrument. A tentative match can be made, provided the spectrum of the compound is found within the library.

The FTIR system is versatile in that it can be used to measure the spectrum of solids, gases, and liquids. One limitation of FTIR analysis is that it does not work well for complicated mixtures, because the resulting spectrum reflects the absorption of all compounds within the mixture.

A typical **FTIR** system. (Perkin Elmer)

8.10 CASE STUDY FINALE: Exploring Biodegradable Polymers

Only 7% of the polymer products that are discarded each year will be recycled. The same properties that make polymers an excellent choice for packaging material give them unusually high stabilities, especially in landfills. The strong intermolecular forces and resistance to chemical reactions keep the polymer chains bonded together rather than breaking down.

While there are seven categories of recyclable plastics, the vast majority of recycled polymers are number 1 (polyester) and number 2 (high density polyethylene). The remaining five classes are sent to landfills, for the most part, except in a larger population center where recycling facilities that accept those plastics are available.

To assist in the recycling effort, a numbering system was adopted for identification and sorting of polymers, as shown in Table 8.6.

The U.S. Department of Energy estimates that 4.6% of the total domestic petroleum consumption (331 million barrels) and 1.5% of natural gas consumption (335 billion cubic feet) were used in the production of polymers within the United States. These figures do not account for polymer products that were imported to the United States. The majority of this material will end up in landfills.

One area of research that could lessen fossil-fuel consumption in the production of polymers is the use of *bioplastics.* Bioplastics are created from such materials as vegetable oils or plant starches. One example that you most likely have come across is packaging peanuts derived from corn starch. It is possible to synthesize many of the same polymers using a renewable resource rather than petroleum, thereby reducing total consumption of fossil fuels.

The current fate of most polymers—ending up in a landfill—is not as easy to address. Improving recycling efforts would have an immediate impact, and there is plenty of room for improvement in that regard. However, scientists are also working on developing *biodegradable plastics* that will degrade under normal environmental

Packaging peanuts can be created from corn starch, a renewable resource. (Christina Micek)

or composting conditions within a reasonable span of time. A biodegradable plastic may be synthesized from petroleum or from a biological source, which creates yet another choice that the educated consumer must make.

In fact, there are biodegradable bioplastics, materials derived from a source of biomass that can degrade when exposed in the environment. You may even be using a product made of one of these materials to take notes for this class. Paper Mate® has designed pens and mechanical pencils using a biodegradable bioplastic that will decompose over time after use. The polymer used to create the biodegradable bioplastic is called polyhydroxybutyrate (PHB) and has the following structure:

$$\left(\!\!\begin{array}{c} CH_3 \\ | \\ CH-CH_2-\overset{\displaystyle O}{\overset{\|}{C}}-O \end{array}\!\!\right)_{\!n}$$

Polyhydroxybutyrate is a polyester polymer formed by a condensation reaction by bacteria that are under stress. The bacteria consume glucose or starch and store that energy as PHB. The commercial usefulness of PHB stems from the fact that under standard conditions, the polymer is rugged and stable, yet when it is exposed to environmental or composting conditions, the bacteria in the soil can break down the polymer into its components.

TABLE 8.6 Recycling Codes for Commonly Used Consumer Products

Identification	Polymer Type	Polymer Structure	Representative Uses
01 PET	Polyester	$\left(\!O\!-\!\overset{\displaystyle O}{\overset{\|}{C}}\!-\!\bigcirc\!-\!\overset{\displaystyle O}{\overset{\|}{C}}\!-\!O\!-\!CH_2\!-\!CH_2\!\right)_{\!n}$	Disposable water bottles, disposable soda bottles, clothing
02 HDPE	High density polyethylene	$\left(\!\!\begin{array}{cc} H & H \\ \| & \| \\ C\!-\!C \\ \| & \| \\ H & H \end{array}\!\!\right)_{\!n}$	Rigid food containers and bottles
03 PVC	Polyvinyl chloride	$\left(\!\!\begin{array}{cc} H & Cl \\ \| & \| \\ C\!-\!C \\ \| & \| \\ H & H \end{array}\!\!\right)_{\!n}$	Plumbing pipes, wire insulation, vinyl records
04 LDPE	Low density polyethylene	$\left(\!\!\begin{array}{cc} H & H \\ \| & \| \\ C\!-\!C \\ \| & \| \\ H & R \end{array}\!\!\right)_{\!n}$ R=H, CH$_3$, CH$_2$CH$_3$, (CH$_2$)$_x$CH$_3$ (x=2-7)	Frozen-food bags, flexible (squeeze) food bottles
05 PP	Polypropylene	$\left(\!\!\begin{array}{cc} CH_3 & H \\ \| & \| \\ C\!-\!C \\ \| & \| \\ H & H \end{array}\!\!\right)_{\!n}$	Yogurt containers, microwave-safe containers
06 PS	Polystyrene	$\left(\!\!-CH_2\!-\!CH\!-\!\!\right)_{\!n}$ (with phenyl ring)	Plastic silverware, insulated coffee cups, egg cartons
07 O	All other polymer products such as polycarbonates and nylon	Polycarbonate: $\left(\!\bigcirc\!\!\overset{\displaystyle CH_3}{\underset{\displaystyle CH_3}{\overset{\|}{\underset{\|}{C}}}}\!\!\bigcirc\!-\!O\!-\!\overset{\displaystyle O}{\overset{\|}{C}}\!-\!O\!\right)_{\!n}$ Nylon: $\left(\!\overset{\displaystyle O}{\overset{\|}{C}}\!-\!CH_2\!-\!\overset{\displaystyle O}{\overset{\|}{C}}\!-\!NH\!-\!CH_2\!-\!NH\!\right)_{\!n}$	Polycarbonates: DVD and CD discs, eyeglass lenses Nylon: clothing, ropes, parachutes

CHAPTER SUMMARY

- Alkanes consist of chains of single-bonded carbon atoms with a sufficient number of hydrogen atoms to satisfy the generic formula C_nH_{2n+2}. The number of carbon atoms in the chain is expressed by beginning the name with *meth-* for one carbon, *eth-* for two, *prop-* for three, and so on, as shown in Table 8.1. The ending *-ane* is added to alkane names. High density polyethylene (HDPE) is a common plastic consisting of long alkane chains.

- Compounds in which a carbon atom forms a double bond to another carbon atom belong to the alkene class of compounds. A closely related class of compounds is the alkynes, in which two carbon atoms form a triple bond.

- A polymer is a long molecule made up of many smaller repeating units. A monomer compound contains a reactive carbon–carbon double bond capable of reacting with other monomers to form a long polymer chain. When a double bond is broken in the monomer, it will form new single bonds to other monomer compounds. This method is called an addition reaction, as all atoms in the monomers are incorporated into the polymer. Other polymers form from a condensation reaction, which is characterized by the formation of water as another product with the polymer.

- The location of the double or triple bond along the carbon chain can create constitutional isomers, which are different molecules that have the same chemical formula but a different structure. Other types of constitutional isomers are created when the carbon chains form carbon atom branches. The names of isomers specify the number of the carbon atom where the double/triple bond or the carbon branch is located.

- Alkane compounds can also form ring structures called cycloalkanes, made entirely from single-bonded carbon atoms. Another type of common cyclic compound is benzene, which is represented by a six-carbon ring with alternating double and single bonds. The benzene ring can be modified by the attachment of an —OH group to form phenol, a —CH_3 group to form toluene, and an —NH_2 group to form aniline.

- Organic compounds fall into various classes based on the presence of functional groups that include the ketone, ether, ester, amine, alcohol, aldehyde, and carboxylic acid groups.

- A plastic is any material that can be shaped or formed. Polymers are simply one type of material that is plastic. Plasticizers are compounds mixed into a polymer to make the resulting bulk physical properties more useful.

- Amines are an example of an organic Brønsted-Lowry base (accepts a hydrogen ion), and carboxylic acids are an example of an organic Brønsted-Lowry acid (donates a hydrogen ion). These properties can be exploited to extract amines or carboxylic acids from mixtures. The ionic forms of the compounds are soluble in water, whereas the neutral forms are soluble in nonpolar organic solvents.

TABLE 8.5 Summary of Organic Compounds

Class	Formula	Example
Alkanes	C_nH_{2n+2}	$CH_3CH_2CH_3$
Alkenes	C_nH_{2n}	$CH_2{=}CH_2CH_3$
Alkynes	C_nH_{2n-2}	$CH{\equiv}CCH_3$
Ethers	$R{-}O{-}R$	$CH_3OCH_2CH_3$
Ketones	$R{-}\overset{\displaystyle O}{\overset{\|}{C}}{-}R$	$CH_3\overset{\displaystyle O}{\overset{\|}{C}}CH_3$
Esters	$R{-}O{-}\overset{\displaystyle O}{\overset{\|}{C}}{-}R$	$CH_3O\overset{\displaystyle O}{\overset{\|}{C}}CH_3$
Amines	R_3N	$CH_3NHCH_2CH_3$
Alcohols	$R{-}OH$	CH_3OH
Aldehydes	$R{-}\overset{\displaystyle O}{\overset{\|}{C}}H$	$CH_3\overset{\displaystyle O}{\overset{\|}{C}}H$
Carboxylic acids	$R{-}\overset{\displaystyle O}{\overset{\|}{C}}{-}OH$	$CH_3\overset{\displaystyle O}{\overset{\|}{C}}OH$

KEY TERMS

alcohol (p. 236)
aldehyde (p. 237)
alkane (p. 219)
alkene (p. 222)
alkyne (p. 224)
amine (p. 234)
aromatic compound (p. 228)
Arrhenius acid (p. 239)
Arrhenius base (p. 239)
benzene (p. 227)
Brønsted-Lowry acid (p. 239)
Brønsted-Lowry base (p. 239)
carboxylic acid (p. 238)

condensation reaction (p. 235)
condensed structural formula
    (p. 221)
congener (p. 236)
cycloalkane (p. 227)
ester (p. 232)
ether (p. 229)
fingerprint region (p. 242)
Fourier Transform Infrared
    Spectrometer (FTIR) (p. 242)
functional group (p. 229)
functional group region (p. 242)
inorganic compound (p. 218)

isomer (p. 222)
ketone (p. 230)
line structure (p. 221)
lone pair electrons (p. 239)
organic compound (p. 218)
plastic (p. 233)
plasticizers (p. 233)
polymer (p. 224)
primary amine (p. 234)
resonance structure (p. 228)
secondary amine (p. 234)
structural formula (p. 220)
tertiary amine (p. 234)

MAKING MORE CONNECTIONS: Additional Readings, Resources, and References

For more information about our consumption and
    recycling of polymers:
    http://www.epa.gov/epawaste/nonhaz/municipal/
    index.htm
    http://www.eia.gov/tools/faqs/faq.cfm?id=34&t=6

For multiple articles regarding the analysis and use of
    alcohol congeners, see J. A. Siegel, ed., *Encyclope-
    dia of Forensic Sciences*, San Diego: Academic
    Press, 2000.
Information on the CAS Registry can be found at
    www.cas.org/cgi-bin/cas/regreport.pl.

REVIEW QUESTIONS AND PROBLEMS

Questions

1. Discuss the historical origins of the terms *organic* and *inorganic* for classifying chemical compounds. (8.1)

2. Discuss the modern definition of the term *organic* for classifying chemical compounds. (8.1)

3. What is the generic formula for all alkanes? What are some common uses of alkanes? (8.2)

4. What is the difference between the alkanes, alkenes, and alkynes? (8.2–8.3)

5. What are the generic formulas of the alkenes and alkynes? (8.3)

6. What is the difference between the chemical formula and the condensed formula of organic compounds? What information does each provide? (8.2)

7. Explain how to interpret the line structures for alkanes, alkenes, and alkynes. (8.2–8.3)

8. What do isomers represent? (8.3)

9. Will two isomers have the same physical properties? Explain why or why not. (8.4)

10. What is a constitutional isomer? Give an example of two compounds that are constitutional isomers. (8.3)

11. What is a stereoisomer? Give an example of two compounds that are constitutional isomers. (8.3)

12. Can you determine the structure of an organic molecule simply by looking at the chemical formula? Why or why not? (8.2–8.3)

13. What is the difference between a monomer and a polymer? (8.3)

14. Using ethene as a monomer, explain how a polymer forms from an addition reaction. (8.3)

15. What is a copolymer? (8.4)

16. What is the structural difference between the high density polyethylene (HDPE) polymer and the low density polyethylene (LDPE) polymer? What effect does this difference have on the physical properties of the polymer? (8.4)

17. What does it mean when a compound has several resonance structures? (8.5)

18. What do the alkanes and cycloalkanes have in common? How do they differ? (8.5)

19. Explain the nature of the single and double bonds found in benzene. (8.5)

20. What is a functional group? What is the difference between an ether, a ketone, and an ester? (8.6)

21. How are ketone, ether, and ester functional groups similar? How are they different? (8.6)
22. What is the difference between a plastic and a polymer? (8.6)
23. What is a plasticizer and why is it added to a polymer? (8.6)
24. How are the aldehyde functional group and the carboxylic acid functional group similar? How are they different? (8.8)
25. What is the amine functional group? What is the difference between a primary, secondary, and tertiary amine? (8.7)
26. What is a condensation reaction? (8.7)
27. Is water produced or consumed in a condensation reaction? (8.7)
28. What is the alcohol functional group? Are all alcohols safe to consume? (8.8)
29. What are congeners? Give several examples of alcohol congeners. (8.8)
30. Explain the difference between a Brønsted-Lowry acid and an Arrhenius acid. Give an example of each. (8.9)
31. Explain the difference between a Brønsted-Lowry base and an Arrhenius base. Give an example of each. (8.9)
32. Explain how you can use the acid-base properties of certain functional groups for extraction of an organic substance from a solution. (8.9)

Problems

33. Name the following straight-chain alkanes. (8.2)
    (a) C_3H_8 (c) CH_4
    (b) C_4H_{10} (d) C_7H_{16}
34. Name the following straight-chain alkanes. (8.2)
    (a) C_8H_{18} (c) C_2H_6
    (b) C_6H_{14} (d) C_5H_{12}
35. What is the chemical formula for each of the following compounds? (8.2)
    (a) Hexane   (c) Pentane
    (b) Butane   (d) Decane
36. What is the chemical formula for each of the following compounds? (8.2)
    (a) Octane   (c) Nonane
    (b) Propane   (d) Heptane
37. Draw the condensed structural formula for the straight-chain alkanes listed in problem 23. (8.2)
38. Draw the condensed structural formula for the straight-chain alkanes listed in problem 24. (8.2)
39. Draw the line structure for each of the compounds listed in problem 25. (8.2)
40. Draw the line structure for each of the following compounds listed in problem 26. (8.2)

41. Write the chemical formula, condensed structural formula, and line structure for each of the following compounds. (8.3)
    (a) Propyne
    (b) 3-nonene
    (c) 2-heptene
    (d) 1-pentyne
42. Write the chemical formula, condensed structural formula, and line structure for each of the following compounds. (8.3)
    (a) 2-hexyne
    (b) 3-octyne
    (c) 3-decene
    (d) 1-butene
43. How many unique isomers can be made by altering the placement of the multiple bond in each of the following compounds? (8.3)
    (a) Butyne
    (b) Octyne
    (c) Pentene
    (d) Propyne
44. How many unique isomers can be made by altering the placement of the multiple bond in each of the following? (8.3)
    (a) Butene
    (b) Decyne
    (c) Pentyne
    (d) Hexene
45. Name the following compounds. (8.4)
    (a)
    (b)
    (c)
    (d)
46. Name the following compounds. (8.4)
    (a)
    (b)
    (c)
    (d)

47. Draw the line structure for each of the following compounds. (8.4)
    (a) 2-methylheptane
    (b) 4-ethyldecane
    (c) 2-methylpentane
    (d) 3-ethyloctane

48. Draw the line structure for each of the following compounds. (8.4)
    (a) 3-methylhexane
    (b) 2-methyloctane
    (c) 3-ethylheptane
    (d) 4-ethyloctane

49. Write the chemical formula and line structure for each of the following compounds. (8.5)
    (a) Benzene   (c) Cyclobutane
    (b) Phenol   (d) Cyclohexane

50. Write the chemical formula and line structure for each of the following compounds. (8.5)
    (a) Aniline   (c) Cyclopentane
    (b) Toluene   (d) Cyclopropane

51. Draw the condensed structural formula for each of the following compounds. (8.6)
    (a) Methoxypropane
    (b) Ethoxyethane
    (c) Butoxyhexane
    (d) Propoxybutane

52. Draw the condensed structural formula for each of the following compounds. (8.6)
    (a) Ethoxypropane
    (b) Methoxypentane
    (c) Propoxypentane
    (d) Butoxybutane

53. Draw the condensed structural formula for each of the following compounds. (8.6)
    (a) 2-hexanone   (c) 3-heptanone
    (b) 2-pentanone   (d) 4-decanone

54. Draw the condensed structural formula for each of the following compounds. (8.6)
    (a) 3-pentanone   (c) 2-heptanone
    (b) Propanone   (d) 3-octanone

55. Draw the condensed structural formula for each of the following compounds. (8.6)
    (a) Methyl propanoate
    (b) Ethyl ethanoate
    (c) Ethyl propanoate
    (d) Propyl ethanoate

56. Draw the condensed structural formula for each of the following compounds. (8.6)
    (a) Ethyl pentanoate
    (b) Butyl propanoate
    (c) Methyl butanoate
    (d) Ethyl butanoate

57. Name the following compounds. (8.6)

(a) $CH_3-CH_2-\overset{\overset{\displaystyle O}{\|}}{C}-CH_2-CH_2-CH_3$

(b) $CH_3-CH_2-O-CH_2-CH_2-CH_2-CH_3$

(c) $CH_3-CH_2-O-\overset{\overset{\displaystyle O}{\|}}{C}-CH_2-CH_2-CH_2-CH_3$

(d) $CH_3-CH_2-CH_2-\overset{\overset{\displaystyle O}{\|}}{C}-CH_3$

58. Name the following compounds. (8.6)

(a) $CH_3-\overset{\overset{\displaystyle O}{\|}}{C}-CH_2-CH_2-CH_2-CH_3$

(b) $CH_3-CH_2-CH_2-CH_2-O-CH_2-CH_3$

(c) $CH_3-CH_2-CH_2-CH_2-CH_2-O-\overset{\overset{\displaystyle O}{\|}}{C}-CH_2-CH_3$

(d) $CH_3-CH_2-\overset{\overset{\displaystyle O}{\|}}{C}-CH_2-CH_3$

59. Draw the condensed structural formula for each of the following compounds. (8.7)
    (a) Pentylamine
    (b) Ethylpropylamine
    (c) Trimethylamine
    (d) Butylmethylpropylamine

60. Draw the condensed structural formula for each of the following compounds. (8.7)
    (a) Propylamine
    (b) Dibutylmethylamine
    (c) Dipropylamine
    (d) Ethylmethylpropylamine

61. Draw the line structure for each of the compounds listed in Problem 49. (8.7)

62. Draw the line structure for each of the compounds listed in Problem 50. (8.7)

63. Draw the condensed structural formula for each of the following compounds. (8.8)
    (a) 2-butanol   (c) 1-pentanol
    (b) 2-propanol   (d) 3-hexanol

64. Draw the condensed structural formula for each of the following compounds. (8.8)
    (a) 2-octanol
    (b) 1-propanol
    (c) Cyclopentanol
    (d) 2-pentanol

65. Draw the condensed structural formula for each of the following compounds. (8.8)
    (a) Butanal   (c) Ethanal
    (b) Hexanal   (d) Octanal

66. Draw the condensed structural formula for each of the following compounds. (8.8)
    (a) Methanal   (c) Propanal
    (b) Pentanal   (d) Heptanal

67. Draw the condensed structural formula for each of the following compounds. (8.8)
    (a) Ethanoic acid   (c) Decanoic acid
    (b) Octanoic acid   (d) Butanoic acid

68. Draw the condensed structural formula for each of the following compounds. (8.8)
    (a) Propanoic acid
    (b) Heptanoic acid
    (c) Nonanoic acid
    (d) Pentanoic acid

69. Name the following compounds. (8.8)

    (a) $CH_3-CH_2-CH_2-CH_2-\overset{\overset{\displaystyle O}{\|}}{CH}$

    (b) $CH_3-CH_2-CH_2-CH_2-CH_2-\overset{\overset{\displaystyle O}{\|}}{C}-OH$

    (c) $CH_3-CH_2-CH_2-\overset{\overset{\displaystyle OH}{|}}{CH}-CH_3$

    (d) $\overset{\overset{\displaystyle O}{\|}}{HC}-CH_2-CH_2-CH_2-CH_2-CH_3$

70. Name the following compounds. (8.8)

    (a) $CH_3-CH_2-\overset{\overset{\displaystyle O}{\|}}{C}-OH$

    (b) $CH_3-CH_2-\overset{\overset{\displaystyle O}{\|}}{CH}$

    (c) $\overset{\overset{\displaystyle O}{\|}}{HC}-CH_2-CH_2-CH_3$

    (d) $CH_3-CH_2-CH_2-\overset{\overset{\displaystyle O}{\|}}{C}-OH$

71. Draw and name all of the isomers of hexanol made by altering only the placement of the alcohol functional group. (8.8)

72. Draw and name all of the isomers of pentanol made by altering only the placement of the alcohol functional group. (8.8)

73. Each of the following names of compounds has an error in it. Explain why the name is incorrect. (8.4)
    (a) 3-methylpropane
    (b) 2-propylethane
    (c) 5-ethylheptane
    (d) 2-methylethane

74. Each of the following names of compounds has an error in it. Explain why the name is incorrect. (8.4)
    (a) 1-methylhexane
    (b) 2-propylbutane
    (c) 4-methylpentane
    (d) 3-methylbutane

75. Each of the following names of compounds has an error in it. Explain why the name is incorrect. (8.6, 8.8)
    (a) 3-propanoic acid
    (b) 2-propanal
    (c) 1-butanone
    (d) 3-butanol

76. Each of the following compound names has an error in it. Explain why the name is incorrect. (8.6, 8.8)
    (a) 2-butanoic acid
    (b) 2-pentanal
    (c) 3-propanone
    (d) 4-butanol

77. Identify each of the following compounds as an acid, base, or neutral compound according to the Brønsted-Lowry concept. (8.8)
    (a) 2-butanone
    (b) Butanoic acid
    (c) 1-propanol
    (d) Methylamine

78. Identify each of the following compounds as an acid, base, or neutral compound according to the Brønsted-Lowry concept. (8.8)
    (a) Methoxyethane
    (b) Methyl propanoate
    (c) Butylamine
    (d) Hexanal

Case Study Problems
Problems 79–85 refer to Table 8.6 on page 245.

79. Which polymer(s) consist of straight-chain alkanes? (8.2)

80. Which polymer(s) consist of branched-chain alkanes? (8.4)

81. Which polymer(s) contain benzene rings? (8.5)

82. Which polymer(s) contain a ketone functional group? (8.6)

83. Which polymer(s) contain an ester functional group? (8.6)

84. Which polymer(s) contain an amine functional group? Identify the compound(s) as a primary, secondary, or tertiary amine. (8.7)

85. Which polymer contains a basic functional group that would make it susceptible to attack by a strong acid? (8.9)

86. The taxes levied against ethanol intended for human consumption are significantly higher than the taxes on ethanol that is intended for use in industry or

laboratories as solvents and reactants. To ensure that ethanol intended for the latter purposes is not being misused, it is *denatured*, a term meaning that methanol has been added to the ethanol. If the denatured alcohol is consumed, the methanol reacts to form the toxic product methanal (formaldehyde). Write the alcohol-aldehyde-carboxylic acid structures that are formed when ethanol and methanol are metabolized. (8.6)

87. The congeners of alcohols present in alcoholic beverages are largely responsible for the effects of a hangover. Write the condensed structural formula for each of the following common congeners found in alcoholic beverages: (8.8)
    (a) Methanol
    (b) 1-propanol
    (c) 1-butanol
    (d) 2-butanol
    (e) 2-methyl-1-propanol
    (f) 2-methyl-1-butanol
    (g) 3-methyl-1-butanol

88. The smell of decaying flesh is due mainly to the production of putrescine and cadaverine from decaying proteins. Putrescine is 1,4-butanediamine and cadaverine is 1,5-pentanediamine. Write the condensed structural formulas for these two compounds. (8.7)

89. The compounds methanal through decanal (1-carbon to 10-carbon aldehydes) are produced during decomposition processes and are currently being studied as possible marker compounds for locating clandestine grave sites. Write the condensed structural formulas for methanal through decanal. (8.8)

90. The possession of 1,4-butanediol is illegal. The compound is a controlled substance that can be abused directly because it is metabolized to form GHB in the body. The following figure shows the FTIR data obtained from analysis of an evidence sample. The defendant claimed the compound was actually the legal compound 1-butanol. Based on the FTIR data here and in Figure 8.15, could this defense work? Explain. (ITL)

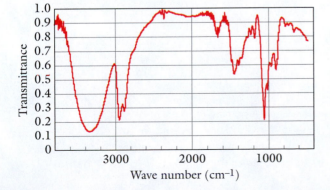

91. During an autopsy, a sample of stomach contents is taken for analysis. According to witnesses, the person was in extreme pain and kept taking aspirin every 5 to 10 minutes. Given the structure of aspirin (below) and your knowledge of the acid–base properties of organic compounds, outline a method for extracting aspirin from the stomach contents.

92. Identify the functional groups present in the digoxin molecule found in the cardiac medicine digitalis shown below.

93. The polymer structure of Kevlar, the material used in bulletproof vests, is shown below. What type of intermolecular forces would exist between chains of the Kevlar polymer when stacked next to one another?

CHAPTER 9

Chemistry of Fire and Heat

CASE STUDY: Exploring the Chemistry of Fire and Arson

Even though this new evidence may establish Mr. Richey's innocence, the Ohio and United States Constitution nonetheless allow him to be executed because the prosecution did not know that the scientific testimony offered at the trial was false and unreliable.

> —*Argument put forth by the prosecution during an appeal of the 1986 murder conviction of Kenny Richey*

The quotation is chilling and contrary to the spirit in which forensic scientists pursue their work. First and foremost, the idea that a person can be executed even though the state realizes serious problems have arisen with the evidence against him clearly violates the rights of the accused. Second, it is disturbing that false and unreliable evidence was allowed to be presented at trial without challenge. Yet in spite of these injustices, an appellate judge apparently agreed with the prosecutor's argument and denied the appeal of Kenny Richey, an inmate on death row in Ohio.

This case started on June 30, 1986, in Columbus Grove, Ohio. Kenny Richey was already living a tough life at the age of 21. He readily admits he was prone to binge drinking, fighting, and petty theft. On the evening in question, Kenny was so inebriated that one witness saw him pass out in the bushes for 10 minutes after talking with the occupants of a car. The prosecutor claimed that Kenny, after waking up in the bushes, proceeded to break into a nearby greenhouse storage shed and steal containers of gasoline and paint thinner. According to the prosecutor, Kenny then climbed on top of a storage shed below the deck of an apartment, hoisted himself onto the deck, and gained access to the living room of Hope Collins. Hope was one of the car's occupants with whom Kenny had been seen speaking earlier. The prosecutor next claimed that Kenny spread the gasoline and paint thinner on the living room floor and deck and set the apartment on fire.

His motive? An ex-girlfriend lived in the apartment below, and he was hoping to burn the apartment building down to kill her and her new boyfriend. Everyone escaped the fire except a two-year-old girl in the apartment where the fire started.

Since Kenny was not only inebriated but also had a broken hand in a cast at the time of the alleged incident, it is unclear how he would have been able to carry out the acts attributed to him. Also, not a single trace of gasoline or paint thinner was found on his hands, clothes, or shoes. In the plea bargain, the prosecutor offered Kenny a 10-year sentence with parole likely after six years, but Kenny refused to plead guilty. The prosecutor apparently used the evidence to make a convincing case to a panel of three judges. The

> **As you read through the chapter, consider the following questions:**
>
> • **What conditions are necessary for a fire to burn?**
>
> • **Is all evidence of arson lost in a fire?**
>
> • **How should evidence of suspected arson be collected and handled?**

inexperienced public defender representing Kenny offered an ineffective and incompetent defense. The panel of judges found Kenny guilty, and he was sentenced to death.

It is very likely that one day you will sit on a jury and listen to experts present opposing opinions on the evidence against the accused. You will have to decide in a criminal case whether or not someone is guilty of a crime, or in a civil case whether or not a person should pay monetary compensation for wrongdoing. The key is to use the critical-thinking skills you are learning throughout your coursework to examine the data and, with the guidance of the expert witnesses, determine if the evidence fits the conclusion presented in court.

Many details of the physical evidence from the Kenny Richey case will be presented in this chapter to illustrate the chemical principles involved in the investigation of arson cases. Then you can make your own decision about whether the physical evidence supports a guilty verdict or not.

MAKE THE CONNECTION

But first, some background information on the chemistry of fire is necessary to understand the evidence in a trial that involves charges of arson . . .

9.1 The Chemistry of Fire

The ability to fight fires effectively and to investigate the origin of a fire relies on having a firm understanding of the chemistry and physics behind flames. A **fire** is a self-sustaining chemical reaction that releases energy in the form of heat and light. A fire is classified as **arson** if it is deliberately set for the purpose of destroying property.

All fires require the same three ingredients: a source of fuel, a source of heat, and oxygen gas. Together these constitute the **fire triangle**. If any one of these three ingredients is missing, a fire will not start. If any one of these components is removed, the fire will be extinguished. This principle is used by firefighters to combat a fire.

For example, if a natural-gas main is ruptured and the gas bursts into flame, the fire is put out by cutting off the fuel supply. Flame-resistant materials in the home provide another method for preventing the spread of a fire. These materials are flame resistant because as they burn they produce a large amount of char, the black crusty material resulting from partial combustion. Char acts as a barrier to prevent oxygen from reaching the potential fuel (the flammable material) beneath it, and it also acts as an insulator that prevents heat from reaching the fuel.

Water is one of the most important tools used in firefighting to cool the area and remove the heat source. But how does water actually cool a fire? One of the properties of water is its ability to absorb a substantial amount of heat from hotter objects. **Heat** is a form of energy that is transferred from hot objects to cold objects that are in contact with one another.

Heat energy is measured in the SI units of **joules (J)** that are the parallel unit to the English system's **calorie (cal)**. The calorie is defined as the amount of energy required to raise the temperature of 1 gram of water from 14.5°C to 15.5°C. To convert from calorie to joule or vice versa, use the conversion factor 1 cal = 4.184 J. The nutritional unit we use to express the caloric content of foods is actually a kilocalorie, or Calorie. The conversion factors are 1 kilocalorie = 1 Calorie = 1000 calories.

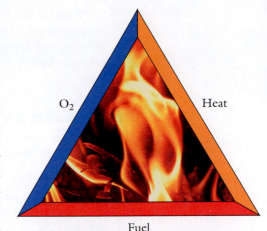

When an object absorbs energy, the kinetic energy (the energy of motion) of its particles increases; this results in an increase in the temperature of the object. **Temperature** is a measurement of the average kinetic energy of the particles in a system, not a measurement of how much heat is in a system. The difference between a piece of wood at 25°C and 250°C is that the internal motion of the molecules is much greater at 250°C.

When two objects that are at different temperatures come in contact, heat is transferred from the warmer object to the colder object, thereby causing the warmer object to cool. The energy added to the colder object increases the kinetic energy of its particles and warms the colder object. **Thermal equilibrium** is achieved when the temperatures of the two objects become equal and remain constant. At this point, the kinetic energy of the particles in both objects is identical. Figure 9.1 illustrates these processes.

The fire triangle. (© Nathan Griffith/Corbis)

(Bill Stormont/Corbis)

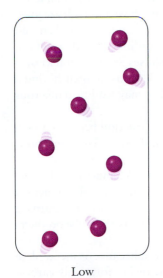

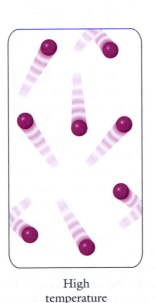

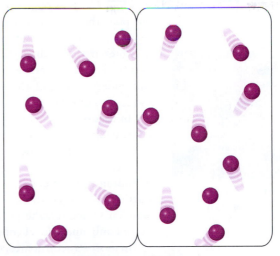

| Low temperature | High temperature | Thermal equilibrium, equal temperatures |

FIGURE 9.1 Temperature and thermal equilibrium. The temperature of an object is a measure of the kinetic energy of the atoms, molecules, or ions that make up the object. When two objects at different temperatures come into contact, energy in the form of heat is transferred until thermal equilibrium is reached.

When water from a fire hose comes into contact with a burning structure, the water absorbs heat from the structure, thus cooling it and removing the heat component of the fire triangle. If the water brings about sufficient cooling, the fire is extinguished.

■ WORKED EXAMPLE 1

What is incorrect about the following statement? *A thermometer measures how much heat is in an object.*

SOLUTION

A thermometer measures the kinetic energy of the particles in a system.
Heat is a form of energy that is transferred from warm objects to cold objects.

Practice 9.1

Using the concept of thermal equilibrium, explain what is incorrect about the following statement. *The tile floor is colder than the carpeted area of the floor.*

Answer

Two objects in thermal contact would have reached thermal equilibrium. Therefore, the temperatures of the tile floor and the carpet are identical. However, the tile floor may *seem* colder to the touch because it draws more heat away from your body than the carpet does. But the temperature of the tile and carpet before you touch them is the same.

9.2 Combustion Reactions

LEARNING OBJECTIVE

Balance the chemical equations for combustion reactions.

A backdraft. (Universal/
The Kobal Collection)

A hidden danger lurks in the fire triangle. When fire heats a poorly ventilated room, the air in the room can fill with flammable compounds created from the incomplete combustion of fuels. The poor ventilation in the room can cause oxygen to be consumed quickly while the heated, flammable vapors build up. Only two of three ingredients for fire (heat and fuel) are present once the oxygen is consumed—until a door is opened or a window is broken and oxygen gas rushes in. Under these circumstances, it is not an ordinary fire that ignites. The presence of such an excess of heated fuel ready to react causes an explosion known as a **backdraft**, or smoke explosion. The safety of firefighters depends on their having a full understanding of the chemical and physical processes that may lead to dangerous conditions during a fire.

In Section 4.9, a *combustion reaction* was first defined as a reaction between an organic compound and oxygen gas to produce water and carbon dioxide. These reactions were presented as a subclass of reduction-oxidation (redox) reactions. However, combustion reactions are not actually confined to organic compounds. We must broaden the definition because combustion reactions can occur with certain metals and nonmetals. Furthermore, even in cases where organic compounds are involved (as in a burning wooden structure), the production of water and carbon dioxide as the only products occurs only under ideal conditions.

In most structural fires, many other compounds are produced because of incomplete combustion; that is, there is insufficient oxygen present for the fuel to form only carbon dioxide and water. The oxygen gas serves as a limiting reactant. The incomplete combustion of fuels produces compounds such as carbon monoxide and solid carbon particles (soot).

The color of the flame and the presence or absence of smoke are actually indicators of how efficient the combustion reaction is. Consider the different fires shown in Figure 9.2.

In the burning of natural gas in a stove, as shown in Figure 9.2a, the flame that is produced is an intense blue with no evidence of smoke. This fire is efficiently converting the fuel into carbon dioxide and water vapor. In Figure 9.2b, there is insufficient oxygen (or excess fuel) for the reaction, resulting in a flame that has a noticeable orange portion due to the production of small carbon particles. The carbon particles glow when heated by the flame. The smoke in Figure 9.2c, produced in a forest fire, is mainly white. This color indicates the condensation of water vapor on small particles of dust and soot, a process much like the formation of clouds. The white color is indicative of a fire in which the burning material is undergoing nearly complete combustion. Finally, Figure 9.2d shows the typical black smoke and flame associated with building fires. The black color is due to heavy production of solid carbon soot particles from incomplete combustion reactions as the amount of oxygen available is restricted. As the smoke escapes the confines of the building, additional oxygen from the atmosphere mixes with the flammable gases and ignites.

(a) (Corbis)

(b) (Corbis)

(c) (Stone/Getty)

(d) (Andrea Comas/Reuters/Corbis)

FIGURE 9.2 (a) The efficient combustion of natural gas to form carbon dioxide and water. (b) The inefficient combustion of natural gas. (c) A typical forest fire has easy access to atmospheric oxygen. (d) Building fires typically have restricted access to oxygen.

WORKED EXAMPLE 2

As illustrated in Figure 9.2, it is common for structural fires to burn with black smoke and for grass or forest fires to have white smoke. Explain why.

SOLUTION

The white smoke indicates a more complete combustion reaction, which occurs when sufficient oxygen gas is present. Because outdoor fires have no barriers between the atmospheric oxygen and the fuel, a more complete combustion reaction can occur. The black smoke from a building is a sign of incomplete combustion due to insufficient oxygen gas reaching the fire because the building walls and ceilings restrict air flow.

Practice 9.2

Firefighters often ventilate a structural fire by cutting holes in ceilings. What processes occurring within a fire would be altered by ventilating a room?

Answer

The conditions for a backdraft require poor ventilation for the buildup of heated, combustible gases within a room. By ventilating the room, firefighters minimize the likelihood of a backdraft. (It also improves visibility and lowers the temperature within the building.)

Balancing Combustion Reactions

The balancing of chemical equations corresponding to the complete combustion of organic materials can sometimes seem complicated. The same method for balancing equations that was discussed in Section 4.6 is also used here. However, it is important for the elements to be balanced in the following order: carbon, hydrogen, oxygen. Balancing equations is illustrated below, using as an example the combustion of propane (C_3H_8) with oxygen gas to form water and carbon dioxide.

Step 1. Write the unbalanced reaction and determine the total number of carbon, hydrogen, and oxygen atoms on each side of the reaction:

$$C_3H_8(g) + O_2(g) \rightarrow CO_2(g) + H_2O(g)$$

C: 3	C: 1
H: 8	H: 2
O: 2	O: 3

Step 2. Balance the carbon atoms:

$$C_3H_8(g) + O_2(g) \rightarrow 3CO_2(g) + H_2O(g)$$

C: 3	C: 1̸ 3
H: 8	H: 2
O: 2	O: 3̸ 7

Step 3. Balance the hydrogen atoms:

$$C_3H_8(g) + O_2(g) \rightarrow 3CO_2(g) + 4H_2O(g)$$

C: 3	C: 1̸ 3
H: 8	H: 2̸ 8
O: 2	O: 3̸ 7̸ 10

Step 4. Balance the oxygen atoms:

$$C_3H_8(g) + 5O_2(g) \rightarrow 3CO_2(g) + 4H_2O(g)$$

C: 3	C: 1̸ 3
H: 8	H: 2̸ 8
O: 2̸ 10	O: 3̸ 7̸ 10

■ WORKED EXAMPLE 3

Balance the equation for the combustion reaction of 1-butanol (C_4H_9OH) with oxygen gas.

SOLUTION

Step 1. $C_4H_9OH(g) + O_2(g) \rightarrow CO_2(g) + H_2O(g)$

       C: 4   C: 1

       H: 10   H: 2

       O: 3   O: 3

Step 2. $C_4H_9OH(g) + O_2(g) \rightarrow 4CO_2(g) + H_2O(g)$

       C: 4   C: ~~1~~ 4

       H: 10   H: 2

       O: 3   O: ~~3~~ 9

Step 3. $C_4H_9OH(g) + O_2(g) \rightarrow 4CO_2(g) + 5H_2O(g)$

       C: 4   C: ~~1~~ 4

       H: 10   H: ~~2~~ 10

       O: 3   O: ~~3~~ ~~9~~ 13

Step 4. $C_4H_9OH(g) + 6O_2(g) \rightarrow 4CO_2(g) + 5H_2O(g)$

       C: 4   C: ~~1~~ 4

       H: 10   H: ~~2~~ 10

       O: ~~3~~ 13   O: ~~3~~ ~~9~~ 13

Practice 9.3

Balance the equation for the combustion reaction of 1-hexene (C_6H_{12}) with oxygen gas.

Answer

$$C_6H_{12}(g) + 9O_2(g) \rightarrow 6CO_2(g) + 6H_2O(g)$$

■ WORKED EXAMPLE 4

Balance the equation for the combustion reaction of butane (C_4H_{10}) with oxygen gas.

SOLUTION

Step 1. $C_4H_{10}(g) + O_2(g) \rightarrow CO_2(g) + H_2O(g)$

       C: 4   C: 1

       H: 10   H: 2

       O: 2   O: 3

Step 2. $C_4H_{10}(g) + O_2(g) \rightarrow 4CO_2(g) + H_2O(g)$

       C: 4   C: ~~1~~ 4

       H: 10   H: 2

       O: 2   O: ~~3~~ 9

Step 3. $C_4H_{10}(g) + O_2(g) \rightarrow 4CO_2(g) + 5H_2O(g)$

       C: 4   C: ~~1~~ 4

       H: 10   H: ~~2~~ 10

       O: 2   O: ~~3~~ ~~9~~ 13

Step 4. $C_4H_{10}(g) + 6.5O_2(g) \rightarrow 4CO_2(g) + 5H_2O(g)$

       C: 4   C: ~~1~~ 4

       H: 10   H: ~~2~~ 10

       O: ~~2~~ 13   O: ~~3~~ ~~9~~ 13

The only way to balance oxygen in this step is to use a coefficient of 6.5. Coefficients, however, should be whole numbers. This can be achieved by doubling all coefficients in the final step.

Step 5. $2C_4H_{10}(g) + 13O_2(g) \rightarrow 8CO_2(g) + 10H_2O(g)$

C: ~~4~~ 8   C: ~~1~~ ~~4~~ 8
H: ~~10~~ 20   H: ~~2~~ ~~10~~ 20
O: ~~2~~ ~~13~~ 26 O: ~~3~~ ~~9~~ ~~13~~ 26

Practice 9.4

Balance the equation for the combustion reaction of octane (C_8H_{18}) with oxygen gas.

Answer

$$2C_8H_{18}(g) + 25O_2(g) \rightarrow 16CO_2(g) + 18H_2O(g)$$

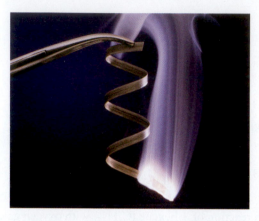

Magnesium (left) and sulfur (right) both undergo combustion reactions, which produce heat and light. Magnesium is often used in mixtures for fireworks displays because of the intense bright light it produces during combustion. Sulfur is one of the three ingredients in gunpowder. (Both: Richard Megna/Fundamental Photos)

Several metals and nonmetals undergo combustion reactions that produce heat and light when ignited in the presence of oxygen gas. The most common examples are the combustion of magnesium ribbon to form magnesium oxide, and the combustion of sulfur to form sulfur dioxide. Other metals such as copper, when heated, react with oxygen gas to form a metal oxide. However, such a reaction is not considered combustion because heat and light are not produced.

9.3 Redox Reactions

The combustion of organic compounds, metals, and nonmetals, and the formation of metal oxides from their elements are all examples of redox reactions involving the gain or loss of electrons. In Section 4.9, *oxidation* was defined as the loss of electrons and *reduction* was defined as the gain of electrons. In this section, we will see how to determine which atoms are oxidized and which are reduced when electrons are not explicitly shown in the reaction.

The concept of **oxidation number** helps in interpreting redox reactions. The oxidation number is merely a value assigned to an atom by the basic rules listed on the following page and should not be interpreted as a true charge on a particular atom. Some exceptions to the rules exist, but for simplicity's sake, they have been omitted here.

Rule 1. All pure elements have an oxidation number of 0. Examples: Oxygen gas (O_2) and iron (Fe) have an oxidation number of 0.

Rule 2. Monatomic ions have an oxidation number equal to their ionic charge. Examples: Mg^{2+} has an oxidation number of $+2$, and S^{2-} has an oxidation number of -2.

Rule 3. Hydrogen always has an oxidation number of $+1$ in a compound.

Rule 4. Oxygen always has an oxidation number of -2 in a compound.

Rule 5. The sum of the oxidation numbers must be 0 for neutral compounds or equal to the ionic charge for polyatomic ions. *First example:* In FeO, the oxygen atom has an oxidation number of -2 according to Rule 4, and there is no rule specific for the iron atom in a compound. Because FeO is a neutral compound, the iron atom must have an oxidation number equal to $+2$. *Second example:* In NO_3^-, the oxygen atoms each have an oxidation number of -2 by Rule 4. Because there are 3 oxygen atoms, they have a total oxidation number of -6. There is no specific rule for nitrogen atoms in a compound, but the nitrate ion must have an overall -1 charge, so the nitrogen atom must have an oxidation number equal to $+5$.

Once the oxidation numbers have been assigned, compare the oxidation number of each element from the reactant side with the same element on the product side. You will notice that one of the elements from the compound being *reduced* has gained electrons, and one of the elements from the compound being *oxidized* has lost electrons. For example, the reaction for the combustion of magnesium is shown below.

$$\overbrace{\underset{0}{2Mg(s)} + \underset{0}{O_2(g)} \rightarrow \underset{+2 \ -2}{2MgO}}$$

Lost 2 e⁻ (top), Gained 2 e⁻ (bottom)

The oxidation numbers of the reactants, Mg and O_2, are both 0 according to Rule 1. The oxidation number of O is -2 in MgO, as indicated by Rule 4. The oxidation number of Mg in MgO is $+2$ according to Rule 5. Therefore, oxygen gas undergoes a change in oxidation number from 0 to -2 by gaining two electrons and thus must have been reduced. Conversely, magnesium undergoes a change in oxidation number from 0 to $+2$ by losing two electrons, which fits the definition of oxidation.

WORKED EXAMPLE 5

Assign oxidation numbers to each atom for the following reactions:

(a) $2Ca(s) + O_2(g) \rightarrow 2CaO(s)$

(b) $CH_4(g) + 2O_2(g) \rightarrow CO_2(g) + 2H_2O(g)$

SOLUTION

Remember that losing an electron makes the oxidation number more positive because of the negative charge on the electron!

(a) $2Ca(s) + O_2(g) \rightarrow 2CaO(s)$

Rule 1: All pure elements have an oxidation number of 0.
$Ca = 0, O_2 = 0$

Rule 4: Oxygen in compounds has an oxidation number of -2.
O in CaO $= -2$

Rule 5: The sum of the oxidation numbers must be 0 for neutral compounds.
Ca in CaO $= +2$

(b) $CH_4(g) + 2O_2(g) \rightarrow CO_2(g) + 2H_2O(g)$

Rule 1: All pure elements have an oxidation number of 0.
$O_2 = 0$

Rule 3: Hydrogen always has an oxidation number of $+1$ in a compound.
H in $CH_4 = +1$
H in $H_2O = +1$

Rule 4: Oxygen always has an oxidation number of -2 in a compound.
O in $CO_2 = -2$
O in $H_2O = -2$

Rule 5: The algebraic sum of the oxidation numbers must be 0 for neutral compounds.
C in $CH_4 = -4$
C in $CO_2 = +4$

Practice 9.5

Determine which elements are oxidized and which elements are reduced in the redox reactions from Worked Example 5.

Answer

(a) Ca went from 0 to $+2$, a loss of 2 e^-. Ca is oxidized.
O went from 0 to -2, a gain of 2 e^- per O atom. O is reduced.

(b) C went from -4 to $+4$, a loss of 8 e^-. C is oxidized.
O went from 0 to -2, a gain of 2 e^- per O atom. O is reduced.

9.4 Thermochemistry of Fire

LEARNING OBJECTIVE
Describe how heat affects both physical and chemical processes.

The fire triangle indicates that a fire needs fuel, oxygen, and heat. But if a fire produces heat, why does it require heat to exist? Heat is needed to initiate the combustion reaction for two reasons. First, liquid or solid fuel must be converted into vapors before the combustion reaction can occur. Second, energy is required to start the reaction of the fuel vapors with oxygen gas. Let's look at each of these requirements.

Energy Needed to Produce Gaseous Reactants

For a liquid fuel, there must be sufficient heat to vaporize the liquid to the gaseous state. Solid fuels must undergo a process called **pyrolysis**, a decomposition reaction that produces small, gaseous compounds capable of undergoing the combustion reaction. Many different chemicals are produced as a result of pyrolysis, but some of the more common products are carbon monoxide, oxygen gas, and small alkanes.

A chemical or physical process that must absorb heat to occur is an **endothermic** process. Endothermic processes cease if the source of the heat is removed. For example, when the copper wiring in a building is heated by a fire, the copper atoms increase in energy, and if sufficient heat is added, the solid wire begins to melt. The copper will continue to melt as long as heat is being added to the copper by the fire. When the firefighters extinguish the fire, the source of heat is removed and the copper will stop melting and solidify.

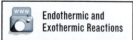
Endothermic and Exothermic Reactions

A related example would be the melting of the plastic housing of a smoke detector, which played a role in analyzing the fire scene in the case of Kenny Richey. There are two critical endothermic processes involved in supplying the fuel to a fire: the vaporization of a liquid fuel and the pyrolysis of solid fuels. Both processes require that heat be transferred into the fuel from the surroundings for the fire to continue.

Any chemical reaction or physical change that produces or releases heat into the surroundings is called an **exothermic** process. A fire is an excellent example of a process that releases heat to the surroundings. Chemical reactions are exothermic if the final energy of the products is lower than the energy of the reactants. The energy that is released in a chemical reaction was stored in the chemical bonds between atoms. The products of an exothermic chemical reaction have less energy stored in the chemical bonds. The excess energy is released into the surroundings.

For the combustion of a fuel to become a self-sustaining chain reaction, the heat produced in the process must be greater than the heat needed to continually convert the fuel to the gas phase. When water is applied to a fire, it absorbs the heat being produced from the exothermic combustion reaction, which in turn makes less heat available for the endothermic vaporization of the liquid fuel or for pyrolysis of the solid fuel. The fire can thus be extinguished.

Activation Energy

What makes a chemical reaction endothermic rather than exothermic? In any chemical reaction, the bonds between atoms in the reactants must be broken before the atoms can be rearranged to form new bonds in the products. The breaking of bonds requires that energy be put into the system, whereas the formation of new bonds releases energy. The energy required to initiate a reaction is called the **activation energy** and represents a barrier that the reactants must overcome before they can react.

If the energy released by bond formation is less than the energy used to break the reactant bonds, the process is endothermic. If the energy released by making new bonds is greater than the energy used to break the reactant bonds, the reaction is exothermic. The energy released in combustion reactions is far greater than the amount of energy required to initiate the reaction. This released energy had been stored within the chemical bonds of the reactants, and it is released because the products have less energy than the reactants. Note, however, that even in an exothermic process, energy must first be absorbed by the reactants to initiate the reaction. Figure 9.3 graphically depicts the energy changes that occur in an exothermic chemical reaction.

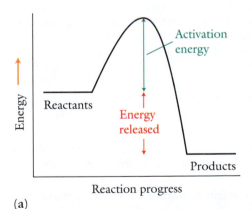

(a)

(b)

FIGURE 9.3 Energy diagram for exothermic reaction. In an exothermic process, the reactant compounds have a higher energy than do the product compounds (a). The activation energy (green arrow) is the amount of energy the reactants must absorb for the reaction to occur. This energy is used to break bonds within the reactant compounds. As new bonds form, energy is released and the amount of energy initially absorbed is recovered. As the reaction continues to release energy, a net release of energy occurs (red arrow). The excess heat of a fire supplies the activation energy to a nearby fuel source such that the exothermic reaction can continue, as illustrated in the combustion of dried leaves (b). (Paul A Souders/Corbis)

The activation energy is the energy required to break the bonds holding the reactants together. If methane gas and oxygen gas are mixed together, combustion does not immediately occur because the molecules do not have sufficient energy to cross the barrier represented by the activation energy. For the methane and oxygen to react, energy must be added to the system, usually by some source of ignition. This added energy gives the reactants enough energy to cross the activation energy barrier, allowing the reaction to proceed. But note that the fact that reactant molecules have sufficient energy to cross the activation energy barrier does not imply that *all* the molecules will form products, a topic explored in more depth in Chapter 11.

In an endothermic reaction, the products have a greater total energy than do the reactant compounds. For example, plants take in carbon dioxide and water, which react to produce oxygen gas and the sugars that will make up cellulose. Photosynthesis is most active in bright sunlight, slows on cloudy days, and shuts down during the night. This is because the sunlight provides the necessary energy for the reactants to cross the activation barrier. The energy changes that occur in an endothermic chemical reaction are shown in Figure 9.4.

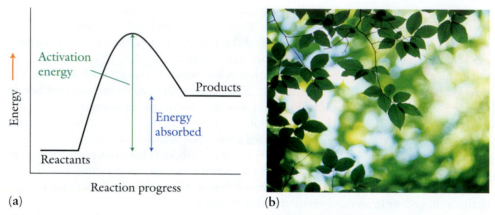

(a)   (b)

FIGURE 9.4 Energy diagram for endothermic reaction. In an endothermic process, the reactant compounds have a lower energy than do the product compounds (a). The activation energy (green arrow) is the amount of energy required for the reactant molecules to successfully form the products. The amount of energy absorbed (blue arrow) is the amount of energy added to the reactants. The energy absorbed from the sun is stored within the chemical bonds of the sugar molecules made during the process of photosynthesis (b). (istockphoto)

The activation energy (green arrow) again represents the energy needed for the reaction to occur. As the reaction proceeds and new bonds are formed, not all of the energy required to cross the activation energy barrier is released. Rather, part of the energy added into the system is stored within the bonds of the newly formed product compounds. The blue arrow in Figure 9.4 represents the amount of energy stored in the product molecules.

A fire can easily spread from its point of origin to involve the entire contents of a room—a process called **flashover**. In a flashover, the heat from a fire in a small area rises to the ceiling and then spreads out across the room. As the fire progresses, the heated layer of smoke and gases just below the ceiling increases its temperature and thickness. Because the heat is restricted by the ceiling, it radiates downward, heating the remaining combustible furniture that has not yet been ignited.

The radiative heating of the furniture continues until the furniture reaches its **autoignition** temperature, the temperature at which material will ignite without a spark or flame directly contacting it. Autoignition can occur because all components of the fire triangle (fuel, oxygen, and heat) are present. The heat radiating down on the furniture is providing the activation energy needed for the combustion reaction. Once the furniture has enough energy to overcome the activation barrier, the combustion reaction proceeds as long as enough oxygen is present in the room. This entire process can occur in less than two minutes from the start of the fire!

WORKED EXAMPLE 6

How does ventilating a room affect the possibility of a flashover?

SOLUTION

Ventilation of a room removes the heated layer of smoke at the ceiling, thereby eliminating the energy source needed for combustible material to reach its autoignition temperature.

Practice 9.6

What happens to the excess energy produced by an exothermic reaction?

Answer

The heat is absorbed by the surroundings (air, beaker, and so forth), which increase in temperature due to the increased kinetic energy of the particles absorbing the energy.

9.5 Heat Capacity and Phase Changes

Reconstructing a fire scene to determine the cause of the fire relies on the investigator's ability to interpret physical evidence left behind and to combine this knowledge with an understanding of the physical and chemical properties of substances in a fire. Once the cause of the fire has been established, the investigator can determine whether the fire was started accidentally or intentionally.

> **LEARNING OBJECTIVE**
> Describe what happens when heat is added to a substance.

One type of physical evidence taken into account is the presence or absence of melted materials. For example, the glass of light bulbs softens in the heat of a fire. Because light bulbs greater than 25 watts are pressurized with inert gases, the bulb bulges outward in the direction of an intense heat source. Plastics and metals used throughout buildings can be indicators of fire temperatures, depending on whether the materials have melted.

Materials can also produce demarcation lines between temperature regions. For example, copper, commonly used for electrical wiring and plumbing, can indicate where along a wall the temperature exceeded 1082°C, the melting point of the metal. Copper that is heated in a fire tends to melt along its surface and flow to lowest points, forming globules and thinned areas.

Not all objects have the same ability to absorb heat. The **specific heat capacity** (C_p), also known as *specific heat*, is a measure of the heat needed to raise the temperature of a 1-gram sample of the substance by 1°C. If a substance has a high specific heat, it takes a large amount of heat to increase the temperature of 1 gram of the substance by 1°C. The same amount of heat could increase the temperature of 1 gram of a different substance by several degrees, depending on the substance.

Water, for example, has a specific heat capacity of 4.184 J/g · °C, indicating that it takes 4.184 joules of heat to raise the temperature of 1 gram of water by 1°C. Water actually has an unusually high specific heat capacity when compared with all other materials. For example, wood construction material has a specific heat capacity of 1.70 J/g · °C. In fire fighting, it takes more energy gram for gram to warm water up than to cool wood down—a tremendous benefit! Table 9.1 provides a representative list of the specific heat capacities associated with materials of interest in fire scene reconstruction.

TABLE 9.1	Specific Heat Capacities
Material	Specific Heat (J/g·°C)
Aluminum	0.90
Concrete	0.80
Copper	0.38
Gasoline	2.01
Gypsum wallboard	1.05
Particle board	1.30
Polyethylene	1.90
PVC (polyvinyl chloride)	1.05
Silica glass	0.75
Water	4.184
Wood	1.70

■ WORKED EXAMPLE 7

What happens to the individual particles of a cold object as it is being heated?

SOLUTION

The particles that make up the object (molecules, atoms, or ions) are absorbing heat, which increases their kinetic energy.

Practice 9.7

The specific heat capacity of water is 4.184 J/g·°C, which is more than twice the value for gasoline at 2.01 J/g·°C. Explain the role of intermolecular forces that cause the specific heat capacity of water to be almost twice that of gasoline.

Answer

The lower specific heat capacity of gasoline compared with water indicates that it is easier to increase the kinetic energy of gasoline particles, which results in an increase in temperature. It is easier to increase the kinetic energy of gasoline particles because the nonpolar organic molecules that make up gasoline have weaker intermolecular forces than the polar water molecules.

The temperature of a typical building fire can easily reach 1300°C or more, well above the boiling point of water (100°C) being used by firefighters to extinguish the fire. Figure 9.5 shows a graph of the temperature of water as heat is added to a system.

Notice that as water absorbs heat, the temperature increases linearly until it reaches a plateau at the boiling point of 100°C. The temperature of the water holds constant at the boiling point until all of the water has been converted to steam (water vapor). Then the temperature of the steam rises above 100°C as more heat is absorbed. The amount of energy required to convert a liquid to a gas is called the **heat of vaporization**. The energy is absorbed by the molecules in the liquid phase, increasing their kinetic energy until the particles can overcome the attractive intermolecular

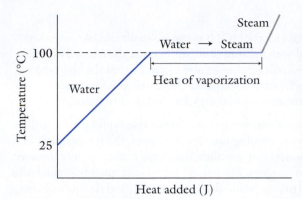

FIGURE 9.5 Temperature of water versus heat added. Heat is added to water at room temperature, causing the temperature of the water to increase. When the water reaches its boiling-point temperature, the heat energy converts the liquid water into steam. The amount of heat required for this conversion is called the *heat of vaporization*. When all the water has been converted to steam, the temperature of the steam increases. (Stocksearch/Alamy)

forces holding them in the liquid state. For water, the heat of vaporization is 2258 J/g. Table 9.2 is a representative list of the heats of vaporization of various liquids that frequently play a role in an arson investigation.

TABLE 9.2	Heat of Vaporization for Various Liquids
Material	**Heat of Vaporization (J/g)**
Diesel fuel	233
Ethanol	921
Gasoline	349
Kerosene	250
Turpentine	293
Water	2258

■ WORKED EXAMPLE 8

Which process cools a fire more effectively—heating water to 100°C or vaporizing the water to steam? Assume the water is initially at 25°C.

SOLUTION

From the heat of vaporization of water, we know that it takes 2258 J to vaporize 1 g of water. The specific heat of water shows that 1.00 g of water requires 4.184 J to increase the temperature by 1.00°C. The water has to be heated from 25.0°C to 100°C, an increase of 75.0°C. Therefore, the water absorbs

$$4.184 \text{ J/g} \cdot °\text{C} \times 1.00 \text{ g} \times 75.0°\text{C} = 314 \text{ J}$$

So, more heat is removed by vaporizing the water than by heating the water.

Practice 9.8

Would you expect the heat of vaporization of ethanol (CH_3CH_2OH) to be different from the heat of vaporization of ethane (CH_3CH_3)? Why?

Answer

The presence of the alcohol functional group (— OH) in the ethanol molecule increases the heat of vaporization as compared with the simple alkane because the ethanol molecule is polar. The alcohol exhibits dipole–dipole intermolecular forces and has the ability to form hydrogen bonds, neither of which is true for ethane. Overcoming the stronger intermolecular forces in the alcohol requires more energy for the liquid to vaporize.

Arsonists often make the mistake of assuming that a flammable liquid used to start a fire will be completely consumed in the ensuing fire. Yet even when the temperature of a fire reaches over 1000°C, the components of gasoline that have boiling points between 40°C and 200°C can still be found in samples. Flammable liquids are quickly soaked into most objects in a room—such as carpeting, wood flooring, cement, and furniture. Rarely does a fire completely destroy all of these materials, so traces of the accelerant can be detected in the debris. Only the vapors of the flammable liquid ignite, and the liquid beneath the flames continues to soak into the absorbent material. Because heat rises, the temperature beneath the flame is considerably lower than the temperature above the flame. Furthermore, as the fire burns upward, debris and ash will fall downward and tend to smother the flames burning the accelerant, thereby preserving it for future analysis. Meanwhile, the fire spreads to other materials and continues burning.

As we saw earlier, fire-scene investigators can estimate the temperature of a fire by determining whether copper wiring in the walls has melted. Other common metals found at a fire scene are aluminum cans (660°C melting point) or steel furniture springs (1100°C to 1600°C melting point, depending on the type of steel).

The amount of heat required to melt a solid completely is referred to as the **heat of fusion** and is analogous to the heat of vaporization discussed at the beginning of this section. When a solid material reaches its melting point, the solid does not melt instantly. It must absorb heat to increase the kinetic energy of the solid particles until they have sufficient energy to overcome the forces holding them in the solid state and break into the liquid phase. Table 9.3 provides a list of heats of fusion for common materials.

TABLE 9.3	Heats of Fusion for Various Materials
Material	**Heat of Fusion (J/g)**
Aluminum	397
Copper	207
High-density polyethylene	296
Iron	247
Low-density polyethylene	98
Water	334

WORKED EXAMPLE 9

The following graph represents the heating curve for copper metal. Estimate the melting point and boiling point of copper.

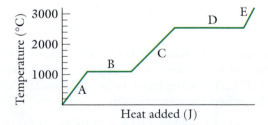

SOLUTION

The melting point corresponds to the temperature of the plateau region B. Extrapolating the line back to the y axis, an estimate of 1050°C is reasonable. (The actual value is 1083°C.) The boiling point corresponds to plateau region D with an estimated value of 2600°C (the actual value is 2595°C).

Practice 9.9

For each distinct region on the heating curve of copper, provide a description of the physical process occurring.

Answer

In region A, the temperature of solid copper increases. In region B, the solid copper changes to liquid copper; the temperature remains constant. In region C, the temperature of liquid (molten) copper increases. In region D, the liquid copper boils and changes to copper in the gaseous state; the temperature remains constant. In region E, the temperature of gaseous copper increases.

9.6 Mathematics of Heat Capacity

The specific heat capacity of a material is most commonly used in calculating the amount of energy required to raise the temperature of an object made of that material. For example, how much energy in joules is needed to heat a 10.0-g copper (Cu) wire from room temperature at 25.0°C to its melting point at 1082°C? The equation for calculating the heat required is

$$q = \underbrace{C_p}_{\substack{\text{Specific heat} \\ \left(\frac{J}{g \cdot °C}\right)}} \times \underbrace{M}_{\substack{\text{Mass} \\ \text{(g)}}} \times \underbrace{\Delta T}_{\substack{\text{Temperature} \\ \text{change} \\ \text{(°C)}}}$$

$\underbrace{q}_{\substack{\text{Heat} \\ \text{(J)}}}$

Look up the specific heat of copper in Table 9.1 and examine the units carefully:

$$0.38 \; \underbrace{\frac{J}{g \cdot °C}}_{\text{Cancel units}} \times 10.0 \, g \times \overbrace{1057 °C}^{1082 - 25} = 4.0 \times 10^3 \; J$$

$\overbrace{\phantom{0.38 \frac{J}{g \cdot °C}}}^{\substack{\text{Desired} \\ \text{final units}}}$

■ WORKED EXAMPLE 10

Will it take more or less energy than that used in the previous example to heat a 10.0-g aluminum (Al) wire from 25°C to its melting point at 660°C, given the specific heat of Al is 0.90 J/g·°C?

SOLUTION

$$q = C \times m \times \Delta T$$
$$C = 0.90 \; J/g \cdot °C$$
$$m = 10.0 \; g$$
$$\Delta T = T_{final} - T_{initial} = 660 - 25 = 635°C$$

The heat required is

$$0.90 \; \frac{J}{g \cdot °C} \times 10.0 \, g \times 635 °C = 5.7 \times 10^3 \; J$$

More energy is needed to heat 10.0 g of Al to 660°C than to heat 10.0 g of Cu to 1082°C.

Practice 9.10

If 3.50×10^3 J of energy is absorbed by a 25.3-g metal object, and the temperature increases from 25°C to 235°C, what is the specific heat of the metal?

Answer

❏ 0.659 J/g·°C

■ WORKED EXAMPLE 11

Calculate the amount of heat necessary to vaporize 1 gallon of gasoline. The density of gasoline is 0.730 g/mL, and the heat of vaporization of gasoline is 349 J/g. One gallon is equal to 3785 mL.

SOLUTION

Step 1. Determine the mass of gasoline present using $D = m/V$ or $m = D \times V$:

$$0.730 \ \frac{g}{mL} \times 3785 \ mL = 2763 \ g$$

Step 2. The heat required to vaporize 73.0 g of gasoline is

$$2763 \ \cancel{g} \times \frac{349 \ J}{\cancel{g}} = 6.943 \times 10^5 \ J$$

Practice 9.11

How many grams of Cu can be melted by heat energy of 7.22×10^3 J?

Answer

❏ 34.9 g

9.7 The First Law of Thermodynamics and Calorimetry

LEARNING OBJECTIVE

Describe how to measure the amount of heat produced in combustion reactions.

As discussed in previous sections of this chapter, combustion reactions require not only activation energy but also energy for phase changes and pyrolysis to convert fuel materials into gaseous reactants. However, once a combustion reaction becomes self-sustaining, more energy is produced by the fire than is required to sustain the reaction. What happens to this excess energy?

The **first law of thermodynamics**, also known as the *law of conservation of energy*, is based on the principle that energy cannot be created or destroyed. The amount of energy released by a process must be equal to the amount of energy absorbed by the rest of the system. For example, the energy released in a fire is absorbed by the air surrounding the fire or by any object in the vicinity of the reaction. The total amount of energy before and after the fire has remained constant.

The law of conservation of energy is used experimentally in the science of **calorimetry**. Calorimetric methods determine the heat released in a combustion reaction by measuring the increase in the temperature of water that surrounds the reaction. A typical calorimeter is shown in Figure 9.6.

The calorimeter consists of a sealed chamber that is completely submerged in a reservoir of water. The chamber contains a fuel source and oxygen gas. The fuel is ignited by an electrical spark from an external power supply. Energy from the combustion reaction is fully absorbed by the water surrounding the chamber, causing the temperature of the water to increase. The heat released can then be determined from the specific heat of water, the temperature change of the water sample, and the volume of water.

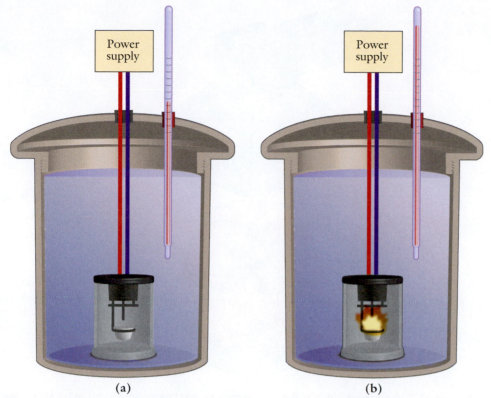

FIGURE 9.6 Calorimeter for measuring energy content. (a) The inner chamber of a calorimeter contains a sample for analysis (fuel), a pure oxygen environment, and an ignition source. (b) The heat produced by the combustion reaction is transferred to the water that completely surrounds the reactions vessel. The heat produced can be measured from the increase in the temperature of the water.

This instrument is often called a **bomb calorimeter** because the fuel is converted to gaseous products that are confined within a sealed container at high pressures. Care is taken to use small amounts of fuel, and the container is typically constructed out of thick stainless steel to withstand the pressure.

Calorimeters are also used to study the amount of heat produced by burning furniture, upholstery, carpet, and various construction materials. Because fire investigators are interested not only in how much heat is produced when such materials burn, but also in how long it takes for these materials to produce their maximum heat output, a slightly modified calorimeter, called a **cone calorimeter**, is used. A cone calorimeter is designed to measure the heat of combustion of ignitable materials by monitoring the decrease in oxygen in the air collected above a burning sample. Figure 9.7 shows a typical cone calorimeter.

A cone calorimeter functions by heating a sample until it ignites, capturing the gases above the flames in a hood and passing the gases by an oxygen sensor. It is known that in a combustion reaction, approximately 13 kJ of heat are produced per gram of oxygen gas consumed. The decrease in oxygen gas concentration above the fire corresponds to greater consumption of oxygen gas and, therefore, to a greater release of heat. Cone calorimeters are used because they rely on atmospheric oxygen to supply the reaction, whereas a bomb calorimeter uses a pure oxygen environment. There are even large-scale calorimeters for studying the combustion of entire pieces of furniture.

Cone calorimeters have changed the way fire investigators evaluate the cause of a fire. Over the last 50 years, our homes have been increasingly filled with foam materials, especially in furniture and mattresses. Because foams are made from petroleum

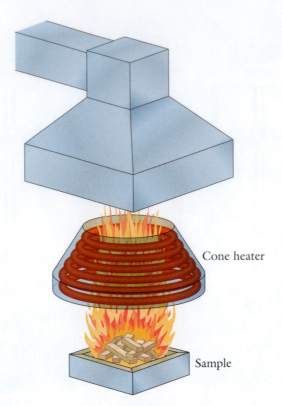

Cone heater

Sample

FIGURE 9.7 A cone calorimeter. A cone calorimeter heats a sample to the point of ignition while measuring the concentration of oxygen in the air above the burning object. The amount of oxygen consumed is used to calculate the amount of heat released by the burning sample. The heat released is recorded as a function of time to provide an accurate depiction of how the sample material would behave in real fire.

products, they tend to make excellent fuel sources and often burn very easily, releasing a large amount of heat back into the fires. Heat is one of the three necessary parts of the fire triangle and one of the key causes for flashovers. Cone calorimeters have helped to show that fires in homes today burn faster and reach higher temperatures sooner than they did 50 years ago.

In the Kenny Richey case, an expert witness testified that the speed at which the fire engulfed the building was corroborating evidence of arson. However, due to the amount of petroleum-based products in a modern home, this is not an accurate method for determining whether a fire was in fact arson.

9.8 Mathematics of Calorimetry

LEARNING OBJECTIVE

Perform calculations in calorimetry.

The heat produced (or absorbed) by a reaction or process must be absorbed (or supplied) by the surroundings. Mathematically, it is necessary to use a sign convention to describe the transfer of heat. If the system under study releases heat through an exothermic process, the amount of heat is indicated by a negative sign. If heat is absorbed in an endothermic process, the amount of heat is indicated by a positive sign.

For example, if a heated metal object is submerged in cold water, heat is transferred from the hot metal to the cold water. The amount of heat leaving the metal is given a negative sign; the amount of heat gained by the water is given a positive sign. According to the law of conservation of energy, the amount of energy lost by the metal must equal the amount of energy gained by the water. The equation for heat transfer becomes

$$q_{lost} = -q_{gained}$$

$$q_{lost} + q_{gained} = 0$$

When the amount of heat lost is added to the amount of heat gained, they must sum to zero. Otherwise, there would be a violation of the law of conservation of energy. Exothermic reactions can also serve as the source of energy in calorimetry problems. In this situation, the energy released by the chemical reaction is absorbed by the water in the calorimeter. For calculations, it is often convenient to summarize the information.

■ WORKED EXAMPLE 12

An unknown amount of copper is heated to 652°C and then plunged into a calorimeter containing 100.0 mL of water at 24.0°C. The temperature of the water increases to 31.3°C. Determine the amount of copper heated. The specific heats are provided in Table 9.1.

SOLUTION

Look up the specific heats of copper and water in Table 9.1 and use the known density of water:

Variable	Copper (lost heat)	Water (gained heat)
Mass	X	$100.0 \text{ mL} \times \dfrac{1.0 \text{ g}}{1 \text{ mL}}$
Specific heat	0.38 J/g·°C	4.184 J/g·°C
Initial temperature	652°C	24.0°C
Final temperature	31.3°C	31.3°C

Calculation 1: The heat gained by the water is calculated using $q = C \times m \times \Delta T$:

$$4.184 \frac{J}{g \cdot °C} \times 100.0 \text{ g} \times (31.3°C - 24.0°C) = 3.05 \times 10^3 \text{ J}$$

Calculation 2: The heat released by the copper is calculated similarly:

$$0.38 \frac{J}{g \cdot °C} \times X \times (31.3°C - 652°C) = \left(-236 \frac{J}{g}\right) X$$

Calculation 3: Since the heat lost plus the heat gained is zero,

$$\left(-236 \frac{J}{g}\right) X + 3.05 \times 10^3 \text{ J} = 0$$

$$X = \frac{-3.05 \times 10^3 \text{ J}}{-236 \frac{J}{g}} = 12.9 \text{ g}$$

Practice 9.12

The heat of combustion of gasoline is -4.40×10^4 J/g. How many grams of gasoline are burned in a calorimeter containing 3000.0 mL of water if the temperature increases from 25.5°C to 46.5°C?

Answer

❑ 5.99 g of gasoline

WORKED EXAMPLE 13

Calculate the specific heat of an unknown gray-colored metal, given that a 24.0 g sample of the metal was heated to 99.8°C and then submerged into 50.0 g of water initially at 23.2°C. The final temperature of the water–metal mixture was 29.2°C.

SOLUTION

Variable	Unknown Metal (lost heat)	Water (gained heat)
Mass	24.0 g	50.0 g
Specific heat	X J/g·°C	4.184 J/g·°C
Initial temperature	99.8°C	23.2°C
Final temperature	29.2°C	29.2°C

Calculation 1: Heat lost by the unknown metal is calculated using $q = C \times m \times \Delta T$:

$$X \text{ J/g·°C} \times 24.0 \text{ g} \times (29.2 \text{ °C} - 99.8°C)$$

Calculation 2: Heat gained by the unknown metal is calculated using $q = C \times m \times \Delta T$:

$$4.184 \text{ J/g·°C} \times 50.0 \text{ g} \times (29.2 \text{ °C} - 23.2°C)$$

Calculation 3: Heat lost + heat gained = 0

$$X \text{ J/g·°C} \times 24.0 \text{ g} \times (29.2°C - 99.8°C) + 4.184 \text{ J/g·°C} \times 50.0 \text{ g} \times (29.2°C - 23.2°C) = 0$$
$$-1.683 \times 10^3 X + 1.255 \times 10^3 = 0$$
$$-1.683 \times 10^3 X = -1.255 \times 10^3$$
$$X = -1.255 \times 10^3 / -1.683 \times 10^3 = 0.746 \text{ J/g·°C}$$

Practice 9.13

A calorimeter was used to determine the specific heat of an unknown metal by submerging an 81.0 g sample of the unknown, initially at 99.7°C, into 75.0 g of water at 24.4°C. The final temperature of the water is 44.1°C. Determine the specific heat of the unknown metal.

Answer

❑ 1.37 J/g·°C

9.9 Petroleum Refinement

Gasoline and paint thinners, the accelerants allegedly used by Kenny Richey, are the most common accelerants used in arson cases. These liquids and a few others, such as kerosene and lighter fluid, comprise a set of highly ignitable and easily obtained liquids that could be used to accelerate a fire in a home or business.

What interests forensic scientists who perform accelerant analysis of these ignitable liquids is that they are homogeneous mixtures made during the processing of crude oil from many different compounds. To understand how these mixtures are analyzed and identified, it is necessary to understand the basics of oil refining. The In the Lab box on page 276 goes into detail about how the flammable petroleum products are analyzed using gas chromatography.

Crude oil contains thousands of compounds. To refine the crude oil into usable products, it is first heated at the base of a tower, as shown in Figure 9.8. The temperature within the tower is lowest at the top and highest at the bottom. As crude oil is heated,

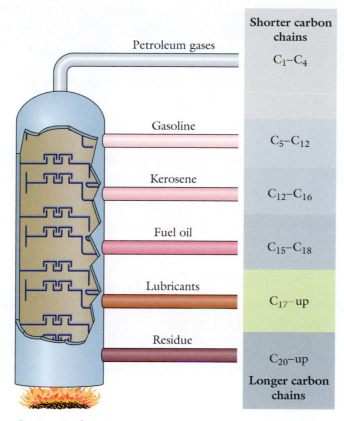

Petroleum gases

Shorter carbon
chains

$C_1–C_4$

Gasoline

$C_5–C_{12}$

Kerosene

$C_{12}–C_{16}$

Fuel oil

$C_{15}–C_{18}$

Lubricants

$C_{17}–up$

Residue

$C_{20}–up$

Longer carbon
chains

FIGURE 9.8 Crude-oil refinement. (Adapted from www.schoolscience.co.uk/content/4/chemistry/fossils/p8.html)

compounds with low boiling points, such as butane and propane, enter the gas phase and proceed to travel upward toward the top of the tower. When a compound reaches a point in the tower where the temperature is lower than its boiling point, the compound condenses back into liquid form. The compounds are in this way grouped by boiling points and removed from different heights along the tower. Gasoline has components with boiling points from approximately 40°C to 220°C. The boiling points of kerosene's components range from 175°C to 270°C.

There are two categories of ignitable liquids—flammable liquids and combustible liquids. The difference between the two is based on their volatility. A **flammable liquid** is defined as a liquid that produces sufficient vapors to support combustion at a temperature lower than 37.8°C. A **combustible liquid** is one that must reach a temperature higher than 37.8°C for sufficient vapors to be produced to support combustion. Gasoline is an example of a flammable liquid. Kerosene is an example of a combustible liquid. If you are wondering why such an odd temperature was chosen, the definition is based on the Fahrenheit scale—37.8°C is the same as 100°F.

Each product made in the refinement of crude oil has a unique set of compounds. The mixture of compounds that make up gasoline is different from the mixture of compounds that make up kerosene. To identify whether an accelerant is kerosene or gasoline, it is necessary to determine what components make up the unknown accelerant and compare those components with standard samples of each accelerant. Chromatography provides an experimental method for separating and detecting multiple compounds within a mixture.

Gas Chromatography

If an accelerant is used to start a fire, trace amounts of liquid recovered after the fire can be used to identify the type of ignitable liquid used. The compounds present in a sample of gasoline are different from the compounds present in a sample of kerosene. To identify the accelerant, the identity and types of compounds present in the sample must be determined. **Gas chromatography (GC)** is the ideal method for analyzing volatile liquids because it separates compounds based on their boiling points and their polarity. A sketch of a typical gas chromatography system is shown in Figure 9.9.

The sample to be analyzed is loaded into a syringe and injected into a heated injector port. The amount of sample needed is very small, as little as several microliters (μL). The heated injector serves two purposes: It vaporizes the liquid sample and loads the sample onto the column that contains the stationary phase. The mobile phase, usually helium gas, carries the sample through a long, narrow glass tube, called a *capillary column*. The capillary column is coated on the inside surface with a thin layer of a nonvolatile liquid that is in the stationary phase. The capillary column is kept inside an oven

so that the temperature can be controlled. The most volatile compounds in the sample, those with the lowest boiling points, are quickly swept through the column by the helium gas. The compounds that have similar polarities to the stationary-phase coating will take longer to reach the detector. The temperature of the column is typically increased throughout the experiment so that those compounds with higher boiling points will be able to travel through the column to the detector in a reasonable amount of time.

When each compound reaches the detector, it triggers a response in which the peak height is proportional to the compound's concentration in the sample. The time from injection of the sample to the time at which the peak response of the compound is detected is called the **retention time**. As long as all the samples are analyzed on the same chromatography system using the same method, the retention time of each compound will not change. A graph of the data, called a **chromatogram**, is then compared against standards made up from known accelerants. Representative chromatograms of gasoline and diesel fuel are shown in Figure 9.10.

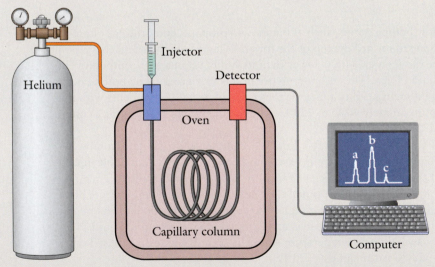

FIGURE 9.9 A typical gas-chromatography system. A mixture of compounds is placed onto a stationary phase, and a mobile phase is passed through the system. Samples injected into the system are vaporized and pass over the stationary phase in a temperature-controlled oven. Each compound spends a different amount of time in the mobile and stationary phases as it passes through the system, based on its physical attraction for the stationary phase and mobile phase. The separation of the compounds is based on the polarity and boiling point of each compound.

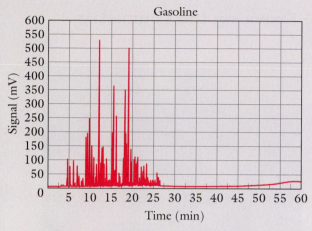

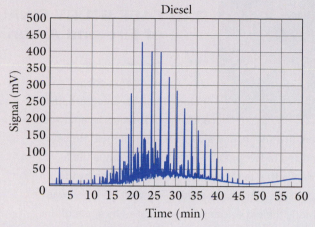

FIGURE 9.10 Each line on the chromatogram represents a unique compound present in gasoline or diesel fuel. Analysis of an arson sample attempts to match the major peaks found in the sample under investigation to peaks obtained from a known sample of accelerant. (www.eti-geochemistry.com/pdf/HRCC-standards.pdf)

If an arson investigator believes that an accelerant was poured on carpeting, the investigator will collect a sample of the carpeting far away from the suspected point of origin of the fire. This is called a *blank sample* because it is used to determine the background levels of organic compounds. Some of the same compounds that make up gasoline and kerosene can be found in household cleaners, insecticides, and adhesives used for carpeting or laminate flooring. These compounds may also be produced as a result of pyrolysis of plastics and polymers. One could get a false positive reading for an accelerant if a blank sample is not evaluated.

Suspected arson samples usually consist of charred wood, clothing, or carpeting taken from the fire scene and can be analyzed using a method called **headspace analysis**, as illustrated in Figure 9.11. To isolate the possible accelerants, the samples are sealed in air-tight vials. If an accelerant such as gasoline or kerosene is present, vapors from these volatile compounds build up inside the sealed container. The air above the charred sample, called the *headspace,* is then sampled with a gas-tight syringe and injected into the gas chromatography system for analysis.

A more sophisticated method, referred to as *gas chromatography-mass spectroscopy* (GC-MS), is used to measure not only the retention time but also the molecular mass of each compound. This

ensures that peaks appearing at a common retention time on two different chromatograms actually represent the same compound. Mass spectroscopy is covered in the In the Lab box in Chapter 10.

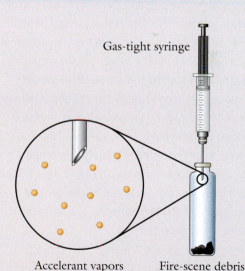

Gas-tight syringe

Accelerant vapors   Fire-scene debris

FIGURE 9.11 Fire debris is analyzed for accelerants by extracting the air (headspace) from a sealed vial containing the debris with a gas-tight syringe. If any accelerant is present, it will be found in the vapor phase and can be detected in the gas chromatograph.

9.10 CASE STUDY FINALE: Exploring the Chemistry of Fire and Arson

The scientific evidence against Kenny Richey included gas chromatography analysis of fire debris, analysis of the fire scene, and a witness who claimed Richey admitted to torching the apartment. What is the new evidence that is referred to in the quotation by the prosecutor in the case study? The new evidence was actually the old evidence as evaluated by some of the leading experts in the world. In Kenny's first trial, his lawyer hired an inept and unqualified expert witness who ended up testifying for the prosecution.

Dr. Andrew Armstrong was the expert responsible for analyzing the debris from the fire that killed 71 people at the Branch Davidian Compound in Waco, Texas in 1993. In the Kenny Richey case, Dr. Armstrong testified about the gas chromatography evidence analysis and determined that not only were the samples improperly taken, they were improperly stored. He had serious doubts about the integrity of the samples because of the way the evidence was handled. The fire marshal initially ruled the fire was an accident and allowed the property owner to remove all of the apartment's contents for disposal at the local landfill. When the police decided to investigate the fire as a crime, officers were sent to the landfill to retrieve the carpeting from the apartment! Most likely because of the carpet's large size, odor, and damaged condition, the police decided to leave the carpet outside on the asphalt next to the gas pump, as shown in the upper right corner of Figure 9.12.

Armstrong also evaluated the analytic methods used by the original forensic scientists and determined that the methods were not accepted by the general scientific community. Furthermore, he concluded that the data were misinterpreted.

According to Armstrong, the evidence actually showed no traces of ignitable fluids at all! Were some petroleum products found on the samples? Yes. But were they the same compounds that would be found in the suspected accelerants? Armstrong testified that the samples were not even close to what would be expected if accelerants were present.

FIGURE 9.12 Improper collection and storage of possible arson evidence. (Courtesy of Goodwin Procter, LLP)

The new evidence also consisted of testimony by a nationally respected expert in the analysis of fire scenes, Richard L. P. Custer. He debunked the initial analysis of the fire scene, which the prosecutors had claimed to show evidence of arson. During the original trial, evidence of wood charring between the edges of wood planks on the deck flooring was presented as proof that a liquid accelerant had been poured and leaked in between the boards. Custer explained that this is in fact a very common pattern caused by air (oxygen) rushing up between the wood decking to feed the fire.

Another piece of evidence used against Kenny Richey in the original trial was that there were pooling marks near the furniture, which were explained as marks left by pools of burning accelerant. Again, an alternative explanation is that modern furniture contains foam made from petroleum ingredients. The foam melts into a pool when ignited, leaving marks identical to those created by accelerants.

It is important to analyze evidence samples from areas believed to contain accelerants as well as from areas that do not. The reason is that many modern household products are derived from petroleum materials, such as foam, glues, and plastics. The presence of petroleum-based chemicals alone is not evidence of arson unless it can be shown that such chemicals are not present in the rest of the home.

Finally, the witness who testified against Richey almost immediately recanted her testimony after the trial, insisting that she was nervous and felt pressured to say what she thought the prosecutor wanted her to say. Yet the prosecutors chose to stay with their original view of the testimony and interpretations given by the laboratory personnel who analyzed the samples.

In an appeal, Richey sought a new trial at which to present the new evidence. During that time, the prosecutor did not challenge any of the arguments debunking the original physical evidence. The prosecution instead decided to focus on the circumstantial evidence and testimony of witnesses to justify the verdict, despite relying heavily on the physical evidence during the original trial.

In 2006, the U.S. Court of Appeals for the Sixth Circuit reversed Kenny Richey's conviction and ordered either his release or a new trial. The state of Ohio appealed to the

Released from Death Row

Supreme Court of the United States, where the decision of the Sixth Circuit Court was overruled and the justices sent the case back to the Sixth Circuit. In the spring of 2007, the Sixth Circuit Court once again ruled in Kenny Richey's favor and ordered either a new trial or his release within 90 days.

The state indicated it would go to trial, yet several months later, Kenny was offered a plea bargain. The murder and arson charges would be dropped if he agreed to plead no contest to charges of breaking and entering and child endangerment. He would then be sentenced to time served and released from prison. While Kenny has always sought to have his name cleared in a court, the plea of no contest was not an admission of guilt. Rather, it meant that he simply would offer no defense to the charges. On January 7, 2008, he was released from prison and returned to Scotland to be reunited with his mother and family.

Freedom, however, presents a new set of challenges for Kenny as he tries to adjust to life outside prison. As is common in such cases, he admits to battling depression and suicidal thoughts because he is overwhelmed by the outside world and trying to figure out how he will fit into a society he left 21 years ago. Kenny has also recently learned that he has cancer of the mouth, most likely due to his years of chain smoking. The fighting spirit he cultivated as a U.S. marine, which helped him survive his years in prison, will now go to fighting cancer and finding peace in his new world.

Richey returned several years ago to the United States to reconnect with his terminally ill father as well as his ex-wife and son. He has recently entered a rehabilitation center to help fight his dependence on alcohol.

Kenny Richey arrives in his home country of Scotland after being freed from death row in an American prison. (David Cheskin/PA Wire URN:5496063 Press Association via AP Images)

CHAPTER SUMMARY

• A fire is a self-sustaining chemical reaction that produces heat and light. Oxygen, heat, and a fuel must simultaneously be present for a fire to exist. If one of these three ingredients is removed, the fire will not burn.

• Heat is a form of energy that is transferred between objects at different temperatures. Temperature is a measure of the average kinetic energy of an object. When heat is transferred from one object to another, the heat will flow from the warmer to the colder object until thermal equilibrium is achieved.

• Solid fuels must undergo pyrolysis to form smaller gaseous molecules. Liquid fuels must be vaporized before they will combust. Pyrolysis and the vaporization process require that energy be consumed (endothermic process). The overall combustion reaction releases energy (exothermic process) because more heat is produced than consumed in the reaction.

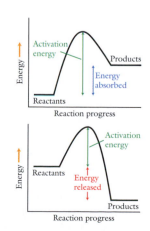

- The activation energy for a reaction is the energy required to initiate the breaking of bonds in the reactant molecules.

- The specific heat capacity of a substance is the amount of energy required to increase the temperature of 1 gram of the material by 1°C and is determined by the intermolecular forces present in the substance.

- Energy is required by a material to overcome intermolecular forces during a phase change from solid to liquid (the heat of fusion) and from liquid to gas (the heat of vaporization).

- The first law of thermodynamics, also called the law of conservation of energy, states that energy is conserved in any physical or chemical process. This law is the basis of calorimetry experiments.

- Gas chromatography is an instrumental method that exploits differences in the boiling points and polarities of compounds to separate the components of a mixture. As the compounds travel the length of the chromatographic column, they interact with a stationary-phase material. Compounds with polarities similar to those of the column's stationary phase tend to be slowed. This results in a unique retention time, the time from injection until the compound is detected at the end of the column, for each compound.

KEY TERMS

activation energy (p. 263)	endothermic (p. 262)	heat of fusion (p. 268)
arson (p. 254)	exothermic (p. 263)	heat of vaporization (p. 266)
autoignition (p. 264)	fire (p. 254)	joule (J) (p. 255)
backdraft (p. 256)	fire triangle (p. 254)	oxidation number (p. 260)
bomb calorimeter (p. 271)	first law of thermodynamics (p. 270)	pyrolysis (p. 262)
calorie (cal) (p. 255)	flammable liquid (p. 275)	retention time (p. 276)
calorimetry (p. 270)	flashover (p. 264)	specific heat capacity (p. 265)
chromatogram (p. 276)	gas chromatography (GC) (p. 276)	temperature (p. 255)
combustible liquid (p. 275)	headspace analysis (p. 277)	thermal equilibrium (p. 255)
cone calorimeter (p. 271)	heat (p. 255)	

MAKING MORE CONNECTIONS: Additional Readings, Resources, and References

National Fire Protection Association, *NFPA 921 Guide for Fire and Explosion Investigations*, Quincy, MA: National Fire Protection Association, 2004.

Richey, Tom. *Death Row Scot—My Brother Kenny's Fight for Justice*, Edinburgh: Black & White Publishing, 2005.

For information about the composition of gasoline and standards in gas chromatography: www.atsdr.cdc.gov/toxprofiles/tp72-c3.pdf

REVIEW QUESTIONS AND PROBLEMS

Questions

1. Draw a molecular view, similar to the one shown in Figure 9.1, for equal masses of Cu at 50°C and at 500°C that come in thermal contact. Explain what would happen if the mass of the Cu at 500°C were twice that of the Cu at 50°C. (9.1)

2. Draw a molecular view, similar to the one shown in Figure 9.1, for equal masses of H_2O at 50°C and at 100°C that are mixed together. Explain what would happen if the mass of the H_2O at 50°C were twice that of the H_2O at 100°C. (9.1)

3. What is the difference between heat and temperature? (9.1)

4. What are the common units for energy and how are they defined? (9.1)

5. Why does a thermometer placed in a solution alter the temperature it is attempting to measure? (9.1)

6. What is the difference between mercury atoms at 20°C and mercury atoms at 50°C? How is this behavior exploited in a mercury-filled thermometer for measuring temperature? (9.1)

7. What combustion conditions can lead to a backdraft? How can a backdraft be prevented? (9.2)

8. How does the absence or presence of smoke, and the color of smoke when it is present, indicate the efficiency of a combustion reaction? Give the main products of the reaction in each example you provide. (9.2)

9. What types of compounds can undergo combustion reactions? Give an example of a specific compound from each type and write a balanced equation for its combustion reaction. (9.2)

10. Does the oxidation number of an atom represent a real charge on the atom? (9.3)

11. How can you determine if an element within a compound is oxidized or reduced in a reaction? What precaution has to be followed in determining which compounds are oxidized and which are reduced? (9.4)

12. What is the nature of the activation energy barrier in a chemical reaction? (9.4)

13. What determines whether a reaction is endothermic or exothermic? (9.4)

14. In which type of reaction, endothermic or exothermic, do the products have more energy than the reactants? (9.4)

15. Explain how the process of a flashover is related to the activation energy of a combustion reaction. How can flashover conditions be prevented? (9.4)

16. Despite the addition of energy to a system during a phase transition, an increase in the temperature does not occur. Explain this observation. (9.5)

17. How is specific heat capacity influenced by intermolecular forces? What types of compounds would tend to have large specific heat capacities? (9.5)

18. How are the heat of fusion and heat of vaporization influenced by intermolecular forces? What types of compounds require very little heat to melt or vaporize? (9.5)

19. Tile floor will feel colder than carpeting despite being in the same room and at the same temperature. Explain the nature of this sensation by discussing relative heat capacities and the flow of heat from one object to another. (9.5)

20. Sketch a temperature-versus-heat curve for an ice cube at −5°C that is converted to steam at 115°C. Label each region with the phases present and the physical processes that are happening. (9.5)

21. Why are accelerants typically not completely consumed in a fire? What process involving the liquid fuel can actually serve to cool it? (9.5)

22. Explain how the first law of thermodynamics is used in calorimetry. When a heated object cools down and releases heat, what happens to that energy? (9.7)

23. Explain how crude oil is refined, and describe the chemical makeup of the resulting petroleum products. (9.9)

24. Describe how a mixture of volatile compounds in gasoline is separated by gas chromatography. What would the difference be between a kerosene sample and a gasoline sample? (ITL)

Problems

25. Convert the following amounts of energy to units of joules. Recall that 1 Cal = 1000 cal. (9.1)
    (a) 65.0 calories (c) 74.6 kilojoules
    (b) 166.0 Calories (d) 82.3 kilocalories

26. Convert the following amounts of energy to units of joules. Recall that 1 Cal = 1000 cal. (9.1)
    (a) 3.58×10^4 Calories (c) 5.36×10^{-3} kilojoules
    (b) 1.00 Calorie (d) 2.00×10^3 calories

27. Heats of combustion for many fuels are reported in units of Cal/g. Convert the following heats of combustion to values of J/g. (9.1)
    (a) Kerosene: 11.0 Cal/g
    (b) Gasoline: 11.5 Cal/g
    (c) Benzene: 10.0 Cal/g
    (d) Paraffin wax: 9.92 Cal/g

28. Heats of combustion for many fuels are reported in units of Cal/g. Convert the following heats of combustion to values of J/g. (9.1)
    (a) Octane: 11.4 Cal/g (c) Fuel oil: 10.7 Cal/g
    (b) Alcohol: 7.4 Cal/g (d) Methanol: 5.42 Cal/g

29. Heats of combustion for many fuels are reported in units of megajoules per kilogram (MJ/kg), mega being the prefix for one million. Convert the following heats of combustion to values of J/g. (9.1)
    (a) Kerosene: 43.3 MJ/kg (c) Gasoline: 43.7 MJ/kg
    (b) Jet fuel: 43.5 MJ/kg (d) Fuel oil: 44.0 MJ/kg

30. Heats of combustion for many fuels are reported in units of megajoules per kilogram (MJ/kg), mega being the prefix for one million. Convert the following heats of combustion to values of J/g. (9.1)
    (a) Propane: 50.0 MJ/kg
    (b) Coal: 30.5 MJ/kg
    (c) Octane: 47.7 MJ/kg
    (d) Wood (average): 20.0 MJ/kg

31. Write and balance equations for combustion of the following compounds. Assume complete combustion to form carbon dioxide and water. (9.2)
    (a) CH_4 (b) C_5H_{12} (c) $C_6H_{14}O$ (d) $C_8H_{16}O$

32. Write and balance equations for combustion of the following compounds. Assume complete combustion to form carbon dioxide and water. (9.2)
    (a) C_3H_8 (b) C_4H_8 (c) C_7H_{12} (d) C_2H_6O

33. Write and balance equations for combustion of the following compounds. Assume complete combustion to form carbon dioxide and water. (9.2)
    (a) Butane (c) 2-pentene
    (b) Octane (d) 1-hexyne

34. Write and balance equations for combustion of the following compounds. Assume complete combustion to form carbon dioxide and water. (9.2)
    (a) Ethanol   (c) 2-butanone
    (b) Ethanal   (d) 3-pentanone

35. Write and balance equations for the combustion of the following compounds. Assume complete combustion to form carbon dioxide and water. (9.2)
    (a) Propane   (c) 1-Propanol
    (b) Pentane   (d) 1-Pentanol

36. Write and balance equations for the combustion of the following compounds. Assume complete combustion to form carbon dioxide and water. (9.2)
    (a) Hexane   (c) Octanal
    (b) 2-octanone   (d) 1-octanol

37. Assign oxidation numbers to each element in the following compounds or ions. (9.3)
    (a) Al^{3+} (b) FeO   (c) H_2S (d) Fe_2O_3

38. Assign oxidation numbers to each element in the following compounds. (9.3)
    (a) $MgSO_4$ (b) TiO_2 (c) NH_3 (d) CH_3OH

39. In the following reactions, determine which element in the reactants is oxidized and which is reduced. (9.3)
    (a) $6CO_2(g) + 6H_2O(l) \rightarrow C_6H_{12}O_6(s) + 6O_2(g)$
    (b) $CH_4(s) + 2O_2(g) \rightarrow CO_2(g) + 2H_2O(l)$

40. In the following reactions, determine which element in the reactants is oxidized and which is reduced. (9.3)
    (a) $Fe(s) + O_2(g) \rightarrow Fe_2O_3(s)$
    (b) $CuCl_2(aq) + Zn(s) \rightarrow Cu(s) + ZnCl_2(aq)$

41. In the following reactions, determine which element in the reactants is oxidized and which is reduced. (9.3)
    (a) $2CO(g) + O_2(g) \rightarrow 2CO_2(g)$
    (b) $2Na(s) + 2H_2O(l) \rightarrow 2NaOH(aq) + H_2(g)$

42. In the following reactions, determine which element in the reactants is oxidized and which is reduced. (9.3)
    (a) $Cu(s) + O_2(g) \rightarrow CuO(s)$
    (b) $C_2H_5OH + 3O_2(g) \rightarrow 2CO_2(g) + 3H_2O(l)$

43. Calculate the specific heat capacity of the metal in a 421 g block that requires 1.00×10^4 J of heat to raise the temperature from 24.7°C to 49.7°C. (9.6)

44. Calculate the specific heat capacity of a concrete cinder block with a mass of 1.8×10^4 g that requires 1.96×10^6 J of heat to raise the temperature from 24°C to 242°C. (9.6)

45. Determine how much heat is required to increase the temperature of a 500.0-g sample of PVC (polyvinyl chloride, a common type of plastic) from 22°C to its autoignition temperature of 507°C, given that the specific heat capacity of PVC is approximately 1.2 J/g·°C. (9.6)

46. Determine how much heat is required to increase the temperature of a 72.1 g sample of No. 2 diesel fuel from 25.0°C to its autoignition temperature of 600.0°C, given that the specific heat capacity of the fuel is 1.8 J/g·°C. (9.6)

47. Determine the mass of copper that would release 488 J of heat while cooling from 284°C to 25°C. (9.6)

48. Determine the mass of water that would release 1.50×10^3 J of heat while cooling from 100°C to 22°C. (9.6)

49. Determine the amount of energy absorbed by 2.00 L of water as it is converted to steam at its boiling point. (9.6)

50. Determine the amount of energy absorbed by 2.00 L of gasoline as it is converted to the vapor phase at its boiling point. (9.6)

51. Calculate the amount of water at its boiling point that would be vaporized by 152.0 Calories. (9.6)

52. Calculate the amount of gasoline at its boiling point that would be vaporized by 152.0 Calories. (9.6)

53. How much heat is required to melt 84.4 g of copper starting at a temperature of 25°C? The melting point of copper is 1082°C. (9.6)

54. How much heat is required to convert 42.6 g of ice at 0°C to steam at 100°C? (*Hint*: (1) Melt the ice. (2) Heat the water. (3) Boil the water to steam.) (9.6)

55. Draw a temperature-versus-heat curve depicting the processes indicated in Problem 53. Label all regions, list all phases present, and indicate the melting point. (9.5)

56. Draw a temperature-versus-heat curve depicting the processes indicated in Problem 54. Label all regions, list all phases present, and indicate the melting point and boiling point. (9.5)

57. A 15.6 g sample of an unknown metal, initially at 225°C, heated 100.0 g of water from 21.2°C to 25.3°C upon being submerged in the water. Determine the specific heat of the unknown metal, given that the specific heat of water is 4.184 J/g·°C. (9.8)

58. A 22.5 g sample of an unknown metal, initially at 189°C, heated 100.0 g of water from 25.0°C to 27.3°C upon being submerged in the water. Determine the specific heat of the unknown metal, given that the specific heat of water is 4.184 J/g·°C. (9.8)

59. What is the initial temperature of a 24.9 g copper slug if it raises the temperature of 50.0 g of water from 20.3°C to 34.1°C upon submersion? The specific heat of copper is 0.38 J/g·°C and of water is 4.184 J/g·°C. (9.8)

60. What is the initial temperature of a 57.4 g concrete brick slug if it raises the temperature of 100.0 g of water from 18.4°C to 78.3°C upon submersion? The specific heat of concrete is 0.80 J/g·°C and of water is 4.184 J/g·°C. (9.8)

61. If the heat of combustion of Teflon is 5×10^3 J/g, how many grams of Teflon are consumed in a bomb calorimeter if 1000.0 mL of water increases in temperature from 25.5°C to 35.8°C? (9.8)

62. If the heat of combustion of gasoline is 47 kJ/g, how many grams of gasoline are required to heat 2500.0 mL of water in a bomb calorimeter from an initial temperature of 21.9°C to a final temperature of 67.8°C? (9.8)

63. If the heat of combustion of newspaper is 1.97×10^4 J/g, what is the final temperature that 500.0 mL of water, initially at 25.0°C, will reach in a bomb calorimeter if 1.20 g of newspaper is burned? (9.8)

64. The heat of combustion of wood is approximately 16 kJ/g. If wood was used in Problem 63 rather than newspaper, what would be the final temperature of the water? (9.8)

65. If 3.35 g of a flammable fuel is ignited in a bomb calorimeter and the temperature of 1.00 L of water increases from 26.8°C to 31.2°C, what is the heat of combustion (J/g) of the fuel? (9.8)

66. If 1.75 g of a flammable fuel is ignited in a bomb calorimeter and the temperature of 1.00 L of water increases from 24.1°C to 31.0°C, what is the heat of combustion (J/g) of the fuel? (9.8)

Case Study Problems

67. Fire sprinklers were first patented in 1872 and operate on a simple principle: Heat from a fire triggers a physical change in a material that is plugging a hole in the water pipe. The plug can be either a piece of metal or a liquid-filled glass vial. What are the desirable properties for (a) a solid plug and (b) a liquid-filled vial? (9.5)

68. The optimal use of water at a fire is to efficiently remove the heat from the fire by completely vaporizing the water to steam. An added benefit is that 4.0 L of water will produce approximately 6.0 m³ of steam capable of displacing oxygen and preventing it from reaching the fuel source. Can too much water be added to a fire to fight it effectively? (CP)

69. If benzene is found in a blank sample of carpeting that has been analyzed by gas chromatography, could the presence of benzene in the suspected sample still be used to determine whether an accelerant was used to ignite the fire? (CP)

70. If benzene is not present in the blank sample of carpeting analyzed by gas chromatography but is

present in the suspected sample, does this constitute proof that a benzene-containing fuel was used to start the fire? (CP)

71. If signs of electrical arcing are present in an electrical box, is this proof that a structural fire was caused by the electric arcing? (CP)

72. What thermal properties are desirable in the equipment a firefighter uses and carries into fires? (CP)

73. Sometimes flames appear to be forming from the smoke that is pouring out of a structural fire. Is this possible? (9.2)

74. In a structural fire, what is most commonly the limiting reagent? When is it advisable for firefighters to ventilate a room? (9.2)

75. Would the heat of combustion calculated in a bomb calorimeter be the same as that observed in a cone calorimeter? Explain why or why not. (9.7)

76. The human body has an approximate specific heat capacity of 3.5 J/g·°C. Calculate the amount of heat that will be released to the surroundings as a body cools from 36.8°C to the outside temperature of 17.5°C, assuming an average adult has a mass of 70 kg. Estimating the time of death based on body temperature is complicated because of the many variables that can dramatically alter the rate at which a body cools down. Discuss what effect the amount of clothing might have on estimating the time of death. Would a body found in water cool faster or slower than a body, similarly clothed, that is found in the woods? Also consider the effects of strong wind. What if the person lay dying for several hours before succumbing to the cold (in winter)?

77. In the Kenny Richey case, a plastic fire detector in the apartment building was found melted and dangling from the ceiling by wires. Is this proof of tampering, or is there a reasonable alternative explanation based on the physical properties of the components and basic principles of fire science? (CS)

78. In the Kenny Richey case, the fact that the apartment building became engulfed so quickly was used as evidence that the fire must have been set with an accelerant. Is this a scientifically sound view? (CS)

79. In the Kenny Richey case, the deck boards were found to have charring between the boards, a fact presented as proof that an accelerant had been poured across the deck's surface and seeped down between the cracks in the boards. Is there another explanation, based on the three requirements of fire, that would explain this observation? (CS)

80. What problems exist in relying on any information obtained from the analysis of the carpet of the apartment fire in the Kenny Richey case? (CS)

CHAPTER 10

Chemistry of Explosions

CASE STUDY: Exploring Airport Security

Whether you are a seasoned traveler with frequent flier miles or a novice, getting through airport security can be the equivalent of an obstacle course: Take off your jacket, your shoes, and all metal objects; pull your laptop out of its case; don't lose your boarding pass; place everything in bins for the X-ray machine; decide whether you prefer a physical pat-down or the advanced imaging system that can see through clothing. When you finally reach the finish line, you gather all your belongings and get moving because more people are pressing behind you. Your carry-on luggage also may be chosen for a random screening process in which a Transportation Security Administration (TSA) agent will take a swab to your luggage and then test the swab for explosives.

Once you have successfully navigated this obstacle course, you arrive at your gate and go through the inevitable gate changes. The amount of security at airports across the world has been continually expanding as the means and the methods of extremist terror groups have changed. Unfortunately, it is usually only after a tragedy that increased safety measures are put in place.

In the 1970s, it was the fear of hijackers using guns that prompted the installation of metal detectors. The 1980s witnessed terrorists using explosive devices in luggage to blow up entire flights in the air, which led to mass screening of checked luggage both by hand and by X-rays. The terrorists of September 11, 2001, were able to hijack the planes used in their attacks by means of small but deadly box cutters that were permitted in the cabin under the regulations of the time.

One of the more unusual parts of the security process is that travelers must remove their shoes and have them pass through the X-ray machine. This regulation came about because of a failed bombing attempt on December 22, 2001, just months after the horrific attacks on the World Trade Center and Pentagon buildings. A man named Richard Reid attempted to detonate explosive materials hidden in his shoes while on a flight from Paris to Miami.

> **As you read through the chapter, consider the following questions:**
>
> • **What happens to the explosive compound when it detonates?**
>
> • **What is the physical state of the products of an explosive reaction?**
>
> • **What physical properties of an explosive compound could be used to detect a bomb before it detonates?**

After you pass through a metal detector, you may also find yourself having to pause inside a second detector in which a puff of air is blown over your body.

MAKE THE CONNECTION

What could that puff of air contain that would warn TSA agents of a potential terror plot to blow up an aircraft?

10.1 Explosives 101

LEARNING OBJECTIVE

Differentiate between high and low explosives.

What are the requirements for a compound or mixture to be explosive? Certainly, the explosive material must be capable of releasing a large amount of energy. However, the combustion of coal actually releases much more energy per gram than nitroglycerin does, yet we do not consider coal to be explosive. The difference between the reactions of coal and of nitroglycerin is that the combustion of coal is a slow process whereas nitroglycerin reacts very rapidly.

In an explosion, substantial amounts of energy and gaseous products are released almost instantaneously. The energy produced in the reaction is released as thermal energy and the kinetic energy of flying debris. There are different mechanisms for initiating explosive reactions and different uses for explosions based on the chemical and physical properties of the explosives. Explosives are separated into two broad classes: low explosives and high explosives. A **low explosive** is a compound that must be confined within a container to produce an explosion and can be triggered by a flame that ignites the explosive compound. The combustion reaction of a low explosive occurs in milliseconds (thousandths of a second) and produces a large volume of gaseous products. Gunpowder is an excellent example of a low explosive. If ignited in the open, it will simply burn and exhaust both heat and gases into the surroundings with little effect. However, when gunpowder is confined within the chamber of a gun, it becomes an explosive compound. Most low explosives serve as propellants for bullets in guns and shells in military artillery.

High explosives produce a violent, shattering effect without having to be confined like low explosives. The detonation is initiated either by heat or by a shockwave, and the explosion occurs in microseconds (millionths of a second). The explosion produces a tremendous change in pressure that shatters any material near the explosive.

High explosive materials must be relatively stable compounds to ensure the safe handling of the explosive until it is needed. However, if the explosive material is too stable, the initiation of the reaction will be difficult. It is common to use several types of explosives with varying degrees of stability in commercial or military applications. The bulk of the explosive is made up of a stable explosive compound. A second explosive is housed in a detonator, which serves to trigger the main charge. The explosive in the detonator is usually very reactive and easily triggered, such as lead azide, $Pb(N_3)_2$. Because of its reactivity, only very small amounts are used, usually less than 0.5 gram. The detonator is never brought in contact with the main explosive charge until just before use. Without the detonator, the bulk explosive material is much safer to handle.

Figure 10.1 shows the chemical structures of several common explosive molecules. Note that the structures are those of organic compounds that contain nitro ($-NO_2$) groups.

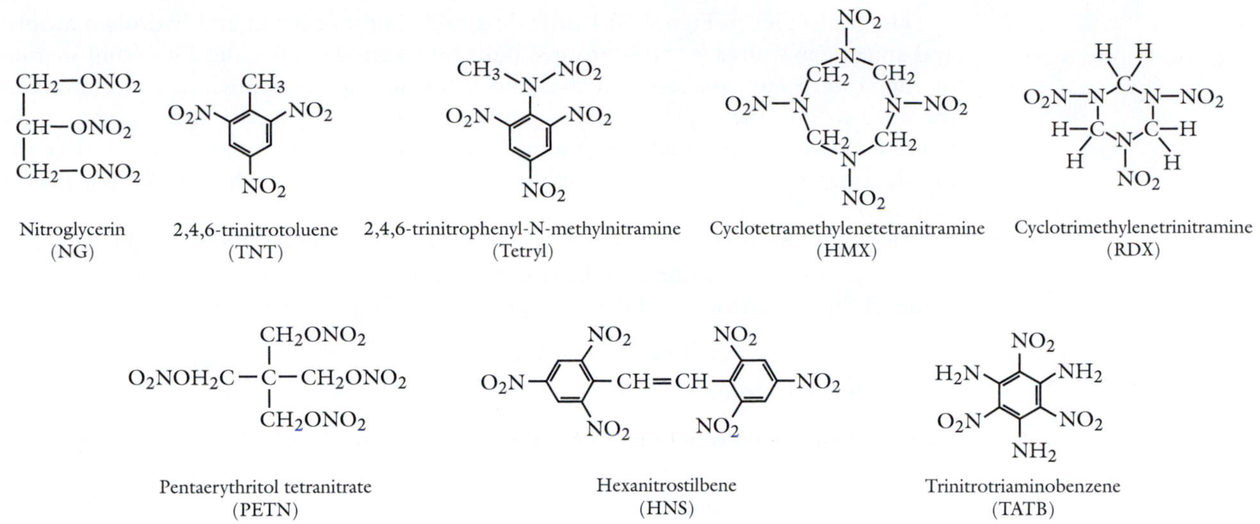

FIGURE 10.1 Structures of explosive molecules. Richard Reid's unsuccessful attempt to blow up an aircraft was done with PETN. However, Pan Am Flight 103 was destroyed by terrorists using RDX on December 21, 1988, over Lockerbie, Scotland.

When these compounds undergo combustion reactions, the products include carbon dioxide, carbon monoxide, nitrogen gas, hydrogen gas, and water vapor. The detonation of these compounds occurs in microseconds, converting nearly all of the solid mass into highly compressed gaseous compounds that expand outward rapidly, creating a blast wave. Consider the detonation of nitroglycerin, as shown in the following reaction:

$$4C_3H_5N_3O_9(l) \rightarrow 12CO_2(g) + 10H_2O(g) + 6N_2(g) + O_2(g)$$

On a molecular level, the reaction of four liquid nitroglycerin molecules will produce 29 total molecules of gas. On the much larger scale of bulk explosives, this reaction produces a tremendous volume of gas. The typical explosion will produce 750 to 1000 liters of gas per kilogram of explosive detonated.

Plastic explosives such as C4 are mixtures of a traditional explosive (RDX) with materials such as rubber or oils that make the mixture moldable. While there are plastics that are true explosives, most plastic explosives are simply moldable mixtures. The plastic explosive that Richard Reid used was a mixture containing PETN as the primary explosive. The Pan Am Flight 103 tragedy was a result of a mixture containing both RDX and PETN.

10.2 Redox Chemistry of Explosives

In Section 9.1 on the chemistry of fire, we discussed combustion reactions and their requirements for fuel and oxygen. Explosions are nothing more than specialized combustion reactions. However, they typically do not rely on oxygen gas from the air. Consider, for example, the reaction of gunpowder:

> **LEARNING OBJECTIVE**
> Determine an explosive molecule's oxygen balance.

$$4KNO_3(s) + 7C(s) + S(s) \rightarrow 3CO_2(g) + 3CO(g) + 2N_2(g) + K_2CO_3(s) + K_2S(s)$$

The carbon component of the reaction is the fuel, and the potassium nitrate serves as the source of oxygen in the reaction. Assigning oxidation numbers to the components of the reaction reveals that carbon is oxidized (loses electrons) and that nitrogen is reduced (gains electrons). Because the fuel source and oxygen source are both solids and must be physically in contact, it is critical for the gunpowder to be a homogenous mixture of fine powders.

The molecules in Figure 10.1 provide a fuel source (carbon and hydrogen atoms) and an oxygen source (—NO$_2$ groups) built into a single molecule. This combination of fuel and oxygen sources increases the efficiency of the explosion and eliminates the need to mix the explosive compounds from Figure 10.1 with other compounds. Some explosive compounds are found in mixtures, but the purpose is to alter the physical or chemical properties of the explosive, as occurs when making a plastic explosive.

The reactions of both nitroglycerin and 2,4,6-trinitrotoluene (TNT) are shown below. Assigning oxidation numbers to the components of each reaction reveals that carbon is oxidized (loses electrons) and that nitrogen is reduced (gains electrons).

Nitroglycerin: $4C_3H_5N_3O_9(l) \rightarrow 12CO_2(g) + 10H_2O(g) + 6N_2(g) + O_2(g)$

2,4,6-trinitrotoluene: $2C_7H_5N_3O_6(l) \rightarrow 7CO(g) + 7C(s) + 5H_2O(g) + 3N_2(g)$

Notice that the detonation of nitroglycerin produces all gaseous compounds, whereas the detonation of TNT produces carbon particles as well as gaseous compounds. To understand the reason for this difference, we must consider a concept called the **oxygen balance** of the explosive.

If there is sufficient oxygen within the molecule to completely oxidize the carbon and hydrogen atoms, the compound is said to have a *neutral* oxygen balance. A *negative* oxygen balance exists when the molecule contains insufficient oxygen atoms for a complete oxidation reaction; a *positive* oxygen balance exists when there is a surplus of oxygen. Nitroglycerin, an example of a compound with a positive oxygen balance, produces a small surplus of oxygen gas. TNT, however, lacks sufficient oxygen to react completely with the carbon present in the explosive and, therefore, has a negative oxygen balance.

The use of explosives for mining or in confined spaces must be done carefully, because carbon monoxide poisoning can occur if explosives with negative oxygen balances are used. Propellants used in firing guns usually have a negative oxygen balance. When the heated, flammable carbon monoxide gas is released at the end of a barrel, it ignites in the muzzle flash.

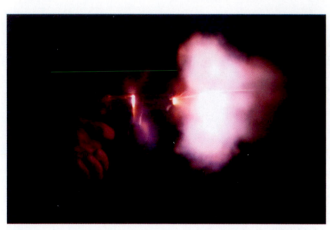

The muzzle flash seen when a gun is fired results from the combustion of hot carbon monoxide gas as it reacts with oxygen gas from the atmosphere. (Scott Doyle/Firearms.com)

■ WORKED EXAMPLE 1

Determine whether the explosive compound RDX from Figure 10.1 has a positive, negative, or neutral oxygen balance. Assume an ideal reaction in which all carbon is converted to carbon dioxide, all hydrogen is converted to water vapor, and all nitrogen is converted to nitrogen gas. Compare the oxygen needed with that present in the molecule.

SOLUTION

The formula of RDX is $C_3H_6N_6O_6$. The complete oxidation of RDX would be:

3 C atoms form 3 CO_2 molecules → requires 6 O atoms

6 H atoms form 3 H_2O molecules → requires 3 O atoms

6 N atoms form 3 N_2 molecules → requires 0 O atoms

Total O atoms needed: 9

RDX contains 6 oxygen atoms, but 9 oxygen atoms are needed. RDX therefore has a negative oxygen balance and would produce CO, not CO_2.

Practice 10.1

Ammonium nitrate (NH_4NO_3) is not typically considered an explosive material. However, when mixed with certain organic compounds, it can form an explosive mixture. This was the explosive used in the Oklahoma City bombing of the Alfred P. Murrah Federal Building in 1995 and also in 2011 to destroy a government building in Oslo, Norway. Determine whether pure ammonium nitrate has a negative or positive oxygen balance upon combustion.

Answer

Ammonium nitrate has a positive oxygen balance.

Explosives are often manufactured as mixtures that benefit from properties of other explosive and nonexplosive compounds. For example, TNT has a negative oxygen balance and ammonium nitrate has a positive oxygen balance. The mixture of the two is called *amatol* and has a neutral oxygen balance. Amatol, first used in World War I, was critical in the war effort, because TNT manufacturing plants could not keep up with the demand for explosives. The amatol mixture reduced the amount of TNT needed in the bombs and provided superior performance.

10.3 Kinetic-Molecular Theory of Gases

Changes in the temperature, volume, and pressure of gases produced in an explosion occur rapidly. Before discussing how these variables are related, we must first look at a molecular model of gases. In the following description of the properties of gases, the term *gas particle* refers to either atoms or molecules.

LEARNING OBJECTIVE

Explain the properties and behaviors of gases using kinetic-molecular theory.

Kinetic-Molecular Theory of Gases

1. Gas particles are extremely small and have relatively large distances between them. *Gases can be compressed easily, but liquids and solids cannot be compressed to any appreciable extent because the particles that make them up are already very close to one another.*

2. Gas particles act independently of one another because there are no significant attractive or repulsive forces between gas particles. *Compared with the energy of the gas particles, the intermolecular forces between gas particles are so weak that they do not have any real effect. Liquid and solid particles are affected by intermolecular forces, as discussed in Chapter 7.*

3. Gas particles are in continuous random, straight-line motion as they collide with one another and with the container walls. *Gas particles move until a collision alters their course, which is why gases fill the entire volume of their container.*

4. The average kinetic energy of gas particles is proportional to the temperature of the gas. *Energy added into the system increases the kinetic energy of the particles, which translates to an increase in temperature of the gas.*

The **kinetic-molecular theory** of gases provides a basis for understanding the observable behaviors of gases. For example, gases can be compressed in containers because of the large distances between gas particles. The particles in liquids and solids cannot be compressed to any great extent because there is very little space between particles in the condensed phases. Collisions of the gas particles against the walls of the container create the pressure of the gas within the container. When a gas is heated, the velocity of the gas particles increases and so does their kinetic energy. Because the temperature of a gas is a measure of the average kinetic energy of all the gas particles present, the temperature of the gas increases.

To understand what happens in explosions, we will use the insights into gas behavior provided by the kinetic-molecular theory and consider how explosive compounds react. For example, if we have a mixture of fuel vapors and oxygen gas from the air, the kinetic-molecular theory tells us that molecules of fuel and oxygen move randomly and collide. Is the energy of the collision sufficient to overcome the activation energy barrier? Detonators or, in the case of fuel-air explosions, a simple flame provides the required activation energy. Yet whether or not the collisions result in a fuel-air explosion depends on several additional factors.

Fuels have a **lower explosive limit** (LEL), the lowest ratio of fuel to air at which the mixture can propagate a flame. Below this level, no explosion will occur because there is not enough fuel present in the mixture for a sufficient number of collisions to sustain a continued reaction. Natural gas, which is mostly methane gas, has an LEL of 3.9%. If the concentration is any lower, natural gas in air will not produce an explosion.

Perhaps surprisingly, fuels also have an **upper explosive limit** (UEL), the highest ratio of fuel to air at which the mixture will support a combustion reaction. Natural gas has a UEL of 15%. At this concentration, the mixture becomes too fuel-rich for the natural gas to react explosively, because there is an insufficient quantity of oxygen present to sustain a reaction. Although such a mixture will not explode, it presents a dangerous situation because gases diffuse quickly. The concentration of fuel vapors may start above the UEL, but the gas will be diffusing at its edges and may quickly fall within the limit that supports combustion and, perhaps, an explosion. Over the last 10 years, airline companies and the FAA have worked to develop *inerting systems* that prevent the air space above the fuel in the aircraft from becoming explosive by reducing the amount of oxygen from 21% found in normal air to 9–12%, which is below the LEL.

Fuel-air explosions are used in military bombs and can also be improvised by terrorists. At the inauguration of President George W. Bush for his second term, for instance, the U.S. Secret Service watched for limousines that could be carrying propane tanks to create a fuel-air explosion. Terrorists had used an improvised explosive device (IED) of this type in the bombing of a Tunisian synagogue in April 2002. Therefore, when terrorists' plans for a limo bomb were discovered before the presidential inauguration, the threat was taken seriously.

10.4 Gas Laws

LEARNING OBJECTIVE

Using the gas laws, describe how gases behave.

Whether an explosion is from a harmless firecracker or an IED built by terrorists, the power of the explosion is due to the almost instantaneous release of energy. In this highly energetic environment, large volumes of gas molecules are produced. Immediately following the explosion, the temperature of the gases reaches several thousand degrees Celsius, the volume of the gases expands at a tremendous rate, and the pressure exerted by the gases can propel solid objects—whether bullets from a gun or fragments from a bomb—at high speeds.

Although explosions are extremely complex, we can comprehend some aspects of what happens by studying a group of principles called the *gas laws*. These laws describe how the temperature, pressure, volume, and number of moles of a gas are related. The gas laws apply to all gases, no matter what their identity.

Avogadro's law states that equal volumes of gases at the same temperature and pressure contain the same number of gas particles. This law means that the volume of a gas is directly proportional to the number of gas particles (and, therefore, the number of moles of gas) in the sample, as long as the pressure and temperature of the gas remain constant. Figure 10.2 illustrates this concept. The gas-filled containers in Figures 10.2a and 10.2b hold different gases but have the same number of gas particles because the containers are equal in volume and have the same temperature and pressure. Avogadro's law is useful because the coefficients in balanced equations that describe chemical reactions of gases can be interpreted as volumes of reactants and products at identical temperatures and pressures.

(a) (b)

FIGURE 10.2 Avogadro's law. Containers of gases at equal temperatures and pressures have the same number of moles.

WORKED EXAMPLE 2

Hydrogen and oxygen gas can form an explosive mixture. Sketch what the missing gas container in the following diagram looks like when 2 L of hydrogen gas react with 1 L of oxygen gas to form water vapor:

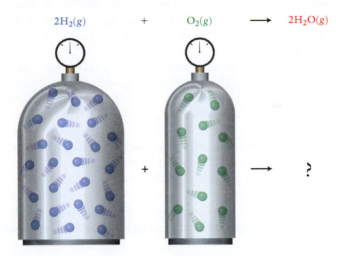

$$2H_2(g) \quad + \quad O_2(g) \quad \longrightarrow \quad 2H_2O(g)$$

SOLUTION

The coefficients of the balanced chemical equation can be interpreted as representing the volumes of the gases that react. Therefore, 2 L of hydrogen gas react with 1 L of oxygen to produce 2 L of water vapor, as shown below:

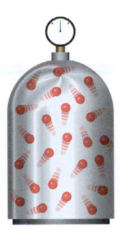

Practice 10.2

If two large, unlabeled industrial gas cylinders of equal volume are filled to the same pressure and temperature, one with hydrogen gas and one with argon gas, is there any way to determine which container has hydrogen and which has argon? The information provided on the periodic table about hydrogen and argon may be useful when thinking about your answer.

Answer

The two containers have exactly the same number of H_2 molecules as Ar atoms. However, because H_2 has a molecular mass of 2.02 amu and Ar has an atomic mass of 39.95 amu, the mass of the argon gas is approximately 20 times greater than the mass of hydrogen gas. Thus, the Ar-filled container would have a much greater mass than the H_2-filled container.

Marshmallow Syringe

For low explosives, the creation of a large amount of gaseous products results in a large pressure increase if the gas is confined in a small space—such as the center of a pipe bomb. The rigid container walls do not allow the volume of the gas to increase as the number of gas particles increases. Because the volume is constant, the explosive reaction causes a tremendous increase in the pressure inside the container. Thus, the container (in this case, the pipe) ruptures, flinging dangerous fragments in all directions. If the explosion occurs in the chamber of a gun, the bullet is propelled out of the barrel to allow the gases to expand.

The relationship between the pressure and volume of a gas is summarized in **Boyle's law**, which states that the pressure of a gas is inversely proportional to its volume. The inverse relationship indicates that if one variable increases, the other must decrease, provided the temperature and number of gas particles stay constant. Figure 10.3 illustrates Boyle's law.

The container in Figure 10.3a holds a fixed number of gas particles. If the volume of the gas is decreased to that shown in Figure 10.3b, the gas particles will not have to travel as far to collide with the container walls and therefore will undergo more collisions. The greater number of collisions results in an increase in pressure. If the volume of the container increases as in Figure 10.3c, the gas particles will have a greater distance to travel between collisions with the container wall, resulting in a decrease in pressure.

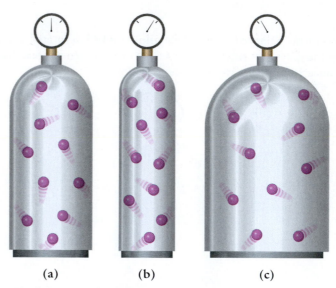

(a)   (b)   (c)

FIGURE 10.3 Boyle's law states that if the volume of a gas decreases, as from (a) to (b), the pressure will increase. Conversely, if the volume increases, as from (a) to (c), the pressure will decrease.

WORKED EXAMPLE 3

At an altitude of 3000 m, the air pressure is approximately 0.35 atm. Most airplane cabins are pressurized to 0.70 atm. If the pressurized cabin of an airplane is breached at an altitude of 3000 m, describe what will happen to the volume of air (doubled, halved, no change) in the lungs of an airplane passenger.

SOLUTION

Boyle's law states that if the pressure decreases, the volume must increase. Because the pressure drops from 0.70 atm to 0.35 atm, the volume must increase by an equal ratio (0.70/0.35). Therefore, the volume of air in a passenger's lungs will double.

Practice 10.3

Does the density of air change as the altitude increases? Explain your answer.

Answer

Air pressure decreases as the height above sea level increases. As the pressure decreases for gases, the volume must increase. The density of a gas is the mass of gas divided by the volume the gas occupies. The equation for density is $D = m/V$, as explained in Chapter 2 (pages 41–42). Because the volume of air increases at higher altitudes, there is less mass within a given volume, and the density of air decreases.

Compressed gas cylinders are used in many different businesses, from welding shops to hospitals, florists to laboratories. By compressing the gases at extremely high pressures, a much larger supply is provided to the consumer within a tank of manageable size. However, the storage temperature of high-pressure gas tanks must be monitored so that the tanks do not become dangerous at elevated temperatures.

Gay-Lussac's law states that the pressure of a gas is directly related to its temperature in Kelvin, provided the volume is constant (enclosed container). Therefore, as the temperature of a gas increases, the pressure increases proportionately.

The kinetic-molecular theory provides an insight into how temperature and pressure are related because the temperature of a gas is a measure of the kinetic energy of the gas molecules. Consider the illustrations in Figure 10.4. The kinetic energy of the gas particles shown in Figure 10.4a is related to the velocity of the gas particles. As the container is heated, the gas particles increase their kinetic energy, which corresponds to a higher average velocity for the gas particles, as shown in Figure 10.4b. The increased velocity of the gas particles increases the pressure because the particles undergo collisions more often with the container walls. The opposite condition occurs if the temperature of the gas particles is decreased, as illustrated in Figure 10.4c.

One of the many dangers that firefighters must confront is explosions that result from the heating of compressed gases in aerosol cans and compressed gas cylinders. If the contents are flammable, they may further react after the explosion. Most compressed gas cylinders are equipped with a safety valve designed to vent the contents if the internal pressure gets too high.

Aerosol cans have temperature warnings for storage because increasing the temperature will increase the internal pressure to potentially dangerous levels. (Michael Dalton/ Fundamental Photographs)

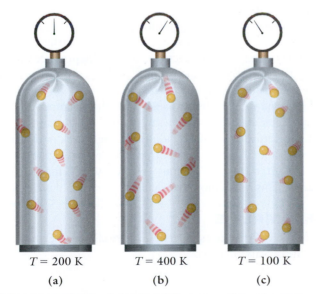

$T = 200$ K $T = 400$ K $T = 100$ K

(a)   (b)   (c)

FIGURE 10.4 Gay-Lussac's law states that if the temperature of a gas increases, as from (a) to (b), the pressure will increase. Conversely, if the temperature decreases, as from (a) to (c), the pressure will decrease.

Compressed hydrogen gas is a nonliquefied gas because it will not form a liquid at room temperatures even when subjected to high pressures. Other gases such as propane will form a liquid at high pressures and are stored in liquefied form. Liquid propane evaporates as the propane is removed from the cylinder to provide a steady flow of gas. When a liquefied gas cylinder is heated, the number of particles in the gas state increases, thus causing the gas pressure to increase until the container walls fail in an explosion.

WORKED EXAMPLE 4

Car tires are heated by friction with the road surface while driving. Is this related to the recommendation by tire manufacturers that the air pressure in the tires be checked before driving a long distance?

SOLUTION

The amount of air in the tire and the volume of the tire are fixed. However, as a result of friction between the tire and the road surface as a car is driven, the temperature of the tires will increase. As the temperature of the air inside the tire rises, so does the pressure. If the tires happen to be overinflated at the outset, the added increase in pressure while driving can cause a tire to rupture.

Practice 10.4

Why do the manufacturers of automobile tires recommend that tire pressures be measured several times per year?

Answer

The pressure of air in the tires changes with seasonal temperature variations. During the winter months, it is necessary to add air to the tires to maintain the proper tire pressure. As the weather warms up, the air pressure in the tire increases and the excess pressure must be released.

Thus far, we have considered gas laws that express the relationship between volume and number of moles (Avogadro's law), between pressure and volume (Boyle's law), and between pressure and temperature (Gay-Lussac's law). The last of the gas laws we will discuss that relate two variables is **Charles's law**, which states that the volume of a gas is directly proportional to its temperature, provided that the pressure and amount of gas remain constant. By increasing the temperature of a gas particle, the kinetic energy and, therefore, the velocity of the gas particle will increase. If the pressure is to remain constant, the volume of the gas must increase. The gases produced in an explosion are initially heated by the exothermic reaction, but as they cool to the temperature of the surroundings, the volume of gas must decrease.

The final volume (the volume after expansion) of the gas produced in an explosion is dictated by the final temperature of the gas, which will be the same as the ambient temperature, or the temperature of the surroundings. A single stick of dynamite has a volume of approximately 400 cm^3 (slightly more than a can of cola), which will upon explosion be converted nearly instantaneously to gaseous products with a final volume of 462 L at a temperature of 25°C. This represents more than a thousandfold increase in volume.

WORKED EXAMPLE 5

Will the density of the gas change as the temperature changes? Consider the variables used to calculate the density of a gas.

SOLUTION

Yes, the density depends on the temperature of the gas. It is calculated by $D = m/V$. Because the volume of a gas increases with an increase in temperature, the density of a hot gas is lower than the density of a cold gas.

Practice 10.5

In Section 9.4, the process of a flashover was discussed. Explain this process in terms of Charles's law.

Answer

A flashover happens when a fire located in one region of the room produces hot gases that become trapped at the ceiling, radiating heat to the remaining contents of a room until the autoignition temperature of the contents is reached. The hot gases produced in the combustion reaction have a lower density than cold gases, due to their increased volume. This is why the less dense gas rises to the ceiling of the room.

The individual gas laws are summarized in Table 10.1. Remember that the individual gas laws assume that all variables, other than those undergoing change according to the law, are held constant. Figure 10.5 focuses on what occurs in a system when the volume changes.

TABLE 10.1	Summary of Gas Laws		
Law	Variables	Constants	Relationship
Avogadro's	n and V	T and P	Directly proportional
Boyle's	P and V	n and T	Inversely proportional
Gay-Lussac's	P and T	n and V	Directly proportional
Charles's	T and V	n and P	Directly proportional

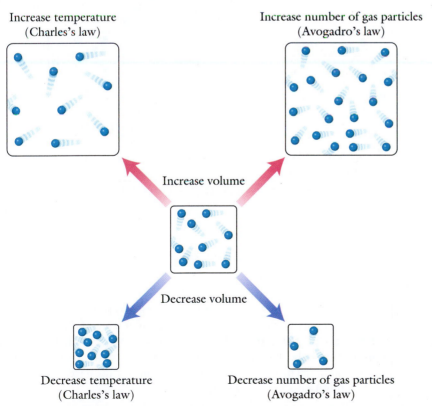

FIGURE 10.5 Summary of gas laws related to volume changes (with pressure held constant).

The increase in volume can be accomplished by either increasing the temperature or the number of gas particles in a system. For the volume of a system to be decreased, either the temperature of the system or the number of gas particles in the system must be decreased.

Figure 10.6 illustrates the variable that must change for the pressure of a system to change. If the pressure increases, the volume of the system must be decreased or the temperature of the system must be increased. If the pressure of a system decreases, either the number of gas particles decreased or the temperature of the system decreased.

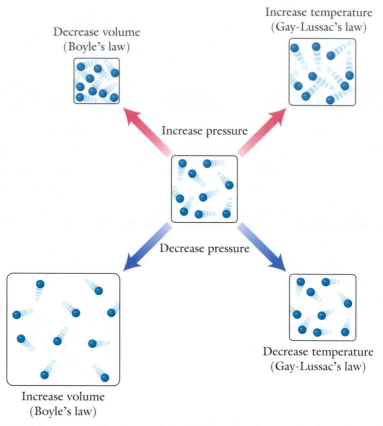

FIGURE 10.6 Summary of gas laws related to pressure changes (with number of moles held constant).

10.5 Mathematics of the Gas Laws

Avogadro's Law

Avogadro's law states that the volume of a gas is proportional to the number of gas particles present at constant temperature and pressure. If additional gas particles are added to a container, the volume must increase proportionately. The new volume can be calculated using the equation shown below. The initial and final conditions for the gases are noted by the subscripts 1 and 2, respectively.

$$\frac{V_1}{n_1} = \frac{V_2}{n_2}$$

The volume occupied by one mole of gas is 22.4 L when the temperature of the gas is 0°C and its pressure is 1 atm. These conditions of temperature and pressure are often referred to as **standard temperature and pressure (STP)**. We can express Avogadro's law at STP as follows:

$$\frac{22.4\ L}{1\ mol} = \frac{V_2}{n_2}$$

■ WORKED EXAMPLE 6

Automobile airbags inflate when sodium azide (NaN_3) contained in the bags explodes during a collision. Calculate the volume of gas produced at STP conditions if 39.0 g of sodium azide are used to inflate a side-impact airbag according to the following equation:

$$2NaN_3(s) \rightarrow 2Na(s) + 3N_2(g)$$

SOLUTION

The first step is to calculate moles of sodium azide reacted:

$$39.0\ \cancel{g} \times \underbrace{\frac{1\ mol}{65.0\ \cancel{g}}}_{\text{Molar mass of } NaN_3} = 0.600\ mol$$

The second step is to calculate how many moles of nitrogen gas are produced from 0.500 mol of NaN_3:

$$0.600\ \cancel{mol\ NaN_3} \times \underbrace{\frac{3\ mol\ N_2}{2\ \cancel{mol\ NaN_3}}}_{\substack{\text{Stoichiometric coefficients} \\ \text{from balanced equation}}} = 0.900\ mol\ N_2$$

The final step is to calculate the volume of 0.900 mol of N_2 at STP:

$$\frac{22.4\ L}{1\ mol\ N_2} = \frac{V_2}{0.900\ mol\ N_2} \Rightarrow V_2 = \frac{22.4\ L}{1\ \cancel{mol\ N_2}} \times 0.900\ \cancel{mol\ N_2} = 20.2\ L$$

Practice 10.6

Calculate the number of moles of nitrogen gas produced and grams of sodium azide consumed if 95.0 L of N_2 at STP results from a sodium azide explosion.

Answer

4.24 mol of N_2 and 184 g of NaN_3

Boyle's Law

Boyle's law states that the pressure of a gas is inversely proportional to the volume of the gas at constant temperature. Compressed gas cylinders are designed with Boyle's law taken into consideration because a large amount of gas is stored in a small volume. This compression leads to extremely high pressures within the cylinder. As the pressure is reduced, the volume of the gases increases. Boyle's law calculations use the following equation:

$$P_1V_1 = P_2V_2$$

Compressed gases have been used to increase the damage done by improvised explosive devices because of the massive increase in volume of gas that occurs when a tank is physically ruptured.

■ **WORKED EXAMPLE 7**

If a compressed gas cylinder has a pressure of 110.0 atm and a volume of 15.0 L, calculate the volume the gas will occupy at 1.00 atm of pressure.

SOLUTION

The first step is to summarize the variables of the problem:

P_1 = 110.0 atm   and V_1 = 15.0 L

P_2 = 1.00 atm   and V_2 = ?

The next step is to use the Boyle equation to solve for V_2:

$$P_1V_1 = P_2V_2 \Rightarrow V_2 = \frac{P_1V_1}{P_2} = \frac{(110.0 \text{ atm})(15.0 \text{ L})}{1.00 \text{ atm}} = 1.65 \times 10^3 \text{ L}$$

Practice 10.7

Calculate the pressure inside a pipe bomb before it ruptures, given that the internal volume of the pipe is 530.0 cm^3 and the contents of the bomb generate exactly 1 mol of gas at STP.

Answer

❏ 42.3 atm

Gay-Lussac's Law

Gay-Lussac's law states that the pressure of a gas is directly proportional to the temperature when the volume of the gas is constant. The equation expressing this relationship between pressure and temperature is given below:

$$\frac{P_1}{T_1} = \frac{P_2}{T_2}$$

An important point about this equation is that the temperatures must be expressed using the Kelvin scale (K) rather than the Celsius scale (°C). The Kelvin temperature of a gas is calculated by adding 273.15 to the Celsius temperature (K = °C + 273.15). For most calculations, in which temperature is expressed only to the nearest degree (as in 25°C), 273.15 is rounded off to 273. Also note that the degree sign is *not* used in the Kelvin units.

■ **WORKED EXAMPLE 8**

The pressure inside a compressed gas cylinder is 134 atm at 25°C. Calculate the new pressure inside the cylinder if it is heated to 85°C.

SOLUTION

The first step is to summarize the variables of the problem:

P_1 = 134 atm   and P_2 = ?

T_1 = 25 + 273 = 298 K   and T_2 = 85 + 273 = 358 K

The next step is to use the Gay-Lussac equation to solve for V_2:

$$\frac{P_1}{T_1} = \frac{P_2}{T_2} \Rightarrow P_2 = \frac{P_1T_2}{T_1} = \frac{(134 \text{ atm})(358 \text{ K})}{298 \text{ K}} = 161 \text{ atm}$$

Practice 10.8

The pressure of CO_2 inside a cola bottle is approximately 1.35 atm at 25°C. What is the pressure inside the bottle if it is chilled to 0°C? What if it is heated to 60°C?

Answer

❏ 1.24 atm, 1.51 atm

Charles's Law

Charles's law states that the volume of a gas is directly proportional to the temperature of the gas at constant pressure. We know that the volume of one mole of gas at 0°C is 22.4 L. But most fires and explosions occur at temperatures much greater than 0°C. Charles's law, shown below in mathematical form, provides a method for calculating the volume of the gas at other temperatures. Temperatures must be expressed in Kelvin, as above.

$$\frac{V_1}{T_1} = \frac{V_2}{T_2}$$

■ WORKED EXAMPLE 9

Calculate the density of 1.00 mol of carbon dioxide at STP. Then calculate the density of the gas if the temperature is increased to 500°C.

SOLUTION

To calculate the density of CO_2, substitute into the formula for density:

$$D = \frac{m}{V} = \frac{\overbrace{44.0 \text{ g}}^{\text{Molar mass of } CO_2}}{\underbrace{22.4 \text{ L}}_{\substack{\text{Volume of} \\ \text{1 mol at STP}}}} = 1.96 \text{ g/L}$$

Before calculating the density of CO_2 at 500°C, we need the volume at that temperature. Summarize the variables in the problem:

$V_1 = 22.4 \text{ L}$

$T_1 = 0°C + 273 = 273 \text{ K}$

$V_2 = ?$

$T_2 = 500°C + 273 = 773 \text{ K}$

Use the Charles equation to solve for V_2:

$$\frac{V_1}{T_1} = \frac{V_2}{T_2} \Rightarrow \frac{22.4 \text{ L}}{273 \text{ K}} = \frac{V_2}{773 \text{ K}} \Rightarrow V_2 = \frac{(773 \text{ K})(22.4 \text{ L})}{273 \text{ K}} = 63.4 \text{ L}$$

Calculate the density at 500°C:

$$D = \frac{m}{V} = \frac{\overbrace{44.0 \text{ g}}^{\text{Molar mass of } CO_2}}{\underbrace{63.4 \text{ L}}_{\substack{\text{Volume of} \\ \text{1 mol at 500°C}}}} = 0.694 \text{ g/L}$$

Practice 10.9

The explosion of 2.00 mol of TNT at 0°C produces a total of 15 mol of gas-state products. Calculate the volume of the gas produced when the final temperature of the gas reaches 25°C. (*Hint:* Use the molar volume of gases at STP to calculate the initial condition for volume and initial temperature.) The balanced equation for the reaction is

$$2C_7H_5N_3O_6(l) \rightarrow 7CO(g) + 7C(s) + 5H_2O(g) + 3N_2(g)$$

Answer

❏ 367 L

10.6 The Combined and Ideal Gas Laws

On Christmas Eve of 2004, the family of Graham Foster received horrible news. Graham had been in a high-speed car accident—he was thrown from the car and suffered fatal injuries. Police interviewed his close friend, David Munn, who said that Graham had dropped him off at his home before the accident occurred. Police would soon learn, however, that the events of that night had not transpired as David described them.

The police evidence that contradicted David's story was found through an investigation of the automobile's airbags, which had inflated during the accident. The collision of the car triggered a mechanism for exploding the sodium azide (NaN_3) contained in the airbags; this produces nitrogen gas. As the nitrogen gas filled the airbags, the occupants' forward momentum was slowed by their impact with the airbags. Investigators looked on the surface of the airbags for skin cells, saliva, blood, and cosmetics from the occupants.

The DNA evidence from the airbag on the driver's side matched that of David Munn, not Graham Foster. When Munn was presented with the evidence, he admitted to being behind the wheel, under the influence of methamphetamine, when he lost control while attempting to navigate a curve at a speed in excess of 100 mph. The explosion that had saved his life provided the evidence needed to prove his guilt. The science behind the airbag mechanism is based on knowledge of the gas laws.

Investigating a car accident in which airbags were deployed. (Age Fotostock/SuperStock)

The detonation of an explosive compound such as sodium azide produces gas-phase products at high pressures that are expanding in volume and simultaneously changing temperature. The gas laws described in the previous section explain how one variable will respond to a change in another, with the assumption that the remaining variables hold constant. In an explosion, however, the pressure, volume, and temperature of the gas *all* are changing. The **combined gas law** is a combination of Boyle's, Charles's, and Gay-Lussac's laws and is written:

$$\frac{P_1 V_1}{T_1} = \frac{P_2 V_2}{T_2}$$

The combined gas law is used when multiple variables of a gas are simultaneously changed.

The **ideal gas law** is based on measurements of the pressure, volume, temperature, and moles of a gas and is given by the equation:

$$PV = nRT$$

R is the universal gas constant and has the value

$$R = 0.08206 \ \frac{\text{L} \cdot \text{atm}}{\text{mol} \cdot \text{K}}$$

The ideal gas law allows calculations of the properties of gases under a specific set of conditions. For example, the ideal gas law can be used to determine the moles of nitrogen gas needed to properly inflate a vehicle airbag by detonation of sodium azide. The combined gas law can be used to calculate the change in volume of the airbag during various phases of deployment.

A typical airbag on the passenger's side can require up to 70 L of gas to inflate completely, whereas an airbag on the driver's side requires only 36 L of gas for full inflation because the airbag is closer to the driver. The side-impact airbags designed to cushion the head are typically 12 L in size. The amount of sodium azide required for each type of airbag differs greatly. The equation for the reaction of sodium azide that fills an airbag is given below. The sodium metal produced in the reaction undergoes further reactions (not shown below) with oxygen gas from the atmosphere to form sodium oxide (Na_2O).

$$2NaN_3(s) \rightarrow 2Na(s) + 3N_2(g)$$

The equation for the explosive decomposition of sodium azide shows that 2 mol of sodium azide will produce 3 mol of nitrogen gas. From the ideal gas law, the number of moles of nitrogen needed to fill a 70-L airbag at 1 atm and 25°C can easily be calculated. Once the amount of nitrogen is determined, the stoichiometry of the reaction is used to determine that it requires approximately 130 g of sodium azide to inflate a passenger side airbag. The airbags on the driver's side contain approximately half this amount, and side-impact bags contain about one-fourth as much sodium azide. Calculations that require the use of the gas laws and stoichiometry are featured in Section 10.8.

Sodium azide is an extremely toxic substance. Environmentalists are greatly concerned about the release of this substance into the environment from automobile junkyards. Many vehicle manufacturers have replaced sodium azide with nitroguanidine, as it is a nontoxic, cost-effective substitute. The switch to nitroguanidine is welcomed by environmentalists, although there is still a generation of used vehicles with toxic sodium azide waiting to end up in junkyards.

10.7 Mathematics of the Combined and Ideal Gas Laws

In Section 10.5, the problems all involved two variables and were solved using either Avogadro's, Boyle's, Charles's, or Gay-Lussac's law. In this section, you will learn to apply the combined gas law and the ideal gas law to solve problems. The combined gas law is used when the pressure, temperature, and volume are all changing in the experiment.

LEARNING OBJECTIVE

Use the gas laws to calculate the variables of a gaseous system.

■ WORKED EXAMPLE 10

A 0.750 L gas container is heated to 95°C and has an internal pressure of 65.0 atm when the container walls undergo a critical failure, releasing the contents. Calculate the final volume the gas will occupy at 25°C and 1.00 atm.

SOLUTION

This problem involves a change in the volume, pressure, and temperature of a compressed gas. Therefore, the combined gas law should be used.

First, summarize the provided data:

$P_1 = 65.0$ atm   and $P_2 = 1.00$ atm

$V_1 = 0.750$ L   and $V_2 = ?$

$T_1 = 95°C + 273 = 368$ K   and $T_2 = 25°C + 273 = 298$ K

Always convert temperature to the Kelvin scale immediately!

Now solve for the missing variable using the combined gas law:

$$\frac{P_1V_1}{T_1} = \frac{P_2V_2}{T_2} \Rightarrow \frac{65.0 \text{ atm} \times 0.750 \text{ L}}{368 \text{ K}} = \frac{1.00 \text{ atm} \times V_2}{298 \text{ K}}$$

$$V_2 = \frac{65.0 \text{ atm} \times 0.750 \text{ L} \times 298 \text{ K}}{368 \text{ K} \times 1.00 \text{ atm}} = 39.5 \text{ L}$$

Practice 10.10

If a 1.00-L container was rated to withstand pressures up to 25.0 atm, calculate what the temperature must have been when it ruptured. The final volume of the gas at 1.00 atm and 25°C was 15.3 L.

Answer

☐ 487 K, or 214°C

Problems that require the ideal gas law equation are different from the previous problem types in that a single set of conditions is provided for three of the four variables (pressure, volume, moles, temperature) and you are asked to solve for the unknown variable. When using the ideal gas law, the unit of the universal gas constant (R) is

$$\frac{L \cdot atm}{mol \cdot K}$$

For your answer to be correct, the units of the variables in the gas law problem must match those used in the constant R. If the units on the variables given in the problem do not match those of the constant R, it is best to convert them immediately to the proper units.

■ WORKED EXAMPLE 11

Mercury fulminate, $Hg(ONC)_2(s)$, is a high explosive compound that historically was used in detonators. Calculate the final volume of gases produced when 2.00 mol of mercury fulminate is detonated at a pressure of 1.05 atm and a temperature of 28°C. The reaction of mercury fulminate is

$$Hg(ONC)_2(s) \rightarrow Hg(l) + N_2(g) + 2CO(g)$$

SOLUTION

List the variables:

$P = 1.05$ atm

$V = ?$

$T = 28°C + 273 = 301$ K

$n =$ The detonation of 1 mol of $Hg(ONC)_2$ produces 1 mol of $N_2(g)$ and 2 mol of $CO(g)$, for a total of 3 mol of gas. Thus, when 2.00 mol of $Hg(ONC)_2$ is detonated, 6.00 mol of gas will be produced. Because there are four variables, use the ideal gas law to calculate V:

$$PV = nRT \Rightarrow V = \frac{nRT}{P} = \frac{(6.00 \text{ mol})(0.08206 \text{ L} \cdot \text{atm/mol} \cdot \text{K})(301 \text{ K})}{1.05 \text{ atm}} = 141 \text{ L}$$

Practice 10.11

Lead azide, $Pb(N_3)_2$, is a high explosive used in detonators. It decomposes to form lead and nitrogen gas. Determine how many moles of nitrogen gas have to be produced for the resulting pressure of nitrogen in a 100.0-mL container to be 74.5 atm at a temperature of 32°C.

Answer

☐ 0.298 mol N_2

10.8 Mathematics of Advanced Ideal Gas Law Problems

The ideal gas law can be used to determine the molar mass of an unknown gas by expanding the calculation of n, moles of gas. The calculation of moles of a substance is simply the mass of the substance divided by the molar mass of the compound. After substituting the calculation for moles of gas into the ideal gas law, the molar mass of the compound can be determined.

$$PV = nRT \text{ where } n = \text{moles} = \frac{\text{mass (g)}}{\text{molar mass (g/mol)}}$$

$$\text{Molar mass} = \frac{(\text{mass})RT}{PV}$$

■ WORKED EXAMPLE 12

A gas-tight syringe is filled to 25.0 mL, at a pressure of 1.34 atm and 25.0°C, with one of the noble gases. If the mass of the empty syringe is 74.206 g and the mass of the filled syringe is 74.321 g, identify the noble gas contained in the syringe.

SOLUTION

The identity of the gas in the syringe can be determined by calculating the molar mass of the compound.

$$\text{Molar mass} = \frac{(\text{mass})RT}{PV} = \frac{\overbrace{(74.321 \text{ g} - 74.206 \text{ g})}^{\text{Mass of gas in syringe}} \times 0.08206 \, \frac{\text{L} \cdot \text{mol}}{\text{atm} \cdot \text{K}} \times 298 \text{ K}}{1.34 \text{ atm} \times 0.0250} = 83.9 \text{ g/mol}$$

The molar mass of krypton is 83.798, which is close to the experimental result of 83.9 g/mol.

Practice 10.12

Identify which of the gaseous diatomic elements has filled a gas-tight syringe to 33.5 mL, at a pressure of 2.41 atm at 21°C, given that the mass of the gas inside the syringe is 0.127 g.

Answer

❏ Fluorine gas, F_2

In stoichiometry problems, the coefficients from the balanced equation are used to calculate either amounts of reactants required or the amount of a specific product that can be theoretically produced. The first step in solving a stoichiometry problem was to calculate moles of the given compound, which typically consisted of a grams-to-moles calculation using the molar mass of the compound. In stoichiometry problems involving either a reactant or product in the gas phase, the ideal gas law can be used as needed to determine the number of moles, n.

■ WORKED EXAMPLE 13

Calculate the amount of sodium azide, NaN_3, that must react to fill a 80.0 L passenger side airbag to a pressure of 1.00 atm at 25°C with N_2 gas. The reaction of sodium azide is $2NaN_3(s) \rightarrow 2Na(s) + 3N_2(g)$.

$$PV = nRT \Rightarrow n = \frac{PV}{RT}$$

$$n = \frac{(1.00 \text{ atm})(80.0 \text{ L})}{\left(0.08206 \frac{\text{L} \cdot \text{atm}}{\text{mol} \cdot \text{K}}\right)(298 \text{ K})} = 3.27 \text{ mol } N_2$$

$$3.27 \text{ mol } N_2 \times \frac{2 \text{ mol } NaN_3}{3 \text{ mol } N_2} \times \frac{65.011 \text{ g}}{1 \text{ mol } NaN_3} = 142 \text{ g } NaN_3$$

Practice 10.13

Calculate how many liters of nitrogen gas would be produced at 1.00 atm and 25°C from 37.0 g sodium azide, based on the equation given in Worked Example 13.

Answer

❑ 20.9 L $N_2(g)$

10.9 Detection of Explosives: Dalton's Law of Partial Pressures

The tragic events surrounding the destruction of Pan Am Flight 103 in 1988 over Lockerbie, Scotland, prompted the U.S. Congress to enact a new law called the Aviation Security Improvement Act of 1990 (H.R. 5732). Section 107 of this act specifically directed the Federal Aviation Administration (FAA) to determine the types and amounts of explosive materials, such as PETN and RDX, that posed the greatest risk for use by terrorists. The FAA was further directed to determine whether current technology could be used to screen passengers for explosives and, if not, then to invest in the development of such technology. If you have traveled by air, you are certainly aware of the X-ray machines used to screen luggage and the metal detectors to screen passengers. However, plastic explosives can easily avoid detection by these means and so require other detection methods.

One method developed in response to this need is based on the detection of very small amounts of gas-phase molecules emitted by solid explosives. A sample of air taken from an area near the explosive material will contain molecules of the material, which can be detected by an electronic instrument designed for such analysis or by the nose of a trained bomb-sniffing dog.

Explosives that have a high vapor pressure will have more molecules that enter the gas phase and are therefore more easily detected than those compounds with a low vapor pressure. Mixtures of gases are governed by **Dalton's law of partial pressures**, which states that the total pressure of a gas mixture is the sum of the pressures of each component in the mixture. Dalton's law of partial pressures is described mathematically as follows:

Luggage being screened for explosives.
(Reuters/Corbis)

$$P_{total} = P_1 + P_2 + P_3 + \ldots$$

Simply put, each gas behaves independently of all other gases in the mixture and acts as if the other gases were not present. For example, the air we breathe is a mixture of nitrogen at 78%, oxygen at 21%, argon at 0.9%, with all other gases making up the remainder. If the total atmospheric pressure is 1.0 atm, then the pressure of nitrogen is 0.78 atm, the pressure of oxygen is 0.21 atm, and the pressure of argon is 0.009 atm.

WORKED EXAMPLE 14

The pressure of a methane-oxygen-nitrogen gas mixture is 2.50 atm. If the partial pressure of oxygen gas is 0.55 atm and of nitrogen gas is 1.72 atm, what is the partial pressure due to methane?

SOLUTION

Substitute into Dalton's law of partial pressures:

$$P_{total} = P_{O_2} + P_{N_2} + P_{CH_4}$$
$$2.50 \text{ atm} = 0.55 \text{ atm} + 1.72 \text{ atm} + P_{CH_4}$$
$$P_{CH_4} = 2.50 \text{ atm} - 0.55 - 1.72 \text{ atm} = 0.23 \text{ atm}$$

Practice 10.14

Given that air is 78% N_2, 21% O_2, and 0.9% Ar, calculate the partial pressure of each gas at an elevation with an atmospheric pressure of 0.90 atm.

Answer

0.70 atm N_2, 0.19 atm O_2, and 0.008 atm Ar

Table 10.2 lists the vapor pressures of some common explosive compounds in air. The first five compounds listed are commonly used high explosives. The vapor pressures of high explosives range widely, but even those like nitroglycerin on the higher end of the range are not easily detected. RDX and PETN have the lowest vapor pressures of the list, and these two compounds are often used to make plastic explosives powerful enough that only a small amount is required to endanger the safety of those on an airplane. Therefore, the detection of plastic explosives has become a priority.

TABLE 10.2	Explosive Compounds and Vapor Pressures
Name	Vapor Pressure at 25°C (1 atm)
Ammonium nitrate (AN)	6.6×10^{-9}
Nitroglycerin (NG)	3.2×10^{-8}
Pentaerythritol tetranitrate (PETN)	5.0×10^{-13}
1,3,5-trinitro-1,3,5-triazacyclohexane (RDX)	1.8×10^{-12}
2,4,6-trinitrotoluene (TNT)	3.9×10^{-9}
2,3-dimethyl-2,3-dinitrobutane (DMNB)	2.8×10^{-6}
Ethylene glycol dinitrate (EGDN)	3.7×10^{-5}
para-Mononitrotoluene (*p*-MNT)	5.4×10^{-5}
ortho-Mononitrotoluene (*o*-MNT)	2.4×10^{-4}

Table adapted from *Containing the Threat from Illegal Bombings*, National Research Council, National Academy Press, Washington DC, 1998.

Mass Spectroscopy

GC-MS

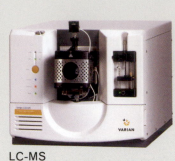

LC-MS

ICP-MS (All: Courtesy of Varian, Inc.)

One of the single most powerful instruments a forensic chemist has in the laboratory to analyze evidence is a **mass spectrometer**, which has the ability to determine the exact molecular mass of an unknown substance. Some uses of a mass spectrometer in the forensic crime laboratory include the confirmation of illegal and prescription drugs and identification of arson accelerants and explosives.

The mass spectrometer can be found coupled to the other methods discussed earlier in this book, such as gas chromatography-mass spectroscopy (GC-MS), liquid chromatography-mass spectroscopy (LC-MS), and inductively coupled plasma-mass spectroscopy (ICP-MS).

Mass spectroscopy is based on the ability of a magnetic field to deflect a moving ion. Figure 10.7 is an illustration of the major components of a mass spectrometer. The ion source allows a very small number of gas-phase ions created from the unknown compound(s) to enter into the vacuum chamber. It is important for the mass

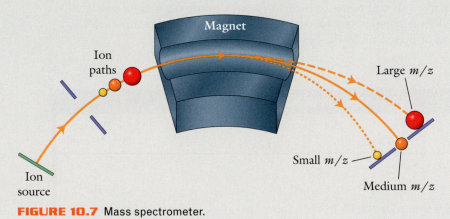

FIGURE 10.7 Mass spectrometer.

In 1991 the United Nations Council of the International Civil Aviation Organization agreed on a standard for the marking of plastic explosives for the purpose of detection. This convention requires that plastic explosives be manufactured with small amounts of **marker compounds** (0.1% to 1.0%). The markers are explosive compounds that have large vapor pressures that ease the detection of the plastic explosives. The bottom four compounds on Table 10.2 are the marker compounds designated for addition to plastic explosives. The chemical markers DMNB, EGDN, spectrometer to be inside a vacuum so the ions do not collide with atmospheric gases that would divert the ions from their paths. Once the ions enter the mass spectrometer, they pass metal plates that have voltages applied such that the ions are repelled by one plate and attracted toward the next. This process increases the velocity of the ions.

The ions next enter a region of the spectrometer where a magnetic field is applied to the ions. The ions are deflected by the magnetic field, much the same way as was first described in the cathode ray experiment from Section 3.4, involving the deflection of electrons. In a mass spectrometer, the ions travel in a curved path, as shown in Figure 10.7. Not all ions will be able to reach the detector, which is on the other side of the narrow slit.

The degree to which each ion is deflected by the magnetic field depends on two variables: the mass of the ion (m) and the charge of the ion (z). The magnetic field of the mass spectrometer is adjusted so that only ions corresponding to a specific mass-to-charge (m/z) ratio can reach the detector. The ions with a larger mass-to-charge ratio are not deflected sufficiently, whereas the ions with a smaller mass-to-charge ratio are deflected too much. The magnetic field of the mass spectrometer is adjusted sequentially so that all possible ions are allowed to reach the detector.

The number of ions at each mass-to-charge ratio is measured for each sample. The relatively simple data obtained for krypton gas are shown in Figure 10.8 as an example. The graph shows that krypton has five isotopes (Kr-80, Kr-82, Kr-83, Kr-84, and Kr-86). For mass spectral data, the peak from the most abundant isotope (in this case, Kr-84) is assigned a relative abundance of 100%, and the ions from all other isotopes are shown as a percentage of that most abundant ion.

The data obtained can be very complex, as shown for cocaine in Figure 10.9. Isotopes alone cannot explain the complexity of the cocaine spectrum. The additional complexity comes from creating ions out of large organic molecules that can break apart into smaller pieces, each forming ions. When chemists analyze an unknown, they compare the experimental results with a known sample of cocaine, because cocaine will always form the same pattern, as shown in Figure 10.9, provided similar experimental conditions are used.

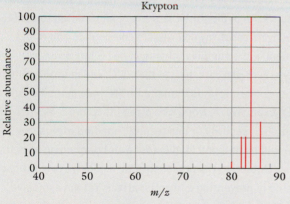

FIGURE 10.8 Mass spectroscopy data for krypton. (Webbook.nist.gov/chemistry)

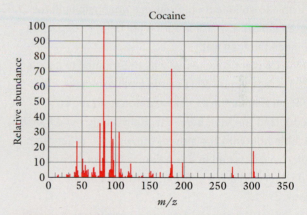

FIGURE 10.9 Mass spectroscopy data for cocaine. (Webbook.nist.gov/chemistry)

p-MNT, and o-MNT have vapor pressures 10,000 to 1 billion times higher than the primary explosive compounds. The high vapor pressures produce a sufficiently large partial pressure in the air samples found near the explosive that they can be detected by either canine or electronic detection systems. Although the marker compounds do not provide any information on the origin of plastic explosives, they do increase the ability of aviation and law enforcement officials to detect the explosives before they can be used.

10.10 CASE STUDY FINALE: Exploring Airport Security

Richard Reid, known as the "shoe bomber," had attempted to board a flight from Paris to Miami on December 21, 2001. His unkempt appearance and incomplete answers to questions from airline screeners raised concern, so they refused him passage. Further questioning by the French police resulted in his being cleared to fly the next day. He passed through the metal detector and security without incident.

While the flight was over the Atlantic Ocean, a flight attendant smelled the odor of a burning match. She scanned the passengers, looking for the person, and a passenger pointed toward Reid. She thought that Reid was attempting to sneak a cigarette aboard the long flight and informed him that there was no smoking. He apologized and waited for the flight attendant to leave before once again attempting to light a fuse that led to 283 grams of C4 plastic explosive hidden in his shoes. The flight attendants and several passengers were able to subdue Reid. Reid was unable to light the fuse, as the rainy weather during his delayed departure had made the fuse wet.

The C4 plastic explosive was determined to consist of PETN (see Figure 10.1) as the main explosive with TATP serving as the explosive detonator (shown below). Had the fuse been dry, Reid would have blown a sizable hole into the side of the airliner, which then would have crashed in the Atlantic Ocean. Instead, Reid is serving a life sentence with no chance of parole.

Reid's failed attempt to blow up an airliner, however, has had a lasting impact on safety standards at airport security in the United States. Since then, all shoes must pass through the carry-on luggage scanner to look for evidence of tampering. X rays can reveal only that an object has been

H₃C   CH₃

Triacetone triperoxide (TATP)

FIGURE 10.10 An airport explosive detector measures the air surrounding an individual for traces of volatile explosive compounds.

modified from its original state, not that a substance is an explosive.

One method used by airports to detect explosives relies on body heat to vaporize traces of explosive compounds. As a person stands inside a scanning booth, as shown in Figure 10.10, puffs of air directed toward the body sweep away any volatile compounds and direct them into a sampling chamber. TATP has a vapor pressure of 6.85×10^{-5} atm at 25°C, which is 137 million times more volatile than the PETN making up the bulk of C4 explosive. Anyone who attempts to bring explosives through security or has recently handled explosives would be stopped long before ever boarding an aircraft.

Detecting the presence of an explosive is a critical step in preventing a terrorist attack. Yet the explosive material itself can provide investigators with valuable intelligence. Consider the example of TNT, which is made by the nitration of toluene using nitric acid. Specifically, the compound produced is 2,4,6-trinitrotoluene, also known as 2,4,6-TNT. In most chemical reactions, unintended compounds are produced, and these include isomers of the main product. The structure of 2,4,6-TNT is illustrated in Figure 10.11. The carbon atoms of the benzene ring are numbered 1 through 6, with number 1 always being the carbon bonded to the methyl group.

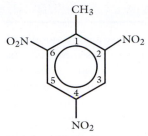

FIGURE 10.11
2,4,6-trinitrotoluene, also
known as TNT.

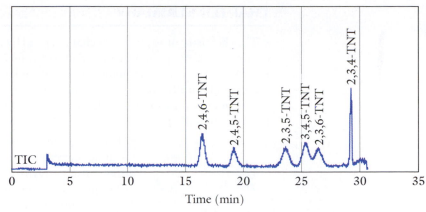

FIGURE 10.12 Separation of TNT isomers through liquid chromatography-
mass spectrometry. (Zhao & Yinon, 2002)

WORKED EXAMPLE 15

There are five isomers of 2,4,6-TNT that occur by
placing the three nitro (—NO$_2$) groups on different
carbon atoms. Draw and name the five isomers.

SOLUTION

2,4,5-TNT   2,3,4-TNT   3,4,5-TNT

2,3,6-TNT   2,3,5-TNT

Practice 10.15

Explain why 3,4,6-TNT is not one of the isomers listed
in Worked Example 15.

Answer

The 3,4,6-TNT isomer is actually the 2,4,5-TNT
isomer, as shown below. 2,4,5-TNT is the correct name
for this structure.

3,4,6-TNT   NO$_2$ 2,4,5-TNT   NO$_2$

The five isomers of 2,4,6-TNT are minor
components that are produced in the manufacturing
process and are present in the final product mix at very
low concentrations. These minor components can
provide clues to where the TNT was produced because
every manufacturer uses slightly different reaction
procedures, equipment, and reaction temperatures
that result in varying amounts of each isomer in the
final product. If law enforcement officials collect a
suspect sample of TNT, the levels of each isomer can
be measured and matched to standards obtained
from each factory. The isomers act as a bar code or
fingerprint for the factory that manufactured the
explosive.

LC-MS is an effective tool for the analysis of
2,4,6-TNT and its isomers, as it is for many mixtures.
The liquid chromatography method separates each
isomer of the mixture, and the mass spectrometer is
able to identify each of the TNT isomers within the
sample. A representative chromatogram of the
separation and detection of a TNT isomer standard
is shown in Figure 10.12. It should be noted that in a
real TNT sample (as opposed to a laboratory mixture
of its isomers used as a standard), the size of the
signal for the isomers other than 2,4,6-TNT would be
extremely small because they would be present in
trace amounts.

CHAPTER SUMMARY

• The chemistry of explosives is characterized by reactions that release a large amount of energy in a short period of time and form substantial quantities of gaseous products. Low explosives are compounds capable of undergoing a rapid combustion reaction and must be physically confined to a small area to become explosive. High explosives are compounds that detonate whether confined or not and undergo extremely rapid decomposition to produce a shock wave that shatters any material near the explosive.

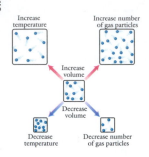

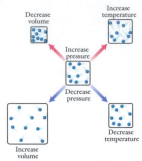

• The kinetic-molecular theory provides an intellectual model for understanding how gases behave and why their properties differ considerably from those of solids or liquids. The effects of temperature, pressure, and volume changes on gas behavior can be understood by using the principles of the kinetic-molecular theory.

• The gas laws allow us to examine the interrelationships of the pressure, volume, temperature, and amount of gas present in a system. We can calculate the effect that a change in one variable has on another. The combination of the individual gas laws leads to the ideal gas law, $PV = nRT$, which shows the relationship of pressure, volume, temperature, and number of moles.

• Dalton's law of partial pressures states that the partial pressures of all gases present in a mixture contribute to the total pressure. One method for the detection of explosives is based on spiking low-vapor-pressure explosives with compounds that have a larger vapor pressure. The vapors from the high-vapor-pressure compounds will produce a measurable partial pressure in the vicinity of the explosive, making them easier to detect.

• One way to identify the source of an explosive is based on measuring trace isomers of the explosive molecules to "fingerprint" the manufacturing facility. The slight variations in the levels of trace isomers reflect varying conditions under which the explosives were manufactured.

KEY TERMS

Avogadro's law (p. 290)
Boyle's law (p. 292)
Charles's law (p. 294)
combined gas law (p. 300)
Dalton's law of partial
  pressures (p. 304)
Gay-Lussac's law (p. 293)

high explosive (p. 286)
ideal gas law (p. 300)
kinetic-molecular theory (p. 289)
low explosive (p. 286)
lower explosive limit (LEL) (p. 290)
marker compound (p. 306)
mass spectrometer (p. 306)

oxygen balance (p. 288)
standard temperature and
  pressure (STP) (p. 297)
upper explosive limit (UEL)
  (p. 290)

MAKING MORE CONNECTIONS: Additional Readings, Resources, and References

"Killer trapped by his DNA," *The Journal* (Newcastle, UK), September 18, 2004.

Witkin, Gorden. "The Debate about Invisible Detectives," *U.S. News & World Report*, September 16, 1996.

Zhao, X., and Yinon, J. "Characterization and Origin Identification of 2,4,6-Trinitrotoluene through Its By-product Isomers by Liquid Chromatography- Atmospheric Pressure Chemical Ionization Mass Spectrometry," *Journal of Chromatography A*, vol. 946, pp. 125–132 (2002).

For information about the Aviation Security Improvement Act of 1990:

http://thomas.loc.gov/cgi-bin/query/D?c101:4:./temp/~c101soI6c1::

For information about bombing incidents in the United States:
www.pbs.org/wgbh/nova/transcripts/2310tbomb.html
For information regarding Richard Reid:
www.timesonline.co.uk/tol/news/world/us_and_americas/article6467828.ece

www.washingtonpost.com/wp-srv/inatl/longterm/panam103/timeline.htm
For information regarding the lower explosive limit of aircraft fuel tanks:
www.faa.gov/news/press_releases/news_story.cfm?newsId=5785

REVIEW QUESTIONS AND PROBLEMS

Questions

1. What are the requirements for a compound to be explosive? (10.1)
2. What is the difference between a low explosive and a high explosive? (10.1)
3. What is the main use of a low explosive? (10.1)
4. Why is a detonator used to trigger the main charge of a high explosive? (10.1)
5. Explosive molecules contain carbon, hydrogen, and oxygen. Why is this important in the detonation process? (10.2)
6. What is the oxygen balance of an explosive? Why is this important information? (10.2)
7. What are the four premises of the kinetic-molecular theory of gases? (10.3)
8. What is the nature of gas pressure within a container? Illustrate your answer with a drawing showing two containers, one with a gas at high pressure and the other with a gas at low pressure. (10.3–10.4)
9. Explain why gases can be compressed to a much greater extent than liquids and solids can. Illustrate your answer with a drawing showing a particle-size view of each physical state. (10.3)
10. Explain why fuel-air mixtures below the LEL will not explode. (10.3)
11. Explain why fuel-air mixtures above the UEL will not explode. (10.3)
12. Explain the physical phenomenon that creates the pressure of a gas. (10.3)
13. State Avogadro's law and explain how the variables are related. Illustrate your answer with a drawing. (10.4)
14. State Boyle's law and explain how the variables are related. Illustrate your answer with a drawing. (10.4)
15. State Gay-Lussac's law and explain how the variables are related. Illustrate your answer with a drawing. (10.4)
16. State Charles's law and explain how the variables are related. Illustrate your answer with a drawing. (10.4)
17. What does STP represent? (10.5)
18. How can you tell which gas law to use in a calculation? (10.7)

19. Explain how the kinetic-molecular theory of gases can be used to understand Dalton's law of partial pressures. (10.9)
20. Discuss the principles behind using taggants in explosive materials. (10.10)
21. Discuss the principles behind using chemical markers in plastic explosives. (10.10)
22. Discuss the use of isomers in tracing the origin of explosives. (10.10)
23. How do taggants, markers, and isomers help security officials? (10.10)
24. What limitations are there on the use of taggants, markers, and isomers? (10.10)

Problems

25. Determine how many moles of oxygen would be necessary for the complete oxidation (combustion) of one mole of each of the following compounds to CO_2 and H_2O. (10.2)
    (a) CH_3CH_3
    (b) $CH_3CH_2CH_2CH_3$
    (c) $CH_2{=}CH_2$
    (d) CH_3CH_2OH
26. Determine how many moles of oxygen would be necessary for the complete oxidation (combustion) of one mole of each of the following compounds to CO_2 and H_2O. (10.2)
    (a) CH_3OH
    (b) $CH_2{=}CHCH_3$
    (c) CH_3OCH_3
    (d) $CH{\equiv}CH$
27. Determine how many moles of oxygen would be necessary for the complete oxidation (combustion) of one mole of each of the following compounds to CO_2 and H_2O. (10.2)
    (a) Ethane   (c) Butane
    (b) Ethanol   (d) Propene
28. Determine how many moles of oxygen would be necessary for the complete oxidation (combustion) of one mole of each of the following compounds to CO_2 and H_2O. (10.2)
    (a) 2-Octene
    (b) Benzene
    (c) 2-methylpentane
    (d) Methane

29. Determine what happens to each variable below (increases, decreases, no change) under the stated conditions. Assume that all other variables not specifically mentioned are constant values. (10.4)
    (a) What happens to the pressure when the volume decreases?
    (b) What happens to the volume when the temperature increases?
    (c) What happens to the pressure when the temperature decreases?
    (d) What happens to the pressure when the amount of gas decreases?

30. Determine what happens to each variable below (increases, decreases, no change) under the stated conditions. Assume that all other variables not specifically mentioned are constant values. (10.4)
    (a) What happens to the volume if the amount of gas doubles?
    (b) What happens to the temperature when the pressure decreases?
    (c) What happens to the temperature when the volume increases?
    (d) What happens to the volume when the pressure increases?

31. Calculate the missing variables in each experiment below using Avogadro's law. (10.5)
    (a) $V_1 = 2.00$ L, $n_1 = 0.651$ mol, $V_2 = 0.575$ L, $n_2 = ?$
    (b) $V_1 = ?$, $n_1 = 1.36$ mol, $V_2 = 750.0$ mL, $n_2 = 1.25$ mol
    (c) $V_1 = 0.334$ L, $n_1 = 0.521$ mol, $V_2 = ?$, $n_2 = 6.77$ mol
    (d) $V_1 = 17.2$ L, $n_1 = 4.71$ mol, $V_2 = 5.86$ L, $n_2 = ?$

32. Calculate the missing variables in each experiment below using Avogadro's law. (10.5)
    (a) $V_1 = 9.15$ L, $n_1 = ?$, $V_2 = 2.23$ L, $n_2 = 3.56$ mol
    (b) $V_1 = ?$, $n_1 = 0.661$ mol, $V_2 = 7.32$ L, $n_2 = 1.45$ mol
    (c) $V_1 = 612$ mL, $n_1 = 9.11$ mol, $V_2 = 123$ mL, $n_2 = ?$
    (d) $V_1 = 5.58$ L, $n_1 = 0.330$ mol, $V_2 = 7.46$ L, $n_2 = ?$

33. Calculate the missing variables in each experiment below using Boyle's law. (10.5)
    (a) $P_1 = 5.44$ atm, $V_1 = 3.74$ L, $P_2 = ?$, $V_2 = 6.64$ L
    (b) $P_1 = 0.177$ atm, $V_1 = 565$ mL, $P_2 = 4.71$ atm, $V_2 = ?$
    (c) $P_1 = 6.03$ atm, $V_1 = ?$, $P_2 = 4.28$ atm, $V_2 = 0.904$ L
    (d) $P_1 = ?$, $V_1 = 7.43$ L, $P_2 = 5.47$ atm, $V_2 = 2.85$ L

34. Calculate the missing variables in each experiment below using Boyle's law. (10.5)
    (a) $P_1 = 6.24$ atm, $V_1 = 5.92$ L, $P_2 = 0.773$ atm, $V_2 = ?$
    (b) $P_1 = ?$, $V_1 = 2.91$ L, $P_2 = 7.74$ atm, $V_2 = 4.02$ L
    (c) $P_1 = 6.99$ atm, $V_1 = ?$, $P_2 = 2.72$ atm, $V_2 = 12.6$ L
    (d) $P_1 = 4.74$ atm, $V_1 = 5.17$ L, $P_2 = ?$, $V_2 = 25.6$ L

35. Calculate the missing variables in each experiment below using Gay-Lussac's law. (10.5)
    (a) $P_1 = 1.02$ atm, $T_1 = 300.0$ K, $P_2 = 7.48$ atm, $T_2 = ?$
    (b) $P_1 = ?$, $T_1 = 266$ K, $P_2 = 8.54$ atm, $T_2 = 607$ K
    (c) $P_1 = 5.74$ atm, $T_1 = ?$, $P_2 = 3.47$ atm, $T_2 = 279$ K
    (d) $P_1 = 6.16$ atm, $T_1 = 287$ K, $P_2 = ?$, $T_2 = 484$ K

36. Calculate the missing variables in each experiment below using Gay-Lussac's law. (10.5)
    (a) $P_1 = 3.11$ atm, $T_1 = 103$ K, $P_2 = 0.216$ atm, $T_2 = ?$
    (b) $P_1 = 4.27$ atm, $T_1 = 334$ K, $P_2 = ?$, $T_2 = 210$ K
    (c) $P_1 = 5.66$ atm, $T_1 = ?$, $P_2 = 7.37$ atm, $T_2 = 797$ K
    (d) $P_1 = ?$, $T_1 = 977$ K, $P_2 = 4.31$ atm, $T_2 = 244$ K

37. Calculate the missing variables in each experiment below using Charles's law. (10.5)
    (a) $V_1 = 1.25$ L, $T_1 = 379$ K, $V_2 = 9.92$ L, $T_2 = ?$
    (b) $V_1 = ?$, $T_1 = 118$ K, $V_2 = 28.3$ L, $T_2 = 612$ K
    (c) $V_1 = ?$, $T_1 = 298$ K, $V_2 = 5.76$ L, $T_2 = 828$ K
    (d) $V_1 = 6.11$ L, $T_1 = 318$ K, $V_2 = 6.83$ L, $T_2 = ?$

38. Calculate the missing variables in each experiment below using Charles's law. (10.5)
    (a) $V_1 = 5.34$ L, $T_1 = 901$ K, $V_2 = 6.98$ L, $T_2 = ?$
    (b) $V_1 = ?$, $T_1 = 659$ K, $V_2 = 5.76$ L, $T_2 = 280$ K
    (c) $V_1 = ?$, $T_1 = 258$ K, $V_2 = 28.8$ L, $T_2 = 818$ K
    (d) $V_1 = 3.17$ L, $T_1 = 408$ K, $V_2 = ?$, $T_2 = 277$ K

39. Calculate the volume of the following amounts of gases at STP. (10.5)
    (a) 5.72 mol of carbon dioxide
    (b) 21.1 mol of hydrogen
    (c) 0.682 mol of sulfur dioxide
    (d) 0.744 mol of hydrogen sulfide

40. Calculate the volume of the following amounts of gases at STP. (10.5)
    (a) 0.312 mol of helium
    (b) 5.72 mol of fluorine
    (c) 36.2 mol of nitrogen
    (d) 8.82 mol of sulfur trioxide

41. Calculate the volume of the following amounts of gases at STP. (10.5)
    (a) 6.57 g of methane
    (b) 3.68 g of oxygen
    (c) 52.2 g of carbon monoxide
    (d) 48.6 g of dinitrogen monoxide
42. Calculate the volume of the following amounts of gases at STP. (10.5)
    (a) 35.8 g of nitrogen
    (b) 31.5 g of argon
    (c) 114 g of propane
    (d) 92.2 g of butane
43. Calculate the density of the gases (g/L) listed in Problem 41 at STP. (10.5)
44. Calculate the density of the gases (g/L) listed in Problem 42 at STP. (10.5)
45. A gas initially occupies 2.00 L at 25°C and 3.00 atm. Calculate its volume after the temperature is increased to 175°C and the pressure is decreased to 1.00 atm. (10.7)
46. A gas initially occupies 25.0 L at 25°C and 3.00 atm. Calculate its volume after the temperature is increased to 175°C and the pressure is decreased to 1.00 atm. (10.7)
47. Calculate the final temperature (K and °C) for 12.0 L of a gas at 28.3°C and 1.10 atm after it expands to a volume of 18.7 L, given the final pressure is 0.625 atm. (10.7)
48. Calculate the final temperature (K and °C) of a gas if 32.0 L of the gas at 25°C and 1.00 atm is compressed into a volume of 3.20 L at a pressure of 5.0 atm. (10.7)
49. What is the pressure of a 18.0-L compressed gas tank that holds 32.4 mol of nitrogen gas at 357 K? (10.7)
50. What is the pressure of 118 g of hydrogen gas stored in a 50.0-L compressed gas cylinder at 276 K? (10.7)
51. What is the volume of a compressed gas tank that contains 2547 g of butane gas at a pressure of 56.4 atm and a temperature of 22.7°C? (10.7)
52. What is the volume of a compressed gas tank that contains 716 g of nitrogen gas at a pressure of 56.3 atm and a temperature of 57.1°C? (10.7)
53. What is the temperature (K and °C) of a 875-L compressed gas tank that contains 64.1 mol of carbon monoxide at a pressure of 38.1 atm? (10.7)
54. What is the temperature (K and °C) of 321 g O_2 gas stored in a 25.0-L gas storage cylinder if pressure is 12.8 atm? (10.7)
55. How many moles of argon gas inside an 12.0-L gas cylinder would have a pressure of 14.4 atm at 25.1°C? (10.7)

56. What is the amount in moles of nitrogen gas found in a 58.0-L compressed gas tank that has a pressure of 89.3 atm at 357 K? (10.7)
57. Calculate the pressure (atm) inside a 5.45-L gas cylinder that contains 72.3 g of carbon dioxide and is stored at 21.4°C. (10.7)
58. Calculate the pressure (atm) inside a 32.0-L gas cylinder that contains 287 g of sulfur hexafluoride and is stored at 26.3°C. (10.7)
59. A gas-tight syringe is filled to 75.0 mL, at a pressure of 3.20 atm and 20.0°C, with a noble gas. If the mass of the syringe increases by 0.201 g when filled, which noble gas is in the syringe? (10.8)
60. A gas-tight syringe is filled to 45.0 mL, at a pressure of 2.33 atm and 25.0°C, with a gaseous diatomic element. If the mass of the syringe increases by 0.137 g when filled, which diatomic element is in the syringe? (10.8)
61. Calculate the mass of sodium azide required to produce 65.0 L of $N_2(g)$ at 0.980 atm and 20.0°C, given $2NaN_3(s) \rightarrow 2Na(s) + 3N_2(g)$. (10.8)
62. Calculate the mass of sodium azide required to produce 125 L of $N_2(g)$ at 1.07 atm and 23.0°C, given $2NaN_3(s) \rightarrow 2Na(s) + 3N_2(g)$. (10.8)
63. Using both the ideal gas law and Dalton's law of partial pressures, calculate the total pressure of a 1.00-L container at 25°C that contains 16.0 g of O_2 and 28.0 g of N_2. (10.7–10.9)
64. Using both the ideal gas law and Dalton's law of partial pressures, calculate the total pressure of a 5.00-L container at 20°C that contains 80 g of argon and 88 g of carbon dioxide gas. (10.7–10.9)

Case Study Problems
65. Determine whether the explosive HMX ($C_4H_8N_8O_8$) has a positive, negative, or neutral oxygen balance. (10.2)
66. Determine whether the explosive nitroglycerin ($C_3H_5N_3O_9$) has a positive, negative, or neutral oxygen balance. (10.2)
67. Would the color of the smoke produced from an explosive differ if it had a positive oxygen balance versus a negative oxygen balance? (CP)
68. A fuel-air explosive is most powerful when there is an exact stoichiometric amount of oxygen present at detonation. Determine how many grams of oxygen gas would be needed to combine exactly with 13.2 kg of propane gas. (10.2)
69. Combustible gases exhibit a concentration range, from the lower explosive limit (LEL) to the upper explosive limit (UEL), within which a mixture becomes explosive. Use the kinetic-molecular theory of gases and Dalton's law of partial pressures to explain why. (CP)

70. In conjunction with Figure 10.1 and the organic functional groups covered throughout Chapter 8, identify compounds that contain the following modified organic functional groups that are present in explosive molecules. (CP)
    (a) Nitroester
    (b) Nitroether
    (c) Nitroamine
    (d) Nitroaromatics

71. Using Boyle's law, explain why a person who has a penetrating chest wound struggles to breathe and risks having a collapsed lung. (CP)

72. If the density of TNT is 1.65 g/cm^3, calculate the volume of a 454-g quantity of the explosive before and after the explosion. Assume that the volume of the $C(s)$ in the product is negligible, and assume STP conditions. See Section 10.5 for the balanced equation. (10.5)

73. During the synthesis of 2,4,6-trinitrotoluene, trace amounts of the dinitrotoluene (DNT) isomers are formed. Draw and name all of the isomers of DNT that could form. (CP)

74. Calculate the partial pressure of each gas and the total pressure inside a 23.0-L container at 500.0°C after the explosion of 100.0 g of TNT. See Section 10.5 for the balanced equation. (10.9)

75. Calculate the partial pressure of each gas and the total pressure inside a 23.0-L container at 500.0°C after the explosion of 100.0 g of nitroglycerin. See Section 10.5 for the balanced equation. (10.9)

76. Cesare Lombroso, a famed criminologist, described a primitive lie detector called a volumetric glove in his 1876 book *Criminal Man*: "The glove is filled with air, and the greater or smaller the pressure exercised on the air by the pulsation of the blood in the veins of the hand acts on an aerial column. . . . [T]his chamber supports a lever carrying an indicator which rises and falls with the greater or slighter flow of blood in the hand." Explain the science (gas laws) of how the "volumetric gloves" function as a lie detector. (10.4)

77. The pressure inside the cabin of an aircraft is kept near one atmosphere, as compared to the external atmosphere of approximately 0.30 atm at an altitude of 31,000 feet. Explain what would have happened when the explosion that occurred in the cargo hold of Pan Am Flight 103 ruptured the integrity of the cabin. (CP)

78. Terrorist plans were uncovered prior to the second inauguration of President George W. Bush that called for filling a limousine with propane tanks and detonating an improvised fuel-air explosion. While it is doubtful this plan would have worked, calculate the volume of carbon dioxide and water vapor that would be created from an improvised fuel-air explosive consisting of 30 propane tanks. Each tank contains approximately 20 lb of propane. Assume STP conditions. (CP)

79. Shown in this figure is the separation of military explosives using thin-layer chromatography. Each experiment used a different mobile phase in an attempt to optimize the separation conditions. For each experiment, evaluate whether each compound is fully distinguishable from the other compounds. Which set of conditions would be best used to screen for the presence of military explosives? (CP)

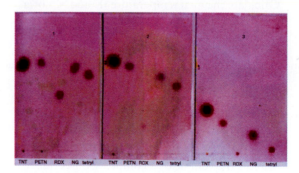

(Encyclopedia of Forensic Sciences, Vol. 2, Ed. Siegel et al., Explosives Analysis, Plate 20, © Elsevier)

80. The confirmatory method of identifying illegal drugs is GC-MS. The suspected drug is analyzed and its spectrum is compared to a library of spectra. Given the following spectra, is it possible to identify the evidence sample? Justify either a positive or negative finding. (CP)

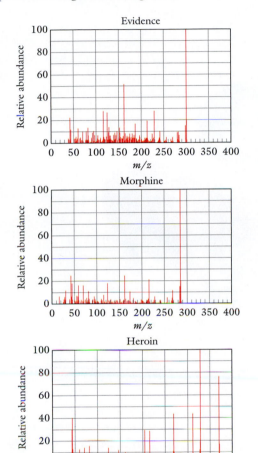

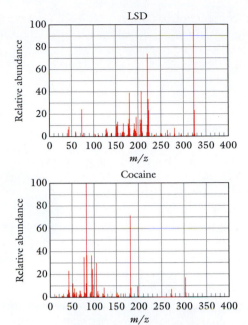

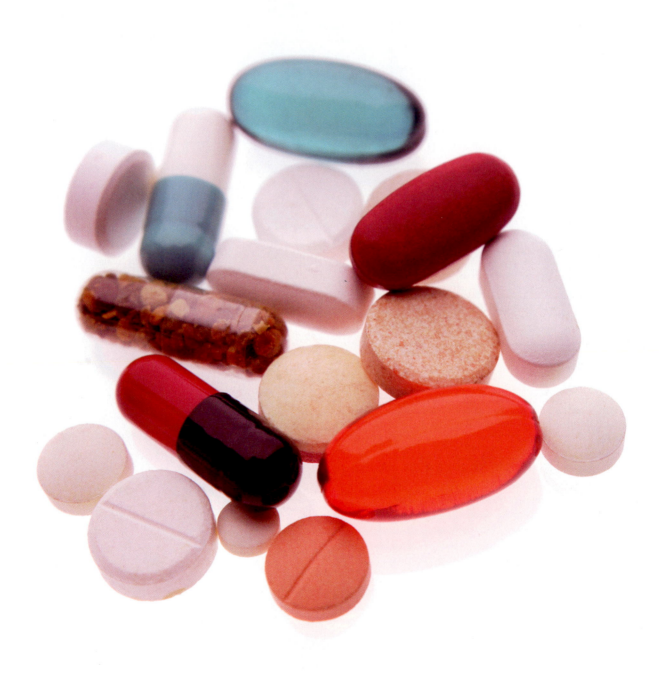

CHAPTER 11

Applications of Chemical Kinetics

CASE STUDY: Investigating Green Chemistry

Pharmaceutical drug manufacturing requires some of the most sophisticated and elaborate chemical reactions to synthesize the complex chemical structures of drug molecules. When complex chemical reactions happen, they often yield only 30% to 50% of the actual chemical compound desired, the rest being a mixture of unwanted compounds. The complex mixtures created during these reactions must be purified using energy-intensive and costly separation methods. Once the desired compound has been isolated from the mixture, the remaining compounds are disposed of.

For example, the drug atorvastatin, which is sold under the name Lipitor®, is used to lower blood serum cholesterol levels and boasts over $10 billion in yearly sales. The production of atorvastatin requires synthetic steps with only 50% yields, extreme alkaline reaction conditions, purification by distillation leading to low yields of the desired product, and extensive amounts of chemical waste that require disposal. The costs of low-yield reactions and waste disposal are significant and also come with increased employee exposure, higher energy use, and significant safety issues.

The chemical industry has responded by shifting its synthetic focus to embrace principles of green chemistry. Green chemistry seeks to prevent pollution and reduce waste while using safer chemicals and less energy. While doing so may seem like common sense, one must remember that synthetic chemists traditionally have been educated in a system that focused only on the goal of creating a complex molecule. Extreme reaction conditions were not viewed as a negative. Now, though, many industrial chemists evaluate their success in terms of green chemistry.

> **As you read through the chapter, consider the following questions:**
>
> • **What dictates how fast a chemical reaction happens?**
>
> • **What role does energy have in making reactions occur?**
>
> • **Is it possible to speed up a reaction?**

MAKE THE CONNECTION
To better understand the principles of green chemistry and how they apply to the creation of atorvastatin, it is necessary to explore how a chemical reaction occurs.

11.1 Introduction to Chemical Kinetics

LEARNING OBJECTIVE

Explain why chemical kinetics is important in understanding reactions.

Chemical kinetics is the study of the rate of chemical reactions. Knowing the rate of a reaction is very important to understanding what occurs during a chemical reaction. For example, explosions are characterized by the release of large amounts of energy and the formation of gaseous products. But these are not the only requirements for an explosion. The combustion of coal yields more energy per gram than nitroglycerin and produces gases, yet there is no explosion. The other factor for an explosion is how quickly the reaction occurs. The danger presented by nitroglycerin lies in the speed of its reaction.

The reaction rate is determined by measuring either the decrease of reactants or the increase of products during a given time period. Consider the formation of sulfur dioxide from the combustion of sulfur with oxygen, as shown below. At the start of the reaction, the S and O_2 are combined and start to form SO_2. The rate of this reaction is determined by measuring how fast the amount of S and O_2 decreases or how fast SO_2 is created.

$$S + O_2 \rightarrow SO_2$$

$$\text{Reaction rate} = \frac{\text{amount of } SO_2 \text{ formed}}{\text{change in time}} = \frac{\text{amount of S reacted}}{\text{change in time}} = \frac{\text{amount of } O_2 \text{ reacted}}{\text{change in time}}$$

Coal contains more energy gram for gram than does nitroglycerin. The rate at which the energy is released, though, is significantly faster for nitroglycerin. (Left: Tim Wright/Corbis; right: Ingram Publishing/Alamy)

■ **WORKED EXAMPLE 1**

Consider the oxidation of carbon monoxide by oxygen:

$$2CO + O_2 \rightarrow 2CO_2$$

Is the rate at which the amount of carbon monoxide decreases the same as the rate at which oxygen decreases? Consider the coefficients of the balanced equation in your explanation.

SOLUTION

No, the rates are not the same. For every 2 mol of CO that react, 1 mol of O_2 is required. Therefore, in a given amount of time, twice as many moles of CO will react as will O_2.

Practice 11.1

Compare the rate of carbon dioxide formation with the rates of loss of both reactants.

Answer

Carbon monoxide will be lost at the same rate at which carbon dioxide forms. The rate of carbon dioxide formation is twice as fast as the rate of oxygen consumption.

11.2 Collision Theory

The observation of many chemical reactions led scientists to some interesting questions about reaction rates. Why do two solids that are mixed together usually react very slowly, whereas if the solids are dissolved in water and mixed, the reaction proceeds quickly? Why is the reaction rate in many instances fastest at the start of a reaction but slower as the reaction continues? Why do most reactions proceed faster at warmer temperatures? To explain these observations about chemical kinetics, scientists use **collision theory**, a model of how chemical reactions occur at a molecular level.

LEARNING OBJECTIVE

Explain what happens on a molecular scale during a reaction.

Collision Theory

1. The molecules of the reactants must collide for a reaction to take place. *However, not all collisions will result in the formation of products.*

2. The collisions must have a sufficiently high energy. *Each reaction has a specific minimum value of energy, called the activation energy, that must be realized for a reaction to occur.*

3. The colliding particles must be properly oriented for a reaction to occur. *Only if the collision has sufficient energy and proper orientation of the colliding particles can the reaction proceed.*

The first point of collision theory—that the molecules of the reactants must collide for a reaction to occur—seems obvious enough. When a reaction in a solution first starts, collisions between reactant molecules happen quite frequently because there is a large amount of reactants in the solution. Having a large number of collisions is one of the prerequisites for a reaction to occur rapidly. However, as the reaction progresses, the reactants are consumed as the products are made. This decreases the amount of reactant available to undergo collisions, so fewer collisions will occur. As the number of collisions decreases, so does the rate of the reaction.

 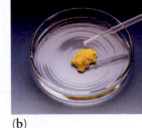

(a)   (b)

Lead(II) iodide, a bright yellow compound, is formed from the reaction of potassium iodide and lead(II) nitrate. Why is the reaction occurring much faster when the reactants are dissolved in solution, as shown in (b), compared with mixing the two solids, as shown in (a)? (Both: Richard Megna/Fundamental Photographs)

Consider Figure 11.1, which illustrates how the concentration of reactants influences the frequency of collisions as the reaction proceeds. At the beginning of a reaction, as shown in Figure 11.1a, the reactants are present at a high concentration and undergo many collisions. As the reaction continues, the concentration of reactants decreases, as shown in Figure 11.1b. This decrease in concentration of reactants causes fewer collisions between reactants and a decrease in the reaction rate.

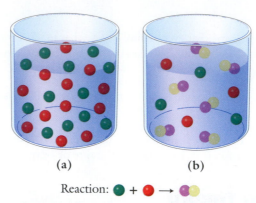

(a)   (b)

Reaction: ● + ● → ●●

FIGURE 11.1 Reaction rates versus concentration. The reaction rate is greatest at the start of the reaction—as is the number of collisions between reactant molecules—because there is a high concentration of reactant molecules, as shown in (a). As the reaction continues, increasing numbers of reactant molecules are consumed in the reaction. The remaining reactants are less likely to collide because fewer molecules are present, as shown in (b), so the reaction rate decreases.

■ WORKED EXAMPLE 2

An experimental procedure calls for mixing reactants that have a concentration of 0.1 M. What will happen to the reaction rate if, by mistake, 1.0 M solutions are mixed? Explain your answer.

SOLUTION

According to collision theory, reactants must collide for a reaction to occur. If the reactants have a higher concentration, they will undergo more collisions and start out with a higher reaction rate.

Practice 11.2

Is it always desirable to have a reaction proceed as quickly as possible? Explain your answer.

Answer

No, there are situations in which a fast reaction is not desirable. For example, explosions are reactions that occur extremely fast, and it is not desirable for all reactions to proceed so quickly. Another example in which a fast reaction rate is not desirable is the corrosion of metal in vehicles or buildings.

The requirement that a collision must occur also explains why reactions with solid reactants tend to have very slow reaction rates, whereas reactions in aqueous solutions are fast in comparison. Not all collisions of reactant molecules result in the formation of products. Only collisions with sufficiently high energy have the ability to bring about a reaction. The kinetic energy of the reactant molecules can vary dramatically, and this distribution of energy among the molecules follows a *bell curve*. For example, Figure 11.2a shows a container that has 36 reactant molecules. The molecules have been assigned a number to represent the amount of energy they contain, from low (1) to high (11). Few molecules have very little energy or very high energy; most are found near the average energy (6). When the number of molecules is graphed against the energy of the molecules, as shown in Figure 11.2b, an outline of the data appears in the shape of a bell.

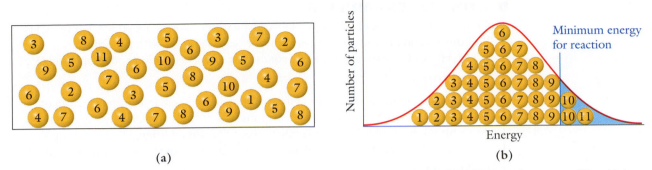

(a)

(b)

FIGURE 11.2 Reactant energy distribution. The energy of reactant molecules varies greatly from low energy (1) to high energy (11), as shown in (a). When the number of molecules of each specific energy is plotted, the data approximate a bell curve (b). Only a few reactants with the highest amount of energy can undergo a reaction.

Why is energy needed for a reaction to start? Recall Section 9.4, in which the concept of an activation energy for a reaction was first discussed. The activation energy is the minimum energy needed in a reaction to break the bonds in the reactants. As new bonds in the products are formed and energy is released, the energy used to overcome the barrier to activation comes from collisions between the reactant molecules. We might expect that all high-energy collisions will lead to formation of reaction products, but this is not the case. An additional factor, the orientation of the colliding molecules, is also critical.

Consider Figure 11.3, showing the reaction of carbon monoxide with nitrogen dioxide to form carbon dioxide and nitrogen monoxide. The energy of the collision needs to be focused between the carbon atom of carbon monoxide and an oxygen atom present in the nitrogen dioxide. If the collision occurs with any other combination of atoms, no

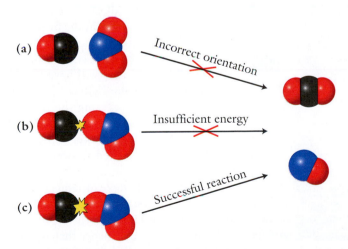

FIGURE 11.3 Reaction collisions. The proper conditions for the successful reaction of carbon monoxide are shown here. Reaction (a) is not successful, as the orientation of the molecules does not lead to the transfer of oxygen from nitrogen dioxide to carbon monoxide. Reaction (b) has proper orientation, yet there is insufficient energy in the collision. Reaction (c) is successful, as it has a collision with proper orientation and sufficient energy.

reaction will happen, even if sufficient energy is present. Figure 11.3a shows a collision that has neither the proper orientation for the needed bonds to break nor sufficient energy despite proper orientation (Figure 11.3b). With sufficient energy and proper orientation, a successful reaction occurs, as in Figure 11.3c.

■ WORKED EXAMPLE 3

Consider the reaction of ethene ($CH_2\!\!=\!\!CH_2$) with H_2 to produce ethane ($CH_3\!\!-\!\!CH_3$). Will this reaction have a greater rate than the reaction of H_2 with propene ($CH_3CH\!\!=\!\!CH_2$), assuming the same conditions of concentration and temperature?

SOLUTION

The reaction of ethene will occur at approximately the same rate as the reaction of propene because the only successful collisions will be those in which hydrogen has a direct hit on the double bond. One might argue that the additional $CH_3\!\!-$ group in propene will reduce the number of successful collisions, but the difference is minimal.

Practice 11.3

In the reaction of hydrogen gas with oxygen gas to form water, will the orientation of the collision be a limiting factor?

Answer

No, there will be no difference in collisions of hydrogen and oxygen gas because of orientation. Every collision will bring the necessary components of the reaction together.

11.3 Kinetics and Temperature

Have you ever wondered how a refrigerator preserves food? You put food into a refrigerator to prevent the food from spoiling; but what is spoiled food and how does a refrigerator prevent it? Spoiled food can mean different things, depending on the food. It can be the growth of mold on leftovers, the curdling of milk, the growth of bacteria on uncooked meat, or flavors being altered by oxidation. All of these processes are chemical reactions, and, as such, the principles of collision theory will apply.

If you lower the temperature of a system (your leftovers), the reactants will have less kinetic energy. This leads to a twofold effect. The number of collisions will decrease, causing the reaction rate to decrease. The collisions that do occur will also have less energy, resulting in fewer collisions that have the minimum energy needed to overcome the activation energy barrier.

An excellent example of temperature's effect on reaction rate can be seen using novelty glow sticks, as shown in Figure 11.4. The chemical reaction in a glow stick produces excess energy that is released as visible light in a process called **chemiluminescence.** If the glow stick is cooled in an ice-water bath before activation, as in Figure 11.4a, the reactants have less kinetic energy, resulting in fewer high-energy collisions. If, however, a glow stick is placed in hot water before activation, as in

FIGURE 11.4 Glow stick kinetics. The brightness of a glow stick depends on the kinetics of the chemiluminescent reaction. The reaction rate can be (a) decreased by placing the glow stick in ice water or (b) increased by placing the glow stick in hot water. (Richard Megna/Fundamental Photographs)

Figure 11.4b, the reactants will have greater kinetic energy, resulting in an increase in high-energy collisions.

What is occurring on a chemical level to slow or increase reaction rates as the temperature changes? Recall that temperature is a measurement of the average kinetic energy of the particles within a system. When heat is added to a system, the energy is absorbed by the particles and their kinetic energy increases, as shown in Figure 11.5. When the kinetic energy increases, the particles travel a greater distance per unit of time and undergo a greater number of collisions per unit of time. Because there are more collisions, there are more opportunities to react successfully.

Glow-Stick Temperature

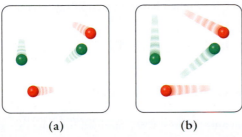

(a)   (b)

FIGURE 11.5 Kinetic energy and collisions. Particles represented in (a) have a lower kinetic energy than those represented in (b). An increase in the kinetic energy of the reactant particles causes them to move a greater distance in the same amount of time, resulting in more collisions and a faster reaction rate.

In addition to increasing the number of collisions, the number of reactant molecules that have the minimum energy to overcome the activation energy barrier increases. Figure 11.6 shows the energy of the molecules of the reactants at two different temperatures. The number of molecules with sufficient energy to react is indicated by the shaded areas.

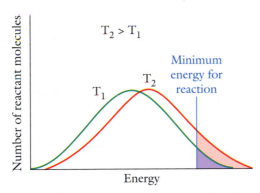

FIGURE 11.6 Energy diagram for two temperatures. Increasing the average kinetic energy of the reactant molecules increases the total number of molecules that have sufficient energy to overcome the activation energy necessary for the reaction. The rate of reaction is approximately doubled for every 10°C increase in temperature or, conversely, halved for every 10°C decrease in temperature.

The faster rate of a reaction is due to both an increase in the number of molecules with sufficient energy to react and an increase in the number of collisions. The rule of thumb is that for every 10°C increase in temperature, the rate of a reaction will double.

■ WORKED EXAMPLE 4

Explain why samples taken by a medical examiner performing an autopsy are routinely refrigerated after collection.

SOLUTION

It is important to preserve the evidence in the same condition in which it was collected. By refrigerating a sample, the rate of any reaction is decreased dramatically.

Practice 11.4

If the temperature of the human body is approximately 37°C and a sample taken from the body is stored at 7°C, how much has the reaction rate changed?

Answer

❑ The rate of the reaction is decreased to 1/8 the original rate.

11.4 Kinetics and Catalysts

LEARNING OBJECTIVE

Describe how a catalyst increases the rate of a chemical reaction.

Luminol lights up the traces of blood left after the commission of a crime, even after attempts are made to clean up the evidence. (Courtesy of Michael E. Stapleton/Stapleton Associates, LLC)

In the previous section, you examined how the rate of a chemical reaction is affected by temperature. It is possible to increase the rate of a chemical reaction by heating the reaction mixture. The elevated temperature increases the number of collisions that have sufficient energy to overcome the activation energy barrier. Another way to increase the reaction rate is to add a substance to the reaction that lowers the activation energy but doesn't interfere with the chemical reaction. Substances that increase the rate of a chemical reaction but are not actually consumed as part of the reaction are called **catalysts.**

If you are a fan of the forensic-science-themed television shows, chances are you have seen a catalyst at work. Trace amounts of blood, invisible to the naked eye, remain behind, despite the best efforts of the guilty to clean up the evidence. The two ingredients needed to reveal this evidence are a darkened room and a luminol solution.

Luminol is a compound that, when oxidized during a chemiluminescent reaction with hydrogen peroxide, releases energy in the form of blue-green light. During the oxidation of luminol, an energetically excited molecule is produced. The excess energy in the molecule is released as a photon of light in the visible region of the electromagnetic spectrum.

When luminol and hydrogen peroxide are combined, the reaction rate is extremely slow. Fortunately for crime-scene investigators, the reaction rate for the oxidation of luminol is increased by iron, which is present in the hemoglobin molecules in blood. The iron-containing portion of hemoglobin catalyzes a luminol oxidation reaction and reveals the location of blood stains, however faint, in a room.

For a better understanding of how a catalyst works, examine the activation energy of a reaction, as illustrated in Figure 11.7. The catalyst in Figure 11.7b lowers the activation-energy requirement as compared with the uncatalyzed reaction by providing an alternate pathway for the reactants to yield the products. The reaction rate increases because more reactants have sufficient energy in their collisions to overcome the lower activation energy barrier.

How exactly does a catalyst provide an alternate pathway? There are two different classes of catalysts, each with its own unique properties and characteristics. Consider the decomposition reaction of hydrogen peroxide to oxygen gas and water:

$$2H_2O_2 \rightarrow 2H_2O + O_2$$

Because the rate of the reaction is extremely slow, H_2O_2 can easily be stored without significant decomposition. However, when hydrogen peroxide is placed on a laceration,

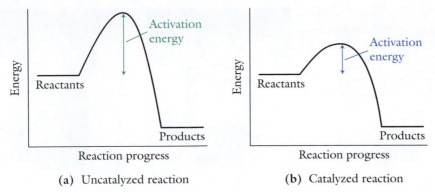

(a) Uncatalyzed reaction

(b) Catalyzed reaction

FIGURE 11.7 Activation energy changes with catalysts. (a) All reactions must overcome the activation energy barrier. The energy for doing this comes from the collisions of reactant molecules. (b) A catalyst increases the rate of a reaction by providing an alternate path for the reaction, which lowers the required activation energy. A larger number of molecules will have sufficient energy to react in the catalyzed system.

the reaction rate increases dramatically. Oxygen gas bubbles can be seen forming because the blood in the wound acts as a catalyst.

A **homogeneous catalyst** is in the same physical state as the reactants, allowing the catalyst to efficiently mix with the reactants. In the case of hydrogen peroxide, a homogeneous catalyst would be another aqueous compound. Figure 11.8a shows the decomposition of hydrogen peroxide by the iodide ion in aqueous solution, which makes the iodide ion a homogeneous catalyst. In the first step of the catalyzed pathway, the hydrogen peroxide reacts with the iodide ion to form the hypoiodite ion (IO^-) and water. In the second step, the hypoiodite ion reacts with another hydrogen peroxide molecule to produce oxygen gas and water.

$$\text{Catalyst} \quad \text{Intermediate product}$$
$$1. \quad H_2O_2 + I^- \rightarrow OI^- + H_2O$$
$$2. \quad OI^- + H_2O_2 \rightarrow O_2 + H_2O + I^- \} \text{Catalyst reformed}$$
$$1+2: \quad 2H_2O_2 \rightarrow 2H_2O + O_2$$

The hypoiodite ion is a **reaction intermediate**, which is a compound formed in the first step of a reaction mechanism but immediately consumed in the following step so that it does not appear in the overall reaction.

Notice that the first step of the catalysis mechanism is the reaction of the iodide ion with hydrogen peroxide. It is a common misconception that catalysts do not undergo chemical reactions themselves. The iodide ion is regenerated in step 2 of the mechanism, so that it appears not to have reacted because there is not a net consumption or production of the ion.

A **heterogeneous catalyst** is typically a solid substance that provides a surface to which liquid or gaseous reactants adhere. The interaction between the reactant and the catalyst weakens the bonds holding the reactants together, thus lowering the activation energy, which in turn increases the reaction rate. Pictured in Figure 11.8b and 11.8c is the catalyzed decomposition of hydrogen peroxide by solid manganese(IV) oxide. A limiting factor for a heterogeneous catalyst can be the accessibility of the reactants to the surface of the catalyst. The figure shows two identical reactions using the same amount of hydrogen peroxide and the same mass of the catalyst MnO_2. The only difference is that MnO_2 is in pellet form in Figure 11.8b and in powdered form in Figure 11.8c.

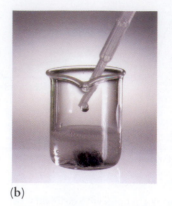

(a)   (b)   (c)

FIGURE 11.8 Homogeneous and heterogeneous catalyzed decomposition of hydrogen peroxide. Hydrogen peroxide decomposes to produce oxygen and water. The reaction is catalyzed by the addition of iodine through a homogenous mechanism (a) that forms the hypoiodite ion, a reaction intermediate. The hypoiodite ion further reacts with hydrogen peroxide to reform the catalyst. The decomposition reaction can be heterogeneously catalyzed (b) and (c) by the addition of solid manganese(IV) oxide. The reaction takes place more rapidly when there is greater surface area (c) for the catalyst to make contact with the reactants. (Richard Megna/Fundamental Photographs)

■ WORKED EXAMPLE 5

Modify the energy diagram (Number of reactant molecules vs. Energy) from Figure 11.2 to show a catalyzed and uncatalyzed reaction.

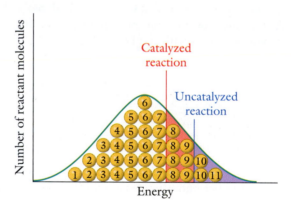

SOLUTION

A catalyst lowers the activation energy barrier, which is the minimum amount of energy required during a collision. A greater number of reactant molecules then possesses sufficient energy to react during collisions.

Practice 11.5

In Figure 11.8, the rate for the decomposition of hydrogen peroxide by MnO_2 is much faster for the powdered MnO_2 than for the pellet MnO_2. Using collision theory, explain why.

Answer

The heterogeneous catalyst lowers the activation energy barrier, which allows more collisions to be successful. The powdered MnO_2 in Figure 11.8c is providing more sites for the reactants to adhere to and ultimately more successful collisions. The smaller the particle size, the larger the surface area available to contact the reactants.

The use of heterogeneous and homogeneous catalysts is commonplace in industry, with each type of catalyst having benefits and drawbacks. For example, homogeneous catalysts will mix thoroughly with the reactant solution, which is desirable. However, recovery of the catalyst involves separating the components of a mixture using chromatography or perhaps distillation. Heterogeneous catalysts have the benefit of simply being filtered from solution after use, but the mixing of the catalyst with the reactants is not nearly as efficient.

11.5 Zero-Order Reactions

The police are charged with pulling over and evaluating any driver suspected of violating the drunk-driving laws. Individuals who fail a field sobriety test are often transported to a hospital or testing laboratory for further determination of their blood alcohol content (BAC) beyond that of any breath test administered by the arresting officer. The analysis of blood alcohol can be done hours after a person has been detained, and an accurate value for the BAC can be determined at the time of arrest. The method for this analysis relies on knowing the rate of oxidation of ethanol to carbon dioxide in the body.

According to collision theory, if the concentration of a reactant is increased, the rate of the chemical reaction should also increase. However, there is a class of chemical reactions in which the rates will not increase, despite an increase in the reactant concentration. These particular reactions are known as **zero-order reactions**—the concentration of a reactant does not influence the reaction rate. The oxidation of ethanol in the human body is an example of a zero-order reaction. The rate of elimination of alcohol does not slow down as the concentration of alcohol decreases, nor does it increase if more alcohol is consumed. The only way for alcohol to be eliminated from the body is to wait until sufficient time passes for the completion of the metabolic processes that remove alcohol. This zero-order dependence is typical of enzyme-catalyzed processes but is not limited to them.

Figure 11.9 shows a typical graph of alcohol levels found in blood after a person consumes alcohol. Once the alcohol is fully absorbed, it is eliminated at a constant rate from the body. On average, a person eliminates from the blood 15 mg/dL of blood per hour.

(Bill Fritsch/Brand X Pictures/Alamy)

Section 8.8 discussed the German practice of using alcohol congener analysis to combat drunk driving. The body oxidizes ethanol at a constant rate until the ethanol has completely reacted. The alcohol congeners build up in a person's system until the ethanol concentration decreases sufficiently to the point at which the congeners will be oxidized. By measuring the concentration of congeners in a person's blood and comparing that

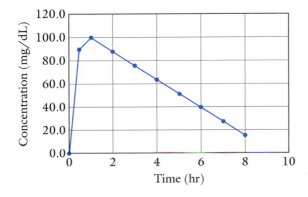

FIGURE 11.9 Graph of blood concentration versus time for alcohol. After the consumption of an alcoholic beverage, the resulting concentration of ethanol in the blood increases dramatically as it is absorbed through the stomach. The concentration then decreases at a constant rate of 15 mg/dL, showing a linear decline.

result with the amount found in alcoholic beverages, the quantity and time of an individual's alcohol consumption can be determined using calculations based on the kinetics of ethanol oxidation.

As stated earlier, it is unusual for a reaction rate to be independent of the concentration, as is the case for zero-order reactions. This unusual behavior can be explained by taking a closer look at how alcohol is oxidized.

An enzyme called *alcohol dehydrogenase* (ADH) catalyzes the oxidation of ethanol in the body. **Enzymes** are large molecules designed to speed up or allow a specific chemical reaction in a living system. In the case of ADH, the consumption of a single alcoholic beverage swamps the capacity of the enzyme to oxidize ethanol. The excess amounts of ethanol cannot reach the limited amount of ADH in the body. Hence, the only way for the body to eliminate the alcohol is to cease intake and wait for the ADH to oxidize the ethanol. The amount of enzyme effectively acts like the mouth of a funnel that limits how much liquid can flow through the funnel. If more liquid is added, it just builds up in the funnel reservoir waiting to pass through. Enzyme-catalyzed reactions are the most common form of zero-order reactions.

11.6 First-Order Reactions

Collision theory predicts that increasing the reactant concentration produces an increase in the reaction rate, and in fact this does occur most of the time. A reaction is said to be a **first-order reaction** when the rate of the reaction mimics changes in the reactant concentration. If the reactant concentration is doubled, so is the reaction rate; if the reactant concentration is halved, the rate of the reaction is also halved.

Consider the elimination of aspirin from the body, as illustrated in Figure 11.10. The aspirin concentration decreases much faster in the beginning of the reaction, as evidenced by the large drop in concentration values in the first hour compared with the decreases in concentration levels later in the reaction. Most common drugs, both legal and illegal, follow first-order reaction rates for elimination.

FIGURE 11.10 Graph of blood concentration versus time for aspirin. The concentration of aspirin in blood shows a dramatic increase at the onset of use. The rate at which aspirin is then eliminated from the body is greatest at the beginning of the reaction and decreases as the reaction progresses.

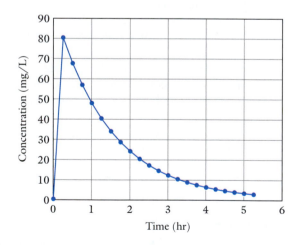

■ WORKED EXAMPLE 6

What is the decrease in concentration of aspirin during the first hour, according to Figure 11.10?

SOLUTION

The aspirin level peaks at 80 mg/L and then decreases to approximately 48 mg/L during the first hour, for a total decrease of 32 mg/L.

Practice 11.6

What is the decrease in concentration of aspirin between the first and second hours, according to Figure 11.10?

Answer

☐ 24 mg/L

A toxicology screening is ordered for any suspicious death. Investigators search the victim's home for signs that the person was taking prescription medications. This information can often provide potential investigative leads. For example, the person might have accidentally overdosed, might have had a deadly reaction to combining illegal drugs or alcohol with the prescription medication, or might have stopped taking a prescribed medicine.

Investigators try to determine what levels of a prescription drug should be in the bloodstream if a person had been taking the prescriptions as directed. Most prescription drugs are partially metabolized by the body while also being excreted in urine. Furthermore, each time a new dose is taken, the pharmaceutical levels in the body increase. It turns out that most pharmaceuticals follow a first-order reaction in their elimination from the body, as seen in Figure 11.10. Still, to determine the proper therapeutic levels that should be in a blood sample, more information—such as the dosage amount and the dosage interval—is necessary.

(Michael Keller/Corbis)

11.7 Half-Life

An approximate time can be calculated using the half-life of aspirin. The **half-life** $(t_{1/2})$ is the amount of time required for one-half of a substance to react. In pharmaceuticals, the half-life is how long it takes for one-half of the drug to be eliminated from the body. The elimination of the drug is the result of its metabolism into other compounds or its physical removal through bodily wastes. Since the half-life of aspirin is approximately three hours, the following concentration profile can be written for the above case:

$$500 \xrightarrow{3\,hr} 250 \xrightarrow{3\,hr} 125 \xrightarrow{3\,hr} 62.5 \xrightarrow{3\,hr} 31.25 \xrightarrow{3\,hr} 15.63 \xrightarrow{3\,hr} 7.81$$

Therefore, the aspirin underwent six half-lives: $6 \times 3\ hr = 18\ hr$.

> **LEARNING OBJECTIVE**
> Explain how the half-life of a drug corresponds to the dosage.

WORKED EXAMPLE 7

Determine the amount of time required for 97% of aspirin to leave a person's system, provided that the person took a single dose of 500 mg and the half-life of aspirin is 3 hours.

SOLUTION

Since 97% of the aspirin has been eliminated,

$$0.97 \times 500\ mg = 485\ mg\ eliminated$$

This corresponds to 15 mg remaining in the person's system. Starting with 500 mg to calculate the number of half-lives,

$$500\ mg \to 250\ mg \to 125\ mg \to 62.5\ mg \to 31.3\ mg \to 15.6\ mg$$

This means it would take approximately five half-lives to eliminate 97% of the aspirin, which corresponds to $5 \times 3\ hr = 15\ hr$.

Practice 11.7

Determine the half-life (in hours) of the pain reliever ibuprofen, given that a person takes a 800 mg dose at 10:00 A.M. and by 6:00 P.M. only 50 mg remain.

Answer

☐ $t_{1/2} = 2$ hr

When you take a prescription drug, there will be instructions for how often per day you should take the drug. It can be important that you take the drug the same time each day or perhaps two or three times per day. Many medications are prescribed so that the dosage interval corresponds to the half-life of the particular drug. This dosage interval leads to a *steady state* in the concentration of the pharmaceutical that is equal to twice the dosage, as illustrated in Figure 11.11.

When an individual starts taking a prescription drug, the maximum level in the body usually does not occur for two to three days, depending on the interval between doses. When the second dose is taken, only half of the original dose remains in the body. This process continues as each dose is taken and gradually increases the concentration in the body. By the seventh dose, the body is within 99% of the final, maximum value that will be attained. It is important to realize that as the drug concentration is being increased with each of these doses, the drug is also being metabolized and excreted. Eventually, a steady state is reached in which the drug concentration stays fairly constant. The values in Figure 11.11 are maximum values.

Dose Number	1	2	3	4	5	6	7	8	9
	400	200	100	50	25	12.5	6.3	3.1	1.6
		400	200	100	50	25	12.5	6.3	3.1
			400	200	100	50	25	12.5	6.3
				400	200	100	50	25	12.5
					400	200	100	50	25
						400	200	100	50
							400	200	100
								400	200
									400
Total	400	600	700	750	775	788	794	797	799

FIGURE 11.11 Steady-state pharmaceutical dosage. This chart gives the total dosage after a dose of 400 mg is administered in intervals equal to the half-life of the drug within a person's body. After a time interval of one half-life, the total dosage is 400 mg plus what remains of the initial dose, and so on. The bottom total line shows that the total concentration of a pharmaceutical taken regularly reaches a near steady-state value after about seven doses.

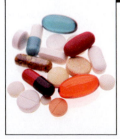

11.8 CASE STUDY FINALE: Investigating Green Chemistry

The synthesis of atorvastatin created a large amount of hazardous waste that required costly disposal, created a hazardous exposure risk for employees, used large amounts of energy, and was a low-yield reaction. The synthesis of atorvastatin was taken on as a challenge by a company called Codexis. The company created a new method of synthesizing the main building block of the drug using the Twelve Principles of Green Chemistry,[*] as follows:

1. Prevent the creation of hazardous waste rather than clean it up afterwards.
2. Use as much of the reactants as possible in the final product to minimize waste.
3. Use fewer hazardous reactants to minimize human and environmental risks.
4. Create safer chemicals with less toxicity.
5. Use safer solvents.
6. Reduce the amount of energy needed through low-temperature reactions.
7. Use renewable sources for raw materials when practical.
8. Simplify reactions to use fewer reagents and therefore produce less waste.
9. Use catalysis when possible.
10. Create products that will biodegrade when no longer needed.
11. Monitor reactions in real time to prevent hazards.
12. Use safer starting materials to prevent accidents.

The synthesis of atorvastatin dramatically changed through the use of enzymes. Recall from Section 11.5 that enzymes are large molecules that catalyze chemical reactions. An enzyme lowers the activation-energy barrier by providing a lower energy route to the product. Enzymes are very large and complex molecules made up of hundreds to thousands of amino acids, which form a specific shape that allows the enzyme to bind in a particular way to a reactant molecule. The interaction of an enzyme with a reactant molecule is often compared to that between a lock and key, as each of these interactions relies on the components having a specific size and shape. Figure 11.12 illustrates the binding of a reactant molecule to a larger enzyme; more information on protein structures is discussed in Chapter 14.

[*]Adapted from Anastas, P.T., and Warner, J. C., *Green Chemistry: Theory and Practice,* Oxford University Press: New York, 1998, p. 30.

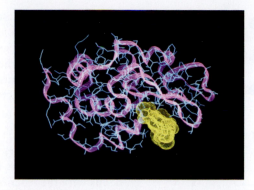

FIGURE 11.12 The large protein molecule forms the "lock" with an opening that is specific for the "key" molecule in yellow. The ribbon is superimposed on the large protein molecule to better illustrate the overall shape of the protein molecule. (Science Photo Library/Photo Researchers, Inc.)

Codexis examined enzymes that, in principle, could catalyze the formation of a key component in the atorvastatin molecule. However, the enzymes were not sufficiently active for the commercial process. The researchers then created variations of the original enzyme to produce a more efficient catalyst for the reaction. The final process uses three enzymes customized by Codexis that have increased the desired product yield from less than 50% to nearly 100% for one of the initial reactions. At the same time, Codexis has eliminated the need for hazardous chemicals, as well as changed the reaction conditions from operating in a highly alkaline environment to a nearly neutral pH at normal atmospheric pressure and room temperatures. In one of the enzyme-catalyzed steps, they increased the reaction rate 100 times while making a second step 4,000 times faster.

Green chemistry illustrates that shifting the emphasis and focus during the development phase of a chemical process can lead to a profitable and environmentally sustainable product.

CHAPTER SUMMARY

• Kinetics is the study of the rate at which chemical reactions occur. Reaction rates can vary from microseconds to years. Variables that affect the rate of a chemical reaction include temperature, reactant concentration, physical state of the reactants, and presence or absence of catalysts.

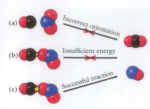

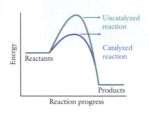

• Collision theory describes the requirements for two reactant molecules to successfully come together and form the new product(s). The three main principles of collision theory are: (1) Reactant molecules must collide to react; (2) they must collide with sufficient energy to overcome the activation energy barrier; and (3) the molecules that collide must have the proper orientation for the reactant bonds to break and the new chemical bonds to form in the products.

• The temperature of a system directly affects the kinetic energy of the reactants, increasing the frequency and energy of collisions.

• A catalyst increases the rate of a chemical reaction by providing an alternative reaction mechanism that has a lower activation energy. Catalysts can be either homogeneous with the reaction solution or a heterogeneous solid material. Enzymes are large protein molecules that serve as biological catalysts to speed up a specific reaction in a living system.

• The reaction rate of a zero-order reaction is independent of concentration. Enzyme-catalyzed reactions tend to be zero-order reactions. First-order reactions have reaction rates that mimic changes in the reactant concentrations. The rate will double when the concentration of a reactant doubles, and the rate is halved when the concentration is halved. In the elimination of most pharmaceuticals, reactions in the body follow first-order reaction rates.

• The time required for the elimination of one-half of a drug in the body is the half-life. Half-life calculations are commonly used to determine the proper dosage interval for pharmaceutical drugs to maintain a therapeutic level of the drug in the body.

KEY TERMS

catalyst (p. 324)
chemical kinetics (p. 318)
chemiluminescence (p. 322)
collision theory (p. 319)

enzyme (p. 328)
first-order reaction (p. 328)
half-life (p. 329)
heterogeneous catalyst (p. 325)

homogeneous catalyst (p. 325)
reaction intermediate (p. 325)
zero-order reaction (p. 327)

MAKING MORE CONNECTIONS: Additional Readings, Resources, and References

For more information on green chemistry:
http://www.epa.gov/greenchemistry/
http://ir.codexis.com/phoenix.zhtml?c=208899&p=irol-newsArticle&ID=1181061
http://www.epa.gov/greenchemistry/pubs/docs/award_recipients_1996_2011.pdf
Anastas, P. T., and Warner, *J. C. Green Chemistry: Theory and Practice,* Oxford University Press: New York, 1998, p. 30.

http://portal.acs.org/portal/acs/corg/content?_nfpb=true&_pageLabel=PP_TRANSITIONMAIN&node_id=830&use_sec=false&sec_url_var=region1&__uuid=edf0a713-99e5-4b35-b829-4948514950ff

REVIEW QUESTIONS AND PROBLEMS

Questions

1. Discuss why it is important to understand the rates of chemical reactions. (11.1)

2. What are the factors that can influence the rate of a chemical reaction? (11.2–11.4)

3. What are the three principles of collision theory? What are the restrictions on each principle? (11.2)

4. Why do many gas-phase reactions occur with rapid kinetics, whereas solid-phase reactions occur slowly? (11.2)

5. Why does the initial concentration of the reactants affect most reaction rates? (11.2)

6. Why do the rates of most chemical reactions slow down over time? (11.2)

7. If bloody clothes are recovered from a crime scene, one of the first things done before storing the evidence is to dry the items out. Why? (11.2)

8. Why do reactants require relatively high-energy collisions to form products? (11.2)

9. Sketch an energy distribution diagram for a reaction indicating the fraction of reactant molecules that have sufficient energy to react. Is the number of molecules with sufficient energy the same as the number of reactant molecules that *will* react? Explain your answer. (11.2)

10. When the temperature of a system is increased, what two factors are changed that lead to an increase in the reaction rate? (11.3)

11. Does increasing the temperature of a reaction system change the amount of energy needed for reactants to cross the activation energy barrier? (11.3)

12. What is the rule of thumb that relates a temperature change to the kinetics of a reaction? How does that rule relate to preventing food from spoiling by placing it in a refrigerator? (11.3)

13. Explain why the amount of surface area of a heterogeneous catalyst can affect the rate of a chemical reaction. (11.4)

14. Sketch the activation energy diagram for a reaction with and without a catalyst present. (11.4)

15. What is the only method of increasing the reaction rate that works by lowering the activation energy? (11.4)

16. Carbon monoxide is the flammable gas that is partially responsible for the muzzle flash seen from a firearm. It is also one of the gases that can cause a backdraft to happen when firefighters open up poorly ventilated rooms. Automobiles produce carbon monoxide as a result of the negative oxygen balance of the fuel-air explosion that powers the engines. Why does the carbon monoxide generated in a gun barrel or in a backdraft ignite, whereas there is no such igniting in the muffler of a car? (CP)

17. Carbon monoxide produced in automobile engines cools at it passes through the exhaust system and, before exiting the vehicle, passes through the catalytic converter for conversion to carbon dioxide. Is the catalytic converter an example of a homogeneous or heterogeneous catalyst? (11.4)

18. Explain the similarities and differences between homogeneous and heterogeneous catalysts. (11.4)

19. What would happen to the rate of an enzyme-catalyzed reaction if the concentration of the enzyme were doubled? Explain your answer. (11.4)

20. Does the rate of an enzyme-catalyzed, zero-order reaction change as a function of time? Explain your answer. (11.5)

21. Explain how the concentration of alcohol in blood, when measured hours after the ingestion of the alcohol, can be extrapolated backwards to the time a suspected drunk driver was pulled over by police. (11.5)

22. Does the rate of a first-order reaction change as a function of time? Explain your answer. (11.6)

Case Study Problems

23. Explain what investigative information can be provided by determining blood serum levels of both legal and illegal drugs and combining this information with the half-lives of those drugs. (11.7)

24. How much aspirin will remain after 12 hours if a person took a single dose of 300 mg? The $t_{1/2}$ of aspirin is 3 hours. (11.7)

25. What is the half-life of a pharmaceutical if after 16 hours a patient had 0.469 mg remaining out of a 30-mg dose? (11.7)

26. The production of nitroglycerin is an exothermic reaction. Explain why scaling up the production of nitroglycerin could lead to an especially dangerous situation. (CP)

27. The injured victim in a pedestrian-vehicle accident states that he walked out into the street because he did not see any headlights. The driver of the car insists the headlights were turned on at the time of the accident. How can a broken headlight conclusively prove whether or not the headlights were turned on? (*Hint:* The temperature of a filament is approximately 3000 K and the headlight is filled with inert gases such as xenon to prevent oxidation of the filament.) (CP)

28. Read the online article from the BBC found at http://news.bbc.co.uk/2/hi/science/nature/3632770.stm and explain the comment about why using the half-life of an isotope that decays is preferred over using insect activity. (CS)

29. From the BBC article referred to in Problem 28, could the use of the isotope lead-210 and its decay be useful for a recent homicide victim? Why would the polonium-210 decay not work for the dating of a victim found buried 30 years after having disappeared? (CS)

30. When illegal drugs are used, many of them will leave trace amounts in the body, because the drugs are taken up by hair follicles and are stored in the hair. What would be the most effective method for detecting the long-term use of illegal drugs by an individual, blood/urine testing or hair analysis? Which method would work best for recent illegal drug use? (CP)

CHAPTER 12

Nuclear Chemistry: Energy, Medicine, Weapons, and Terrorism

CASE STUDY: Exploring Nuclear Power

The United States is a nation that consumes resources, and at a disproportionate rate as compared to the rest of the world. The population of the United States is approximately 5% of the world's population, but we consume nearly 25% of the world's energy resources. As the economies of China and India continue to expand, their level of energy consumption will increase dramatically because they have a significant proportion of the world's population.

Competition for fossil fuels will increase and have a direct impact on national and world economics.

Industry and economic development thrive with abundant and relatively inexpensive energy. Historically, little thought was given to the pollution created by burning fossil fuels. China has spurred its economic development in this fashion, as became evident when the 2008 Olympic Games were held in Beijing, China. The Chinese government forced large numbers of factories and power plants to be shuttered for extended periods of time before the games began, in an attempt to clean the air.

The United States uses a variety of sources to generate electricity. Twenty percent of the electrical power generated in the United States comes from 65 nuclear power plants. The remaining 80% of the nation's power comes from 3,450 power plants that are fueled by natural gas, coal, and oil. Another 2,650 power plants are based on hydroelectric, wind, solar, and other renewable energy sources. Nuclear power plants are generally found near large population centers, as they can provide a much larger amount of energy. However, proximity to large population centers also poses potential risks.

As you read through the chapter, consider the following questions:

• **What is radiation and why is it dangerous?**

• **How can radiation be used to diagnose and treat disease?**

• **How is energy generated in a nuclear power plant and what risks are associated with nuclear power plants?**

On March 11, 2011, an earthquake occurred in the Pacific Ocean 230 miles northeast of Tokyo, Japan, with a magnitude of 9.0 on the Richter scale. This earthquake was the fifth most powerful earthquake ever measured. The earthquake produced a tsunami with a 30-foot-high wave of water that devastated the Japanese coast, including the Fukushima Daiichi nuclear power plant.

MAKE THE CONNECTION
What role should nuclear power play in meeting the needs of our nation? To answer this question, it is first necessary to understand the nature of radiation and nuclear chemistry.

12.1 The Discovery of Natural Radioactivity

LEARNING OBJECTIVE

Trace Becquerel's experiments that led to the discovery of radioactivity.

Glow-in-the-dark toys contain a phosphorescent substance. (Doug Steley/Alamy)

Nuclear medicine, radiation therapy, atomic weapons, and nuclear power plants all trace their origins back to the discovery of natural radioactivity one cloudy day in 1896. Henri Becquerel, a French scientist, was studying the phosphorescence of minerals after exposure to sunlight. **Phosphorescence** is the spontaneous emission of light from a substance after it has been activated by exposure to light. This is the process by which glow-in-the-dark toys function.

Becquerel hypothesized that the process of phosphorescence was related to X-rays, which had only recently been discovered. Although Becquerel was incorrect about a link between X-rays and phosphorescence, his experiments set in motion a chain of events that would usher in the nuclear age.

Becquerel's experiment consisted of wrapping a photographic plate in black paper to prevent its exposure to light and placing it next to a mineral crystal made of salts of uranium. The next step in the experiment was to expose the mineral-photographic plate combination to sunlight to induce phosphorescence in the mineral. When the photographic plate was developed, the resulting image corresponded to the silhouette of the mineral. Becquerel incorrectly assumed that the image was due to the emission of X-rays during phosphorescence.

Becquerel intended to repeat this experiment but was delayed several days by cloudy weather. He placed his mineral-photographic plate combination in a desk drawer. When sunshine returned later that week, he decided he should start over with a fresh photographic plate. Becquerel removed the mineral sample and unused plate from his desk drawer and, by chance, developed it. To his surprise, the photographic plate showed exactly the same pattern of the mineral's shape as had been

(a) Henri Becquerel discovered rays of energy being emitted spontaneously from the phosphorescent uranium-containing minerals that he was studying. (b) Marie Curie won two Nobel Prizes, one in chemistry and one in physics, for her pioneering research in nuclear chemistry. (Left: Time & Life Pictures/Getty Images; right: Hulton-Deutsch Collection/CORBIS)

obtained from the previous experiments. Because there was no phosphorescence in this case, it was clear that something else caused the pattern to appear on the photographic plate in his desk drawer.

Becquerel continued his experiments with the mineral and soon determined that the uranium present in the mineral sample was causing the exposure of the photographic plate. He postulated that the uranium was spontaneously emitting rays of some sort.

A young student by the name of Maria Sklodowska, who was looking for a thesis project to earn her doctorate, took up the study of these "uranic rays." Better known today by her married name, Madame Marie Curie, she is responsible for coining the term **radioactivity** to describe the spontaneous emission of high-energy rays and particles from an atomic nucleus.

Marie Curie is also responsible for the discovery of two previously unknown radioactive elements, radium and polonium. Her isolation of radium required the processing of eight tons of ore to obtain a single gram of radium! During her distinguished career, she became the first person ever to win two Nobel Prizes—one in physics and one in chemistry. The element with the atomic number 96 is named *curium*, in honor of her work.

12.2 Radiation Types and Hazards

The danger of radioactivity is that the rays and particles emitted from the radioactive source collide with molecules in body cells. The collision ionizes the molecules by breaking their internal bonds. The ionized molecules can react with other molecules in the cell and adversely affect bodily processes. Extended exposure to elevated levels of ionizing radiation can increase the odds of developing cancer; exposure to high levels of ionizing radiation can cause severe illness or even death.

> **LEARNING OBJECTIVE**
> **Explain the risks in handling radioactive material.**

We are all exposed to low levels of ionizing radiation, called **background radiation**, that occurs naturally in the environment from a variety of sources. Figure 12.1 illustrates the various sources of background radiation. The most prevalent is radon gas, a product of the radioactive decay of trace amounts of uranium naturally found in the soil. The U.S. Environmental Protection Agency estimates that one out of 15 homes has elevated radon gas levels. This is of concern because the second leading cause of lung cancer is elevated radon levels.

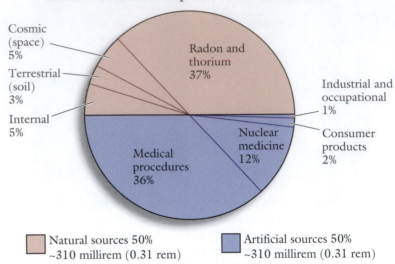

Sources of Radiation Exposure in the United States

Cosmic (space) 5%

Terrestrial (soil) 3%

Internal 5%

Radon and thorium 37%

Industrial and occupational 1%

Nuclear medicine 12%

Consumer products 2%

Medical procedures 36%

Natural sources 50% ~310 millirem (0.31 rem)

Artificial sources 50% ~310 millirem (0.31 rem)

FIGURE 12.1 Sources of background radiation. The major source of radon gas comes from the natural decay of trace amounts of uranium occurring naturally in soil. Ionizing radiation from medical X-rays and nuclear medicine is used to diagnose and treat medical disorders. Cosmic radiation is produced by the sun, and a small percentage passes through the earth's atmosphere and reaches the surface. Terrestrial radiation comes from the earth's natural radioactive isotopes. Some of the naturally occurring radioactive isotopes are also present in our bodies and serve as internal sources of radioactivity. (www.nrc.gov/reading-rm/basic-ref/glossary/exposure.html)

Medical X-rays and nuclear medicine are another source of background radiation, but the benefit in diagnosing and treating medical conditions outweighs the risk involved in exposure to the radiation. Yet another form of background radiation comes from outer space and passes through our atmosphere; the sun is the main producer of the cosmic radiation that reaches earth. Terrestrial radiation, on the other hand, refers to the naturally occurring radioactive elements found in the earth's crust. Radioactive materials are also intentionally placed in certain consumer products such as smoke detectors and static eliminators.

The final source of background radiation is our own bodies—radioactive carbon, lead, and potassium are among the most prominent radioactive isotopes that are natural components of our body and are present even at birth. Even though we are continually bombarded with background radiation, it is prudent to limit unnecessary exposure to ionizing radiation.

Many elements have naturally occurring radioactive isotopes. Uranium is probably the best known, but we will encounter other examples in this chapter. What makes some isotopes radioactive and others stable? Hydrogen consists of three isotopes—$_1^1H$, $_1^2H$, $_1^3H$—of which the first two are stable and the third is radioactive. To understand why a particular isotope may be unstable, we have to look more in depth at the nucleus of an atom, which was originally discussed in Chapter 3.

The nucleus makes up a minute fraction of the atom's total volume. Although it is approximately 1/100,000 the size of the hydrogen atom, the nucleus contains all of the positively charged protons of the atom. Recall that like charges repel each other. The repulsive forces between the protons within the nucleus are stabilized by the neutrons, which can be thought of as the glue holding the nucleus together.

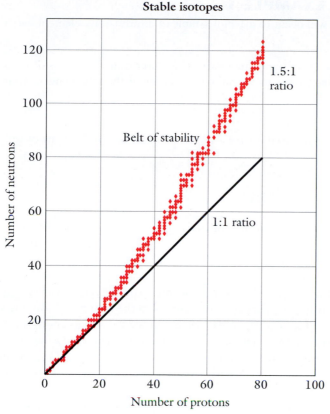

Stable isotopes

FIGURE 12.2 Isotope stability. The belt of stability represents isotopes that have a stable ratio of neutrons to protons within the nucleus. Radioactive isotopes have a ratio with either too many neutrons as compared with protons (above the belt of stability), or too many protons as compared with neutrons (below the belt of stability).

Figure 12.2 shows a graph that compares the number of neutrons with the number of protons of all stable (nonradioactive) isotopes. For lighter elements, stable isotopes have one neutron for every proton. Heavier elements, however, require more neutrons to stabilize the increased number of protons present in the nucleus. The data in Figure 12.2 form a dense elongated cluster referred to as the **belt of stability** because there are no radioactive isotopes that appear in this region. Radioactive isotopes have neutron-to-proton ratios that fall outside of the belt of stability. By emitting radioactive particles from their nucleus, unstable isotopes will ultimately form isotopes or elements that are found within the belt of stability.

Consider the periodic table on the inside front cover of this book. Most elements have radioactive isotopes, but any element with an atomic number greater than 83 consists entirely of radioactive isotopes. Notice that the atomic masses listed for most of these elements are in parentheses. This number is the mass of the most stable isotope. Thorium (Th), protactinium (Pa), and uranium (U), however, have normal atomic masses because their most stable isotopes have an extremely long half-life that lasts millions to billions of years. Two elements with atomic numbers lower than 83 also consist of only radioactive isotopes, technetium (Tc) and promethium (Pm).

Uranium, with atomic number 92, is the heaviest known naturally occurring element. Most of the elements with an atomic number greater than 92 are artificially created in nuclear research facilities.

■ **WORKED EXAMPLE 1**

What changes will occur to a 1-mol sample of a radioactive element over time?

SOLUTION

The decay process will generate new isotopes over time. The mass of the sample is constantly changing due to the radioactive decay of the atomic nuclei to produce lighter elements.

Practice 12.1

How many elements do not have a single stable isotope and consist of only radioactive isotopes?

Answer

❑ 30

How does the emission of radioactive particles change the number of protons and neutrons within the nucleus to create completely different elements? Three types of radiation are emitted, and the emission of each type has a different effect on the original nucleus.

Alpha particles (α), the heaviest of the three particles, consist of two protons and two neutrons. Because alpha particles are identical to the helium +2 ion, the symbol for the alpha particle is ^4_2He. The alpha particle is emitted to decrease both the number of protons and neutrons in the nucleus, as the nucleus is too heavy and the new, lighter isotope that is formed will be more stable. Note that the ionic charge is generally not shown in the symbol for the alpha particle because the charge is not necessary for balancing nuclear equations.

The danger presented by alpha particles must be gauged on two different scales. The **ionizing power** of a radioactive particle refers to its ability to ionize another molecule. The ionizing power of the alpha particle is extremely high because it is massive in size (compared with the other two types of radioactive particle) and has a +2 charge, which easily attracts electrons away from other molecules.

The **penetrating power** of ionizing radiation is a measure of how far the particles can penetrate into a material and cause damage. Alpha particles have very weak penetrating power because of their large size. A sheet of paper or a layer of clothing is sufficient to stop alpha particles. The location of the source of the alpha radiation is therefore important. An alpha emitter within a person's body does tremendous damage, yet an alpha emitter outside the body doesn't pose as much of an immediate threat.

An isotope undergoes alpha decay when it emits an alpha particle from the nucleus of an atom. The new isotope formed has two fewer protons and two fewer neutrons and will be shifted closer to the belt of stability. Often, a radioactive isotope undergoes many decay processes before achieving a stable isotope. A uranium nucleus, for example, undergoes 14 separate decay processes before it finally forms a stable isotope of lead ($^{206}_{82}\text{Pb}$). In the transformation of uranium to lead, not all of the decay steps involve alpha particles. A second type of decay occurs when a neutron is converted to a proton by emitting a **beta particle** (β).

A beta particle, often written as $^{0}_{-1}\text{e}$, is identical to an electron. Beta particles tend to have less ionizing power than alpha particles because the mass of the beta particle is much smaller and transfers less kinetic energy in a collision. However, the penetrating power of beta particles is greater than that of alpha particles. Several millimeters of metal sheeting or very dense wood is required to stop the flow of beta particles. Beta decay occurs when the nucleus has too many neutrons. A neutron is converted into a proton and the ejected beta particle. The new isotope has the same mass as the original parent isotope, but its atomic number has increased by one proton. The mass lost due to the beta particle is too small to be

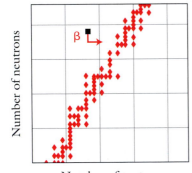

Number of neutrons

β

Number of protons

A distinct pattern can be seen when the number of neutrons is plotted against the number of protons for stable isotopes. Radioactive isotopes have ratios of neutrons to protons that fall outside of this stable region (belt of stability).

noticeable. When a beta particle is emitted, the newly created nucleus is closer to the belt of stability, because the net change is a gain of one proton and a loss of one neutron.

The third type of radiation, the **gamma ray** ($\boldsymbol{\gamma}$), is emitted from a nucleus during the emission of alpha and beta particles. A gamma ray has no mass and no charge, but it is an extremely high-energy photon—a discrete quantity of electromagnetic energy, or a packet of light. Gamma rays have low ionizing power but very high penetrating power. To shield personnel from gamma rays requires either several centimeters of lead or thick cement walls.

■ WORKED EXAMPLE 2

If an unstable isotope situated just above the belt of stability undergoes beta decay, will it be closer to or farther away from the belt of stability?

SOLUTION

Beta decay involves a nucleus converting a neutron into a proton and giving off a beta particle. Beta decay decreases the number of neutrons by 1 and increases the number of protons by 1, moving the nucleus down and to the right, as shown in the figure at left. Therefore, the nucleus located just above the belt of stability will shift toward the stable region.

Practice 12.2

If a radioactive isotope undergoes alpha decay, which location will the new isotope have with respect to the original isotope on the belt of stability graph?

Answer

Alpha decay involves a nucleus decreasing the total number of protons and neutrons by 2 each, which would move the nuclei down and to the left.

12.3 Balancing Nuclear Equations

What becomes of an atom of uranium-238 ($^{238}_{92}$U) once it has emitted an alpha particle? We can analyze this change by writing and balancing a nuclear equation—a process that is different from balancing a traditional chemical equation. In a traditional chemical equation, we make sure to account for the total number of atoms of each element on both sides of the equation. If two atoms of magnesium appear on the left side of an equation, two atoms of magnesium must appear on the right.

> **LEARNING OBJECTIVE**
>
> **Write balanced reactions for the radioactive decay of an isotope.**

In contrast, a nuclear equation involves an atom emitting a radioactive particle that can change the identity of the element. The result of emitting a radioactive particle is that an atom of a radioactive element such as uranium can appear on the left side of the equation, but not on the right. Nuclear equations are balanced by making sure that the total number of neutrons and protons on both sides of the equation is equal. The directions for balancing a nuclear equation are listed below and include as an example U-238 emitting an alpha particle.

Balancing Nuclear Equations

Step 1. Write all elements and particles using isotope notation, with the mass numbers written as superscripts and the atomic numbers written as subscripts.

$$^{238}_{92}\text{U} \rightarrow {}^{4}_{2}\text{He} + ?$$

Step 2. Since the total number of neutrons and protons does not change, the mass numbers must add up to the same value on both sides of the equation.

$$238 = 4 + \underline{234}$$

Step 3. The atomic numbers must also add up to the same value on both sides of the equation.

$$92 = 2 + \underline{90}$$

Step 4. The identity of the missing element or particle is determined by the atomic number calculated in the previous step. In the example, the element with the atomic number of 90 is thorium.

Step 5. Using information from steps 1 to 4, construct the balanced nuclear equation:

$$^{238}_{92}\text{U} \rightarrow {}^{4}_{2}\text{He} + {}^{234}_{90}\text{Th}$$

■ WORKED EXAMPLE 3

Radioactive isotopes are used to diagnose and treat various medical disorders. Balance the following nuclear equations for (a) alpha emission of radium-226 and (b) beta emission of iodine-131.

SOLUTION

(a) Step 1. The atomic number of radium is 88, and the problem states that the mass number is 226. Therefore, the symbol is $^{226}_{88}\text{Ra}$. The alpha particle is $^{4}_{2}\text{He}$. The unbalanced equation is

$$^{226}_{88}\text{Ra} \rightarrow {}^{4}_{2}\text{He} + \text{?}$$

Step 2. Determine the mass number: $226 = 4 + 222$

Step 3. Determine the atomic number: $88 = 2 + 86$

Step 4. Identify the missing element by its atomic number: $86 = \text{radon (Rn)}$.

Step 5. Write the balanced nuclear equation:

$$^{226}_{88}\text{Ra} \rightarrow {}^{4}_{2}\text{He} + {}^{222}_{86}\text{Rn}$$

(b) Step 1. The atomic number of iodine is 53, and the problem states that the mass number is 131. Therefore, the symbol is $^{131}_{53}\text{I}$. The symbol for the beta particle is $^{0}_{-1}\text{e}$. The unbalanced equation is

$$^{131}_{53}\text{I} \rightarrow {}^{0}_{-1}\text{e} + \text{?}$$

Step 2. Determine the mass number: $131 = 0 + 131$

Step 3. Determine the atomic number: $53 = -1 + 54$

Step 4. Identify the missing element by its atomic number: $54 = \text{xenon (Xe)}$.

Step 5. Write the balanced nuclear equation:

$$^{131}_{53}\text{I} \rightarrow {}^{0}_{-1}\text{e} + {}^{131}_{54}\text{Xe}$$

Practice 12.3

Determine what isotope would be produced when (a) curium-243 undergoes alpha decay, and (b) sodium-24 undergoes beta emission.

Answer

(a) $^{239}_{94}\text{Pu}$ (b) $^{24}_{12}\text{Mg}$

12.4 Half-Lives and Risk Assessment

In 1987, the third largest accidental release of radioactivity in history occurred at a junkyard in Goiânia, Brazil. Salvagers who had targeted an abandoned hospital clinic for scrap metal chanced upon a radiation therapy machine and took it for scrap. Several days later, an employee pried open the machine, found an unusual lead container, and forcibly opened it to reveal a powdery material that glowed blue. The workers shared this novel but deadly discovery with friends, family, and neighborhood children. The powdery material was cesium-137.

By the time the first person was diagnosed with radiation sickness nearly a week later, 244 people had been exposed to the cesium-137 and several city blocks contaminated. Radiation sickness consists of nausea, diarrhea, skin burns, hair loss, and bleeding from the mouth, nose, and gums. The severity of radiation sickness depends on the amount of radiation that enters the body.

Entire homes and layers of soil had to be removed and placed in cement-lined storage barrels for disposal at a nuclear landfill. Figure 12.3 shows a contaminated home that has been demolished; the rubble is being prepared for storage at the landfill. Within the next week, four persons died from their acute exposure. The long-term chronic effects of the exposure are still being determined. Because cesium-137 has a half-life of 30.2 years, trace amounts that could not be removed are still present in the area.

Radiation Sickness in Goiânia

FIGURE 12.3 Demolition of a contaminated home. Due to the extensive contamination by radioactive cesium-137, this home had to be demolished. The entire contents and the layer of topsoil were taken to a landfill designed to accept radioactive materials. (International Atomic Energy Agency)

Which source of exposure is more harmful—short-term exposure to high levels of radiation or long-term exposure to low levels of radiation? In Section 11.7, the concept of a half-life was introduced to measure the length of time a pharmaceutical or illicit drug remains in a person's body. The half-life of radioactive isotopes is the time required for one-half the original amount of isotope to undergo radioactive decay.

Popcorn Half-Life

One might think that radioisotopes with long half-lives are safer than those with short half-lives (see Table 12.1). For example, uranium-238 has a half-life of 4.5 billion years, whereas iodine-131 has a half life of eight days. If a person is exposed to equal molar amounts of each isotope for the same period of time, fewer total uranium-238 atoms will have spontaneously undergone radioactive decay in that time than iodine-131. But iodine-131 decays by beta emission and uranium-238 decays by alpha emission. The danger presented by radioactive isotopes has to be gauged by the length of time of exposure, the type of radiation, whether the radiation source is external or internal, and the dose of radiation received.

The half-life of a radioactive isotope determines the method required for disposal. It is common for radioactive medical waste to be collected and buried in fairly shallow landfills. The landfills are carefully monitored, but as long as the radioisotopes have short half-lives, the natural decay process will eliminate low-level radioactive materials in a short time. However, the same cannot be said for all radioisotopes used in medicine. Some, like cesium-137, have long half-lives and require special disposal procedures.

The disposal of waste from nuclear power facilities is another problem, especially the spent fuel rods from nuclear reactors. These materials have long half-lives and emit dangerous levels of radiation. For example, the half-life for uranium-235 is 710 million years and the half-life for uranium-238 is 4.5 billion years, as mentioned earlier. Most radioactive nuclear waste materials are stored on site at a power plant.

TABLE 12.1	Half-Lives of Selected Radioisotopes		
Names	Symbol	Half-Life ($t_{1/2}$)	Radiation
Einsteinium-243	$^{243}_{99}\text{Es}$	21 seconds	α
Rhodium-106	$^{106}_{45}\text{Rh}$	29.8 seconds	β
Nobelium-259	$^{259}_{102}\text{No}$	1.0 hours	α
Sodium-24	$^{24}_{11}\text{Na}$	14.7 hours	β
Americium-240	$^{240}_{95}\text{Am}$	2.1 days	α
Radon-222	$^{222}_{86}\text{Rn}$	3.8 days	α
Iodine-131	$^{131}_{53}\text{I}$	8.0 days	β
Phosphorus-32	$^{32}_{15}\text{P}$	14.3 days	β
Polonium-210	$^{210}_{84}\text{Po}$	138 days	α
Curium-243	$^{243}_{96}\text{Cm}$	28.5 years	α
Cesium-137	$^{137}_{55}\text{Cs}$	30.2 years	β
Radium-226	$^{226}_{88}\text{Ra}$	1.6×10^3 years	α
Lead-202	$^{202}_{82}\text{Pb}$	5.3×10^4 years	α
Iron-60	$^{60}_{26}\text{Fe}$	1×10^5 years	β
Uranium-235	$^{235}_{92}\text{U}$	7.6×10^6 years	α
Uranium-238	$^{238}_{92}\text{U}$	4.5×10^9 years	α

■ WORKED EXAMPLE 4

On March 11, 2011, an earthquake occurred in the Pacific Ocean 230 miles northeast of Tokyo, Japan, with a magnitude of 9.0 on the Richter scale. This earthquake was the fifth most powerful earthquake ever measured. The earthquake produced a tsunami with a 30-foot-high wave of water that devastated the Japanese coast, including the Fukushima Daiichi nuclear power plant. The breach in nuclear containment released iodine-131, which has a half-life of 8.02 days. Calculate the percent of iodine that would have been remaining on April 12, 2011.

SOLUTION

March 11 → March 19 → March 27 → April 4 → April 12

  100%   →   50%   →   25%   → 12.5% → 6.25%

Practice 12.4

The accident at the Fukushima Daiichi nuclear power plant also released cesium-137, which has a half-life of 30.2 years. However, the biological half-life of this isotope is approximately 70 days. (a) How long will it take for the amount of cesium-137 to decline to less than 1% of its original value? (b) How long would it take to decline to less than 1% if a person ingested a single dose of the radioactive isotope?

Answer

(a) 211 yrs   (b) 490 days, based on 7 half-lives

12.5 Medical Applications of Nuclear Isotopes

Since the discovery of radioactivity, the field of nuclear medicine has provided for the diagnosis and treatment of a vast array of diseases and disorders. From the earliest days of the discovery of radioactivity, Marie Curie knew that radiation could be used to treat cancer. This application alone could have made the Curie family very wealthy, had they patented the processes.

> Physicists always publish their results completely. If our discovery has a commercial future, that is an accident by which we must not profit. And radium is going to be of use in treating disease. . . . It seems to me impossible to take advantage of that.
>
> —*Madame Curie: A Biography,* by Eve Curie

LEARNING OBJECTIVE

Explain how the properties of radioisotopes are used for medical diagnosis and treatment of disease.

In general, nuclear medicine has two roles. The first role is to diagnose diseases and disorders. The second role is to treat cancerous tumors by exposing the tumors to radiation. The isotopes chosen for diagnostic purposes have an affinity for a particular organ or target a specific system within the body.

Suppose, for example, that a patient is having problems absorbing vitamin B_{12}. The structure of vitamin B_{12}, shown in Figure 12.4, consists of a large organic molecule with a cobalt atom near the center. Radioactive vitamin B_{12} is made with cobalt-60 and is ingested by the patient. The radioactive vitamin B_{12} will act the same as normal vitamin B_{12} in the body because the chemistry of molecules is unaffected by the identity of the isotope. However, the radioactive cobalt-60 atom will emit gamma rays that are easily detected by medical sensors. If the radioactive vitamin B_{12} is being evenly absorbed by the intestines, the image produced by the radiation will be even and bright. However, if a portion of the intestines is not absorbing vitamin B_{12}, a corresponding dark spot on the image will result from a lack of gamma rays originating from that region.

Most isotopes intended for diagnosing disorders emit gamma rays. Recall that gamma rays have a large penetrating power that allows the gamma rays to escape through a

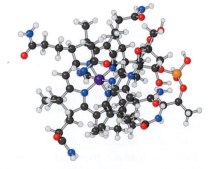

FIGURE 12.4 Vitamin B_{12} ($C_{63}H_{88}CoN_{14}O_{14}P$). Vitamin B_{12} contains a single cobalt atom (purple) located within the center of the structures. By substituting a radioactive isotope for the normal cobalt atom, the efficiency of the body's absorption of vitamin B_{12} can be determined.

person's body to reach the detector. Secondly, they produce the lowest amount of ionization of the three radioactive particles, minimizing damage to the body's tissues.

The second major use of radioisotopes in nuclear medicine is to destroy tumors. Radiation interferes with a cell's ability to reproduce and affects both healthy and cancerous cells. However, because the cancer cells are growing very rapidly, they are affected to a greater extent. Healthy cells that are damaged can often repair themselves.

Because the ionizing nature of radiation is being exploited to cause damage to the cancer cells, all forms of radioactive emission can be used. Gamma rays are generally used when the radiation is provided externally by focusing the emitted gamma rays onto the location of the tumor. Since beta rays do not have the penetration power of gamma rays, beta emitters are implanted into a tumor to directly expose the cancer cells to the radioisotope. Alpha emitters are also employed in a similar way. Table 12.2 lists a representative sampling of the many radioisotopes used in nuclear medicine. The second column of the table lists each radioisotope's half-life—the time required for one-half the original amount of isotope to undergo radioactive decay.

The half-lives of radioactive isotopes used for imaging range from hours to days, which serves to minimize a patient's total exposure to radiation because the source is quickly eliminated. Isotopes that are used for the treatment of tumors have a longer half-life and emit large doses of radiation.

TABLE 12.2	Medical Uses of Radioactive Isotopes			
Isotope	Half-Life	Decay	Form	Use
^{241}Am	432.7 years	α	Americium metal	Treatment of malignancies
^{60}Co	5.271 years	β, γ	Radioactive vitamin B_{12}	Diagnostic for defects of intestinal vitamin B_{12} absorption
^{113}In	1.658 hours	γ	Indium-labeled red blood cells	Determination of blood volume
^{131}I	8.040 days	β, γ	Sodium iodide	Imaging and function studies of the thyroid gland
^{32}P	14.282 days	β	Sodium phosphate	Localization of eye, brain, and skin tumors
^{42}K	12.360 hours	β, γ	Potassium chloride	Tumor detection; determination of total exchangeable potassium
^{87}Sr	2.795 hours	γ	Strontium nitrate	Bone imaging
^{133}Xe	5.245 days	β, γ	Xenon	Lung ventilation studies

Adapted from *Merck Index*, 13th ed.

■ WORKED EXAMPLE 5

Why is Sr-87 useful for bone imaging? Why is the short half-life of Sr-87 desirable? (*Hint:* Consider what bones are made of, and use the periodic table!)

SOLUTION

Recall that all elements found within a group in the periodic table undergo similar reactions. Because strontium is among the alkaline earth metals, it will undergo reactions similar to those of calcium, the main elemental ingredient in bone. The radioactive strontium replaces some of the calcium atoms that make up the bone material and becomes a permanent part of the skeleton. It would be undesirable to have elements with a long half-life become a permanent source of internal radioactive exposure. The short half-life ensures that the radiation does not last long in a person's body.

Practice 12.5

Why does xenon-133 work well for studying the function of the lungs?

Answer

Xenon is a noble gas and, as such, is inert. The xenon gas will fill the lungs upon inhalation but will not react with any components of the respiratory system. The xenon gas can be expelled from the lungs, reducing exposure to the radioisotope.

When you take biology or psychology courses, it is common to find textbook images of the brain that show the location of various brain functions or the difference between the brain of a healthy individual and a person suffering from a mental illness. These images are made using **positron emission tomography** (PET). Figure 12.5 is one such image that shows the effect of cocaine on the human brain. The first row of images is from a healthy individual, the second row shows the brain of an individual who had last used cocaine 10 days prior, and the bottom row of images corresponds to a person who had last used cocaine 100 days earlier.

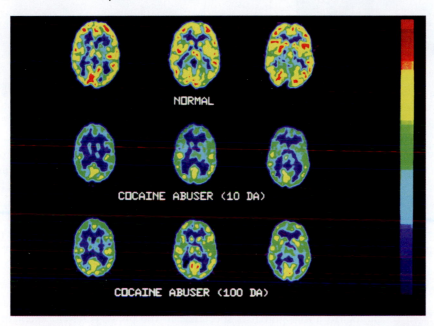

FIGURE 12.5 The effect of cocaine use on normal brain function as shown by positron emission tomography. The top row shows normal brain function, followed by brain function ten days after cocaine use, and finally 100 days after the last use of cocaine. (Dr. Nora Volkow, Brookhaven National Laboratory/Photo Researchers)

A **positron** is a radioactive particle emitted from a nucleus that contains too many protons, and one of the protons is converted into a neutron. The positron is sometimes called an *antielectron*, as it has the same properties as an electron, but with a charge of +1 instead of −1. Positrons do not last long after being emitted, as they will collide with an electron and the two opposite particles will *annihilate* each other. In this process of positron–electron annihilation, two gamma rays are created that travel in opposite directions, 180° apart. These gamma rays are then used to map brain function.

Figure 12.6a shows a PET scan instrument in which the patient lies down with his head surrounded by a circular detector. The patient has been given a sugar solution to drink that contains the carbon-11 isotope, a positron emitter. He is then given different mental tasks to complete, depending on the nature of the tests. The radioactive sugar will concentrate itself in the portion of the brain being used. Figure 12.6b shows a patient

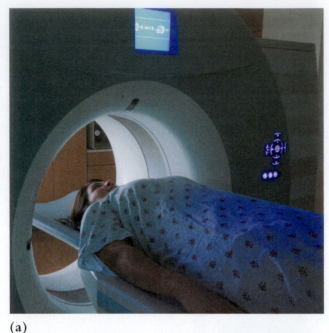
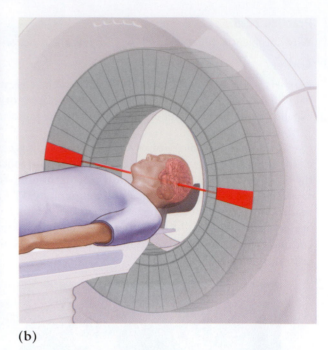

(a) (b)

FIGURE 12.6 Positron emission tomography. (a) A typical PET scan instrument functions by (b) detecting the pairs of gamma rays released during positron–electron pair annihilation. (Mark Kostich/Getty Images)

using the front lobe of the brain. When a positron is emitted, the annihilation process releases two gamma rays, traveling in opposite directions. If the gamma rays are emitted from the exact center of the circle, they will reach the detectors simultaneously. If, however, as in Figure 12.6b, the gamma rays are not in the exact center, one gamma ray will strike a detector first, and the time is measured for the second gamma ray to strike the detector on the opposite side. This time difference allows the annihilation location to be mapped.

12.6 Investigative Application of Carbon-14

LEARNING OBJECTIVE

Explain how the half-life of carbon-14 can be used in forensic investigations.

Biological anthropologists specialize in the physical development of the human species and are able to analyze skeletal human remains to determine such information as the sex, age, race, and, at times, the manner of death of a person. However, estimating how long a person has been dead is usually beyond the scope of an anthropologist because a skeleton might be from a victim murdered several weeks prior or might belong to a burial site of an indigenous people from hundreds of years ago.

If there is any question as to the age of the skeleton, investigators can use **carbon-14 dating**. Carbon-14 is a radioactive isotope created in the atmosphere, as shown in Figure 12.7. Cosmic rays (protons, alpha particles, and beta particles) from the sun collide with atoms in our upper atmosphere and dislodge neutrons. The neutrons then collide with stable nitrogen-14 atoms to create the radioactive carbon-14 atoms. The carbon-14 atoms then react with oxygen gas to form radioactive carbon dioxide, absorbed by plants during photosynthesis and incorporated into the plant. Carbon-14 is then present in the plants and thus in any animal that ingests plant material. As long as plants or animals are alive, there is a continual uptake of carbon-14. Once the plant or animal dies, the existing carbon-14, which is approximately equal to the atmospheric concentration of carbon-14, is lost through beta decay to form nitrogen-14 and is no longer replenished.

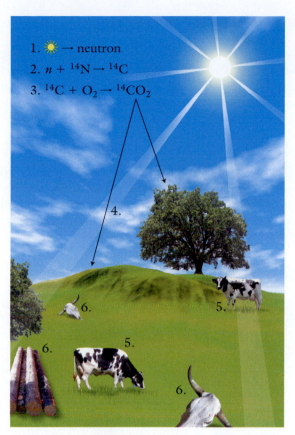

1. ☀ → neutron
2. $n + {}^{14}N \rightarrow {}^{14}C$
3. ${}^{14}C + O_2 \rightarrow {}^{14}CO_2$
4.
5.
6.

FIGURE 12.7 Creation and intake of carbon-14. Carbon-14 is generated by the interaction of cosmic rays with the atmosphere (steps 1–3). Radioactive carbon-14 is taken up by plant materials in photosynthesis (step 4) and enters the food chain (step 5). With the death of the plant or animal, the intake of carbon-14 halts, and the amount of carbon-14 starts to decrease by beta decay to form nitrogen-14.

The age of the sample is determined by how much carbon-14 is present in the sample and by comparing it to the half-life of carbon-14. For example, if the sample has half the original carbon-14 left, it has gone through one half-life, which for carbon-14 is equal to 5730 years. If the sample has one-fourth the original amount of carbon-14, its age would be two half-lives, or 11,460 years.

Another use of carbon-14 dating is verifying the age of materials present in ancient artifacts. Authentic artifacts can be worth tens of thousands of dollars. Therefore, forgery of such items is always a concern. Carbon-14 dating is only one tool, however, in investigating such forgeries, because counterfeiters will forge artifacts out of old materials to make them appear as something more valuable.

Current research is underway to use two different radioactive isotopes to measure the time of death. The lead-210 isotope has a half-life of 22.3 years, making this measure useful for determining whether a discovered skeleton is of modern or ancient origins. The shorter half-life would also be helpful in linking skeletal remains to decades-old crimes. The other isotope being studied is polonium-210, which has a half-life of 138 days, making it useful for identifying skeletons of people who died within the previous several months. Both lead-210 and polonium-210 are found in trace levels in food because they are two of the natural sources contributing to background radiation, and they are continually incorporated into human bodies until death.

12.7 Nuclear Power

There are two nuclear processes that produce excess energy for use by power plants. The first method, nuclear fission, powers nuclear reactors across the world. The second method, nuclear fusion, is currently being developed by scientists across the world as a potential future energy source. Each process will be explored further in this section.

Nuclear Power Plant Security

Fission

Nuclear reactors generate heat by nuclear **fission**, a reaction in which a large nucleus is split into smaller components. The reaction that powers most nuclear power plants is the fission of uranium-235 ($^{235}_{92}U$), as shown in Figure 12.8. A neutron striking the uranium-235 nucleus destabilizes it, causing it to split into barium-142 ($^{142}_{36}Ba$), krypton-91 ($^{91}_{56}Kr$), and three additional neutrons. The reaction produces a large amount of heat.

$$^{1}_{0}n + {}^{235}_{92}U \longrightarrow {}^{142}_{36}Ba + {}^{91}_{56}Kr + 3{}^{1}_{0}n$$

Atomic Fission

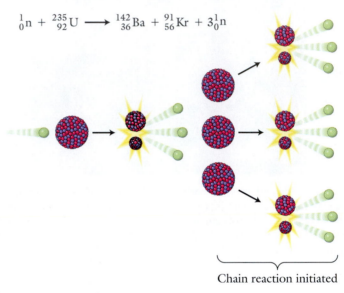

Chain reaction initiated

FIGURE 12.8 Fission of uranium-235.

The fission of a single uranium-235 atom produces three neutrons. Notice that each of the three neutrons created in the first fission reaction strikes another uranium-235 atom, which in turn releases three neutrons per atom. If the concentration of uranium-235 atoms is great enough, the reaction can become a self-sustaining **chain reaction** within the reactor core.

The fission reaction inside a nuclear power plant generates heat. If left unregulated, the amount of heat generated by a chain reaction would be too great for the reactor core to handle—resulting in a **meltdown**. To prevent this, the number of uranium atoms undergoing fission must be regulated. **Control rods** within a nuclear reactor are made out of boron or hafnium, both of which will absorb the neutrons produced during fission. If the control rods are lowered into the reactor, they absorb the neutrons and less uranium undergoes fission. If more heat is desired in the reactor core, the control rods are raised. In the Fukushima Daiichi disaster, the cooling tanks that stored the fuel rods ruptured and drained. The temperature of the fuel rods increased to the point of a meltdown.

The actual products of uranium-235 fission are more numerous than those illustrated in Figure 12.8. There are hundreds of radioactive isotopes that can be produced. All of the waste products from the nuclear reactor are radioactive and present a long-term storage problem for nuclear facilities. Most facilities have highly secured waste storage on site.

The security of nuclear wastes, both domestically and internationally, is a high priority to the United States because terrorist organizations are known to be actively searching for sources of radioactive materials. The federal government constructed a centralized storage facility in Yucca Mountain, Nevada, where the radioactive waste could be buried over 500 meters deep in the mountain. The site was chosen because it is not located near any population center and has little likelihood of being struck by earthquakes. The site is also in an extremely dry area, which inhibits the corrosion of storage containers. However, in 2011, continued political controversy over shipping nuclear waste cross-country and budgetary concerns caused the de-funding of the Yucca Mountain storage site. Radioactive waste will continue to be stored on-site at each nuclear power plant facility.

The use of nuclear power plants to supply large amounts of electricity had decreased in popularity in the 1970s and 1980s. However, with the current concern over global warming due to the release of carbon dioxide into the atmosphere, combined with U.S. dependence on foreign oil for energy, the use of nuclear power is being reevaluated.

Perhaps one of the more surprising supporters of nuclear power is Patrick Moore, co-founder of Greenpeace. In a newspaper article he wrote in 2006, he states that 10% of the carbon dioxide released globally is from the 600-plus coal-burning power plants found in the United States and that the use of nuclear power could reduce emission of carbon dioxide by 36%. He notes that reactors are much safer than they had been and that, by reprocessing nuclear fuel, the storage of nuclear waste is reduced.

On-site storage of spent fuel rods. (Roger Ressmeyer/ Corbis)

Fusion

Research into a new type of nuclear reactor that may one day provide electrical power without the production of highly radioactive wastes is currently being conducted. A nuclear **fusion** reactor produces energy by joining two light nuclei to form a heavier nucleus. The amount of energy released by fusion reactions is much greater than that produced by fission reactions. However, temperatures in excess of 10 million degrees Celsius are needed to initiate a fusion reaction. Since no material can withstand such a high temperature, it is easy to see the problem of constructing a fusion reactor. Current research is directed toward using magnetic fields as the "container" for a fusion reaction and triggering the reaction with lasers.

The most notable fusion reactors are stars, our sun being the closest example. The energy of the sun comes from the fusion of hydrogen nuclei to form helium atoms as follows:

$$\,^1_1H + \,^1_1H \rightarrow \,^2_1H + \,^0_{+1}e$$

$$\,^2_1H + \,^1_1H \rightarrow \,^3_2He + \gamma$$

$$\,^3_2He + \,^3_2He \rightarrow \,^4_2He + 2\,^1_1H$$

The first reaction is the fusion of two hydrogen nuclei together to form the deuterium isotope of hydrogen and a positron ($\,^0_{+1}e$). The second step of the reaction is the fusion of a deuterium nucleus with a hydrogen nucleus to form the helium-3 isotope and a gamma particle. The final reaction is the combination of two helium-3 isotopes to form a helium-4 isotope and two protons.

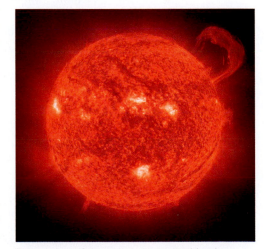

The energy of the sun is generated by the fusion of hydrogen atoms to form helium atoms (NASA/JPL-Caltech/Corbis)

12.8 Military Uses of Nuclear Isotopes

Today, the physicists who participate in watching the most formidable and dangerous weapon of all time . . . cannot desist from warning and warning again: we cannot and should not slacken in our efforts to make the nations of the world and especially their governments aware of the unspeakable disaster they are certain to provoke unless they change their attitude towards each other and towards the task of shaping the future. We helped in creating this new weapon in order to prevent the enemies of mankind from achieving it ahead of us. Which, given the mentality of the Nazis, would have meant inconceivable destruction, and the enslavement of the rest of the world . . .

—Albert Einstein

Nuclear weapons were first developed in World War II as part of the top-secret Manhattan Project. Initiated at the urging of Albert Einstein out of fear that Nazi Germany was working toward the deployment of such weapons, Einstein later opposed the use of nuclear weapons and warned of the potential chaos that would ensue if all nations pursued nuclear weapons programs. The Germans never built a fully operational nuclear weapon and surrendered before the United States had readied its nuclear weapons for use. However, the United States unleashed nuclear weapons on Japan to end the war in the Pacific.

Nuclear weapons proliferation is still a matter of international concern that challenges the stability and peace of the world. One reason is that the technology used for the production of electricity in nuclear power plants can be used to produce material for nuclear weapons and is politically referred to as *dual-use technology*. The existence of such facilities in countries that support the efforts of international terrorists is a cause for concern. Could terrorists obtain radioactive material overseas and transport it into the United States?

Uranium has three isotopes, all of which are radioactive: uranium-238 (99.2745%), uranium-235 (0.720%), and uranium-234 (0.0055%). Uranium in its natural composition is not usable for either weapons or power plants. Only uranium-235 will undergo the fission reaction. Uranium metal is processed to increase the concentration of uranium-235 for use in either nuclear power plants or nuclear weapons. The resulting uranium is referred to as *enriched* uranium, and the level of enrichment determines its use. If the uranium is enriched to contain 3 to 5% uranium-235, it can be used in a nuclear reactor. If the concentration is 20%, the uranium becomes suitable for a crude nuclear weapon, although modern nuclear weapons use uranium enriched to contain 85% uranium-235. Note that the concentration of uranium-235 in a fuel rod is too low to support a nuclear detonation, so nuclear power plants will not under any circumstance produce a nuclear explosion.

The mushroom cloud from the detonation of a fusion-based hydrogen bomb, designated Castle Romeo, on March 27, 1954, in the South Pacific. The nuclear fallout from the weapons testing would poison nearby islanders and the crew of a Japanese fishing boat. (Corbis)

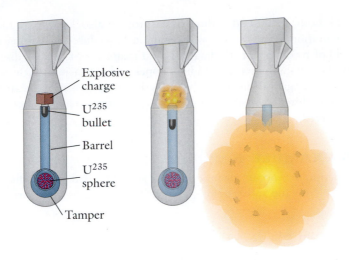

FIGURE 12.9 How nuclear weapons operate. (Adapted from Howstuffworks.com)

The amount of uranium-235 needed to produce a chain reaction is called the **critical mass**. The basic nuclear weapon, illustrated in Figure 12.9, functions by combining two subcritical masses of uranium-235 at the time of detonation to create a critical mass. At the same moment the two subcritical masses are joined, a neutron emitter initiates the fission of the uranium-235 nuclei. The reaction is allowed to accelerate with no constraints, and a nuclear explosion results. The actual mass of uranium needed to achieve a critical state depends on the level of enrichment.

Armor-piercing missiles are routinely made from *depleted* uranium metal, which is uranium that has had the uranium-235 isotope extracted from it. The depleted uranium has a density of 19.0 g/cm^3, which is much greater than the density of steel at 7.85 g/cm^3. This high density is significant because it enables depleted uranium to easily penetrate steel armor when used as a projectile for antitank weapons. The depleted uranium is also used as armor plating to make some tanks resistant to the use of antitank weapons made of depleted uranium.

There is some controversy about using depleted uranium because the uranium metal, now mostly uranium-238, still is a radioactive alpha emitter. Recall that the alpha particles produce high levels of ionization but have low penetrating power. An alpha emitter is most dangerous when it can enter a person's body. When a depleted uranium shell strikes an armored vehicle, it pierces a hole through the target but is vaporized in the process. The greatest risk to personnel would be near a battlefield in which a large amount of depleted uranium was used, because the depleted uranium might be breathed in with the air and dust.

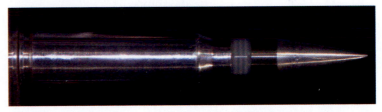

Standard 30-mm antitank shell tipped with depleted uranium. (Reuters/Corbis)

12.9 Nuclear Transmutations

In a nuclear **transmutation** reaction, a stable isotope is converted into a different element or isotope through a nuclear reaction. Ernest Rutherford conducted the first nuclear transmutation experiment in 1919 when he used alpha particles emitted from a radioactive source to bombard a sample of nitrogen-14. The two nuclei fused together and formed one oxygen-17 atom plus one hydrogen atom, as shown below.

LEARNING OBJECTIVE

Determine the isotope formed during a nuclear transmutation reaction.

$$^{14}_{7}N + ^{4}_{2}He \rightarrow ^{17}_{8}O + ^{1}_{1}He$$

Creation of both stable and radioactive isotopes, also known as **radioisotopes**, through transmutation reactions is now a common procedure for making the isotopes used in the field of medicine for the diagnosis and treatment of various diseases. Major hospitals have the capability to create radioisotopes and use them directly at the facility. Can the same technology used in hospitals across the world be used to create the material for a dirty bomb? More information on nuclear chemistry is required before that question can be answered.

■ WORKED EXAMPLE 6

Balance the reaction that occurs when sodium-23 is bombarded with alpha particles to produce a proton (^1_1H) and another isotope. Identify the other isotope formed.

SOLUTION

Step 1. Write the equation:

$$^{23}_{11}Na + ^4_2He \rightarrow ^1_1H + ?$$

Step 2. Determine the mass number of the unknown:

$$23 + 4 = 1 + \underline{26}$$

Step 3. Determine the atomic number of the unknown:

$$11 + 2 = 1 + \underline{12}$$

Step 4. Determine the unknown element from the atomic number:

$$12 = magnesium\ (Mg)$$

Step 5. Write the balanced equation:

$$^{23}_{11}Na + ^4_2He \rightarrow ^1_1H + ^{26}_{12}Mg$$

Practice 12.6

Balance the following nuclear transmutation reactions:

(a) Fluorine-19 bombarded with neutrons to produce an alpha particle and another radioisotope.

(b) Uranium-238 bombarded with neutrons to produce a beta particle and another radioisotope.

Answer

(a) $^{16}_7N$

(b) $^{239}_{93}Np$

12.10 Nuclear Terrorism

LEARNING OBJECTIVE

Explain the nature and potential danger to public safety posed by a dirty bomb.

Several possible uses of nuclear technology by terrorists exist. One issue involves the security of nuclear weapons in countries that are politically and economically unstable. A second issue involves the security around nuclear power stations both domestically and internationally as potential targets for attack. These risks are geographically limited to a relatively small number of sites. One potential act of nuclear terrorism that could strike any target location is the use of a **radiological dispersion device**, also known as a *dirty bomb*. A dirty bomb is a traditional explosive that has radioactive materials packed around it. Upon detonation, the radioactive material is spread over a large region. It is important to note that there is no *nuclear explosion*, that is, a dirty bomb is not a nuclear weapon.

What purposes might a terrorist organization have for exploding a dirty bomb? And what would be the real effects of the detonation of such a device? Government officials must address these issues in developing response plans in case of an emergency. Below is a summary of questions that will arise if there is a dirty bomb incident.

Dirty Bomb Issues

- What type of radioactive material was used and where did it come from?
- What are the casualties and destruction from the explosion?
- What is the exposure risk to the first responders on the crime scene?
- How will the public respond to the explosion?
- How will the area be decontaminated and individuals evacuated?
- What will be the long-range economic damage?

By all estimates, the initial deaths from the detonation of a dirty bomb will be relatively few, particularly when compared with the impact of a nuclear weapon. Casualties might number in the hundreds. The type of radioactive material contained in the bomb would be of great importance in assessing the hazards at the scene. For example, most medical radioisotopes will not present the long-term contamination issues that would result if radioisotopes from a nuclear reactor were used. Risk estimates have shown that the first responders to the scene, who would not be aware of the radiological nature of the explosion, will most likely not be exposed to life-threatening levels of radiation. The amounts and types of radiological materials that could be used will not present an imminent danger after detonation.

Even though a dirty bomb is not likely to create a large number of casualties, it can, nevertheless, wreak havoc in other ways. A potent psychological weapon, it might demoralize and exhaust the resources of the citizens who must deal with a large-scale decontamination and evacuation of a region of their community. The United States Nuclear Regulatory Commission refers to a dirty bomb as a "weapon of mass disruption." Terrorists undoubtedly hope that the public's first response, when news of a radioactive bomb detonation is announced, will be mass panic and chaos.

Terrorists will be likely to choose a target that would cause economic hardship or make uninhabitable an area of real and symbolic importance. Imagine if the White House, the U.S. Capitol, or the Wall Street financial district were evacuated for long-term decontamination. The costs of the cleanup and decontamination could quickly escalate, especially since many radioactive isotopes would be permanently absorbed into concrete or asphalt, requiring the complete removal of those materials.

In the planning and preparation by state and federal government agencies for the response to a nuclear incident, the purchase and stockpiling of **radiation pills** has been recommended. However, it is important to understand the limitations of these pills. The radiation pill is simply a dose of potassium iodide that serves to protect the thyroid gland, and only the thyroid gland, from absorbing radioactive iodine ($^{131}_{53}I$). The thyroid gland serves as a regulator of bodily functions and growth, which is why protecting it is desirable.

Iodine-131 has a half-life of 8.0 days. Therefore, it is doubtful that terrorists would use iodine-131 in a dirty bomb because it would be transformed into stable isotopes before use or soon after. The only real danger for exposure to iodine-131 comes from two sources: fallout from the explosion of a nuclear weapon and a catastrophic failure at a nuclear power plant. If either of these two events should happen, countless radioactive isotopes as well as iodine-131 would be present, and there are no pills to counteract the other isotopes. Potassium iodide was used in Poland following the Chernobyl disaster and in the Fukushima Prefecture, when a large amount of iodine-131 was released into the atmosphere. The pills proved to be a safe and effective method for protecting the thyroid gland until the radioactive iodine naturally decayed.

Dirty Bombs

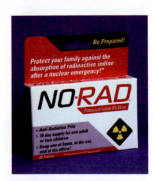

Radiation pills used for thyroid protection from the radioactive iodine-131. (Leonard Lessin/Photo Researchers, Inc.)

Neutron Activation Analysis

Neutron activation analysis (**NAA**) is an experimental method based on the nuclear transmutation reactions of a sample that is placed inside a nuclear reactor and bombarded by neutrons.

President John F. Kennedy
(Ted Spiegel/Corbis)

President Zachary Taylor
(Bettmann/Corbis)

Sir Francis Drake (Hulton Archive/Getty Image)

NAA has been used to gather data in some interesting forensic cases. Bullet fragments from the assassination of President John F. Kennedy on November 22, 1963, were analyzed by NAA, as were hair and nail samples from President Zachary Taylor's remains to determine whether he might have died of arsenic poisoning. A brass plaque purported to have been left by Sir Francis Drake in 1579, commemorating his landing on the California coastline, was analyzed to see if the plaque was genuine or a forgery. Even though the samples in these cases were obtained from 40 to 425 years ago, NAA provided useful information.

Neutron activation analysis is based on the creation of isotopes of an element by bombarding the element with neutrons (see Figure 12.10). A neutron combines with the nucleus of the sample atom to create an energetically excited isotope, as

shown in Reaction 12.1 for the bombardment of arsenic-75.

$$^{75}_{33}\text{As} + ^1_0\text{n} + \rightarrow ^{76}_{33}\text{As*} \quad \textbf{(Reaction 12.1)}$$

In Reaction 12.2, the energetically excited arsenic atom (arsenic-76) releases excess energy as a gamma ray. Recall that gamma rays are photons—packets of light. The emission of the gamma particle enables a forensic scientist to determine what the excited isotope is because the energy is unique to the isotope that generated it.

$$^{76}_{33}\text{As*} \rightarrow ^0_0\gamma + ^{76}_{33}\text{As} \quad \textbf{(Reaction 12.2)}$$

The first gamma ray is released almost instantly, and the gamma ray detector must be in the nuclear reactor with the sample. It is common for a third, slower reaction to occur that releases a second gamma ray during a beta decay of the nucleus. The beta decay process can take place anywhere from minutes to days later, depending on the half-life of the newly created isotope. The half-life for arsenic-76 is 1.1 days. Therefore, the sample can be removed from the nuclear reactor, and the energy of the second gamma ray measured, as shown in Reaction 12.3.

$$^{76}_{33}\text{As} \rightarrow ^0_{-1}\beta + ^0_0\gamma + ^{76}_{34}\text{Se} \quad \textbf{(Reaction 12.3)}$$

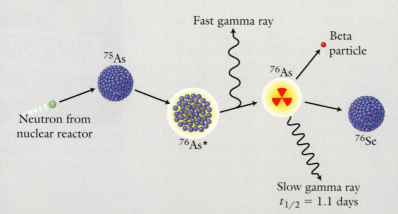

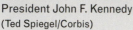

FIGURE 12.10 Neutron activation analysis. The atoms are bombarded by neutrons in the center of a nuclear reactor. The neutrons will combine with the nucleus of an atom to create an energetically excited isotope of the same element. The excited nucleus will relax by emitting a gamma ray photon with an energy unique to the element of origin. The relaxed nucleus may still undergo further radioactive decay through beta emission, which is also accompanied by gamma emission. (Adapted from www.missouri.edu/~glassock/naa_over.htm).

Fast gamma ray

Beta particle

^{76}As

^{75}As

^{76}As*

Neutron from nuclear reactor

^{76}Se

Slow gamma ray
$t_{1/2} = 1.1$ days

TABLE 12.3	Neutron Activation Analysis Results for the Bullet Fragments			
Sample	Description	Silver ppm	Antimony ppm	Conclusion
Q1	Bullet from Governor Connally's stretcher	8.8 ± 0.5	833 ± 9	Bullet No. 1
Q9	Governor Connally's wrist	9.8 ± 0.5	797 ± 7	
Q2	Large fragment recovered from car	8.1 ± 0.6	602 ± 4	Bullet No. 2
Q4	Fragment from JFK's skull	7.9 ± 0.3	621 ± 4	
Q14	Small fragment recovered from car	8.2 ± 0.4	642 ± 6	

Guinn, Vincent P. "JFK Assassination: Bullet Analyses," *Analytical Chemistry*, vol. 51, pp. 484A–493A, 1979.

The bullet fragments recovered from the assassination of President Kennedy were analyzed by NAA to measure the amounts of the silver and antimony impurities in the lead fragments. Impurity levels that are close in value would indicate that the fragments *could* have come from the same bullet. Levels of trace impurities that vary greatly would suggest that the fragments most likely came from different bullets, assuming that the impurities are distributed homogeneously within the bullet.

Critics of the Warren Commission Report of the assassination have long questioned the number of bullets that struck President Kennedy and surmised that a second gunman was involved. The NAA results showed that two bullets struck the presidential motorcade that day—there was no evidence to support the existence of a third bullet. This analysis does not contradict the Warren Commission Report of the assassination, which stated that *only* two bullets struck President Kennedy. The experimental results of the NAA are summarized in Table 12.3. When examining the results, recall that distribution of trace metals within the lead bullet will vary slightly if they come from the same bullet but differ greatly if they are from different sources.

Was President Zachary Taylor assassinated by arsenic poisoning? Some historians believed that the symptoms associated with President Taylor's illness matched those of arsenic poisoning. In 1991, the descendants of President Taylor agreed to allow his body to be exhumed and to have samples taken for NAA at the Oak Ridge National Laboratory. The samples proved that President Taylor did not have arsenic levels in his body beyond the natural trace levels.

The World Encompassed, published in 1628, describes a plaque that Sir Francis Drake allegedly left on the coastline of California to commemorate his landing there in 1579. Because the plaque had never been located, it was of interest to many historians. When it was discovered in 1936, it soon became the most significant archeological find in the history of California. However, it was proven conclusively in 1977 that the plaque was a forgery constructed at the time of its discovery.

NAA of the brass, conducted by the Lawrence Berkeley National Laboratory, revealed a high purity of the metal alloy and an elemental composition that would not have been possible using technology from 1579. The found plaque was created as a practical joke among a group of historians. Before they had a chance to reveal the joke, a friend and unwitting victim had publicly declared the plaque's authenticity, and a media frenzy ensued. The perpetrators of the joke decided to avoid publicly humiliating their friend and colleague and let the hoax go unchallenged.

■ WORKED EXAMPLE 7

Does the NAA evidence prove that President Taylor died of natural causes? Explain your answer.

SOLUTION

No, the NAA results simply prove that President Taylor did not die from arsenic poisoning. He may or may not have died from another poisonous substance that was not detected by the NAA procedure.

Practice 12.7

Does the NAA evidence in the Kennedy assassination case prove definitively that only two shots struck the President? Explain your answer.

Answer

The evidence suggests that two bullets struck President Kennedy. It eliminates the possibility that only one bullet struck the president. If a third bullet struck President Kennedy, either it had an elemental composition identical to one of the other bullets or it was never recovered and analyzed.

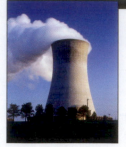

12.11 CASE STUDY FINALE: Exploring Nuclear Power

There are two main designs for nuclear reactors. Figure 12.11 is an illustration of the **pressurized water reactor.** This nuclear reactor design has three independent heat transfer systems. The heat is generated in the reactor core, which contains the uranium-235 fuel rods and is shown in orange. As the uranium-235 undergoes fission, the excess heat is absorbed by the water and transferred to the steam-generation system.

The steam-generation system (shown in blue) absorbs the heat from the reactor core and converts water into steam. The steam is then used to turn the blades of a turbine, which generates electricity. The electricity is directed to the power grid. The excess heat contained in the steam then needs to be removed by the cooling system.

The cooling system consists of the large tower that dominates the nuclear power-plant facility along with a large body of water such as a river or lake. A common misconception is that the reactor core is located in this

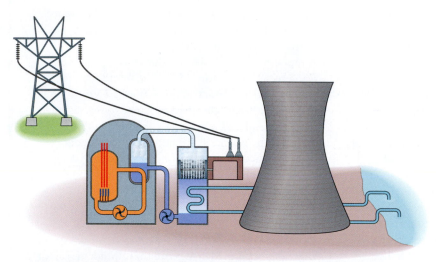

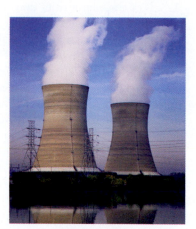

Cooling towers

Reactor core

Fuel rods

Steam lines

Steam-powered generators

FIGURE 12.11 Electrical power generated from a nuclear power plant. Uranium fuel rods release large amounts of heat while undergoing fission reactions in the reactor core. The heat from the reactor core is used to convert water into steam, which is used to turn a steam turbine that generates the electricity. The steam is then cooled and converted back into liquid water by fresh water pumped in from a lake or river. The fresh water is then cooled in the cooling towers before it is returned to the lake or river. (Illustration adapted from Howstuffworks.com; photos, top right: W. Cody/Corbis; left to right: Yann Arthus-Bertrand/Corbis; Tim Wright/Corbis; Corbis; Charles E. Rotkin/Corbis.)

structure and that the substance coming out of the top of the cooling towers is smoke. What is actually coming out of the cooling towers is steam. The water in the cooling tower absorbs the excess heat from the steam and allows the water in the steam-generation cycle to condense, producing liquid water for reuse. Any heat absorbed by the river or lake water must be released into the atmosphere before the water is returned, or else it could wreak havoc on the natural ecosystem. The large tower is used to cool the natural waters before they return to their source.

Each system is linked to the others in much the same way as three links in a chain are attached—each system crosses into another, but the contents of each system never mix under normal circumstances. The most recent nuclear accident at a power plant did not happen under normal circumstances.

The Fukushima Daiichi nuclear power plant was a **boiling water reactor**, which contains only two heat transfer systems. In this type of reactor, the water in the reactor core is heated until it boils. The boiling water forms steam, which then directly turns the turbines to generate electricity. The natural water supply, in this case the Pacific Ocean, then absorbs the excess heat from the steam and converts it back to liquid water for return to the reactor core.

The Fukushima Daiichi disaster occurred as a result of a magnitude 9.0 earthquake that forced an emergency shutdown of the reactor. Because the fuel rods in a boiling water reactor cannot be removed from the reactor core, the control rods that absorb the neutrons that initiate fission reactions were inserted between the fuel rods. This stopped the active nuclear-fission process; however, the reactor core still required cooling because of the tremendous amount of heat it contained from its normal use and because the radioactive

products of the fission reaction continued to release energy as they underwent radioactive decay.

The pumps that circulated the coolant malfunctioned because of the earthquake. A backup diesel-fueled emergency pump supplying coolant then malfunctioned when the 30-foot-tall wave from the resulting tsunami struck the power plant. A final backup cooling system would fail when, it is believed, a leak developed in the core, allowing water to escape and the fuel rods to be exposed. The temperature of the fuel rods then exceeded the melting point, creating a nuclear meltdown. Further problems struck the power plant when the storage tanks containing old fuel rods would empty of coolant, allowing a similar situation to happen.

As the future energy policy of the United States is being planned, the fate of nuclear power must be decided. Most power plants (in fact, all but one) were built in the 1960s and 1970s with a 40-year life expectancy. Most plants have been given a license to operate for an additional 20 years. Consider the following points as you evaluate the pros and cons of nuclear power for the future energy needs of the United States. Alternative energy sources (wind, solar, hydroelectric) cannot yet replace the energy-generation demands of the United States cost-effectively. Nuclear power plants also release no carbon dioxide into the atmosphere. A typical coal-fired power plant releases more radioactive material into the atmosphere than does a nuclear power plant because a large coal power plant burns 10,000 tons of coal per day that is contaminated with low levels of radioactive material. This is in addition to the carbon dioxide, nitrogen oxides, and sulfur oxides that are released by burning coal that contribute to global climate change and acid rain.

CHAPTER SUMMARY

• The discovery of radioactivity by Henri Becquerel and the subsequent development of the field of nuclear chemistry by Marie Curie have created a new world in which radio-isotopes have cured countless numbers of people from disease, provided the allies with a victory in World War II, and generated electricity to power millions of homes.

• Atoms that lie outside of the belt of stability adjust their ratio of protons to neutrons by releasing one or more of the three main forms of radiation: the alpha particle, beta particle, or gamma ray.

• The alpha particle consists of two protons and two neutrons and is identical to the helium nucleus. The beta particle is an energetic electron produced when a neutron is converted to a proton within the nucleus. The gamma ray is a high-energy photon that is emitted by a nucleus in an elevated energy state.

• The danger presented by radiation comes in two forms: the power of the radiation to penetrate matter and the ionizing power of the radiation. Ionizing power is the ability to ionize the matter through which the radiation passes. The alpha particle has the greatest ionizing power but the lowest penetrating power. The gamma ray is at the other extreme, requiring several centimeters of lead to be absorbed, but it has the lowest ionizing power. The properties of beta particles are intermediate between the alpha and gamma rays.

• Balancing a nuclear equation is based on the law of conservation of mass. The total mass of protons and neutrons on each side of the equation is conserved, but the identity of the elements changes.

• Radioactive isotopes can be used in the diagnosis of medical conditions and for the treatment of cancerous tumors.

• Carbon-14 dating determines the age of a plant or animal artifact by measuring the amount of carbon-14 remaining in the object and using the half-life of 5730 years for carbon-14.

• The fission reaction of uranium-235 provides the fuel source for both atomic weapons and nuclear power plants, although the fuel in a power plant cannot detonate like a nuclear weapon. Nuclear power plants generate electricity by using the heat of the fission reaction to convert water into steam and turn a traditional steam turbine.

• The dangers of nuclear power lie in the extremely long half-life of the waste products, which must be stored indefinitely. The waste produced by most medical isotopes generally does not pose a long-term storage problem because of the relatively short half-life of those isotopes.

• A radiological dispersion device, or dirty bomb, is actually geared to inflict economic and psychological damage rather than mass casualties.

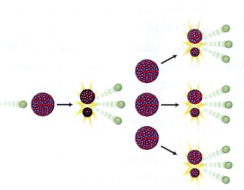

Fission of Uranium-235

KEY TERMS

alpha particle (α) (p. 340)
background radiation (p. 337)
belt of stability (p. 339)
beta particle (β) (p. 340)
boiling water reactor (p. 359)
carbon-14 dating (p. 348)
chain reaction (p. 350)
control rod (p. 350)
critical mass (p. 353)

fission (p. 350)
fusion (p. 351)
gamma ray (γ) (p. 341)
ionizing power (p. 340)
meltdown (p. 350)
neutron activation analysis
  (NAA) (p. 356)
penetrating power (p. 340)
phosphorescence (p. 336)

positron (p. 347)
positron emission tomography
  (PET) (p. 347)
pressurized water reactor (p. 358)
radiation pill (p. 355)
radioactivity (p. 337)
radioisotope (p. 354)
radiological dispersion device (p. 354)
transmutation (p. 353)

MAKING MORE CONNECTIONS Additional Readings, Resources, And References

Curie, Eve. *Madame Curie*, New York: Doubleday, 1939.

Guinn, Vincent P. "JFK Assassination: Bullet Analyses," *Analytical Chemistry* 51, 1979: 484A–493A.

For Patrick Moore's article about the benefits of nuclear power: www.washingtonpost.com/wp-dyn/content/article/2006/04/14/AR2006041401209.html

For more information on Drake's plaque: www.lbl.gov/Publications/Currents/Archive/Mar-07-2003.html#Drake

For a comprehensive article on nuclear forensic science: www.llnl.gov/str/March05/Hutcheon.html

U.S. Environmental Protection Agency, *A Citizen's Guide to Radon: The Guide to Protecting Yourself and Your Family From Radon*, Washington DC: U.S. Environmental Protection Agency, 2005. www.epa.gov/radon/pubs/citguide.html

For information on which consumer products contain radioisotopes: www.orau.org/ptp/collection/consumer%20products/consumer.htm

For use of half-lives in forensic anthropology: news.bbc.co.uk/2/hi/science/nature/3632770.stm

For more on neutron activation analysis and the cases mentioned: www.ornl.gov/info/ornlreview/rev27-12/text/ansside6.html

A list of commonly found medical radioisotopes can be found in the *miscellaneous tables* section of the Merck Index.

For information on Fukushima Daiichi: www.yomiuri.co.jp/dy/national/T110321004250.htm www.asahi.com/english/TKY201108180318.html

REVIEW QUESTIONS AND PROBLEMS

Questions

1. Discuss Henri Becquerel's original hypothesis about phosphorescence and how, using the scientific method, he disproved it. (12.1)

2. What is phosphorescence? (12.1)

3. How would history have changed had Becquerel been using a mineral ore other than one that contained uranium? (12.1)

4. Would history have significantly changed if Becquerel had had sunny weather throughout his experiments? (12.1)

5. Explain the role of the neutron in the nucleus. (12.2)

6. Explain why some isotopes are stable and others are radioactive. (12.2)

7. What happens to the total number of neutrons and protons of a nucleus during alpha, beta, and gamma decays? (12.2)

8. What happens to the total charge of a nucleus during alpha, beta, and gamma decays? (12.2)

9. Which radioactive decay particle has the greatest ionizing power? Which has the least ionizing power? (12.2)

10. Which radioactive decay particle has the greatest penetrating power? Which has the least penetrating power? (12.2)

11. Which radioactive decay particle poses the greatest risk if present inside the body? Why? (12.2)

12. Which radioactive decay particle poses the greatest risk if it is external to the body? Why? (12.2)

13. Explain how the conservation of the total number of neutrons and protons applies to balancing nuclear equations. (12.3)

14. Does the ionic charge on a radioisotope affect the type of radioactive decay it will undergo? (12.3)

15. Which is more stable, an isotope with a short half-life or one with a long half-life? (12.4)

16. Which is safer, an isotope with a short half-life or one with a long half-life? (12.4)

17. What are some desirable characteristics of radioisotopes for medical diagnosis? For cancer treatment? (12.5)

18. Explain how a PET scan identifies the section of the brain being used by a patient. (12.5)

19. How is carbon-14 created and how does it become part of the food cycle? (12.6)

20. Can carbon-14 dating be used to determine the age of a clay pot? (12.6)

21. Sketch a nuclear power plant and describe how the three independent heat exchange systems generate electricity. (12.11)

22. How does a fission reaction differ from a fusion reaction? Which is more desirable for generating electricity and why? (12.7)

23. Describe how a nuclear chain reaction occurs. Include a sketch of U-235 undergoing a chain reaction. What is the role of control rods in a nuclear power plant in relation to a chain reaction? (12.7)

24. What are the limitations on creating a fusion reactor? (12.7)

25. What is the difference between enriched and depleted uranium? What are the uses of each form? (12.8)

26. Describe two military uses of radioisotopes. (12.8)

27. One advantage of neutron activation analysis is that it does not destroy the sample being tested. Why is this desirable for forensic analysis of evidence? (ITL)

28. What is one limitation on the widespread use of neutron activation analysis? (ITL)

29. Because mass casualties from the explosion of a radiological dispersion device are highly unlikely, what are the true goals of a terrorist organization using such a device? What are the scientific issues that would have to be addressed in case of such an incident? (12.10)

30. Why are dirty weapons referred to as *weapons of mass disruption*? (12.10)

31. Historically, the testing of nuclear weapons technology moved from above-ground or underwater detonations to underground testing. Why would underground testing of nuclear weapons be considered a safer alternative? (CP)

32. What is a nuclear transmutation reaction and how does it differ from radioactive decay? (CP)

Problems

33. Write a balanced equation for the following radioactive decay systems. (12.3)
    (a) Uranium-239 undergoing beta decay
    (b) Polonium-210 undergoing alpha decay
    (c) Strontium-90 undergoing beta decay
    (d) Gold-174 undergoing alpha decay

34. Write a balanced equation for the following radioactive decay systems. (12.3)
    (a) Radium-202 undergoing alpha decay
    (b) Rubidium-90 undergoing beta decay
    (c) Sodium-27 undergoing beta decay
    (d) Copper-73 undergoing alpha decay

35. Provide the missing information to balance the following radioactive decay systems. (12.3)
    (a) $^{198}_{79}\text{Au} \rightarrow {}^{0}_{-1}\text{e} + \underline{\hspace{1cm}}$
    (b) $^{239}_{94}\text{Pu} \rightarrow \underline{\hspace{1cm}} + {}^{235}_{92}\text{U}$
    (c) $^{3}_{1}\text{H} \rightarrow \underline{\hspace{1cm}} + {}^{3}_{2}\text{He}$
    (d) $^{235}_{92}\text{U} \rightarrow \underline{\hspace{1cm}} + {}^{231}_{90}\text{Th}$

36. Provide the missing information to balance the following radioactive decay systems. (12.3)
    (a) $^{233}_{92}\text{U} \rightarrow \alpha + \underline{\hspace{1cm}}$
    (b) $^{137}_{55}\text{Cs} \rightarrow \underline{\hspace{1cm}} + {}^{137}_{56}\text{Ba}$
    (c) $^{232}_{90}\text{Th} \rightarrow \underline{\hspace{1cm}} + {}^{228}_{88}\text{Ra}$
    (d) $^{42}_{19}\text{K} \rightarrow \beta + \underline{\hspace{1cm}}$

37. Determine the amount of radioisotope under the following conditions. Refer to Table 12.1 for half-life information. (12.4)
    (a) The original amount of iodine-131 if 31.3 g remains after four half-lives.
    (b) The original amount of iodine-131 if 4.62 kg remains after 40.0 days.
    (c) The amount of einsteinium-243 if 47.1 mg undergoes three half-lives.
    (d) The amount of einsteinium-243 if 0.147 g is observed for 105 seconds.

38. Determine the amount of radioisotope under the following conditions. Refer to Table 12.1 for half-life information. (12.4)
    (a) The original amount of phosphorus-32 if 6.32 kg remain after three half-lives.
    (b) The original amount of phosphorus-32 if 5.83 g remain after 28.3 days.
    (c) The amount of nobelium-259 if 181 mg undergoes six half-lives.
    (d) The amount of nobelium-259 if 0.268 g is observed for 5 hrs.

39. Provide the missing information to balance the following nuclear transmutation reactions. (12.8)
    (a) $\underline{\hspace{1cm}} + {}^{0}_{-1}\text{e} \rightarrow {}^{57}_{26}\text{Fe} + \gamma$
    (b) $\underline{\hspace{1cm}} {}^{1}_{0}\text{n} \rightarrow {}^{60}_{27}\text{Co} + \gamma$
    (c) $^{111}_{50}\text{Sn} + {}^{0}_{-1}\text{e} \rightarrow \underline{\hspace{1cm}}$
    (d) $\underline{\hspace{1cm}} + \alpha \rightarrow {}^{36}_{18}\text{Ar}$

40. Provide the missing information to balance the following nuclear transmutation reactions. (12.8)
    (a) $^{14}_{7}\text{N} + {}^{4}_{2}\text{He} \rightarrow \underline{\hspace{1cm}} + {}^{1}_{1}\text{H}$
    (b) $\underline{\hspace{1cm}} + {}^{4}_{2}\text{He} \rightarrow 3{}^{1}_{0}\text{n} + {}^{239}_{94}\text{Pu}$
    (c) $^{4}_{2}\text{He} + {}^{239}_{94}\text{Pu} \rightarrow \underline{\hspace{1cm}}$
    (d) $^{238}_{92}\text{U} + \underline{\hspace{1cm}} \rightarrow {}^{239}_{93}\text{Np} + {}^{0}_{-1}\text{e}$

Case Study Problems

41. Neutron activation analysis of a hair sample can be used to determine a person's history of drug use. Explain how hair analysis could provide data on a person's drug use six months earlier. (CP)

42. One suggested method for tagging explosives involves using varying amounts of isotopes from rare elements. Why wouldn't radioactive isotopes be used? (CP)

43. If analysis of skeletal remains shows that polonium-210 has undergone 5.25 half-lives, how old is the skeleton? The half-life of polonium-210 is 138 days. (12.4)

44. If analysis of skeletal remains shows that lead-210 has undergone 2.5 half-lives, how old is the skeleton? The half-life of lead-210 is 22.3 years. (12.4)

A clock manufactured by Westclox that has glow-in-the-dark numbers and hands. The radioactive paint used was called Undark. (Ray Chipault/Underwood Archives)

45. Radium was first discovered by Madame Marie Curie in 1898. It has a half-life of 1602 years and is an alpha emitter. In the early 1900s, small amounts of it were mixed into paint and used to paint the numbers on clocks so the glow from the radium could be seen at night. One of the largest clock factories, Westclox in Ottawa, Illinois, employed hundreds of young women to paint the dials. The young women were instructed to keep a fine point on their paint brushes by using their lips. Many of these young women would start to suffer from radiation poisoning, yet the labor laws of the time left few options for recourse against the company.

One of the Chicago newspapers called this group of women the "Ottawa Society for the Living Dead." The company defended the safety of the product by saying that Madame Curie had been exposed to far greater amounts of radium and had not suffered radiation poisoning. Why did these young women suffer whereas Madame Curie did not? Are these antique clocks dangerous?

46. Mark Hofmann was an expert forger who sold hundreds of fake documents. He is best known for duping the Mormon Church into purchasing many fake writings from its early history. As his web of lies started to unravel, Hofmann turned to murder to cover his crimes, killing two individuals in two separate incidents. One of his last sales was to the Library of Congress of the only known copy of the Oath of a Freeman, a loyalty oath taken by the original settlers of the Massachusetts Bay colony. Carbon-14 dating of the document verified both the age of the paper and the ink, despite its modern origin. How could this be? (CP)

47. On April 26, 1986, an explosion at the Chernobyl nuclear power plant in the former Soviet Union sent large amounts of radioactive materials into the atmosphere, which then were carried by winds across Europe and beyond. One of the radioisotopes released into the atmosphere was tellurium-132, which has a half-life of 3 days. By May 2, the radioactive plume of dust had reached the United Kingdom. Calculate the percent of tellurium-132 remaining in the plume on May 2. (CP)

48. The Chernobyl accident also released cesium-134, which has a half-life of 2.07 years. How many years would it take for the amount of cesium to decline to less than 1% of the original? (CP)

CHAPTER 13

Chemical Equilibrium

CASE STUDY: Exploring Athletic Performance

Whether you are a collegiate athlete, an intramural sports star, or someone going to the gym to lose weight and stay fit, you are bombarded constantly with advertisements for products that claim to give you an edge over the competition. To understand athletic endurance and if the claims made by commercial products have any validity, we must first understand the biochemical reactions that produce energy for the body to use.

The mitochondrion is the cellular component responsible for energy production in our bodies. The source of that energy is the reaction of oxygen with glucose. When you start to exercise, you burn more

energy and require more oxygen, which means that you begin to breathe faster to get more oxygen into your lungs. If the delivery of oxygen to the cells is sufficient to provide the energy being used, the process is called *aerobic respiration.* If the expenditure of energy outpaces the supply of oxygen, however, the body will continue to supply energy using *anaerobic respiration,* which produces the lactate ion as a byproduct.

During exercise, the body generates excess heat and a person begins to sweat as a cooling mechanism to prevent overheating. Electrolytes are required in the cell for proper function. One risk that has been linked to the loss of electrolytes and dehydration is muscle cramping during exercise.

Misinformation abounds about how to exercise optimally and how to prevent muscle burn and muscle cramps. Lactic acid and lowering of blood pH have often been linked to the muscle burn that occurs during workouts and to the muscle soreness during the recovery period over the next day or two. Many sports drink manufacturers continually reinforce the perception that their products can prevent muscle cramps and increase an athlete's performance during competition. Sports drinks are now specializing to the degree that they have separate versions for before, during, and after a strenuous workout or competition.

As you read through the chapter, consider the following questions:

• **How is oxygen delivered from the lungs to the muscle tissue that requires it?**

• **What does it mean if a system is in equilibrium?**

• **Can equilibrium be disturbed, and if so, how can it be restored?**

Does lactic acid cause muscle burn? Will sports drinks eliminate muscle cramps? In order to evaluate such claims it is necessary to understand how equilibrium processes work. The delivery of oxygen and electrolytes to muscle tissue are complex processes regulated by equilibrium processes that will be discussed within this chapter.

MAKE THE CONNECTION
At the conclusion of the chapter, we will reconsider these questions as well as look into a report that links dark chocolate with enhanced athletic performance.

13.1 Chemical Equilibrium

LEARNING OBJECTIVE

Describe chemical equilibrium.

A common misconception about chemical reactions is that the reactants are completely changed to the products. This notion is conveyed by the simplified description of chemical reactions in which a single arrow ($\rightarrow$) is used to indicate that a reaction takes place. The reaction arrow is translated as "to form," without any qualifier. In reality, most reactants do not completely react to form the product even if sufficient quantities of the reactants are present in the exact molar proportions needed for a complete reaction. One possibility that you may not have considered is that the products react together to re-form the reactants. Consider the following generic reaction:

$$A + B \rightarrow C + D$$

Recall from Section 11.2 the requirements for a chemical reaction according to collision theory: Molecular collisions of sufficient energy and orientation will be successful. Since there will be successful collisions between the products, the **reverse reaction** of products to form reactants can occur:

$$C + D \rightarrow A + B$$

When the rate of the forward reaction is equal to the rate of the reverse reaction, the system is in **chemical equilibrium.** Rather than writing two equations for each reaction, a double arrow notation is used to represent the equilibrium system:

$$A + B \rightleftarrows C + D$$

When we breathe, oxygen is taken into our lungs, enters the alveoli, and diffuses into the pulmonary capillaries. The oxygen gas then reacts with hemoglobin (Hb) to form oxyhemoglobin (HbO_2), as shown in the following reaction:

$$Hb + O_2 \rightleftarrows HbO_2$$

Note that this reaction is reversible, as the oxyhemoglobin (product) can react to form hemoglobin and oxygen gas (reactants), which is how it delivers oxygen to cells in the body. This process is explained in greater depth in Figure 13.6.

If the reverse reaction of C and D coming together to remake the reactants A and B occurs, can the reaction ever be finished? As soon as A and B are formed, they can react together to form the products C and D once again. To understand the extent to which the forward and reverse reactions take place, and what it means for a reaction to be "finished," we need a more complete understanding of collision theory.

The magnitude of both the forward and reverse reactions is controlled by the restrictions of collision theory and especially the activation energy barrier. Consider the reaction progress in relation to the activation energy, as shown in Figure 13.1.

The forward reaction is limited to those molecules that have enough energy to overcome the activation energy barrier illustrated by the blue arrow in Figure 13.1. The reverse reaction occurs to a lesser extent because its activation energy, illustrated by the green arrow, is much greater than that of the forward reaction.

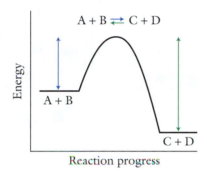

FIGURE 13.1 Activation energy barrier. The energy required for the forward reaction (A + B → C + D) is much less than the energy required for the reverse reaction (C + D → A + B).

13.2 Dynamic Equilibrium

A chemical reaction is at equilibrium when the rate of the forward reaction is equal to the rate of the reverse reaction. Under these circumstances, for every set of reactants that form products, a set of products will undergo the reverse reaction to re-form the reactants. The energy diagram in Figure 13.1 might easily be misinterpreted to mean that the rate of the forward reaction will always be greater than the rate of the reverse reaction due to the differences in activation energy. Recall, though, that activation energy is not the only factor influencing a reaction—the frequency of collisions and the orientation of the molecules are also important.

When compounds A and B first react, as shown in Figure 13.2a, no reverse reaction takes place. As compounds C and D are formed, they are initially far apart and seldom collide, as illustrated in Figure 13.2b. Because compounds A and B are present in high concentrations at the start of the reaction, the forward reaction occurs rapidly. As the reaction continues, the concentrations of A and B decrease, causing fewer collisions between A and B and a slower rate in the forward direction. Simultaneously, the concentrations of C and D increase, causing more collisions between C and D and a higher rate in the reverse direction, as shown in Figure 13.2c.

LEARNING OBJECTIVE
Explain how dynamic equilibrium is established and the effect of adding a catalyst.

At any time, a juggler keeps a constant number of balls in the air, although the specific balls change locations. (Art Vandalay/ Getty Images)

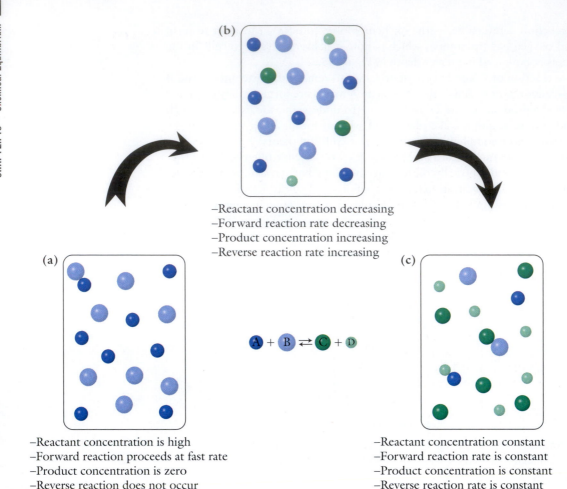

(b)

–Reactant concentration decreasing
–Forward reaction rate decreasing
–Product concentration increasing
–Reverse reaction rate increasing

(a)

$A + B \rightleftharpoons C + D$

–Reactant concentration is high
–Forward reaction proceeds at fast rate
–Product concentration is zero
–Reverse reaction does not occur

(c)

–Reactant concentration constant
–Forward reaction rate is constant
–Product concentration is constant
–Reverse reaction rate is constant
*Reactant concentration ≠ product concentration
*Forward reaction rate = reverse reaction rate

FIGURE 13.2 Establishment of equilibrium for A + B $\rightleftharpoons$ C + D. (a) At the start of the reaction, only the forward reaction is possible because only compounds A and B collide. (b) As the reaction progresses, the number of collisions between compounds A and B decreases because their concentrations are lower. The number of collisions between compounds C and D increases because they are now both present in solution. (c) When the reaction reaches equilibrium, the number of successful collisions between reactants in the forward reaction is equal to the number of successful collisions between products in the reverse reaction.

Figure 13.3a illustrates the changes in reaction rates until the point at which equilibrium is established. The blue line indicates the rate of the forward reaction, and the green line represents the rate of the reverse reaction. At equilibrium, both the forward and reverse reactions occur at the same rate, and the concentrations of reactants and products do not change over time. We often refer to this as *dynamic equilibrium* because the reactions continue but do not produce any overall change in the concentrations of reactants and products. It is important to note that just because the concentrations are no longer changing does not mean that the concentrations are equal at equilibrium. A good analogy for dynamic equilibrium is juggling. At any point in time, there is a constant number of balls in the air, but the specific balls are in fact continually changing positions.

Figure 13.3b shows a concentration profile for the generic reaction of

$$A + B \rightleftharpoons C + D$$

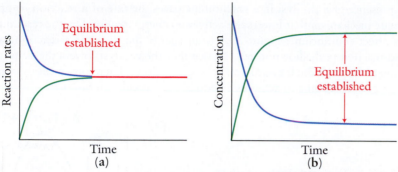

FIGURE 13.3 Change in reaction rates and concentrations during establishment of equilibrium. (a) The rate of the forward reaction (blue) decreases as the reaction consumes reactant molecules. The rate of the reverse reaction (green) increases as the reaction progresses because the newly created product molecules may now react. Equilibrium is established when the rates of the forward and reverse reactions are the same. (b) The concentration of the reactants (blue) decreases and the concentration of the products (green) increases as the reaction progresses. Equilibrium is established when the concentrations of reactants and products become constant.

The concentration profile differs from the kinetic graphs shown in Figure 13.3a because the concentrations become constant but not equal to one another. Figure 13.3b shows that the final concentration of reactants (in blue) is less than the final concentration of products (in green). This is not always the case, because some reactions do not create a significant amount of the products.

WORKED EXAMPLE 1

Does the following sketch of a reaction vessel represent a reaction at equilibrium, given that the vessel in Figure 13.2c represents the same reaction at equilibrium? Explain your answer.

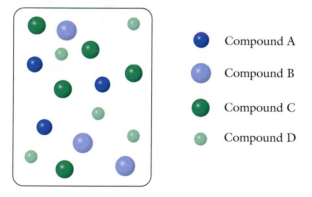

SOLUTION

This system is not in equilibrium with the one shown in Figure 13.2c. The concentration of the reactants in this vessel is higher, and the concentration of the products is lower, than shown for the vessel in Figure 13.2c.

Practice 13.1

Draw a reaction vessel that is in equilibrium with that shown in Figure 13.2c.

Answer

The vessel should have six units each of compounds C and D, and two units each of compounds A and B.

The presence of a catalyst in a reaction increases the rate of a reaction by providing an alternate mechanism that lowers the activation energy required by the reactants. How will this affect equilibrium? Figure 13.4 is an energy diagram for a reaction, with the catalyzed reaction path shown in red. Notice that the catalyst lowers the activation energy required for both the forward and reverse reactions. Therefore, a catalyst increases the rate of each reaction and achieves an equilibrium condition faster.

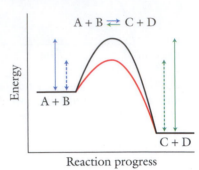

FIGURE 13.4 Catalyzed reaction pathway. A catalyst increases the rate of a reaction because the activation energy barrier is decreased, which allows a greater number of collisions to be successful. Notice that the catalyzed reaction path (red) lowers the activation energy barrier for both the forward and reverse reactions.

13.3 Values of the Equilibrium Constant

LEARNING OBJECTIVE

Determine concentrations of reactants and products using the equilibrium constant.

There are many reasons not to smoke cigarettes, including the fact that smoking reduces the ability of hemoglobin to carry oxygen to the cells, disrupting chemical equilibrium. Cigarette smoke contains carbon monoxide created from the incomplete combustion of the tobacco. The carbon monoxide molecule directly competes for the same binding site on hemoglobin as oxygen, forming carboxyhemoglobin (HbCO). However, the carbon monoxide binds to the hemoglobin 250 times more strongly than oxygen does. A nonsmoker typically will have HbCO levels of 0.5% or less from ambient sources of carbon monoxide, whereas smokers may have up to 13% HbCO, severely decreasing the capacity of their blood to deliver oxygen.

The ability of a reaction to form stable products is measured by the **equilibrium constant** (K). The equilibrium constant is equal to the concentrations of the products raised to their stoichiometric coefficients, divided by the concentrations of the reactants raised to their stoichiometric coefficients at equilibrium. For the hemoglobin-carbon monoxide example, the reaction and equilibrium constant would be as follows:

$$Hb + CO \rightarrow HbCO \qquad K = \frac{[HbCO]^1}{[Hb]^1[CO]^1} = \frac{[HbCO]}{[Hb]\,[CO]}$$

The numerical value of the equilibrium constant for a chemical reaction is always the same, provided the temperature is held constant. The actual concentration of reactants and products may vary under different conditions, but their ratio is always the same. The equilibrium constant is also called the *stability constant*, the *association constant*, or the *binding constant*, especially when discussing toxins.

The magnitude of the equilibrium constant provides information about the relative amounts of reactants and products found in solution at equilibrium. Consider the following symbolic equation for the equilibrium constant:

$$K = \frac{[\text{Product C}]\,[\text{Product D}]}{[\text{Reactant A}][\text{Reactant B}]}$$

If you have a high concentration of products and a low concentration of reactants,

$$\frac{[\text{large number}]}{[\text{small number}]} = \text{very large number} \Rightarrow \text{mostly products in solution}$$

If you have a low concentration of products and a high concentration of reactants,

$$\frac{[\text{small number}]}{[\text{large number}]} = \text{very small number} \Rightarrow \text{mostly reactants in solution}$$

Therefore, a large equilibrium constant indicates a solution consisting primarily of products. A small equilibrium constant indicates that mostly reactants are present in solution.

Table 13.1 lists the binding constants for the mercury(II) ion (Hg^{2+}), a toxin that can be fatal if ingested. The mercury(II) ion binds to the amino acids methionine and cysteine, found in protein molecules such as albumin. Hg^{2+} forms a strong bond to the sulfur atom located in the methionine molecule. The equilibrium constant of 3.2×10^6 indicates that the bound form of the ion predominates at equilibrium. The bonding of Hg^{2+} to the sulfur atom in cysteine is even stronger, as reflected by an equilibrium constant of 1.6×10^{14} that is 50 million times larger! The structures for cysteine and methionine are provided in Table 13.1, and it is evident that the mercury(II) ion has a stronger bonding preference for the R—SH group of cysteine rather than the R—S—R group of methionine.

TABLE 13.1	Equilibrium Constants for Hg^{2+} Binding	
Methionine	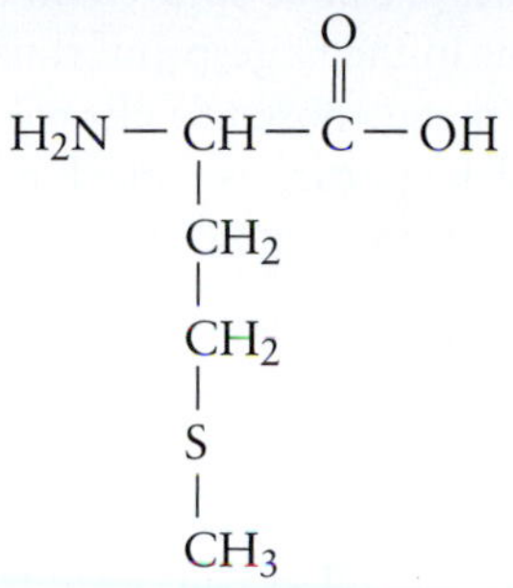 $H_2N - CH - C - OH$ (with C=O above, and chain CH₂–CH₂–S–CH₃ below)	3.2×10^6
Cysteine	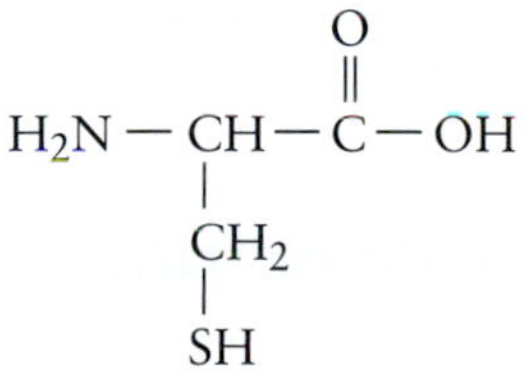 $H_2N - CH - C - OH$ (with C=O above, and chain CH₂–SH below)	1.6×10^{14}
Dimercaprol	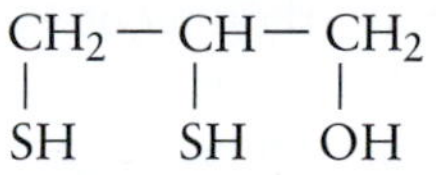 $CH_2 - CH - CH_2$ with SH, SH, OH below	5.0×10^{25}

The last compound listed in Table 13.1, dimercaprol, is the antidote for mercury poisoning because it exploits the mercury(II) ion's affinity for the —SH functional group. Dimercaprol can also be used to treat poisoning by heavy metals other than mercury. In fact, dimercaprol is often referred to as BAL, which stands for British Anti-Lewisite. BAL was the antidote developed in World War I to combat the chemical warfare agent Lewisite, an arsenic-based compound. The binding constant for the mercury(II) ion with the —SH group of cysteine is extremely large, indicating that at equilibrium there is very little free mercury(II) ion. The antidote dimercaprol takes advantage of the strong bond between toxic metals and the —SH functional group by having two —SH groups within the same molecule. Both —SH groups can interact with a single metallic ion, yielding an extremely large binding constant.

■ **WORKED EXAMPLE 2**

Weak acids are in equilibrium with a hydrogen ion and the corresponding anion according to the reaction:

$$HA \rightleftarrows H^+ + A^-$$

Which of the following weak acid compounds will have the most acidic (that is, the lowest) pH value?

Weak Acid	Formula	Equilibrium Constant
Hydrofluoric acid	HF	7.1×10^{-4}
Acetic acid	$HC_2H_3O_2$	1.8×10^{-5}

SOLUTION

The most acidic pH value will correspond to the compound with an equilibrium constant that most favors the product side of the reaction ($HA \rightleftarrows H^+ + A^-$). Because hydrofluoric acid has a higher equilibrium constant than acetic acid, it will have the most acidic pH value.

Practice 13.2

An aqueous solution of silver nitrate can be sprayed on a fingerprint to develop it by reacting with the chloride ions in the fingerprint, thus forming solid AgCl. The precipitate-coated fingerprint is then exposed to ultraviolet radiation to decompose the AgCl to elemental silver, which appears as a black powder. Using the following information, explain why silver is a good choice for this application.

$$Ag^+ + Cl^- \rightleftarrows AgCl(s) \qquad (K = 5.6 \times 10^9)$$

Answer

Silver ion will react with chloride ion to precipitate silver chloride. The very large equilibrium constant value indicates that at equilibrium there is mostly product and very little reactants. This is beneficial for the development of fingerprints because it would be of little use to apply a reagent that did not form a stable compound.

13.4 Le Chatelier's Principle

LEARNING OBJECTIVE

Apply Le Chatelier's principle to reactants and products.

Henry Louis Le Chatelier (1850–1936). (Roger Viollet/Getty Images)

Carbon monoxide is a colorless, odorless poison that incapacitates a person before death. Accidental carbon monoxide poisonings claim hundreds of lives each year. Many of these deaths can be prevented with the purchase and use of a carbon monoxide detector, now mandatory for residences in many states. Carbon monoxide poisoning is common among victims of residential fires and is also a hazard for firefighters. Carbon monoxide poisoning can be treated if caught in time.

To understand the mechanism of carbon monoxide poisoning, we must first know how equilibrium systems can be influenced by outside factors and a few facts about the chemical basis of respiration. When a system is at equilibrium, the rate of the forward reaction is equal to the rate of the reverse reaction, and the concentration of all reactants and products is constant. If conditions are changed—for example, if reactants are added to the reaction mixture—the system is no longer at equilibrium. How does the system respond to this disruption? The response is summarized by a general rule called **Le Chatelier's principle**, which states:

When a system at equilibrium is disturbed, the system will shift in such a manner as to counteract the disturbance and reestablish an equilibrium state.

Consider Figure 13.5a, in which equilibrium has been established and the concentration of all reactants and products is constant. In Figure 13.5b, reactant (yellow) has been added to the system, disrupting the equilibrium. As the concentration of reactant increases, the rate of the forward reaction increases and consumes some of the additional reactants, producing more of the product (green); this counteracts the original disturbance. This proceeds until equilibrium has been reestablished, as shown in Figure 13.5c. Note that when equilibrium is reestablished, the concentration of reactants and products is not the same as in Figure 13.5a.

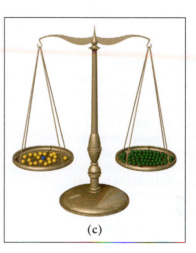

(a)   (b)   (c)

Reaction: 🟡 + 🔵 ⇌ 🟢

FIGURE 13.5 If it were possible to measure on a balance the rate of the forward and reverse reactions, (a) the two pans would be equal at equilibrium. If (b) more reactant were added, Le Chatelier's principle states that a system will counteract the disturbance in such a manner as to (c) reestablish equilibrium. Note that both systems (a) and (c) would have the same equilibrium constant.

Recall that the equilibrium constant K has a value that remains constant but that there is no specific value for the reactant and product concentrations associated with achieving equilibrium. A summary of the effects of changing the concentration of reactant(s) or product(s) is given in Table 13.2.

TABLE 13.2	Le Chatelier's Principle Applied	
Action	Response to Action	Change in Equilibrium Constant
Increase reactant concentration	Shifts equilibrium to the right: Consume reactants, produce products	None
Decrease reactant concentration	Shifts equilibrium to the left: Consume products, produce reactants	None
Increase product concentration	Shifts equilibrium to the left: Consume products, produce reactants	None
Decrease product concentration	Shifts equilibrium to the right: Consume reactants, produce products	None

FIGURE 13.6 The respiratory system. The transportation of oxygen through the body is regulated by the Hb + O_2 ⇌ HbO_2 equilibrium system. Hemoglobin in the blood enters the lungs and passes through the alveoli, which are small air sacs in the lungs that provide a large surface area for oxygen to pass into the bloodstream and become bound with hemoglobin. The high concentration of O_2 shifts the equilibrium to produce a large amount of HbO_2. As the oxygenated hemoglobin passes into the body, the unbound oxygen is removed by cells. When a reactant is removed, the equilibrium shifts to counteract this change, releasing additional oxygen gas. When the hemoglobin is depleted of oxygen, it returns to the lungs to be replenished.

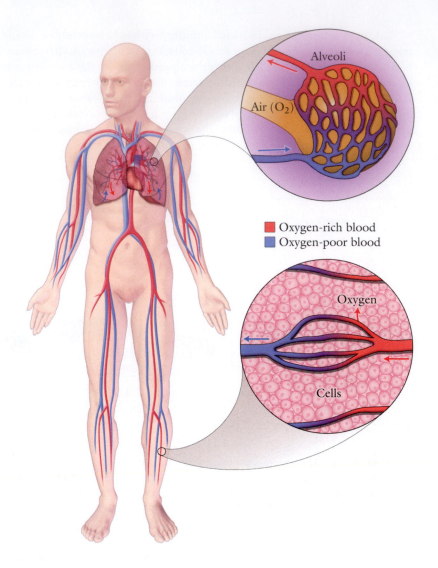

Alveoli

Air (O_2)

■ Oxygen-rich blood
■ Oxygen-poor blood

Oxygen

Cells

The protein hemoglobin, found in red blood cells, is responsible for transporting oxygen gas from the lungs to tissues throughout the body. Recall that the equilibrium reaction established between hemoglobin, oxygen gas, and oxyhemoglobin (HbO_2) is

$$Hb + O_2 \rightleftarrows HbO_2$$

The respiratory system is illustrated in Figure 13.6, with emphasis placed on the regions where the Hb–HbO_2 equilibrium is altered. The oxygen diffuses through alveoli into the pulmonary capillaries and is present at a high concentration, thus shifting the equilibrium to favor the formation of HbO_2. The blood is then pumped by the heart throughout the body to the extremities. As the blood circulates, a small amount of free O_2 is in equilibrium with the bound HbO_2. This free oxygen is available for use by the cells and is removed from the blood as needed.

The response to the removal of O_2 is to produce more O_2 to replace the amount that was consumed by the cells. The hemoglobin–oxygen system replaces the oxygen by increasing the rate of the reverse reaction. The production of O_2 in the reverse reaction will continue until equilibrium conditions are reestablished.

The process of supplying more oxygen to the blood by adjusting the rates of the reactions repeats itself as the newly freed O_2 is taken in by more tissue cells. Eventually the hemoglobin can no longer supply O_2, and the deoxygenated hemoglobin returns to the lungs.

Upon reaching the capillaries surrounding the alveoli, the hemoglobin is exposed to a high concentration of O_2. According to Le Chatelier's principle, exposure to a high concentration of a reactant will drive the forward reaction to produce the bound HbO_2, thus resupplying the blood with a source of oxygen and starting the process over again. An important fact to remember is that even though the concentration of reactants and products has changed when equilibrium is reestablished, the ratio of the concentration of products to reactants has not changed.

When a person is exposed to carbon monoxide, the hemoglobin–oxygen equilibrium is disturbed because carbon monoxide forms a stronger bond to hemoglobin than oxygen does. Once carbon monoxide is bound to hemoglobin to form carboxyhemoglobin (HbCO), the hemoglobin can no longer carry oxygen to the tissues. Also, hemoglobin does not release carbon monoxide through the lungs to be exhaled, as it does with carbon dioxide.

Treatment for carbon monoxide poisoning is based on Le Chatelier's principle. The reaction, which favors the production of the products, is

$$HbO_2 + CO \rightleftharpoons HbCO + O_2$$

The undesirable product is carboxyhemoglobin, and the desirable substance is oxyhemoglobin. To shift the equilibrium to favor the production of oxyhemoglobin, excess oxygen is supplied to the system. In cases of minor carbon monoxide exposure, this can be achieved by escorting the individual to a region of fresh air, which contains approximately 21% oxygen gas, or providing a breathing mask attached to a source of pure oxygen.

In severe cases of exposure to carbon monoxide, a hyperbaric oxygen chamber is commonly used, as shown in Figure 13.7. A hyperbaric chamber is a pressurized chamber filled with pure oxygen gas. The high oxygen pressure forces the equilibrium to shift to the left side, favoring the production of oxyhemoglobin and minimizing the production of carboxyhemoglobin.

The preceding discussion of hemoglobin–oxygen equilibrium was greatly simplified by omitting the role of the transport of carbon dioxide through the body. When hemoglobin transports oxygen to the cells, it also carries away carbon dioxide, a waste product of the cells, back to the lungs to be exhaled. The carbon dioxide–hemoglobin equilibrium is influenced by another system involving carbonic acid and the bicarbonate ion. Figure 13.8 illustrates all of the equilibrium processes involved in normal respiration.

FIGURE 13.7 Hyperbaric oxygen chamber. A hyperbaric oxygen chamber is used to provide oxygen gas under high pressure to victims of carbon monoxide poisoning. The large amount of oxygen forces the hemoglobin bound with carbon monoxide to release the carbon monoxide and become bound to the oxygen. (Kike Calvo/V & W/The Image Works)

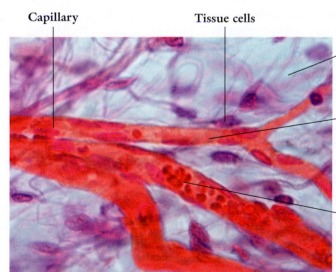

Capillary   **Tissue cells**

Interstitial fluid
- O_2 partial pressure 30 torr
- CO_2 partial pressure 50 torr

Plasma
- O_2 partial pressure 100 to 40 torr
- CO_2 partial pressure 40 to 46 torr
- $CO_2 + H_2O \rightleftharpoons H_2CO_3 \rightleftharpoons HCO_3^- + H^+$
- $H^+ + Protein^- \rightleftharpoons HProtein$

Red blood cell
- $CO_2 + H_2O \rightleftharpoons HCO_3^- + H^+$
- $H^+ + HbO_2^- \rightleftharpoons HHb + O_2$
- $CO_2 + HbO_2^- \rightleftharpoons HbCO_2^- + O_2$

FIGURE 13.8 The equilibrium systems involved in respiration. A complex set of equilibrium systems within the human body is involved in the transfer of oxygen and carbon dioxide (measured in torr where 760 torr = 1 atmosphere) between hemoglobin, plasma, interstitial fluid between cells, and tissue cells. (Warren Rosenberg/Fundamental Photographs)

■ WORKED EXAMPLE 3

What will happen to the pH of blood if a person is exposed to elevated carbon dioxide levels?

SOLUTION

According to the equilibrium reactions listed in Figure 13.8, increasing carbon dioxide (CO_2) will push the equilibrium position to favor the production of carbonic acid (H_2CO_3). Increasing the amount of carbonic acid shifts a second equilibrium, the dissociation of carbonic acid, to the product side, increasing the concentration of hydrogen ions (H^+). The increase in the concentration of H^+ corresponds to a decrease in pH.

Practice 13.3

What effect will breathing in a higher concentration of carbon dioxide have on the ability of hemoglobin to carry oxygen?

Answer

The increase in carbon dioxide concentration can have several effects. When the deoxygenated hemoglobin reaches the lungs, it is carrying carbon dioxide. The higher concentration of carbon dioxide in the lungs results in a lower amount being released from the hemoglobin. Also, the amount of oxygen in the lungs will be less than normal, and less oxyhemoglobin will be formed. The result is a decrease in the ability of hemoglobin to carry oxygen.

13.5 Solubility Equilibrium

LEARNING OBJECTIVE

Extend the concept of equilibrium to insoluble compounds.

In Chapter 6, the solubility rules were first discussed and compounds were labeled as being either soluble or insoluble in water. That was an oversimplification, as solubility ranges from one extreme to the other. Solubility is an equilibrium process that occurs between the solid compound and the aqueous-phase ions that have dissolved from the solid.

For example, calcium oxalate (CaC_2O_4) is a highly insoluble compound that produces a small number of calcium ions and oxalate ions when exposed to water. The extremely small value for the equilibrium constant K indicates that very few products (ions) are present at equilibrium, so the compound is quite insoluble. The equilibrium constant for the solubility of a compound is denoted as K_{sp}.

$$CaC_2O_4(s) \rightarrow Ca^{2+} + C_2O_4^{2-} \qquad K_{sp} = 2 \times 10^{-9}$$

Calcium oxalate is the compound responsible for the formation of kidney stones. The kidneys process excess electrolytes as well as regulate pH and water balance. As a result, sufficiently high concentrations of calcium ions and oxalate ions can occur in the kidneys so as to precipitate them out as insoluble crystals. If the crystals continue to grow, they can form a kidney stone. Although calcium oxalate is one of the more common compounds that form kidney stones, they can also be formed by insoluble calcium hydrogen phosphate crystals, according to the following equilibrium:

$$CaHPO_4(s) \rightarrow Ca^{2+} + HPO_4^{2-} \qquad K_{sp} = 3 \times 10^{-7}$$

One method to reduce the likelihood of developing a kidney stone is by consuming a large amount of water daily. Water will reduce the concentration of the ions and shift the equilibrium toward the formation of more aqueous-phase ions, thereby reducing the size of the stone.

Minerals, for the most part, are highly insoluble compounds that have formed over the geological time scale. However, as you have just learned, insoluble compounds can be in equilibrium with their constituent ions when exposed to water. The United States

Kidney stones are most commonly formed by insoluble precipitates of calcium oxalate and calcium hydrogen phosphate. (© Nathan Griffith/Corbis)

Geological Survey has been monitoring the amount of arsenic in ground water across the United States. Arsenic occurs naturally in the minerals realgar (AsS), orpiment (As_2S_3), and arsenopyrite (FeAsS), to name just a few. These natural sources of arsenic have been a cause for concern as a possible source of unintended poisoning. The arsenic concentrations shown in Figure 13.9 are of samples taken directly from wells without prior treatment.

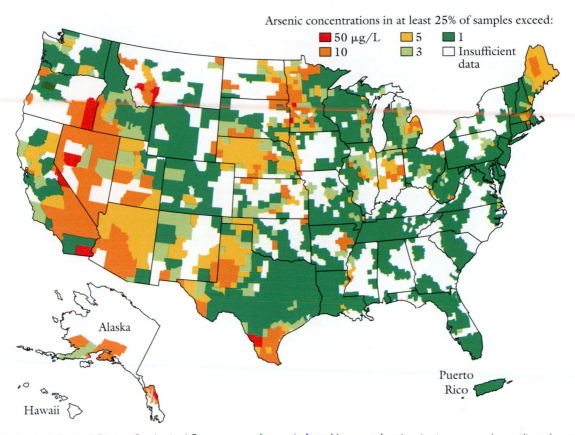

Arsenic concentrations in at least 25% of samples exceed:

- 🟥 50 µg/L
- 🟧 10
- 🟨 5
- 🟩 3
- 🟩 1
- ⬜ Insufficient data

FIGURE 13.9 United States Geological Survey map of arsenic found in ground water. (water.usgs.gov/nawqa/trace/arsenic)

The EPA-recommended level for arsenic in water is not to exceed 10 µg/L. People living in a region with elevated arsenic levels in ground water are most likely drinking water that has been treated. Arsenic would be removed in water treatment facilities to a safe level. It is also important to realize that even well water in an area with low reported arsenic levels is not necessarily safe to drink. Only analysis of a water sample, conducted by a certified laboratory, can provide information about the quality of ground water that is being directly used without treatment.

How does arsenic get into ground water? The arsenic-containing minerals are in equilibrium with their dissolved ions. Using the mineral orpiment (As_2S_3) as an example, the equilibrium reaction is shown by the following equation:

$$As_2S_3(s) \rightleftarrows 2As^{3+}(aq) + 3S^{2-}(aq)$$

At equilibrium, the arsenic(III) and sulfide ions will re-form the solid orpiment mineral at the same rate at which the orpiment dissolves. The magnitude of the equilibrium constant is extremely small, so there is very little arsenic dissolved by the ground waters. Yet the dangers of arsenic are great enough that the amount is worth monitoring. If a solution contains a high level of free arsenic(III) ions, the addition of

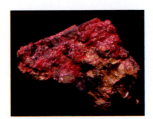

One possible source of accidental arsenic poisoning is the presence of naturally occurring arsenic minerals in contact with waters used for human consumption. Pictured here is the mineral realgar (AsS). (Maurice Nimmo/Frank Lane Picture Agency/Corbis)

sulfide ion will trigger the precipitation of the insoluble compound As_2S_3. The equilibrium reactions of arsenic minerals are actually more complicated than written above because other dissolved ions present in solution have a direct effect on the solubility of the mineral.

13.6 Mathematics of Equilibrium

The previous sections of this chapter discussed the nature of chemical equilibrium, how to interpret the value of equilibrium constants, and several different kinds of equilibrium systems. This section will focus on two types of equilibrium constant calculations: first, the determination of equilibrium constants from equilibrium data; then, the use of equilibrium constants to determine the concentration of a compound.

The value of the equilibrium constant can be determined, provided that the concentrations of all the reactants and products are known for a system that has achieved equilibrium. For example, consider a solution of acetic acid, a weak acid present in vinegar, which will partially dissociate according to the equation:

$$HC_2H_3O_2(aq) \rightleftarrows H^+(aq) + C_2H_3O_2^-(aq)$$

The equilibrium constant can be calculated by multiplying the concentrations of the products and then dividing by the concentrations of the reactants. The coefficients of all the values are 1.

$$K = \frac{[H^+][C_2H_3O_2^-]}{[HC_2H_3O_2]}$$

For example, if the concentrations of the components in a solution of acetic acid at equilibrium were measured to be: $[H^+] = 0.0010$ M, $[C_2H_3O_2^-] = 0.0010$ M, and $[HC_2H_3O_2] = 0.055$ M, the equilibrium constant would be calculated as follows:

$$K = \frac{[0.0010][0.0010]}{[0.055]} = 1.8 \times 10^{-5}$$

The equilibrium constant (K) does not change as long as the temperature of the system remains constant. However, the individual concentrations of the component ions may take on an infinite number of combinations that will produce the single equilibrium constant value. For example, if $[H^+] = 0.0020$ M and $[HC_2H_3O_2] = 0.0020$ M under a new set of equilibrium conditions, the equilibrium concentration of the acetate ion is calculated as follows:

$$K = \frac{[0.0020][0.0020]}{[HC_2H_3O_2]} = 1.8 \times 10^{-5}$$

$$[HC_2H_3O_2] = \frac{(0.0020)(0.0020)}{(1.8 \times 10^{-5})} = 0.222$$

When writing the equilibrium expressions, care should be taken to note the physical state of the reactants and products. For heterogeneous systems, such as the solubility equilibrium from section 13.5 in which relatively insoluble compounds in the solid state dissolve slightly to produce aqueous ions as the system reaches equilibrium, the solid compounds are omitted from the equilibrium expression because their concentrations cannot change. For example, the solubility equilibrium expression for lead(II) chloride is given by the reaction:

$$Pb(Cl)_2(s) \rightleftarrows Pb^{2+}(aq) + 2Cl^-(aq)$$

Note that the stoichiometric coefficient of the chloride ion is 2, which leads to squaring the chloride ion concentration in the equilibrium expression, and the solid lead(II) chloride compound is omitted as shown:

$$K = [Pb^{2+}][Cl^-]^2$$

Equilibrium constants are often given subscripts to help denote the type of equilibrium system they represent. The equilibrium constant for a weak acid system is denoted as K_a.

WORKED EXAMPLE 4

Determine the equilibrium constant K_a for hydrofluoric acid (HF) given the equation $HF(aq) \rightleftarrows H^+(aq) + F^-(aq)$ and that under equilibrium conditions $[HF] = 0.0395$ M, $[H^+] = 0.00530$ M, and $[F^-] = 0.00530$ M.

Step 1: Write the equation for the equilibrium system.

$$HF(aq) \rightleftarrows H^+(aq) + F^-(aq)$$

Step 2: Write the equilibrium-constant expression based on the equation.

$$K_a = \frac{[H^+][F^-]}{[HF]}$$

Step 3: Substitute the given values into the expression and solve.

$$K_a = \frac{[0.00530][0.00530]}{[0.0395]} = 7.11 \times 10^{-4}$$

Practice 13.4

Determine the equilibrium concentration of the hydrogen ion for a weak acid solution containing $[HF] = 0.0850$ M, $[F^-] = 0.0320$, and $K_a = 7.11 \times 10^{-4}$.

Answer
❑ 0.00189 M

WORKED EXAMPLE 5

Determine the equilibrium constant (K_{sp}) for the solubility of lead(II) chloride given the equation: $Pb(Cl)_2(s) \rightleftarrows Pb^{2+}(aq) + 2Cl^-(aq)$, and at equilibrium $[Pb^{2+}] = 0.016$ M and $[Cl^-] = 0.032$ M.

Step 1: Write the equation for the equilibrium system.

$$Pb(Cl)_2(s) \rightleftarrows Pb^{2+}(aq) + 2Cl^-(aq)$$

Step 2: Write the equilibrium-constant expression based on the equation.

$$K = [Pb^{2+}][Cl^-]^2$$

Step 3: Substitute the given values into the expression and solve.

$$K = [0.016][0.032]^2 = 1.6 \times 10^{-5}$$

Practice 13.5

Determine the equilibrium concentration of the barium ion for a solution containing solid barium hydroxide given that the $[OH^-] = 0.250$ M, the equation is $Ba(OH)_2(s) \rightleftarrows Ba^{2+}(aq) + 2OH^-(aq)$, and $K_{sp} = 5.0 \times 10^{-3}$.

Answer
❑ 0.080 M

13.7 Chemistry of Poisons

Alle Dinge sind Gift, und nichts ist ohne Gift.
Allein die Dosis macht, dass ein Ding kein Gift ist.
All things are poison, and nothing is without poison.
The dosage alone determines that a thing isn't poison.

—Paracelsus (1493–1541)

A **poison** is defined as any compound that injures or harms a living organism. As Paracelsus suggested, all things have the potential to be a poison—only the dose dictates whether a compound is a poison or not. The term **toxic compound** is used to indicate that a compound is dangerous even in small amounts.

Sodium chloride and potassium cyanide can both be poisonous. Potassium cyanide is a very toxic compound, however, whereas sodium chloride is not. Different precautions are needed when handling sodium chloride as compared to potassium cyanide. It is evident that a method is needed to measure the toxicity of compounds.

The **lethal dose** is a measurement of the toxicity of a compound. The name indicates the actual experimental procedure, as illustrated in Figure 13.10. A compound is given to a population of lab rats (or other test species) until the dose is reached that will be lethal to a percentage of the population, the most common value being 50% (LD_{50}). The dose is typically measured as milligrams of compound per kilogram of body mass (mg/kg rat). The smaller the LD_{50} value, the greater the toxicity of the compound. Table 13.3 illustrates the LD_{50} for a range of compounds.

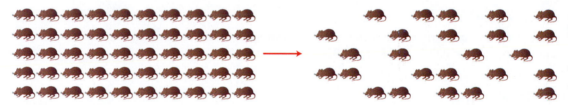

FIGURE 13.10 The LD_{50} is a measurement of a compound's toxicity by determining the dose that will be lethal to 50% of a test population.

Many poisons work by interfering with an organism's normal cell functions in a way that results in cell death. Poisons can be in the form of pure elements (mercury), inorganic compounds (carbon monoxide), or organic compounds (strychnine). The term **toxin** usually refers to a naturally occurring poisonous substance produced by a living organism.

TABLE 13.3	The LD_{50} of Various Compounds		
Substance	LD_{50} (mg/kg)	Substance	LD_{50} (mg/kg)
Mercury(II) chloride	1	Calcium chloride	1940
Potassium cyanide	10	Sodium chloride	3000
Arsenic(III) oxide	15	Methanol	5628
Barium chloride	118	Ethanol	7060

Values reported correspond to oral dose administered to rats.

WORKED EXAMPLE 6

Can a person suffer from water poisoning?

SOLUTION

Yes. If excessive amounts of water are consumed, it can act as a poison by seriously disrupting the body's balance of electrolytes. News reports indicate that several college students in the last few years have died from hazing incidents that involve drinking excessive amounts of water.

Practice 13.6

Iron is a common element that is included in many over-the-counter multivitamins. Are iron supplements a toxic compound? Search the Centers for Disease Control and Prevention's Web site at www.cdc.gov for information regarding the safety of iron supplements.

Answer

As with other substances that are usually benign when ingested at recommended levels, iron supplements are safe for adults. However, iron can be fatal if ingested by children.

> [Ghost:] With juice of cursed hebona in a vial,
> And in the porches of my ears did pour
> The leperous distilment; whose effect
> Holds such an enmity with blood of man
> That swift as quicksilver it courses through
> The natural gates and alleys of the body,
> And with a sudden vigour it doth posset
> And curd, like eager droppings into milk,
> The thin and wholesome blood. So did it mine;
>
> —*Hamlet,* Act I, Scene 5, by William Shakespeare

William Shakespeare used several methods of poisoning in his telling of *Hamlet:* The ghost of Hamlet's father reveals that a poison was poured into his ear; Hamlet's mother dies from drinking poisoned wine; and Hamlet himself dies from being stabbed with the poison-coated tip of a foil during a duel. As illustrated by the play, the means by which an individual is exposed to a poison is a crucial factor in the danger presented by the compound.

Some poisons are extremely hazardous by inhalation, but contact with skin is not particularly dangerous. Other poisons may have multiple routes of exposure and entry. The LD_{50} is normally reported for an oral dose, although for some compounds, the LD_{50} may be reported for a subcutaneous or intravenous dose. If a poison is discovered during an autopsy, the medical examiner will attempt to determine how the poison entered the victim's system. This information can help investigators determine whether the case is a homicide or an accidental death.

Warning signs for containers that are poisons, corrosive, or a health hazard (carcinogen, mutagen, teratogen).

The LD_{50} measurement is for **acute poisoning**, that is, a single exposure to a poison or an exposure over a relatively short time period. In an acute poisoning, the effects are noticeable almost immediately. There is also danger from **chronic poisoning**, that is, exposure over a longer and more continuous interval to a lesser amount of the poison. For example, lead compounds had been a common component of paints before the 1970s and are a source of chronic exposure to lead in lead poisoning.

Some other chemical substances are considered dangerous, even deadly, although they do not fit the traditional definition of a poison. Acids and bases are **corrosive** to human flesh; that is, they decompose the protein molecules that make up your body. Acids and bases were used in the past to disfigure victims' faces and to remove victims' fingerprints to prevent their identification. DNA analysis, however, has largely ended this practice.

There are several other sets of dangerous chemical compounds, such as carcinogens, teratogens, and mutagens. **Carcinogens** are compounds that interfere with normal cell growth and result in rapid, uncontrolled growth and tumor formation. **Teratogens** cause abnormal growth of fetal cells and birth defects in the developing fetus. A **mutagen** alters the DNA of cells in the parent, which can cause future disease in offspring. Teratogens and mutagens may also be carcinogens, but not all teratogens and mutagens cause cancer in exposed individuals.

13.8 CASE STUDY FINALE: Exploring Athletic Performance

Does lactic acid cause muscle burn, or does it supply the fuel to power muscles? Do sport drinks help relieve muscle cramps? Perhaps most intriguing is some recent research that suggests dark chocolate can increase athletic performance. Let's examine these questions with a more complete understanding of equilibrium.

For years, it has been said that the burning sensation in your muscles during an intense workout is due to lactic acid build-up. Even soreness that lingers a day or more after the workout during the recovery phase was attributed to lactic acid. Energy is created by the reaction of glucose, shown in Figure 13.11a, to form the pyruvate ion, shown in Figure 13.11b. In an anaerobic setting, the pyruvate ion is converted into the lactate ion, shown in Figure 13.11c.

Lactic acid is formed by the reaction of lactate ion with a hydrogen ion, however, lactic acid will not form, as blood is buffered at a pH of 7.4 and does not contain a sufficient amount of the hydrogen ion. Furthermore, the lactate ion concentration drops rapidly when oxygen levels are restored. As lactic acid never forms to any appreciable amount and the lactate ion is converted back to the pyruvate ion when aerobic conditions are reestablished, it cannot be the cause of muscle pain and soreness.

Many sports drink manufacturers would like potential customers to believe that their product provides near-instant energy and restores electrolyte

(a) Glucose   (b) Pyruvate   (c) Lactate

FIGURE 13.11 Aerobic respiration of glucose (a) produces pyruvate (b) in the first step. The pyruvate ion further reacts if sufficient oxygen is available. Under anaerobic conditions, when there is not a sufficient amount of oxygen, the pyruvate is converted to lactate (c).

balance. When athletes get a muscle cramp during practice or competition, it is often attributed to dehydration and electrolyte loss due to sweating. However, your body is continually adjusting a complex set of dynamic equilibria to regulate pH, hydration, blood-sugar levels, and electrolyte balance. Muscle cramps are localized in a single muscle group, and they do not indicate a systemwide problem such as dehydration or electrolyte loss which would cause widespread symptoms. The electrolytes lost during exercise are replenished continuously from the digestive system as the loss of electrolytes would shift equilibrium according to Le Chatelier's principle. Furthermore, it would take at least 30 minutes for electrolytes in a sports drink to reach the affected muscle group.

The final question to consider, then, is whether dark chocolate improves athletic endurance. One of the components found in cocoa is a chemical compound called *epicatechin*. It has been shown in mice that *epicatechin* can enhance athletic endurance by over 50% when administered in small doses over a 15-day period. The compound appears to trigger a significant amount of new capillary growth within muscle tissues. It also appears to trigger an increase in the number of mitochondria within the cells. Increasing the number of mitochondria while also increasing the ability of blood to circulate oxygen to the cells would increase the

overall ability of the muscles to provide aerobic power, enhancing athletic endurance.

Epicatechin belongs to a group of compounds called the *flavonoids*, which are found in dark chocolate but are removed in the process of making other forms of chocolate like milk chocolate. Because chocolate is high in fat, though, it is best to keep your intake to a minimum. One study on human beings that cited improved vascular health used 1.6-ounce bars of dark chocolate, the amount shown in the margin. However, increased caloric intake from the chocolate was compensated by decreasing the total calories consumed.

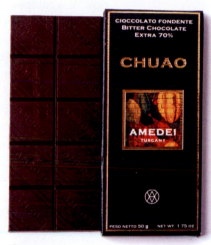

(Superstock)

CHAPTER SUMMARY

• A chemical system achieves equilibrium when the rate of a forward reaction is equal to the rate of the reverse reaction. In an equilibrium state, reactants form products at the same rate that products regenerate reactants.

• Equilibrium has been reached when there are no further changes in the concentrations of the reactants or products. A reaction in equilibrium appears to have stopped, but in reality, the reaction continues with the concentrations of products and reactants remaining constant.

• The ratio of the concentrations of products to the concentrations of reactants in a system at equilibrium is indicated by the equilibrium constant. A very large value for the equilibrium constant means the concentration of products is much greater than the concentration of reactants at equilibrium. If the equilibrium constant has a small value, the concentrations of reactants are much greater than the concentrations of products at equilibrium.

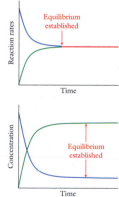

• Although the ratio of product concentrations to reactant concentrations at equilibrium is constant for a particular reaction, the concentrations of individual species can vary dramatically from one equilibrium situation to another.

• Equilibrium can be lost by the addition or removal of either reactants or products because changes in concentration levels disrupt equilibrium.

• According to Le Chatelier's principle, when a system at equilibrium is disturbed, the system will shift in such a manner as to counteract the disturbance and reestablish an equilibrium state. If a reactant is removed, the reverse reaction attempts to restore the equilibrium status by preferentially producing the reactant until the equilibrium ratio is once again achieved.

• While all compounds can be toxic to living organisms, the term *poison* is usually reserved for those that cause harm in relatively small doses. A toxin is a poisonous compound that originates from either a plant or an animal.

• The LD_{50} is a measure of the toxicity of a compound, determined by establishing the dose of a compound that will kill 50% of a test animal population. LD_{50} values are commonly reported for oral doses given to rats, and the units are typically mg/kg rat.

• Compounds can also present health hazards other than being poisonous if, for instance, they are corrosive (dissolve flesh), carcinogenic (cause tumor growth), teratogenic (cause birth defects), or mutagenic (alter DNA). Compounds may be classified under multiple health hazards.

• The ability of a poison to enter into and disrupt the normal function of cells is dependent on the chemical equilibrium of the poison with biological compounds.

KEY TERMS

acute poisoning (p. 382)
carcinogen (p. 382)
chemical equilibrium (p. 366)
chronic poisoning (p. 382)
corrosive (p. 382)

equilibrium constant (K) (p. 370)
Le Chatelier's principle (p. 372)
lethal dose (LD_{50}) (p. 380)
mutagen (p. 382)
poison (p. 380)

reverse reaction (p. 366)
teratogen (p. 382)
toxic compound (p. 380)
toxin (p. 380)

MAKING MORE CONNECTIONS Additional Readings, Resources, and References

For more information about carbon monoxide poisoning: http://www.ncbi.nlm.nih.gov/pmc/articles/PMC1740215/pdf/v059p00708.pdf

For more information on the health benefits of dark chocolate: http://www.webmd.com/diet/news/20040601/dark-chocolate-day-keeps-doctor-away

Epicatechin enhances fatigue resistance and oxidative capacity in mouse muscle, Leonardo Nogiueira, et al, J Physiol 589.18 (2011) pp4615-4631. http://www.chocolate.org/health/epicatechin.html http://jp.physoc.org/content/589/19/4643.full

May, Meredith. "Fraternity Pledge Died of Water Poisoning. Forced Drinking Can Disastrously Dilute Blood's Salt Content," *San Francisco Chronicle*, February 4, 2005.

Smith, R. M., Martell, A. E., and Motekaitis, R. J. NIST Critically Selected Stability Constants of Metal Complexes, *NIST Standard Reference Database*, no. 46, version 3.0, Gaithersberg: U.S. Department of Commerce, 1997.

Trestrail, John Harris. *Criminal Poisoning: Investigational Guide for Law Enforcement, Toxicologists, Forensic Scientists, and Attorneys*, Totowa, NJ: Humana Press Inc., 2000.

For more on the fraternity pledge and water poisoning: www.nbcdfw.com/news/local/Prairie-View-Frat-Suspended-After-Hazing-Death-91592214.html and www.nbcdfw.com/news/local/Parents-of-Prairie-View-Student-Sue-Allege-Hazing-69019122.html

For additional information on arsenic in ground water: http://water.usgs.gov/nawqa/trace/arsenic

REVIEW QUESTIONS AND PROBLEMS

Questions

1. What are the conditions needed for the reverse reaction of a system to occur? (13.1)

2. What is meant by a dynamic equilibrium? (13.1)

3. Why does the rate of the forward reaction decrease as a function of reaction time? (13.2)

4. Why does the rate of the reverse reaction increase as a function of reaction time? (13.2)

5. Why does it *appear* that a reaction has stopped when it reaches equilibrium? (13.2)

6. Will the concentrations of reactants and products always be the same at equilibrium? (13.2)

7. How does a catalyst influence the equilibrium process? (13.2)

8. In a style similar to Figure 13.2c, sketch (on a molecular scale) what the equilibrium system would look like for a system with a very small equilibrium constant. (13.3)

9. In a style similar to Figure 13.2c, sketch (on a molecular scale) what the equilibrium system would look like for a system with a very large equilibrium constant. (13.3)

10. The containers shown in parts (a) and (b) of the following figure represent snapshots of the same chemical solution taken at different times. Has the solution reached equilibrium? Explain your answer. (13.3)

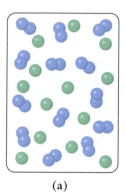 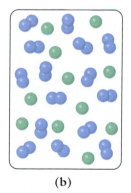

(a)   (b)

11. What does the magnitude of the equilibrium constant represent? (13.3)

12. When an equilibrium constant has a large value, is it the reactants or products that predominate in a solution? How about when an equilibrium constant has a small value? (13.3)

13. Given the following K_{sp} values, determine which equilibrium solution has the greatest concentration of Cd^{2+} ion. (13.3)
    (a) $Cd(OH)_2(s)$, $K_{sp} = 7.2 \times 10^{-15}$
    (b) $CdS(s)$, $K_{sp} = 1.0 \times 10^{-27}$
    (c) $CdCO_3(s)$, $K_{sp} = 1.0 \times 10^{-12}$
    (d) $CdF_2(s)$, $K_{sp} = 6.4 \times 10^{-3}$

14. Given the following K_{sp} values, determine which equilibrium solution has the greatest concentration of Cu^{2+} ion. (13.3)
    (a) $Cu(OH)_2(s)$, $K_{sp} = 4.8 \times 10^{-20}$
    (b) $CuC_2O_4(s)$, $K_{sp} = 4.4 \times 10^{-10}$
    (c) $CuS(s)$, $K_{sp} = 8.0 \times 10^{-37}$
    (d) $Cu(IO_3)_2(s)$, $K_{sp} = 6.9 \times 10^{-8}$

15. Given the following list of K_a values, determine which weak acid solution would have the highest concentration of $[H^+]$ (lowest pH value) at equilibrium assuming all acids have the same initial concentration. (13.3)
    (a) $HC_2H_3O_2(aq)$, $K_a = 1.8 \times 10^{-5}$
    (b) $HCN(aq)$, $K_a = 6.2 \times 10^{-10}$
    (c) $HCHO_2(aq)$, $K_a = 1.8 \times 10^{-4}$
    (d) $HC_3H_5O_2(aq)$, $K_a = 1.3 \times 10^{-5}$

16. Given the following list of K_a values, determine which weak acid solution would have the highest concentration of $[H^+]$ (lowest pH value) at equilibrium assuming all acids have the same initial concentration. (13.3)
    (a) $HF(aq)$, $K_a = 6.3 \times 10^{-4}$
    (b) $HOCl(aq)$, $K_a = 3.5 \times 10^{-8}$
    (c) $HC_7H_5O_2(aq)$, $K_a = 6.4 \times 10^{-5}$
    (d) $HN_3(aq)$, $K_a = 1.9 \times 10^{-5}$

17. Summarize the four types of stress that can be applied to a system in equilibrium, and the response to each, according to Le Chatelier's principle. (13.4)

18. When a chemical reaction is carried out in an industrial setting, it is very common to have one of the reactants present in an excess amount above the stoichiometric quantity that would be needed for the reaction. Why? (13.4)

19. Describe how oxyhemoglobin delivers oxygen gas for use by cells, as it relates to Le Chatelier's principle. (13.4)

20. Describe how hemoglobin is converted to oxyhemoglobin in the lungs, as it relates to Le Chatelier's principle. (13.4)

21. How does carbon monoxide interfere with the mechanism of hemoglobin? (13.4)

22. What is a hyperbaric oxygen chamber, and why is it used to treat individuals with carbon monoxide poisoning? (13.4)

23. Using Le Chatelier's principle, predict what will happen to the concentration (increase, decrease, no change) of each remaining component of the equilibrium system:
$H_2CO_3(aq) \rightleftarrows H^+(aq) + HCO_3^-(aq)$ (13.4)
   (a) Increase H^+ concentration
   (b) Decrease H^+ concentration
   (c) Increase H_2CO_3 concentration
   (d) increase HCO_3^- concentration

24. Using Le Chatelier's principle, predict what will happen to the concentration (increase, decrease, no change) of each remaining component of the equilibrium system:
$HC_2H_3O_2(aq) \rightleftarrows H^+(aq) + C_2H_3O_2^-(aq)$ (13.4)
   (a) Increase $HC_2H_3O_2$ concentration
   (b) Decrease H^+ concentration
   (c) Decrease $C_2H_3O_2^-$ concentration
   (d) Increase $C_2H_3O_2^-$ concentration

25. If a sample of the arsenic-containing mineral realgar is in equilibrium with a solution, what could be done to decrease the amount of dissolved arsenic? (13.5)

26. If a sample of the arsenic-containing mineral realgar is in equilibrium with a solution, what could be done to increase the amount of dissolved arsenic? (13.5)

27. Determine the solubility equilibrium constant (K_{sp}) for lead(II) fluoride given that the equilibrium concentrations of $[Pb^{2+}] = 0.0043$ M and $[F^-] = 0.0022$ M. (13.6)

28. Determine the weak-acid equilibrium constant (K_a) for nitrous acid, HNO_2. The equilibrium concentration of $[H^+] = 0.020$ M and $[NO_2^-] = 0.020$ M. (13.6)

29. Using the K_{sp} value calculated in problem 35, determine the missing value for each set of values below. (13.6)
   (a) $[Pb^{2+}] = 0.0017$ M and $[F^-] = $ _____
   (b) $[Pb^{2+}] = $ _____ M and $[F^-] = 0.0012$ M
   (c) $[Pb^{2+}] = 0.031$ M and $[F^-] = $ _____ M
   (d) $[Pb^{2+}] = $ _____ M and $[F^-] = 0.079$ M

30. Using the K_a for nitrous acid calculated in problem 36, determine the missing value for each set of values below. (13.6)
   (a) $[H^+] = 0.013$ M and $[NO_2^-] = $ _____ M
   (b) $[H^+] = $ _____ M and $[NO_2^-] = 0.0750$ M
   (c) $[H^+] = 0.059$ M and $[NO_2^-] = $ _____ M
   (d) $[H^+] = $ _____ M and $[NO_2^-] = 0.0081$ M

31. What is the difference between a toxin and a poison? (13.7)

32. Are all poisons toxins? Are all toxins poisons? (13.7)

33. Explain the quote by Paracelsus on the nature of poisons. (13.7)

34. All substances, even water, can be poisonous. But what property is shared by most substances that we consider to be "poisons"? (13.7)

35. How is the relative toxicity of a compound measured? (13.7)

36. What is an acute poisoning? (13.7)

37. What is a chronic poisoning? (13.7)

38. What health risk does each of the following classifications represent? (13.7)
   (a) Corrosive   (c) Teratogen
   (b) Carcinogen   (d) Mutagen

Case Study Problems

39. Using the Web site www.nlm.nih.gov/medlineplus/encyclopedia.html as a resource, investigate the medical condition known as *acidosis*. What are some of the possible causes for acidosis? How does this condition affect the ability of the body to transport oxygen gas? (CP)

40. Using the Web site www.nlm.nih.gov/medlineplus/encyclopedia.html as a resource, investigate the medical condition known as *alkalosis*. What are some of the possible causes for alkalosis? How does this condition affect the ability of the body to transport oxygen gas? (CP)

41. Explain why dimercaprol is an effective antidote for mercury poisoning. (13.3)

42. Dimercaptosuccinic acid (DMSA) is used to treat lead poisoning. Draw the structure of DMSA, given that it is an organic compound with a 4-C chain, two —SH groups, and two carboxylic acid groups. (CP)

43. When you exhale into a breathalyzer, the air from your lungs provides an accurate measure of the alcohol in your blood. Explain why, based on equilibrium principles. (CP)

44. Table 13.1 lists the LD_{50} for several compounds, as determined by using oral doses given to rats. Assuming that the toxicity for these compounds is the same in humans (average mass 70 kg), calculate the amounts of sodium chloride and potassium cyanide that a human would have to ingest in a single dose to reach the LD_{50}. (13.7)

45. Section 13.5 discussed the contamination of water supplies with arsenic from arsenic-containing minerals. If a person lived in the area and consumed this water, is it an example of an acute poisoning or chronic poisoning? Would the danger level be the same for an infant drinking formula prepared using this water as for an adult drinking the water? (13.5)

46. When a firefighter discovers a body at the scene of a fire, a complete investigation into the cause and manner of death of the victim will follow. Recall from Section 9.2 that carbon monoxide is often produced as a result of incomplete combustion. What information would analysis of the hemoglobin for carbon monoxide provide investigators? Could it be the sole piece of evidence to determine if the death was suicide, homicide, or accidental? Explain the limits of the information that could be provided from the lab. (CS)

47. Cyanide functions by binding very strongly with the Fe^{3+} ion in hemoglobin, which prevents the cells from utilizing the oxygen and results in suffocation. What antidotes are there for cyanide poisoning and what do they do chemically to counteract the effect of cyanide? Is oxygen present in the blood despite the suffocation? Consult the *Antidotes & Other Treatments and Laboratory Tests* sections at www.atsdr.cdc.gov/mmg/mmg. asp?id=1073&tid=19 for further information in constructing your answer. (CP)

48. During exercise, carbon dioxide is generated at a greater rate than it can be removed from your body, resulting in elevated levels of carbon dioxide. Given the carbonate-based equilibrium below, what will happen to the blood pH? (CS)

$$CO_2 + H_2O \rightleftarrows H_2CO_3 \rightleftarrows HCO_3^- + H^+$$

49. Blood will typically contain calcium ion concentration of 0.00115 M. Assuming this is the same concentration of calcium ions present in the kidneys, calculate the concentration of oxalate ions that would be in equilibrium with a calcium oxalate kidney stone. (CP)

$$CaC_2O_4(s) \rightleftarrows Ca^{2+} + C_2O_4^{2-} \qquad K_{sp} = 2 \times 10^{-9}$$

CHAPTER 14

Introduction to Biochemistry

CASE STUDY: Exploring Genetically Modified Food

The first traces of agriculture appeared with the end of the last ice age approximately 10,000 years ago. Archeologists believe that in most parts of the world, a gradual transition occurred from hunter-gatherer societies to the more permanent settlements of agricultural society. The transition is quite logical: Whenever a surplus of seeds was gathered, the surplus seeds could be planted, providing a larger yield to be gathered the next season.

The planting of wild seeds marked the earliest form of agriculture, which would later give rise to domesticated grain crops. As a surplus of wild grains developed, it enabled the selection of the best grains for planting in the following season. These seeds would come from the larger, healthy, more rugged plants that produced larger seeds and fruits. Humans had started to influence which genetic traits would be passed on and created new, *domesticated* varieties that could not have developed without human selection. The ability of human beings to domesticate crops such as corn, wheat, rye, barley, and oats as well as animals like cows, goats, sheep, and horses created a food supply that could sustain larger numbers and that coincided with a population boom across the world.

The population of the earth was 3.5 billion in 1968, and in 2011 it topped 7 billion and continues to grow at an approximate rate of 160 people per minute. The unprecedented number of people on earth has created a demand for food that has strained agricultural resources in some regions of the world, even as other areas have abundant resources.

Modern agriculture met the demand for increased food production in the 1900s by using fertilizers, insecticides, and pesticides to optimize growing conditions for crops. A new chapter in agriculture started in the 1990s with the first plants that had been optimized in a laboratory. Scientists were able to locate specific genes that produced a desirable trait on the DNA of one plant and then inserted those genes into a separate plant,

As you read through the chapter, consider the following questions:

• **How are proteins assembled by cells?**

• **What is a gene?**

• **How does DNA function as a blueprint for proteins?**

creating a *genetically modified (GM)* plant. There is a great deal of concern among environmental groups about the safety of genetically modified plants and the unintended consequences that they may cause.

MAKE THE CONNECTION

To understand the fundamental issues related to genetically modified plants or animals, it is best to first understand the science behind the ability to create such organisms.

14.1 Lipids: Fats, Waxes, and Oils

In Section 5.7, the properties of fats and oils were first discussed, and the health risks of trans fats were explained. Fats and oils are two types of molecules that, along with wax compounds, are collectively referred to as **lipids.** Lipids are molecules that are insoluble in water but will dissolve in a nonpolar solvent such as hexane. This section will explore the chemical differences between the fats, oils, and wax molecules.

You are already familiar with the physical differences between waxes (hard solids), fats (soft semisolids), and oils (liquids). But what are the *chemical* differences between these subclasses of lipid molecules? **Waxes** are made from a **fatty acid**, which is a long-chain carboxylic acid molecule, and a long-chain alcohol, as shown in Figure 14.1. Waxes are hard solids at room temperature because the large molecules have strong dispersion forces between them. Recall that a high degree of intermolecular forces is reflected in high melting and boiling points.

Fats and oils are examples of **triglycerides**, compounds that are made by combining a glycerol molecule with three fatty acid molecules, as shown in Figure 14.2. The distinction between fats and oils

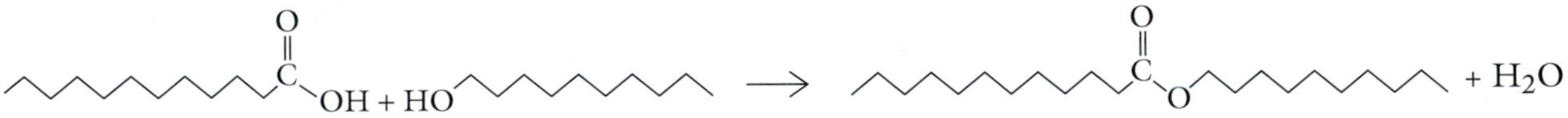

Three examples of lipids shown above are wax candles, olive oil and butter. (Sally Johll)

FIGURE 14.1 The structure of wax. A wax is an ester compound formed when a fatty acid reacts with an alcohol.

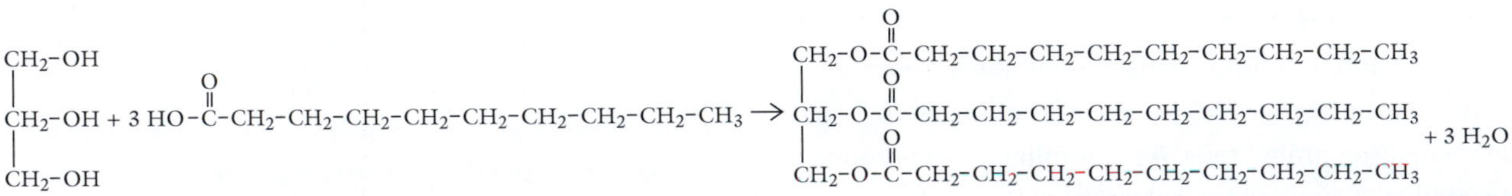

$$CH_2-OH$$
$$CH_2-OH + 3\ HO-\overset{\overset{\displaystyle O}{\|}}{C}-CH_2-CH_2-CH_2-CH_2-CH_2-CH_2-CH_2-CH_2-CH_3 \rightarrow$$
$$CH_2-OH$$

Glycerol   Fatty acid

$$CH_2-O-\overset{\overset{\displaystyle O}{\|}}{C}-CH_2-CH_2-CH_2-CH_2-CH_2-CH_2-CH_2-CH_2-CH_3$$
$$CH_2-O-\overset{\overset{\displaystyle O}{\|}}{C}-CH_2-CH_2-CH_2-CH_2-CH_2-CH_2-CH_2-CH_2-CH_3$$
$$CH_2-O-\overset{\overset{\displaystyle O}{\|}}{C}-CH_2-CH_2-CH_2-CH_2-CH_2-CH_2-CH_2-CH_2-CH_3$$
$$+ 3\ H_2O$$

Triglyceride   Water

FIGURE 14.2 Formation of triglycerides.

lies in the nature of the fatty acid molecules composing the triglycerides. **Fats**, which are semisolids at room temperature, have fatty acids whose carbon chains contain only single bonds between carbon atoms. **Oils**, which are liquids at room temperature, have fatty acids whose carbon chains contain one or more carbon–carbon double bonds. Nutritional labels classify fats and oils together as fats, however they do distinguish between **saturated** (all single bonded carbons) and **unsaturated** (contains one or more double bonds between carbon atoms).

Why does the presence or absence of double bonds have an effect on whether a compound is solid or liquid at room temperature? Consider Figure 14.3, which shows the structure of several fatty acids. When the fatty acid constituents consist of only single bonds, as in stearic acid, the molecule is linear. Oleic acid, on the other hand, has a double bond that puts a kink in the chain of atoms. The linoleic acid molecule with two double bonds has a bent shape with a 90° corner in the molecule. Each of the molecules in Figure 14.3 contains the same number of carbon atoms.

How does the shape of the fatty acid molecule, caused by the presence or absence of C=C atoms, influence the physical properties of the compounds? Table 14.1 lists the chain length, the number of C=C groups in the molecule, and the melting points of the compounds. The first three fatty acids listed in the table all have 18 carbon atoms in the chain. Yet the presence of a single C=C lowers the melting point from 69°C to 13°C, and the presence of two double bonds further lowers the melting point to −5°C. Stearic acid, with its linear shape, has the strongest intermolecular forces; linoleic acid, with a significantly bent shape, has the weakest intermolecular forces. The linear hydrocarbon portions of stearic acid molecules are better able to align so that dispersion forces have greater effect. The substantial bend in the linoleic acid molecule decreases the contact area between molecules and, therefore, decreases the melting point.

One interesting use of wax and oils is in the production of lipstick. Chemists are heavily involved in the development of cosmetics, and their efforts focus largely on the physical properties and intermolecular forces of the compounds that are to be mixed together for the desired effect. A quick examination of the ingredient list in the photo of lipstick shows a large number of wax and oil compounds. Lipstick is a homogeneous mixture of many compounds, each of which has a role in creating the desired physical properties of the lipstick. The wax compounds provide the rigid structural support for lipstick and are insoluble in water, so the lipstick doesn't dissolve. Yet wax is too

Stearic acid,
all C—C

Oleic acid,
one C=C

Linoleic acid,
two C=C

FIGURE 14.3 Effect of single and double bonds on the molecular shape of a fatty acid. Each molecule has the same number of carbon atoms.

Common ingredients in lipstick include: coconut oil, sunflower oil, castor oil, jojoba oil, palmarosa oil, geranium oil, olive oil, orange oil, lanolin oil, soybean oil, synthetic wax, ceresin wax, carnauba wax, paraffin wax, palm wax, beeswax, and candelilla wax. (iStockphoto/Thinkstock)

TABLE 14.1	Properties of Saturated and Unsaturated Fatty Acids		
Name	Length of Carbon Chain	Number of (C=C Bonds)	Melting Point (°C)
Stearic acid	18	0	69
Oleic acid	18	1	13
Linoleic acid	18	2	−5
Arachidic acid	20	0	77
Arachidonic acid	20	4	−49

rigid and would not spread easily or evenly across lips—imagine using a candle as lipstick. Therefore, lipstick has to be softened by adding oils. The oils also help disperse the colored pigments, fragrances, and flavoring compounds throughout the lipstick. The pigments often are solid particles that are finely ground and dispersed in the oils, while the fragrances and flavoring compounds tend to dissolve in the oils. A high-gloss finish tends to have more oils to produce a shiny appearance.

■ WORKED EXAMPLE 1

Lip balm is used to protect lips from chapping and dryness, especially during the winter months. How do the physical properties of waxes and oils in lip balm help prevent lips from drying out when exposed to dry air?

SOLUTION

Both waxes and oils are nonpolar compounds. A lip balm creates a thin barrier layer of the nonpolar substance across the lips. Water is a polar substance, so it will not penetrate the barrier layer.

Practice 14.1

Margarine is produced from vegetable oil in a process called *hydrogenation*. Based on the different properties of fats and oils, what must happen to the vegetable oil during hydrogenation?

Answer

Vegetable oil is a liquid due to the double bonds in the fatty acid molecules. For oil to become a semisolid, the C=C bonds must be converted to C—C bonds through hydrogenation.

Nutrition Facts

Serving Size 1 Bar (85g)
Servings Per Container 4

Amount Per Serving	
Calories 170	Calories from Fat 50

	% Daily Value *
Total Fat 6g	**9%**
Saturated Fat 4g	**19%**
Trans Fat 0g	
Polyunsaturated Fat 0.5g	
Monounsaturated Fat 1g	
Cholesterol 13mg	**4%**
Sodium 83mg	**3%**
Total Carbohydrate 33g	**11%**
Dietary Fiber 4g	**16%**
Sugar 25g	
Protein 3g	

Vitamin A 110%	•	Vitamin C 2%
Calcium 10%	•	Iron 3%

*Percent Daily Values are based on a 2,000 calorie diet. Your daily values may be higher or lower depending on your calorie needs.

		Calories	2,000	2,500
Total Fat	Less than		65g	80g
Sat Fat	Less than		20g	25g
Cholesterol	Less than		300mg	300mg
Sodium	Less than		2,400mg	2,400mg
Total Carbohydrate			300g	375g
Dietary Fiber			25g	30g

Calories per gram:
Fat 9   •   Carbohydrate 4   •   Protein 4

(Hemera/Thinkstock)

14.2 Carbohydrates

> **LEARNING OBJECTIVE**
>
> **Describe the structure of carbohydrates.**

When you explore the Nutrition Facts label on food items, you will encounter the Total Carbohydrate section, which has two subcategories: Sugars and Fiber. The term **sugar** is commonly used to refer to sucrose, also known as *table sugar,* but actually represents any sweet-tasting carbohydrate compound. *Fiber,* also referred to as **dietary fiber,** is any carbohydrate that resists digestion.

Carbohydrates, also known as *saccharides,* are literally the hydrates of carbon and have a carbon–hydrogen–oxygen ratio of 1:2:1. A **monosaccharide** is a small carbohydrate molecule, also referred to as a *simple sugar*. Examples include glucose, fructose, galactose, ribose, and deoxyribose. (Deoxyribose will be discussed further in Section 14.4.) Two monosaccharides can bond together to form a **disaccharide,** or double sugar. For instance, lactose (milk sugar) is made up of galactose bonded to glucose, sucrose (table sugar) is glucose bonded to fructose, and maltose (malt sugar) is two glucose molecules bonded together. Monosaccharide and disaccharide compounds are classified as sugars because they have a sweet taste.

The term **oligosaccharide** is sometimes used to describe a carbohydrate containing a small number of monosaccharides—between three and ten. (The prefix *oligo-* means "a few.") Oligosaccharides played an important role in forensic science before DNA analysis because of their significance in the ABO blood typing system. For type A and type B blood (but not type O blood), red cells have different oligosaccharide units on the outer cell wall that enable the cell to differentiate between foreign and non-foreign material. Blood typing was based on the reactions of these oligosaccharides and was the primary way investigators attempted to match to a specific person the blood found at a crime scene. DNA analysis of blood, though, can now provide much more specific identification.

Carbohydrates can exist as very large molecules made by linking together many individual sugar molecules. The term **polysaccharide** is used to characterize large carbohydrates that can range in size from hundreds to thousands of individual sugar units. In this sense, polysaccharides are a type of polymer.

We usually think of plastics when we hear the term **polymer**—a long-chain molecule made by linking together smaller molecules called **monomers**. A carbohydrate is a polymer made up of sugar monomers. The sugar glucose ($C_6H_{12}O_6$), shown in Figure 14.4, is the basic monomer found in cotton fibers. The glucose monomers link up to form **cellulose**, a polysaccharide of glucose that is used in plants to make up the structural material of the cell walls. Cotton fibers are 92% cellulose.

Cellulose is also a component of foods derived from plants. Because humans are unable to digest cellulose, it is often referred to as dietary fiber. Dietary fiber supplements containing cellulose usually state on the label that sufficient amounts of water should be consumed when taking them. The reason is that cellulose is capable of absorbing a substantial quantity of water through hydrogen bonding. Consider the number of hydroxyl groups (—OH) on a glucose unit in Figure 14.4, all of which are capable of forming hydrogen bonds with water. The cellulose molecule consists of hundreds of glucose units, so the amount of water that can be associated with cellulose is considerable.

Glucose, a monosaccharide.

Glucose	Galactose	Fructose	Ribose	Deoxyribose

FIGURE 14.4 Chemical structures of several common monosaccharides.

Green plants produce another polysaccharide of glucose called **starch**, a substance that serves as a source of energy for the body. Potatoes and rice are common sources, although all plants that undergo photosynthesis produce starch. Enzymes in the human digestive system are able to cleave off the individual glucose monomers for use in metabolism.

What is it about starch that makes it digestible by humans whereas cellulose is not? Both are polymers of glucose, but the key to understanding the difference in their digestibility lies in how the glucose units are bonded together. Figure 14.5 shows the bonding of glucose units in cellulose and starch. Notice how the glucose units in cellulose link in a way that results in a linear structure for the overall molecule. The linkages between glucose molecules in starch are oriented such that an angular molecule results.

The enzymes in the body are designed to cleave starch molecules apart, but the shape of the enzymes will not allow cellulose to be cleaved. Some animals such as cattle contain cellulose-digesting bacteria in their digestive tracts and are thus able to use cellulose as an energy source.

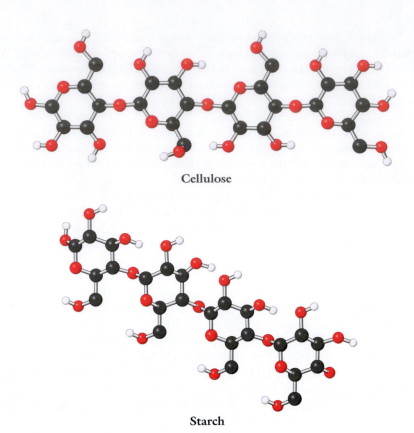

Cellulose

Starch

FIGURE 14.5 Cellulose and starch polysaccharides.

14.3 Proteins

Proteins are a group of biological compounds that are a required component of nearly every aspect of life, such as the digestion of food; transportation of oxygen, food, and waste products; cell wall structure; reproduction; immune system; creation of energy; and the storage of energy. Each process requires a unique protein, which can be created by the cells as the body requires a particular protein.

A **protein** is a long polymer chain of amino acids that typically contains several hundred to several thousand amino-acid monomer units. The largest protein contains over 34,000 amino acids. Estimates of the number of different proteins in the human body range from 24,000 to more than 100,000. The exact number is still in question, although data from the Human Genome Project suggest it is near the lower value. The key to the function of a protein is its shape, which is dictated by the sequence of amino acids in the polymer chain. The blueprint for each protein is stored as a *gene* on DNA.

The **amino acids** are a group of 20 naturally occurring compounds that are similar in structure. Each molecule contains an amino functional group ($-NH_2$) and a carboxylic acid functional group ($-COOH$), separated by a carbon atom. An organic group ($-R$) bonded to the middle carbon atom is what distinguishes one amino acid from another. Figure 14.6 shows the 20 naturally occurring amino acids and the standard abbreviations used when writing out the constituents of protein molecules. The top structure in Figure 14.6 is the generic amino acid backbone with the purple-colored sphere depicting the $-R$ group. For the specific amino acid structure, replace the purple sphere on the amino acid backbone with the group listed beneath each name. The human body is capable of

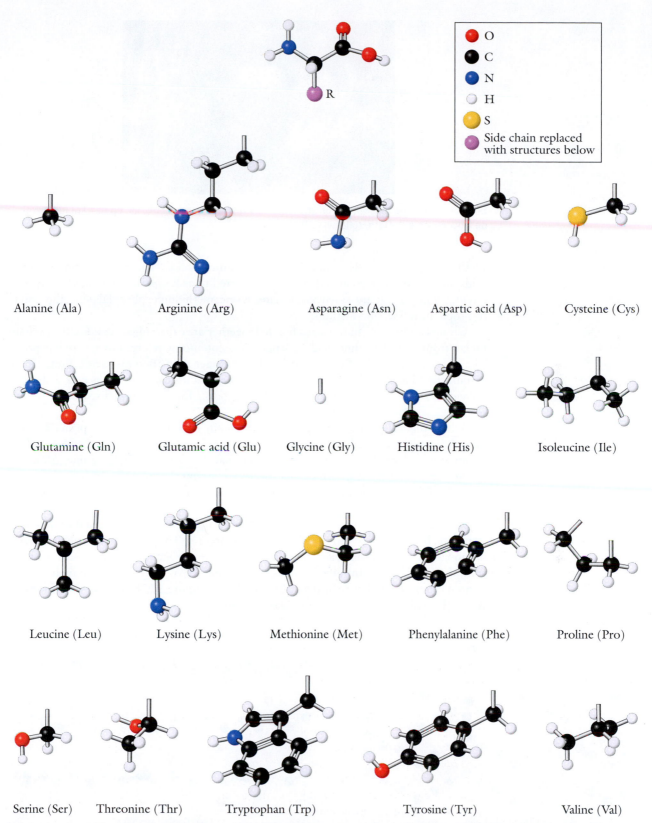

FIGURE 14.6 The 20 amino acids. The top structure is the backbone, with the purple sphere depicting the —R group. Replace the purple sphere on the amino acid backbone with the group listed beneath each name for the structure of each amino acid. It should be noted that the proline side chain actually bonds back to the nitrogen atom of the backbone.

Silkworm cocoons that will be unraveled. The individual strands of silk, up to 1 km long, will be woven into silk cloth. (blickwinkel/Alamy)

synthesizing 11 of the 20 naturally occurring amino acids. The remaining nine amino acids must be present in one's diet and are referred to as *essential amino acids*. The nine essential amino acids are isoleucine, leucine, lysine, methionine, phenylalanine, threonine, tryptophan, valine, and histidine.

Groups of 10 to 20 amino acids bonded together are called **oligopeptides**. If the chain ranges from 20 to 50 amino acids in length, it is commonly referred to as a **polypeptide**. The term *protein* is usually reserved for chains longer than 50 amino acids. Proteins are manufactured in the cells as needed, and the blueprint for each protein is encoded on an individual's DNA. The mechanism of DNA encoding for proteins is discussed in a later section.

The sequence of amino acids within a protein gives the protein its **primary structure**. For a silk protein fiber, the primary structure is [Gly-Ala-Gly-Ala-Gly-Ser]$_n$. This notation includes abbreviations for three amino acids—glycine (Gly), alanine (Ala), and serine (Ser)—and shows how they link together in a unit that repeats *n* times. Strands of silk, which can reach over a kilometer in length when spun from the silkworm, are composed of this protein.

Beyond the primary structure, proteins assume different shapes, called the **secondary structure**, based on how the amino acid chains interact. The secondary structure of silk is known as a **pleated sheet**, shown in Figure 14.7a. (Hydrogen atoms have been removed from the structure to simplify the image.) The distinctive shape is created by hydrogen bonding that occurs between two adjacent chains. Figure 14.7b is a top-down view in which the hydrogen bonding (green dashes) can be seen.

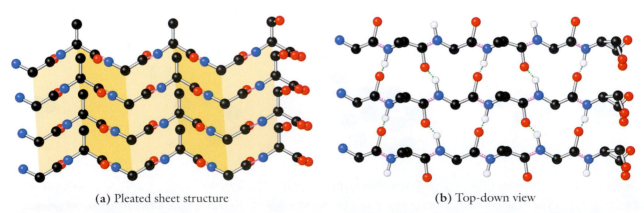

(a) Pleated sheet structure   (b) Top-down view

FIGURE 14.7 (a) The pleated sheet secondary structure of silk. Shading has been added to help illustrate the pattern. (b) The individual strands of protein are held in place by hydrogen bonding (green dashes).

Animal hair consists of a protein called *keratin* that does not have a fixed amino acid sequence. The primary structure is simply the order of amino acids that happened to be used in the creation of the molecule. The secondary structure of keratin is quite different from that of silk. Keratin assumes an **alpha helix** form, much like a spring. The coils of the spring are held in place by hydrogen bonds that form between every fourth amino acid. Figure 14.8a illustrates an alpha helix structure. The ribbon has been added to aid in visualization. Figure 14.8b provides a close-up view of the hydrogen bonding (green dashes) present in the alpha helix structure.

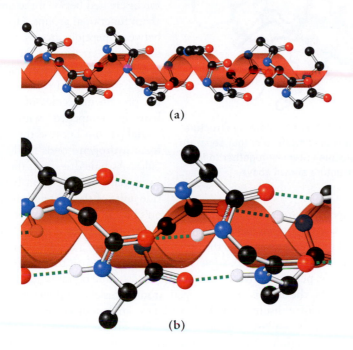

(a)

(b)

FIGURE 14.8 The structure of hair. (a) An alpha helix structure. The ribbon has been added to highlight the structure. (b) A close-up view of the hydrogen bonding (green dashes) in the alpha helix structure.

One example of an important protein with an alpha helix structure is hemoglobin, the protein in blood that transports oxygen from the lungs to the rest of the body. However, proteins such as hemoglobin are much more complex than a simple alpha helix. The helix bends and twists upon itself, creating a **tertiary structure** that forms a larger three-dimensional shape. Figure 14.9 illustrates the protein structure present in hemoglobin.

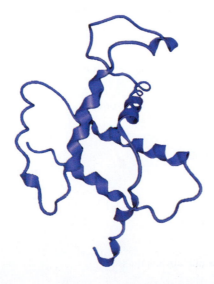

FIGURE 14.9 A single protein molecule from hemoglobin folds back upon itself in a distinct three-dimensional pattern that comprises its tertiary structure.

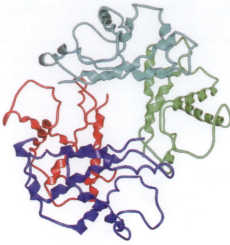

FIGURE 14.10 The quaternary structure of hemoglobin is comprised of four separate protein molecules that link together to form the superstructure shown above.

Protein molecules can also join with one or more different protein molecules to form what is known as a **quaternary structure**. The full structure of hemoglobin consists of four separate protein molecules, as illustrated in Figure 14.10.

Protein molecules maintain their three-dimensional shape as a result of attractive forces created between the amino acid side-chain functional groups. Hydrogen bonding between peptide groups is the dominant force in the creation of the secondary structure of protein molecules. For example, curly hair is due to hydrogen bonding within the protein molecules of hair. When curly hair is dampened with water, it will straighten out because the water molecules form hydrogen bonds to the protein molecules. As hair dries out, the curls return as hydrogen bonds re-form between the peptide units of the protein molecule itself.

Curls in hair come from a combination of hydrogen bonding and disulfide bonds that form along the protein fibers. (M. Thompsen/zefa/Corbis)

Also responsible for the curliness of a person's hair is a second type of bond due to the amino acid cysteine, which has a terminal —SH group. Cysteine in the hair can bond to another —SH group farther down the protein chain to form a **disulfide bond**, —S—S—. If a person has straight hair, the existing disulfide bonds are such that the protein molecules can lie flat. If the person receives a permanent, the disulfide bonds are first reduced to form the —SH bonds, the hair is then wrapped around curlers, and the —SH is ultimately oxidized back to form disulfide bonds that are now angled, producing a curl in the hair.

Attractions between large nonpolar functional groups on the amino acids are called **hydrophobic interactions**. Nonpolar functional groups tend to form in the tertiary and quaternary structures of a molecule. Ionic forces also influence the structure of proteins through an **ionic bridge** that forms between an amine functional group and a carboxylic acid functional group. At normal physiological pH, the amine group gains a hydrogen ion to form —NH_3^+, and the carboxylic acid group loses a hydrogen ion to form —COO^-. These interactions are illustrated in Figure 14.11.

$$-(CH_2)_4 \overset{+}{N}H_3 ---- \overset{-}{O}OC-CH_2CH_2-$$

Ionic bridge

CH_2

Hydrophobic interaction

CH_3

CH_2

OH

Hydrogen bond

O

$CH_2CH_2-\overset{O}{\overset{\|}{C}}-NH_3$

FIGURE 14.11 Protein molecule linkages.

14.4 DNA Basics

DNA is the abbreviation for **deoxyribonucleic acid.** It is the molecule responsible for transmitting all hereditary information from one generation to the next. As our knowledge of DNA has grown over the last several decades, applications such as forensic identification of biological samples and the ability to genetically modify living organisms have become commonplace. To better understand how these applications work, it is necessary to first explore DNA further.

DNA has a double-stranded alpha helix shape in which each strand of DNA is a polymer made up of subunits, or building blocks, called **nucleotides.** A nucleotide consists of a phosphoric acid molecule bonded to a deoxyribose sugar and to one of four possible nitrogen-containing compounds: cytosine, thymine, adenine, or guanine. The six substances that combine to create nucleotide building blocks of DNA are shown in Figure 14.12.

Phosphoric acid and deoxyribose sugar form the backbone of the DNA polymer, with the nitrogen base group extending across toward the nitrogen base on the complementary strand of DNA. The DNA molecule is held together in this configuration by hydrogen bonding between the nitrogen bases, represented by the green dots in Figure 14.12.

The interaction between nitrogen bases is quite specific: Cytosine always bonds to guanine (C-G), and adenine always bonds to thymine (A-T), to form the classic double helix structure pictured at the start of this section. The combination of C-G or A-T is referred to as a **base pair** and is often used to refer to the length of a DNA molecule or portion of DNA. If the entire DNA material were extracted from one cell and extended full length, it would be one meter long and would consist of three billion base pairs.

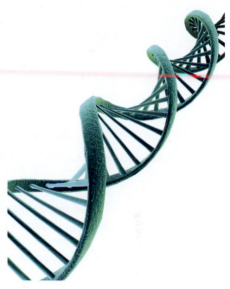

DNA forms a double helix shape, held in place by hydrogen bonding.

www DNA on a Stirring Rod

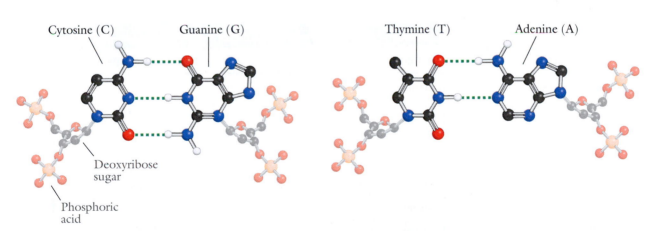

FIGURE 14.12 DNA components. The DNA molecule is a polymer in which phosphoric acid and deoxyribose sugar molecules make up the backbone of the DNA strand. Each DNA strand is linked to a second one by center structures consisting of the nitrogen-containing compounds guanine (G), cytosine (C), thymine (T), and adenine (A). Due to their shapes and the hydrogen bonding that occurs between the side branches, C always binds to G with three hydrogen bonds, and A always binds to T with two hydrogen bonds.

WORKED EXAMPLE 2

What is the complementary strand of DNA for the strand given below?

A A C A G G C T T

SOLUTION

The nucleotide adenine always associates opposite to thymine (A-T or T-A), and cytosine to guanine (C-G or G-C). Therefore, the second complementary strand of the DNA is

T T G T C C G A A
A A C A G G C T T

Practice 14.2

What is the complementary strand of DNA for the strand given below?

T C A C C A G C T

Answer

A G T G G T C G A

DNA provides hereditary information by supplying the blueprints for the creation of the estimated 100,000 different protein structures found within the human body. The blueprint for each protein is referred to as a **gene** and provides instructions for the order in which the amino acids bind together in the protein. The order of each amino acid in the protein is stored by a series of three nucleotides called a **codon**. If a protein molecule has 100 amino acids, the gene would be $100 \times 3 = 300$ nucleotides long.

In the process by which a protein is created in a cell, the DNA strands untwist near the gene, and a copy of the blueprint is made by creating the complementary strand to the gene. The blueprint is then sent to the ribosome, which synthesizes protein molecules. The job of the ribosome is to assemble the individual amino acids into a protein in the order specified by the gene. If the nucleotide sequence AAA is detected at the ribosome, it is the specific code for the amino acid lysine. If the next amino acid in the protein is glutamate, the nucleotide pattern is GAA. Table 14.2 is a partial list of the nucleotide sequences used by the ribosome to assemble amino acids into proteins.

TABLE 14.2	Nucleotide Code for Selected Amino Acids
Amino Acid	**Nucleotide Sequence**
Alanine (Ala)	GCC
Arginine (Arg)	CGC
Asparagine (Asn)	AAC
Aspartate (Asp)	GAC
Glutamine (Gln)	CAA
Glutamate (Glu)	GAA
Glycine (Gly)	GGC
Histidine (His)	CAC
Lysine (Lys)	AAA
Proline (Pro)	CCC
Threonine (Thr)	ACC

WORKED EXAMPLE 3

What is the amino acid sequence indicated by the sequence of codons below?

C A A A A G G C

SOLUTION

The first codon is CAA, the unique pattern for the amino acid Gln, given in Table 14.2. Similarly, the second codon is AAA, which codes for Lys. The final codon, GGC, codes for Gly.

Practice 14.3

What is the nucleotide sequence for the section of a protein consisting of Pro-Gln-Asn-Asp?

Answer

CCCCAAAACGAC

14.5 DNA Analysis

DNA analysis is most closely linked with forensic science, as in real cases such as the O. J. Simpson and Casey Anthony trials that captivated the headlines. The use of DNA analysis has also freed hundreds of prisoners convicted before the common use of DNA analysis, by proving their innocence through later analysis of the evidence. Many of these same prisoners were awaiting the death penalty.

> **LEARNING OBJECTIVE**
>
> Describe the techniques used to take a sample of DNA and analyze its genetic information.

The forensic value of DNA lies in the fact that each individual has DNA unlike that of any other person, and samples of DNA can be obtained from common types of evidence such as hair, skin, sweat, blood, and saliva. DNA analysis is a relatively new technique, used for the first time in a criminal case only in 1986. Since that time, the technology has improved, and it is now routine to detect one-billionth of a gram of DNA.

DNA is packaged in a unit called a **chromosome**, found inside body cells. Humans have 23 pairs of chromosomes, totaling 46 chromosomes. The dominant means of DNA analysis today is a method called **short tandem repeat (STR) analysis**. This method makes use of the fact that 95% of the DNA in a chromosome is noncoding information—long stretches of DNA that don't code for any particular gene. Within these unused portions of DNA, patterns of nucleotides that repeat one after another can be identified.

Humans have 23 pairs of chromosomes stored in each cell. One chromosome in each pair is inherited from each parent. The chromosome structure consists of protein and DNA molecules. (Scott Camazine & Sue Trainor/Photo Researchers, Inc.)

The short tandem repeat (STR) sections of DNA appear as dark lines during analysis. Each line represents an STR of a different length. (Mauro Fermariello/Photo Researchers, Inc.)

For example, on chromosome number 2, a section of noncoding DNA called TPOX has the short tandem repeat sequence [AATG], which occurs anywhere from eight to twelve times. Each individual has two copies of chromosome number 2, one from each parent. By studying a large sample of the population, it has been determined that 28.5% of humans have eight repeats of [AATG] on each chromosome, but only 0.24% of humans have 10 repeats of [AATG] on each chromosome. The population frequency for each possible repeat combination of [AATG] for the TPOX location of chromosome number 2 is provided in Table 14.3.

Obviously, if an individual who committed a crime had left DNA at the crime scene, having the 12,12 (0.24%) repeat sequence would limit the possible number of suspects with matching DNA better than the 8,8 (28.5%) repeat sequence. The FBI has chosen a standard set of 13 different regions, or *loci*, of noncoding DNA in which to analyze DNA. After comparing the statistics of a person's genetic makeup at all 13 loci, the odds of two people sharing the same pattern usually becomes one in a billion or more. It is important to understand that even if two individuals are found to match at all 13 loci, both individuals still have unique DNA but just happen to share the same exact pattern of repeats at each of the 13 loci. Table 14.3 also shows data for the D16S539 region of chromosome number 16. At this location, the sequence [GATA] can repeat itself eight to 15 times.

To find the probability of having multiple patterns of repeats, the percentage of each pattern is multiplied together. The population size required to find an individual with a matching pattern is determined by taking the inverse of the probability.

TABLE 14.3	STR Probabilities for Two of the 13 FBI CODIS Loci						
Chromosome #2 TPOX STR	$[AATC]_n$ Repeats	Probability of Matching (%)	Chromosome #16 D16S539 STR	$[GATA]_n$ Repeats	Probability of Matching (%)	$[GATA]_n$ Repeats	Probability of Matching (%)
	8,8	28.5		8,8	0.023	10,13	1.94
	8,9	12.7		8,9	0.38	10,14	0.29
	8,10	4.91		8,10	0.18	10,15	0.037
	8,11	2.70		8,11	0.93	11,11	9.67
	8,12	5.23		8,12	0.90	11,12	18.6
	9,9	1.41		8,13	0.48	11,13	9.89
	9,10	1.09		8,14	0.072	11,14	1.49
	9,11	6.02		8,15	0.009	11,15	0.19
	9,12	1.17		9,9	1.64	12,12	8.94
	10,10	0.212		9,10	1.56	12,13	9.51
	10,11	2.33		9,11	7.96	12,14	1.44
	10,12	0.451		9,12	7.65	12,15	0.180
	11,11	6.40		9,13	4.07	13,13	2.53
	11,12	2.48		9,14	0.61	13,14	7.63
	12,12	0.24		9,15	0.077	13,15	0.095
				10,10	0.37	14,14	0.058
				10,11	3.79	14,15	0.014
				10,12	3.65	15,15	0.0009

The FBI has chosen 13 STR locations on the various chromosomes for use in STR analysis. While 28.5% of people have the TPOX 8,8 pattern, only one out of 15,500 people will have the TPOX 8,8 pattern *and* the D16S539 8,8 pattern. The chance that two individuals will have a matching pattern of STRs at all 13 locations is about one in a billion.

WORKED EXAMPLE 4

What percent of the population has the 9,11 pattern on TPOX and the 10,13 pattern on D16S539?

SOLUTION

From Table 14.3, we see that:

1. The TPOX 9,11 pattern is present in 6.02% of the population.
2. The D16S539 10,13 pattern is present in 1.94% of the population.

The percent of the general population that has both patterns is

$$0.0602 \times 0.0194 = 0.00117 \text{ or } 0.117\%$$

To find out how many people are needed to find one person with this pattern, take the inverse of 0.00117:

$$\frac{1}{0.00117} = 855$$

Practice 14.4

How large a sample of the human population would be required to find one individual with the TPOX pattern 11,11 and the D16S539 pattern 11,12?

Answer

84

14.6 Mitochondrial DNA

When you think of DNA, it is usually the chromosomal DNA inherited from both parents, which is found in the nucleus of the cell. However, a second form of DNA, called *mitochondrial DNA (mtDNA)*, is found within the body and is inherited only from the mother. Mitochondrial DNA is found within the mitochondrion, a cell structure that serves as the source of power for the cell. Mitochondrial DNA is inherited only

> **LEARNING OBJECTIVE**
>
> Describe how mitochondrial DNA can be used to identify samples.

from a person's mother because sperm cells do not have mitochondria, whereas egg cells do. The structure of a cell and the location of the DNA are illustrated in Figure 14.13.

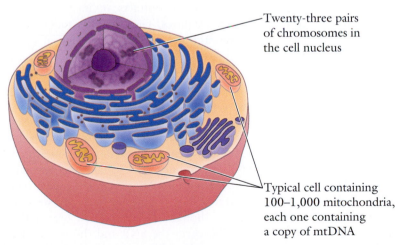

Twenty-three pairs of chromosomes in the cell nucleus

Typical cell containing 100–1,000 mitochondria, each one containing a copy of mtDNA

FIGURE 14.13 Cell structure and DNA. Each cell contains two sets of chromosomal DNA found as the 23 chromosomes located in the nucleus of the cell. The mitochondrial DNA is found within the mitochondrion. There can be several hundred to over one thousand mitochondria per cell.

The differences between mitochondrial DNA and nuclear DNA are significant, one being that mitochondrial DNA has a total of approximately 16,569 base pairs compared with the three billion base pairs found in nuclear DNA. This difference in size is related to the number of genes for which each molecule codes. Nuclear DNA maintains the code for some 100,000 proteins whereas the mtDNA codes for less than one thousand proteins. There are only two copies of nuclear DNA corresponding to the two chromosome pairs stored in the nucleus of the cell. The number of mtDNA copies can range from the hundreds to thousands for a single cell and are located outside the nucleus of the cell. Another difference is that mtDNA comes only from the mother's side of the family, whereas DNA in the nucleus is contained in 23 chromosomes from the maternal side and 23 chromosomes from the paternal side.

The sheer number of mtDNA copies in each cell makes recovery of a usable sample much easier than does the relatively scarce amount of nuclear DNA. The mtDNA does not decompose as easily as nuclear DNA, which increases the odds of obtaining a usable sample. If a set of human remains is located, you would need a known source of nuclear DNA from a missing individual in order to compare the two sets of nuclear DNA. However, the mtDNA of a sibling, mother, maternal aunt, or child of a maternal aunt can be compared with the mtDNA of the remains to establish whether there is a family connection, as all maternally related individuals share the same mtDNA. Because all maternally related individuals share the same mtDNA, it cannot be used to distinguish between those individuals in cases where several members of the same family are missing.

14.7 Genetic Genealogy

Did your ancestors come to the United States during colonial times or, perhaps, in one of the later waves of immigration? Do you celebrate St. Patrick's Day, Oktoberfest, Cinco de Mayo, a powwow, Yom Kippur, Mardi Gras, Chinese New Year, or Ramadan? Or perhaps you celebrate several of those holidays, as you learn of the great diversity within your own family heritage and genealogy. Tracing your family tree can provide insight and, in many cases, some surprises about the people and lives that have led to your own existence.

Traditional genealogy is conducted by searching through a paper trail of birth records, marriage certificates, census data, legal contracts, military records, and death certificates stored in government or church archives. There is a new tool that can provide insight into our family history, however, and its archives are in each of our cells. Our DNA is a record of our ancestors, who passed their genes down from one generation to the next. As we learn to decipher the information contained within DNA, we learn more about who we are.

The Y chromosome is the sex chromosome present in all males and is passed down unchanged from one generation to another. Over time, since our earliest ancestors started their trek out of Africa nearly 60,000 years ago, small mutations in the short tandem repeat sections of the Y chromosome have occurred. Each mutation is called a **single nucleotide polymorphism (SNP)**, pronounced "snip," and happens when one of the pairs of nucleotide bases is switched within a short tandem repeat section. SNP mutations are very rare. It is estimated that they occur once every few hundred generations. Each time a SNP mutation occurs, though, it creates a new branch that can differentiate among those descendants who carry the SNP and those who may carry a different SNP mutation from a separate occurrence at a different time and location during human migration. Each unique pattern of SNP mutations on the Y chromosome creates a *haplogroup* in which all members share a common ancestor who first exhibited that mutation.

SNP mutations can also occur on mtDNA, which then allows one to look into the deep genealogical roots of the maternal heritage. Each haplogroup created by a mutation on the mtDNA links our maternal families back to a common ancestor.

Genetic genealogy predates surname use and, therefore, has some limitations in tracing back your family line. Consider what happens over 10 generations. You are the descendant of 1024 grandparents, but your Y chromosome and mtDNA are shared only by the direct grandfather and direct grandmother on opposite parental lineages. Nonetheless, interesting information can be found: The data provided by genetic genealogy go back to the earliest migrations of humans.

People are often surprised to learn of multiethnic backgrounds in their DNA, such as Native American, African, Asian, Middle Eastern, and European. The author of this book was surprised to learn that his family originated in Scandinavia, most likely during the great wanderings of the Vikings. Traditional genealogical research had located the author's ancestors in the tiny country of Liechtenstein between Switzerland and Austria near the southern border of Germany. The haplogroup map corresponding to the Y-chromosome I1 haplogroup and the U mtDNA haplogroup, of which the author is a descendant, is shown in Figure 14.14. The colored circles on the maps represent points in time when a SNP mutation occurred, creating a new haplogroup. Fewer SNP mutations have occurred on the mtDNA than on the Y chromosome.

The author's lineage follows the colored arrows starting at Y-chromosome Adam and mitochondrial Eve. The terms *Adam* and *Eve* are used in human genetics to represent the most recent common ancestor from which all humans have descended. It is estimated that mitochondrial Eve lived 200,000 years ago while Y-chromosome Adam lived 60,000 years ago.

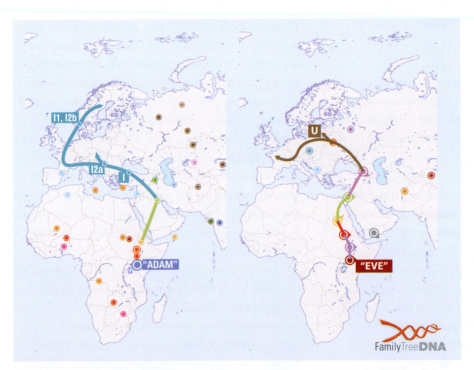

FIGURE 14.14 The haplogroup designation for the Y chromosome and mtDNA of the author reveals the migration path of his ancient ancestors.

14.8 Case Study Finale: Exploring Genetically Modified Food

Are genetically modified foods safe? This debate is mired in fear and a misunderstanding of how genetically engineered foods are created. All domesticated crops are genetically modified, since they contain genes not found in their wild counterparts. The new twist that genetic engineering introduced was the ability to insert genes from different species into a target species.

The basis for genetic engineering lies in the nucleotide code, also called the *genetic code*. The genetic code that represents the 20 amino acids used in the construction of a protein is universal to all organisms. All living organisms have the necessary genetic tools to make any protein and it is the instructions (a.k.a. genes) contained in the DNA that determine which proteins are actually made. When scientists identify a gene that can produce a protein with desirable effects, it is possible to transfer that gene into another organism that lacks the gene. The genetically modified organism will now have the ability to produce the desired protein.

For example, scientists identified a soil bacterium called *Bacillus thuringiensis* (*Bt*) that produces its own insecticide compounds. The genes that produced those compounds were transferred into corn plants so that the new genetically modified corn, called *Bt-corn,* can produce its own insecticide. The use of Bt-corn reduces the use of insecticides, diminishes insecticide exposure to farmers and consumers, and improves the crop yield.

The first step in genetic engineering is to identify the gene that produces the protein responsible for the desired trait such as insect resistance, drought resistance, faster growth rate, or larger fruit size, as shown in Figure 14.15a. The next step is to cut out that section of DNA containing the gene from the donor organism using a *restriction enzyme*. The restriction enzyme cleaves the sugar-phosphate bond asymmetrically across the two strands of DNA, as shown in Figure 14.15b.

The next step in creating a genetically modified crop is to modify the DNA of a species of bacterium called *Agrobacterium* that produces tumors in plants. *Agrobacterium* contains a circular section of DNA called a *plasmid.* The plasmid contains *transfer DNA* (*t-DNA*), which inserts the bacterial DNA into a plant's DNA, causing a tumor to grow. The plasmid is altered by removing the section that codes for the tumor growth using restriction enzymes.

The desired gene is transferred to the *Agrobacterium* plasmid by joining the fragmented ends of DNA back together. The asymmetric cleavage of DNA exposes a single strand of base pairs (A, G, C, T) on both strands of the DNA, as shown in Figure 14.15c. The exposed bases are capable of hydrogen bonding to their respective counterparts on the asymmetrically cut plasmid DNA, as shown in Figure 14.15d. The sugar-phosphate bonds of the DNA backbone are repaired by the addition of a compound called *ligase*.

The final step is the exposure of the host plant to the genetically modified *Agrobacterium,* which will insert the new gene into the DNA of the host. The genetically modified host plant now contains the new genes from the donor DNA to code for the protein.

Genetic engineering entails some risks, and one attempt to pass a gene from the Brazil nut to soybeans resulted in the passing of an allergen in Brazil nuts to the soybeans. The counterpoint to this criticism is that plants that often produce allergic reactions could be genetically engineered so that they no longer contain the proteins that trigger the allergic response. Thus, a hypoallergenic food could be engineered that does not trigger allergic reactions.

The debate over the safety and role of genetically modified foods will continue as the world population continues to increase at an unprecedented pace and as demand for food increases. A fundamental understanding of the process on the part of policy makers and consumers is required so that informed decisions can be made as to what regulations should be placed on these foods.

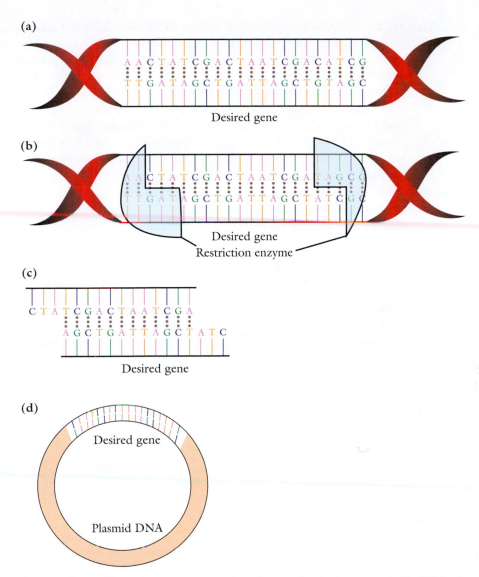

FIGURE 14.15 Genetically modified plants are created by selecting the desired gene in the DNA of an organism (a) and then removing that DNA using a restriction enzyme that cleaves the DNA backbone asymmetrically across the two strands (b). The desired DNA fragment (c) is then placed into the circular DNA of bacteria, called a *plasmid* (d), and the sugar-phosphate bond on the DNA is repaired. The plasmid has the ability to inject its DNA into the DNA of the target organism.

CHAPTER SUMMARY

• Lipids are a large class of organic compounds that are soluble in nonpolar solvents but not in water. The main subclasses of lipids are fats, oils, and waxes. Fats and oils are triglycerides made up of a glycerol molecule bonded to three fatty acid molecules. A wax consists of a long-chain fatty acid bonded to a long-chain alcohol.

• Fatty acids, which are long-chain carboxylic acids, contain double bonds (unsaturated) between adjacent carbon atoms in oils, whereas fats contain only single bonds (saturated).

• Carbohydrates are compounds such as sugars, cellulose, and starch that provide both an energy source and the structural material of cell walls. Cellulose and starch are polysaccharides of glucose. Due to a difference in the bonding of the glucose units in cellulose as compared with starch, only starch is usable by humans as a food source. Cellulose is indigestible and is referred to as dietary fiber.

• Proteins are polymers of amino acids. The primary structure of a protein is simply the order in which the amino acids appear in the molecule. The secondary structure of an amino acid describes the pattern that short segments of a protein may adopt—such as the pleated sheet arrangement of silk or the alpha helix shape of wool, cashmere, angora, and hair.

• Protein molecules may have tertiary structures in which the long chains fold back on one another, forming a larger three-dimensional shape. Several protein molecules may join together to form a superstructure referred to as a quaternary structure.

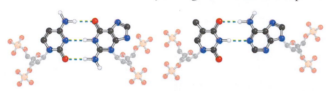

• DNA molecules are long polymer chains of nucleotides that serve as the blueprint for all proteins. The basic nucleotide consists of a phosphoric acid molecule and a deoxyribose sugar unit that together form the backbone of the polymer, and one of four nitrogen base compounds. The cross-linking interaction between chains of DNA occurs because of hydrogen bonding between the bases adenine and thymine (A-T or T-A) or between cytosine and guanine (C-G or G-C).

• The pattern of base pairs serves as the blueprint for DNA. Three base pairs, called a codon, provide the code for one of the 20 amino acids. The group of codons that produces a specific protein is referred to as a gene.

• Mitochondrial DNA is used in another form of DNA analysis. However, such analysis is limited in that all maternal relatives will share identical mitochondrial DNA.

• Genetic mutations called *single nucleotide polymorphism* (*SNP*) mutations, found in the short tandem repeat (STR) sections of the Y chromosome and mtDNA, can be used to link individuals with the migration routes their ancestors followed out of Africa as human beings spread across the world.

KEY TERMS

alpha helix (p. 397)
amino acid (p. 394)
base pair (p. 399)
carbohydrate (p. 392)
cellulose (p. 393)
chromosome (p. 401)
codon (p. 400)
deoxyribonucleic acid (DNA) (p. 399)
dietary fiber (p. 392)
disaccharide (p. 392)
disulfide bond (p. 398)
fat (p. 391)
fatty acid (p. 390)
gene (p. 400)
hydrophobic interaction (p. 398)

ionic bridge (p. 398)
lipid (p. 390)
mitochondrial DNA
　(mtDNA) (p. 403)
monomer (p. 393)
monosaccharide (p. 392)
nucleotide (p. 399)
oil (p. 391)
oligopeptide (p. 396)
oligosaccharide (p. 392)
pleated sheet (p. 396)
polymer (p. 393)
polypeptide (p. 396)
polysaccharide (p. 393)
primary structure (p. 396)

protein (p. 394)
quaternary structure (p. 398)
saturated (p. 391)
secondary structure (p. 396)
short tandem repeat (STR)
　analysis (p. 412)
single nucleotide
　polymorphism (SNP)
　(p. 404)
starch (p. 393)
sugar (p. 392)
tertiary structure (p. 397)
triglyceride (p. 390)
unsaturated (p. 391)
wax (p. 390)

MAKING MORE CONNECTIONS Additional Readings, Resources, and References

For more information on the history of agriculture: http://news.nationalgeographic.com/news/2006/06/060601-agriculture.html

For more information on world population: http://www.prb.org/

For more information on the proteins: http://www.ornl.gov/sci/techresources/Human_Genome/faq/gene-number.shtml#sixth

For more information on genetic genealogy: http://www.familytreedna.com https://genographic.nationalgeographic.com/genographic/lan/en/index.html

REVIEW QUESTIONS AND PROBLEMS

Questions

1. Explain what properties lipid molecules share and how a lipid molecule is defined. (14.1)

2. Is there a restriction on what types of organic molecules can be considered a lipid? (14.1)

3. What is the difference between a fatty acid and a wax? (14.1)

4. What are the components of a triglyceride? What are the two classes of triglycerides? (14.1)

5. What properties determine whether a triglyceride is a fat or an oil? (14.1)

6. Is a saturated triglyceride most likely to be a fat or an oil? An unsaturated triglyceride? (14.1)

7. If a fatty acid has a 24-carbon chain that is completely saturated, is it most likely a solid or liquid at room temperature? (14.1)

8. What is a polymer? A monomer? (14.2)

9. What is the monomer in a starch polymer? In a cellulose polymer? (14.2)

10. The term *sugar* commonly refers to sucrose. However, in chemistry what type of compounds can be given the generic title of sugar? (14.2)

11. Draw a sketch of both cellulose and starch to highlight the differences between the two molecules. (14.2)

12. Why are proteins classified as polymers? What are the monomers? (14.3)

13. What are the three distinct regions of an amino acid? What distinguishes one amino acid from another? (14.3)

14. What is the difference between an oligopeptide and a polypeptide? (14.3)

15. What is the primary structure of a protein molecule? How does that differ from the secondary structure of a protein? (14.3)

16. What are the two secondary protein structures? Sketch each. (14.3)

17. How are the secondary structures held together? (14.3)

18. Discuss the importance of the tertiary and quaternary structures of proteins. Do all proteins have primary, secondary, tertiary, and quaternary structures? (14.3)

19. What are the four types of bonds that hold the tertiary and quaternary structures together? (14.3)

20. Why is DNA classified as a polymer? What are the monomers? (14.4)

21. What force holds together the strands of DNA? Why would covalent bonding between strands be problematic? (14.4)

22. What is a gene? How does it function? What is it responsible for creating? (14.4)

23. What is the complementary DNA strand to the section with the sequence [GACTTAGGG]? (14.4)

24. What is the complementary DNA strand to the section with the sequence [GTGTTTGCG]? (14.4)

25. What is the complementary DNA strand to the section with the sequence [CCGTGGTTGCTTGGGCCGGCG]? (14.4)

26. What is the complementary DNA strand to the section with the sequence [TTTGGGCCGCTGTGGGTTGTG]? (14.4)

27. What is the sequence of amino acids represented by the nucleotides [CACCCCGCGCGCACAAAAAC]? (14.4)

28. What is the sequence of amino acids represented by the nucleotides [AAACACACCAAACGCCACGCC]? (14.4)

29. What is the sequence of amino acids represented by the nucleotides [GAAAAACCCCAAAACAACACC]? (14.4)

30. What is the sequence of amino acids represented by the nucleotides [GAACAACACAACCCCACCGCC]? (14.4)

31. What is the nucleotide sequence for the section of a protein consisting of [Glu-Arg-Pro-His-Gly-Asp-Thr]? (14.4)

32. What is the nucleotide sequence for the section of a protein consisting of [Glu-Ala-His-Asn-Pro-Thr-Lys]? (14.4)

33. What is the nucleotide sequence for the section of a protein consisting of [Thr-Pro-Pro-Asn-Asp-Gln-Gln]? (14.4)

34. What is the nucleotide sequence for the section of a protein consisting of [His-His-Lys-Asp-Gln-Glu-Lys]? (14.4)

35. Explain the advantages and disadvantages of using mtDNA for identification purposes. (14.6)

36. Explain how the STR sections in the noncoding portion of DNA allow for identification of a person. (14.5)

37. Explain how a single nucleotide polymorphism (SNP) can be used in genetic genealogy. (14.7)

38. Why are the Y chromosome and mtDNA singled out for use in genetic genealogy? (14.7)

Case Study Problems

39. What is the sample of people statistically needed to find a person with both the TPOX 10,10 repeat pattern and the D16S539 8,15 repeat pattern? (14.5)

40. One of the raw ingredients for the production of alcohol is glucose. Discuss why blood alcohol levels in postmortem samples can be skewed by the presence of starches in the person's system, whereas the presence of cellulose does not affect blood alcohol levels. (14.2)

41. Two cousins are implicated in an assault case where mtDNA has been recovered. However, only one of them is believed to have committed the crime. Could mtDNA be used to identify the guilty person? (14.6)

42. It is estimated that only 5% of DNA actually constitutes genes, the rest being noncoding sections of nucleotides. Could DNA be used for identification purposes if it contained only genes? (14.5)

43. It is known that the fingerprints of children do not last on objects as long as those of adults. Considering the nature of fingerprints, discuss why this might be. For further information, go to www.ornl.gov/info/ornlreview/rev28-4/text/tech.htm. (CP)

44. Ninhydrin is a chemical reagent capable of forming a bright purple compound when reacting with primary amines. Secondary amines are known to form an orange-colored compound. Which of the 20 amino acids are secondary amines? Would ninhydrin give a positive result for small peptide chains? For starches or cellulose? (CS)

45. Judy Buenoano, nicknamed the Black Widow, was convicted of murdering several husbands, boyfriends, and one of her children to collect insurance money. She would use arsenic to poison them, and local doctors who were unfamiliar with the symptoms of arsenic poisoning ruled the deaths as due to natural causes. A person who has been exposed to arsenic or heavy metal poisons such as lead or mercury will retain trace amounts of the poison in the hair. Each exposure to the poison is recorded in the victim's hair as the hair continues to grow. Based on what you learned in Chapter 13 about poisons and from the primary protein structure of hair, explain why these poisons will show up in a person's hair. (CP)

46. If a woman would like to have her DNA analyzed for genetic genealogy, she would be able to submit mtDNA for maternal lineage, but women have two X chromosomes and no Y chromosome. What options would she have to learn of her paternal lineage? (CS)

47. Police suspect a person has been repeatedly poisoned with arsenic and have taken samples of the victim's hair. The hair sample was cut into 1.0-cm lengths, as shown below, and analyzed for arsenic. The first, fifth, and eighth samples tested positive for arsenic. Based on the fact that hair grows at a rate of 0.5 mm per day, determine the time intervals for which the victim was exposed to arsenic. (CP)

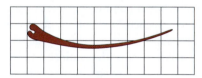

1 cm × 1 cm

APPENDIX A

Conversion Factors

the number of . . .	To convert from into the number of . . .	multiply by . . .
Mass:		
pounds	kilograms	0.454
kilograms	pounds	2.20
tons	tonnes	0.907
tonnes	tons	1.102
kilograms	tonnes	0.001
tonnes	kilograms	1000
Volume:		
quart	liter	0.946
liter	quart	1.057
cm^3 or mL	liters	0.001
liters	cm^3 or mL	1000
m^3	liters	1000
liters	m^3	0.001
Length:		
inches	centimeters	2.54
centimeters	inches	0.394
yards	meters	0.914
meters	yards	1.094
miles	kilometers	1.61
kilometers	miles	0.621

Temperature

To convert temperatures in °F to those in °C, subtract 32 and then divide the result by 1.80.

To convert temperatures in °C to those in °F, multiply by 1.8 and then add 32 to the result.

To convert temperatures in °C to Kelvin scale, add 273.

To convert Kelvin temperature to °C, subtract 273.

The two other electron orbital types, the *d* and **f orbitals,** are shown in the figure below. As you can see, the number and complexity of the orbitals increase dramatically and are provided only for completeness; this book does not delve further into the orbital discussion.

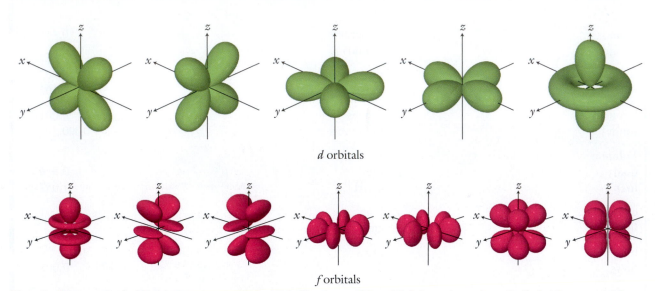

d orbitals

f orbitals

There are five separate *d* orbitals and seven separate *f* orbitals. The shapes of the orbitals in a set vary dramatically, but the energy of the orbitals in a set is equivalent.

ANSWERS TO ODD-NUMBERED REVIEW QUESTIONS AND PROBLEMS

Chapter 1

1. Chemistry has a seemingly infinite number of impacts on our daily lives. Basic respiration is a chemical process, as is metabolism. The paper on which this answer is printed was produced using chemical processes and is itself made of a composite of chemical compounds. There are numerous potential correct answers to this question.

3. The substances that compose a homogeneous mixture are so evenly distributed that all portions of the mixture are identical. A heterogeneous mixture is not evenly distributed. If the substance has any uneven distribution of its components (i.e., it is "chunky" in some way), then it is a heterogeneous mixture. If its composition is distributed evenly throughout, it is a homogeneous mixture.

5. Pure substances and homogeneous mixtures are both homogeneous, meaning that they both have uniform composition throughout. The method for determining whether a sample is a pure substance or a homogeneous mixture depends on the phase (solid, liquid, or gas) of the sample. Generally, the best method is to determine whether the substance can be separated into different components by using a physical change. If the sample is a solid, heating the solid to its melting point may determine whether the sample is a pure substance or a mixture. If a portion of the solid melts to liquid and another portion is left behind, then it is a mixture. If the sample is a liquid, boiling it (and then condensing it into another container) will show if the sample is a homogeneous mixture when a solid residue is left after boiling or if the boiling temperature changes as more of the sample boils away. If the sample is a gas, it can be very difficult to determine whether it is a homogeneous mixture or a pure substance. One might try condensing the gas to a liquid, but a better method is gas chromatography (discussed in a later chapter).

7. The type of mixture determines the method used for separating it into its pure components. For example, a solution (homogeneous mixture) composed of salt dissolved in water is best separated by distillation, where the liquid component (water) is evaporated from the mixture and then condensed into another container, leaving the solid component (salt) behind as a residue. A mixture of two solids is more difficult to separate; one might employ the technique of extraction, where one solid is dissolved into a solvent, leaving behind the other solid. Other, more complex mixtures will most likely require more involved methods to separate them into their components.

9. Elements are rarely found as pure substances because most elements are reactive under the conditions within earth's atmosphere, in the oceans, and in the crust. Only elements that are unreactive can be found in their pure form under normal conditions.

11. Some atomic symbols are based on the Latin name for the elements rather than on their contemporary names, often due to historical reasons (and sometimes because a symbol based on the contemporary name was already taken by another element). (For example, tungsten has the symbol "W" because of its Germanic name, *wolfram*, and iron has the symbol "Fe" because of its Latin name, *ferrum*.)

13. It is essential for a scientist to be able to make observations because all branches of science involve testing hypotheses through experimentation. The observations made by a scientist are helpful in providing evidence for or against a particular hypothesis. Without observations, there would be no way of providing support for the validity of a scientific law or theory. Put in the terms of forensic science, observations are required to make a judgment between the guilt or innocence of a suspect (or often to determine whether a crime has taken place at all).

15. The failure to prove a hypothesis is not a failure of the scientific method. Disproving a hypothesis is just as valid a result of the scientific method as is proving a hypothesis; both can require modification or refinement of the hypothesis and further experimental testing.

17. The container shows a mixture of two elements as well as a compound made from those two elements. A yellow sphere represents the atoms of one element, a green sphere represents the atoms of another element, and a yellow sphere connected to a green sphere represents the compound.

19. a. Gasoline—mixture
    b. Air—mixture
    c. Water—pure substance
    d. Steel—mixture

21. a. Silicon—element
    b. Carbon dioxide—compound (CO_2)
    c. Arsenic—element
    d. Water—compound (H_2O)

23. a Soil—heterogeneous
    b. Air—homogeneous
    c. Diesel fuel—homogeneous
    d. Concrete—heterogeneous

25. a. magnesium
    b. krypton
    c. phosphorus
    d. germanium

27. a. manganese
    b. beryllium
    c. cadmium
    d. rubidium

29. a. Ne
    b. Zn
    c. Rb
    d. I

31. a. Ba
    b. Cs
    c. Ag
    d. Ir

33. a. Cadmium, Cd
    b. Potassium, K
    c. Fluorine, F
    d. correct

35. a. metal
    b. nonmetal
    c. metalloid
    d. metal

37. a. metalloid
    b. nonmetal
    c. metalloid
    d. nonmetal

39. d. $MgSO_4$ contains one magnesium atom, one sulfur atom, and four oxygen atoms.

41. a. CaF_2
    b. Na_2SO_4
    c. H_2O
    d. Mg_3P_2

43. Hematite's formula is Fe_2O_3.

45. b. Validating the original hypothesis is not part of the scientific method. Once a hypothesis has been revised due to the results of experimental testing, it is not necessary to validate the original hypothesis when using the scientific method.

47. a. In this quotation, Sherlock Holmes is following the scientific method because he is collecting all relevant data before formulating his hypothesis. He is correct in not forming any theories at that stage of his investigation.
    b. Holmes is completely correct. Forming a theory is done only after testing a hypothesis that has been formulated and revised as the result of many observations.
    c. This is not a correct statement. Until a hypothesis or theory has been validated, it is not considered to be correct simply because all other possibilities have been eliminated.

49. Testing the risks associated with low-level pharmaceuticals and personal-care product chemicals in natural waters could be accomplished in several ways. Using the scientific method, careful experiments would need to be constructed to determine whether the low-level concentrations cause adverse effects. A comparison group should be exposed to identical natural waters that do not have the low-level pharmaceuticals and personal-care products in them to allow accurate conclusions to be reached.

51. The thin-layer chromatography (TLC) results indicate that the evidence may have been methamphetamine ("meth") because the pink spot for the evidence has the same elution distance as the sample containing pure methamphetamine. However, this is not absolute proof that the person was in possession of the illegal drug. TLC is a screening test and not a confirmatory test for controlled substances; further testing (such as gas chromatography-mass spectrometry, GC-MS) is necessary for proper identification.

Chapter 2

1. The key difference between chemical changes and physical changes is that in physical changes, the chemical properties do not change. Chemical changes alter the chemical identity of the substance.

3. Physical properties are useful in identifying an unknown substance because they can be measured without altering the chemical identity of the substance, and it is often possible to pinpoint the identity of the substance by using a combination of several different physical properties. When comparing two or more chemical substances, physical properties can often be unique to a specific chemical substance and permit identification.

5. Mass is the amount of substance, measured in grams. Weight is the force of gravity acting on an object.

7. The standard SI unit for measuring mass is kilograms (abbreviated kg). The standard unit for measuring distance is meters (abbreviated m).

9. Tera, T (1,000,000,000,000); giga, G (1,000,000,000); mega, M (1,000,000); kilo, k (1000); deci, d (0.1); centi, c (0.01); milli, m (0.001); micro, μ (0.000 001); nano, n (0.000 000 001).

11. 2.54 cm = 1 in.

13. The accuracy of a measurement is limited by the precision of the measuring instrument if the operator uses only the measurement that is known with a high degree of certainty. By estimating the last digit on a measurement, the operator is including a more precise (and potentially more accurate) value for the measurement.

15. a. Physical change
    b. Chemical change
    c. Physical change
    d. Chemical change

17. a. Physical property
    b. Chemical property
    c. Chemical property
    d. Physical property

19. a. Weight depends on gravity.
    b. Mass is measured in grams.
    c. Mass measures the amount of matter.
    d. Weight is measured with a scale.

21. a. micro, μ, 0.000 001
    b. kilo, k, 1000
    c. centi, c, 0.01
    d. nano, n, 0.000 000 001

23. a. 0.001 = m (milli)
    b. 1000 = k (kilo)
    c. 0.1 = d (deci)
    d. 0.000 000 001 = n (nano)

25. a. 25 mg
    b. 3.525 L
    c. 7.8 dm
    d. 0.433 m

27. a. 4.5 ft.
    b. 12.0 yd.
    c. 820 mm
    d. 23.9 cm

29. a. 3.0 m^2
    b. 0.34 ft.^{2}
    c. 0.224 dm^3
    d. 1.3 × 10^6 cm^3

31. a. 6540 ft./hr
    b. 6.7 lb/day
    c. 24 ft./s
    d. 20 km/L

33. a. 4
    b. 3
    c. 4
    d. 1

35. a. 2.3 × 10^3
    b. 1.0 × 10^{-3}
    c. 1.75 × 10^4
    d. 2.40 × 10^{-5}

37. a. 0.00614
    b. 259,000
    c. 10,025
    d. 0.02226

39. a. 5.12 mL
    b. 3.55 mL
    c. 5.70 mL
    d. 2.70 mL

41. a. 3 significant figures
    b. 1.20 × 10^3
    c. 0.0500
    d. 2.50 × 10^2

43. a. 3.520 × 10^4
    b. 8.71 × 10^{-3}
    c. 1.9 × 10^3
    d. 2.2 × 10^{-2}

45. a. 5.10 × 10^1
    b. 7.82 × 10^1
    c. 3.3 × 10^{-1}
    d. 2.5 × 10^2

47. a. 7.4
    b. 373.5
    c. 17.8
    d. 3.0 × 10^1

49. a. 280 = 2.8 × 10^2
    b. 0.024
    c. 2300 = 2.3 × 10^3
    d. 2.71

51. a. 5.0 × 10^1
    b. 6.0
    c. −35.8
    d. −127

53. Set (c) is the most precise and accurate.

55. a. 1.45 g/cc
    b. 3.56 g/mL
    c. 1.45 g/cm^3
    d. 0.847 g/mL

57. a. 14.7 mL
    b. 3.10 mL
    c. 3.51 cc
    d. 4.02 cm^3

59. a. 21.8 g
    b. 17.7 g
    c. 127 g
    d. 12.6 g

61. *Determine the nature of the problem:* The investigators believed that Pitera was linked to the mass grave, but they needed evidence to prove that link. *Collect and analyze all relevant data:* The investigators collected soil samples from all over Staten Island for comparison. *Form an educated guess, called a hypothesis, as to what happened:* The hypothesis was that there was a unique soil type at the mass grave and that Pitera would have soil matching that unique type to link him to the crime. *Test the hypothesis:* Bruce Hall examined the color, texture, and composition of the soil in each of the reference samples and in the sample taken from Pitera's shovel. The soil on the shovel matched only the burial site and none of the alibi sites proposed by the defense. This linked Pitera to the crime and helped lead to a conviction in the case.

63. Soil on the blade of a shovel is easily brushed off or mixed with other soils as the shovel is used repeatedly. The soil in the rounded-over flange of the shovel has been compacted and doesn't mix with soil as the shovel is used at new sites.

65. 279 mL

67. 0.00025 g/(dL · min)

69. 36.76 mL

71. If a suspected arson sample sent to the laboratory comes back negative, it means accelerants were not found in the sample. This indicates that the accelerants were not detected, not that they weren't used nor that the fire wasn't deliberately set. If a sample comes back positive for petroleum-based accelerants, it means accelerants were present in the fire. It could indicate that the fire was deliberately set or it could mean that the fire burned in a location that already contained accelerants (such as a garage or storage shed). The investigator can determine only if accelerants were detected in the arson sample. How that information is used in the investigation will depend on the location and type of fire and the suspects being considered.

73. Borosilicate (Pyrex®)

75. 5×10^3 m traveled in the first 15 s after being fired

77. 11.1 cigarettes; 167 cans of Red Bull

79.

Solution 4 (25 mL, 1.80 g/mL) floats on top

Mineral 4 (2.16 g/mL) floats at the interface of Solutions 2 and 4

Solution 2 (25 mL, 2.45 g/mL)

Mineral 1 (2.58 g/mL) floats at the interface of Solutions 3 and 2

Solution 3 (50 mL, 3.34 g/mL)

Mineral 2 (3.56 g/mL) floats at the interface of Solutions 1 and 3

Solution 1 (50 mL, 4.25 g/mL)

Mineral 3 (4.35 g/mL) sinks to the bottom

81. It was not reasonable to convict the suspect based on the evidence given. The scientific method was only applied to the second case.

83. The recovered bullet and case are consistent with the .22 LR pistol and not the .25 automatic.

Chapter 3

1. Atoms were described by Leucippus and Democritus as small, hard, indivisible particles that come in various sizes, shapes, and weights. They were thought to be in constant motion and to combine to make up all the various forms of matter. The observable properties of matter were thought to be a direct result of the type of atoms it contained.

3. Aristotle believed that matter could be divided infinitely (a direct contradiction of the theory proposed by the atomists).

5. The law of conservation of mass says that matter changes form and is neither created nor destroyed in a chemical reaction.

7. While many of his contemporaries were making careful measurements of mass, the difference in Lavoisier's experiments was that he was using closed systems. By making careful measurements of the mass of a closed system, Lavoisier was able to demonstrate the law of conservation of mass.

9. The law of definite proportions (law of constant composition) helped scientists understand that atoms of different elements can combine in specific ratios to make different compounds. It is not just the identity of the elements that matters but how much of each one is present.

11. In science, a theory is the best current explanation of a phenomenon. In popular use, it is often an opinion that can be easily swayed by argument.

13. The four principles of Dalton's atomic theory: (1) All matter is made up of tiny, indivisible particles called atoms. (2) Atoms cannot be created, destroyed, or transformed into other atoms in a chemical reaction. (3) All atoms of a given element are identical. (4) Atoms combine in simple, whole-number ratios to form compounds.

15. The law of definite proportions is a restatement of the fourth principle of Dalton's atomic theory: Atoms combine in simple, whole-number ratios to form compounds.

17. We will represent carbon atoms as black circles and oxygen atoms as white circles:

19. The law of multiple proportions shows that different ratios of the same elements yield different compounds. One carbon and one oxygen atom bonded together makes carbon monoxide, while one carbon atom bonded to two oxygen atoms makes carbon dioxide.

21. There are actually more than three subatomic particles, but these three are the most important subatomic particles from the perspective of chemistry: electrons, protons, and neutrons.

23. Rutherford compared his results to seeing an artillery shell reflected back by having it hit a piece of paper. He called the dense positive region in the gold foil the nucleus and subsequently explored the structure of the atom by other techniques to arrive at a physical model to account for the observed behavior.

25. Isotopes of the same element are different because they have a different mass per atom. This difference in mass results from a differing number of neutrons in the nucleus. (If the atoms had differing numbers of protons in the nucleus, they wouldn't be the same element. If they had differing numbers of electrons, they would be ions with different overall charges.)

27. The atomic mass for einsteinium (Es) is listed as (252) because the isotope ^{252}Es is the longest-lived isotope of the element. The isotopes of einsteinium have short half-lives that make it impossible to measure an accurate average atomic mass for the element.

29. A continuous spectrum is produced by a source that produces a large number of emissions at varying wavelengths. A good example is an incandescent light bulb, where the high temperature of the tungsten filament causes the emission of light at wavelengths throughout the visible spectrum. A line spectrum is produced by a source that is based on atomic emission from a single element. High voltage discharge tubes containing samples of elements in the gas phase produce such a spectrum, as do flame emissions of many inorganic salts.

31. The line spectrum from an excited lithium atom results from electronic transitions of the three electrons within the atomic orbitals of the lithium atom. The line spectrum from an excited cesium atom results from electronic transitions of the 55 electrons within the atomic orbitals of the cesium atom. The higher number of electrons results in a more complex line spectrum.

33. Radio waves have the longest wavelength and gamma rays have the shortest wavelength of the forms of electromagnetic radiation.

35. As the wavelength of light increases, the energy decreases. As the wavelength of light decreases, the energy increases. (Energy and wavelength are inversely proportional.)

37. According to the Heisenberg uncertainty principle, if the location of an electron is known precisely, then at that instant the energy cannot be known precisely, and vice versa.

39. An s-orbital is a sphere centered at the nucleus of the atom. A p-orbital has two spherical lobes (similar to a peanut or dumbbell) with the two lobes meeting at the nucleus of the atom.

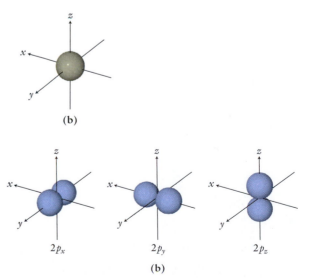

(b)

$2p_x$ $2p_y$ $2p_z$

(b)

41. The order is: Democritus (440 B.C.), Lavoisier (1785), Dalton (1803), and Thomson (1897).

43. Statement (c) is false because the reaction vessel was a closed system. By definition, a closed system cannot exchange any matter with its surroundings.

45.
a. 46.5 g carbonic acid
b. 60.0 g sodium hydroxide
c. 37.0 g calcium hydroxide
d. 8.8 g carbon dioxide

47. Observation (b) is a direct conclusion, and (a) and (d) are valid by extension.

49.

Particle	Charge	Mass (amu)	Symbol
Electron	−1	0.0005486	e^-
Proton	+1	1.0073	p or p^+ or H^+
Neutron	0	1.0087	n or n^0

51.

Protons	Neutrons	Electrons	Isotope
28	30	28	$^{58}_{28}Ni$
22	25	22	$^{47}_{22}Ti$
6	6	6	^{12}C
18	20	18	$^{38}_{18}Ar$

53.
a. $^{24}_{12}Mg$
b. $^{41}_{19}K$
c. $^{74}_{34}Se$
d. $^{136}_{56}Ba$

55. The atomic mass of magnesium on the periodic table is 24.3050. Therefore, the ^{24}Mg isotope is probably the most abundant. Option (a) is the correct answer.

57. B: 10.811 amu
Br: 79.903 amu
Rb: 85.468 amu
Sb: 121.760 amu

59. Statement (b) is false.

61.
a. 1.224×10^{15} Hz
b. 4.399×10^{14} Hz
c. 3.529×10^{14} Hz
d. 1.064×10^{15} Hz

63.
a. 128 nm
b. 52.8 nm
c. 35.7 nm
d. 99.0 nm

65.
a. 4.677×10^{-19} J
b. 2.915×10^{-19} J
c. 2.339×10^{-19} J
d. 7.049×10^{-19} J

67.
a. 1.56×10^{-18} J
b. 3.76×10^{-18} J
c. 5.57×10^{-18} J
d. 2.01×10^{-18} J

69.
a. $1s^22s^22p^63s^23p^64s^23d^8$
b. $1s^22s^22p^4$
c. $1s^22s^22p^63s^23p^5$
d. $1s^22s^22p^63s^23p^64s^23d^5$

71.
a. $[He]2s^22p^2$
b. $[Ar]4s^23d^{10}4p^2$
c. $[Kr]5s^24d^{10}5p^3$
d. $[Ar]4s^23d^{10}4p^3$

73.
a. Rh
b. P
c. Sr
d. V

75.
a. P
b. Te
c. Na
d. I

77.
a. Y: $[Kr]5s^24d^1$—The error is in the noble gas core: Kr not Ar.
b. Sc: $[Ar]4s^23d^1$—The error is in the principal quantum number for the d orbitals: 3d not 4d.
c. Fe: $1s^22s^22p^63s^23p^64s^23d^6$—The principal quantum numbers do not go up for every orbital.
d. F: $1s^2 2s^2 2p^5$—Fluorine has only 5 electrons in its 2p orbital.

79. ^{88}Sr: 38 protons, 50 neutrons, 38 electrons
^{87}Sr: 38 protons, 49 neutrons, 38 electrons
^{86}Sr: 38 protons, 48 neutrons, 38 electrons
^{84}Sr: 38 protons, 46 neutrons, 38 electrons

81. Blue: 447 nm; $\nu = 6.71 \times 10^{14}$ Hz;
E = 4.45×10^{-19} J
Green: 502 nm; $\nu = 5.96 \times 10^{14}$ Hz;
E = 3.95×10^{-19} J
Yellow: 597 nm; $\nu = 5.02 \times 10^{14}$ Hz;
E = 3.33×10^{-19} J
Red: 668 nm; $\nu = 4.49 \times 10^{14}$ Hz;
E = 2.98×10^{-19} J

83. Because strontium is continually replaced in bone but not in teeth, the ratio of the amount of radioactive strontium isotope in the bones to the amount of the same isotope in the teeth can be used in conjunction with the half-life of that isotope to determine age at death, given that most permanent teeth are formed within a few years during childhood.

85. The match between the evidence recovered from the burn victim and the elemental profile of the explosives used by the neighbors is compelling evidence, but it alone does not prove the neighbors guilty. It is still possible that someone else might have used another lot of the same batch of explosives (with the same elemental profile). They may even have stolen some of the neighbors' explosives and used them, or perhaps the victim stole some and was accidentally killed.

Chapter 4

1. The periodic table is printed inside the front cover of your textbook. Here is a smaller version with the appropriate sections labeled:

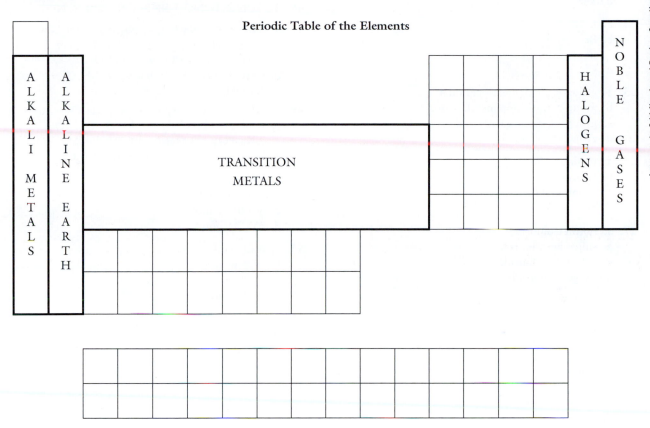

Periodic Table of the Elements

3. Metals tend to form cations (they tend to lose electrons, which leaves them with a positive charge); they are located on the left-hand side of the periodic table. Nonmetals tend to form anions (they tend to gain electrons, which leaves them with a negative charge); they are located on the right-hand side of the periodic table.

5. Covalent bonds typically form between nonmetallic elements.

7. Cations are formed when an atom loses one or more electrons. Anions are formed when an atom gains one or more electrons.

9. Polyatomic ions are ions composed of more than one atom bonded together. Examples include NH_4^+, NO_3^-, SO_4^{2-}, PO_4^{3-}, and $CH_3CO_2^-$.

11. In MgS, there are Mg^{2+} cations and S^{2-} anions. In NaF, there are Na^+ cations and F^- anions. The divalent ions (Mg^{2+} and S^{2-}) have stronger electrostatic attraction than do the monovalent ions (Na^+ and F^-).

13. Balanced chemical equations are important because they reflect the fact that all chemical reactions obey the law of conservation of mass. Since matter is neither created nor destroyed in a chemical reaction, the chemical equation should be balanced to reflect that fact.

15. The concept of the mole is necessary to make a connection between the world we operate in and the abstract world of atoms. Without it, scientists would be unable to relate the mass of a substance to the number of atoms present in the substance.

17. The limiting reactant is the one that determines the maximum theoretical amount of product formed. If the limiting reactant is unknown, it is impossible to predict how much product is formed as a result of a chemical reaction.

19. A solution containing a greater amount of FD&C Red Dye No. 40 is more concentrated than a solution containing a lower amount of the dye. The more

concentrated solution contains more dye molecules in the same volume of solution. The higher the concentration of a colored compound, the less light passes through the solution, and the solution appears darker.

21. a. I
    b. Br
    c. F
    d. Cl

23. a. group 1, alkali metal, period 4
    b. group 8, transition metal, period 5
    c. group 18, noble gas, period 2
    d. group 15, period 4

25. a. platinum, Pt
    b. fluorine, F
    c. krypton, Kr
    d. arsenic, As

27. a. boron trifluoride
    b. sulfur hexafluoride
    c. chlorine trifluoride
    d. oxygen difluoride

29. a. S_2Cl_2
    b. N_2S_5
    c. SF_4
    d. SO_3

31. a. +1
    b. −3
    c. multiple (+2, +4)
    d. −1

33. a. $AlCl_3$
    b. CsF
    c. CdO
    d. SrI_2

35. a. Nitrate ion: NO_3^-, −1
    b. Hydroxide ion: OH^-, −1
    c. Cyanide ion: CN^-, −1
    d. Phosphate ion: PO_4^{3-}, −3

37. a. $CsC_2H_3O_2$
    b. K_2CO_3
    c. $Ba_3(PO_4)_2$
    d. $Al(MnO_4)_3$

39. a. Cu^{2+}
    b. Co^{2+}
    c. Cu^+
    d. Cr^{2+}

41. a. aluminum sulfide
    b. magnesium oxide
    c. potassium hydroxide
    d. aluminum phosphate

43. a. copper(II) chloride
    b. cobalt(II) nitrate
    c. copper(I) chloride
    d. chromium(II) oxide

45. $3\,NaCN + Fe(NO_3)_3 \rightarrow Fe(CN)_3 + 3\,NaNO_3$

47. a. $3\,BaCl_2 + 2\,Na_3PO_4 \rightarrow Ba_3(PO_4)_2 + 6\,NaCl$
    b. $Na_2S + Fe(NO_3)_2 \rightarrow 2\,NaNO_3 + FeS$
    c. $C_3H_8 + 5\,O_2 \rightarrow 3\,CO_2 + 4\,H_2O$
    d. $Ca(C_2H_3O_2)_2 + 2\,KOH \rightarrow 2\,KC_2H_3O_2 + Ca(OH)_2$

49. a. $NH_4Br + AgC_2H_3O_2 \rightarrow NH_4C_2H_3O_2 + AgBr$
    b. $K_2SO_4 + Pb(NO_3)_2 \rightarrow 2\,KNO_3 + PbSO_4$
    c. $2\,Na_3PO_4 + 3\,MgCl_2 \rightarrow 6\,NaCl + Mg_3(PO_4)_2$
    d. $Li_2CO_3 + Cu_2SO_4 \rightarrow Li_2SO_4 + Cu_2CO_3$

51. a. 191.956 g/mol
    b. 138.206 g/mol
    c. 601.933 g/mol
    d. 383.789 g/mol

53. a. 0.123 mol
    b. 0.00669 mol
    c. 0.139 mol
    d. 0.0932 mol

55. a. 114 g
    b. 26.6 g
    c. 162 g
    d. 103 g

57. 31.2 g $BaSO_4$

59. 2.50×10^2 g $NaCl$

61. 264 g CO_2

63. a. The action of hydrochloric acid on zinc is a redox (reduction–oxidation) reaction. The zinc is oxidized (it loses electrons to go from a neutral charge to a +2 charge) and the hydrogen is reduced (it gains electrons to go from a +1 charge to a neutral charge).
    b. The burning of gun cotton is a combustion reaction. The statement that it leaves almost no ash after ignition implies that its combustion is nearly complete—the chemical composition is such that the reaction with oxygen produces only carbon dioxide gas, water vapor, and nitrogen gas.
    c. The production of a solid from two solutions is a precipitate reaction. The reaction of chloride ion with silver nitrate solution to produce solid silver chloride is an excellent forensic science application of this chemical principle.
    d. The key word in this question is "acid." That gives the hint that this is an acid–base reaction. In fact, it is a neutralization reaction where the milk

of magnesia is the base and stomach acid (as it refluxes into the esophagus) is the acid.

65. The blue diatomic molecules are the limiting reactant and the theoretical yield is four product molecules.

67. 15.9 g $Fe(OH)_3$

69. Iodine ($I_2(s)$) is a halogen (Group 17) in Period 5. Phosphorus ($P_4(s)$) is in Group 15 and Period 3. Lithium metal ($Li(s)$) is an alkali metal (Group 1) in Period 2.

71. 23.13 g N_2 produced

73. 150 mg/L

75. Some spectrophotometers simply yield an absorbance reading at one particular wavelength. If that is the case, the spectrophotometer should be set to a wavelength corresponding to the complementary color of yellow, which is violet. A good wavelength would be in the 450–500 nm range.

77. The killer's fingerprints would have to be found on the inside of the Tylenol bottle. The average customer does not open over-the-counter medicine bottles in the store, especially not those that have the "tamper resistant" seals.

79. 1.88 g $BaSO_4$ theoretical yield

81. Scientifically speaking, it is important to use the phrase "was consistent with" rather than "was" because the scientist cannot travel backward in time to determine what took place in the past. Instead, it is possible only to do specific measurements that can indicate whether the evidence is consistent with or inconsistent with the proposed hypothesis within the context of the particular measurements being performed.

Chapter 5

1. Most single atoms do not have a full shell (sometimes called an octet) of valence (outermost) electrons. To obtain a full shell, they can either exchange (donate or accept) or share electrons. Two atoms can share a pair of electrons to form a bond:

$$H\cdot \; + \; H\cdot \longrightarrow H\cdot\cdot H \longrightarrow H-H$$

This can also be represented as shown in Worked Example 1 on page 138.

3. Atoms form covalent bonds when they share electrons equally. Nonmetals tend to form covalent bonds when bonding with other nonmetals. Atoms form ionic bonds when they exchange (donate or accept) electrons. Metals tend to form ionic bonds with nonmetals.

5. When atoms exchange or share electrons, they achieve a full valence shell (sometimes called an octet). This is a stable configuration that minimizes the energy of the atoms present. This energy minimization is the reason why many chemical reactions proceed the way they do.

7. A double bond will form between two atoms when both atoms need the second bond to achieve a full valence shell. For example, oxygen is two electrons short of an octet. If each oxygen atom shares two electrons with the other, a double bond is formed.

9. Resonance structures show equivalent possible structures based on the placement of electrons in multiple bonds or lone pairs. They do not represent accurate bonding within the molecule because the actual bonding is some combination of the resonance structures. Crudely speaking, the actual structure is a sort of "average" of all possible resonance structures.

11. According to the VSEPR theory, electron regions around an atom will move until they are as far apart as possible. The number of electron regions determines the angle required between regions to allow maximum separation. The angle describes the electron geometry of the molecule.

13. The nature of the electron region does not affect the overall electron geometry of the molecule. However, the "size" of an electron region does depend on its nature (single bond, double bond, triple bond, lone pair). Lone pairs take up more space than bonds do, so they cause a distortion of the bond angles to make more space for themselves. This slightly alters the electron geometry.

15. As described in Question 13, the nature of electron regions slightly alters electron geometry. Since electron geometry determines the molecular geometry, the molecular geometry is also affected (albeit slightly) by the nature of the electron regions in the molecule.

17. According to our textbook, stereoisomers occur when two compounds share the same chemical formula and the same connections between atoms, but exhibit differences in the way their atoms are arranged three-dimensionally. This is extremely significant in drug chemistry—different stereoisomers can have very different properties. The example given in the text is that l-methamphetamine is an ingredient in (legal) vapor-rub products while d-methamphetamine is an extremely addictive illegal drug.

19. The molecular shape of neurotransmitters is the key function that controls their uptake into the uptake channel of the neuron. According to our textbook, the tunnel walls are strands of protein molecules arranged so that neurotransmitters of a particular shape and size may pass through.

21. Cocaine and other illegal drugs cause a permanent change in brain function by causing destruction of uptake channels. The resulting damage to neurons is not repaired, often leading to depression and other symptoms.

23. a. $\dot{S}r\cdot$

b. $\cdot\dot{S}i\cdot$

c. $:\dot{O}\cdot$

d. $:\ddot{B}r\cdot$

25. a. Na^+ Na^+ $:\ddot{S}:^{2-}$

b. Mg^{2+} $:\ddot{C}l:^-$ $:\ddot{C}l:^-$

c. Sr^{2+} $:\ddot{O}:^{2-}$

d. Li^+ Li^+ Li^+ $:\ddot{P}:^{3-}$

27. $:\ddot{I}-\ddot{N}-\ddot{I}:$
       $\quad\quad\;|$
       $\quad\quad:\ddot{I}:$

29. a. $[:C\equiv N:]^-$

b.
$$\left[\begin{array}{c} :\ddot{O}: \\ | \\ :\ddot{O}-P-\ddot{O}: \\ | \\ :\ddot{O}: \end{array}\right]^{3-}$$

c.
$$\left[\begin{array}{c} :\ddot{O}: \\ C \\ \ddot{O}.\quad.\ddot{O}: \end{array}\right]^{2-} \longleftrightarrow \left[\begin{array}{c} :O: \\ \| \\ C \\ \ddot{O}.\quad.\ddot{O}: \end{array}\right]^{2-} \longleftrightarrow \left[\begin{array}{c} :\ddot{O}: \\ C \\ :\ddot{O}.\quad.\dot{O} \end{array}\right]^{2-}$$

d. $[:\ddot{O}-H]^-$

31. a. $:\ddot{F}-\ddot{P}-\ddot{F}:$
          $\quad\quad\;|$
          $\quad\quad:\ddot{F}:$

b. $:\ddot{O}=\ddot{N}-\ddot{O}\cdot$
   $\quad\quad\updownarrow$
   $\cdot\ddot{O}-\ddot{N}=\ddot{O}:$

c. $:\ddot{C}l:$
   $\quad|$
   $:\ddot{C}l-C-\ddot{C}l:$
   $\quad|$
   $:\ddot{C}l:$

d. $H-\ddot{A}s-H$
   $\quad\quad|$
   $\quad\quad H$

33. a. Linear electron geometry
b. Tetrahedral electron geometry
c. Trigonal planar electron geometry
d. Tetrahedral electron geometry

35. a. tetrahedral
b. trigonal planar
c. tetrahedral
d. tetrahedral

37. a. Linear molecular geometry
b. Tetrahedral molecular geometry
c. Trigonal planar molecular geometry
d. Linear molecular geometry

39. a. trigonal pyramidal
b. bent (120°)
c. tetrahedral
d. trigonal pyramidal

41.

Electron Regions	Number of Lone Pairs	Electron Geometry	Molecular Geometry
2	0	linear	linear
3	1	trigonal planar	bent (120°)
4	0	tetrahedral	tetrahedral

43. a. Trigonal planar electron geometry, bent molecular geometry

$$\left[\begin{array}{c} \ddot{N} \\ \ddot{O}.\quad\quad\ddot{O} \end{array}\right]^- \longleftrightarrow \left[\begin{array}{c} \ddot{N} \\ \ddot{O}.\quad\quad\ddot{O}. \end{array}\right]^-$$

b. Tetrahedral electron geometry, tetrahedral molecular geometry

$$\left[\begin{array}{c} :\ddot{O}: \\ | \\ :\ddot{O}-Cl-\ddot{O}: \\ | \\ :\ddot{O}: \end{array}\right]^-$$

c. Tetrahedral electron geometry, bent molecular geometry

$$[:\ddot{O}-Cl-\ddot{O}:]^-$$

d. Tetrahedral electron geometry, trigonal pyramidal molecular geometry

$H-\ddot{N}-H$
$\quad\quad|$
$\quad\quad H$

45. a. polar covalent ←
b. polar covalent →
c. nonpolar covalent
d. nonpolar covalent

47. a. polar covalent ←
b. nonpolar covalent
c. nonpolar covalent
d. nonpolar covalent

49. a. polar
b. polar
c. nonpolar
d. polar

51. Immunoassays are a screening method because they do not provide definite identification of a chemical compound. Because molecular geometry is an important part of how the immunoassay functions, it is possible for a molecule that has similar shape to an antigen (or its antibody) to cause the same reaction in the immunoassay as the target molecule. This could lead to a false positive if a confirmative test was not administered.

53. Oxygen at top of ring (tetrahedral, four electron regions); oxygen to right side of ring, double-bonded to carbon (trigonal planar, three electron regions)

55. The negative results from the GC-MS test could provide sufficient "reasonable doubt" that a conviction may not have been secured if the defense had properly challenged the evidence. There is significant possibility for cross-reactivity in RIA, so it is possible that the tissue samples did not really have LSD present.

57. Physical evidence should always be more important than circumstantial evidence in a criminal trial. Because the physical evidence in the Martin case was not convincing, it may be possible for the prosecution (and the defense) to use circumstantial evidence to try and bolster their case. However, circumstantial evidence is never stronger than physical (scientific) evidence.

Chapter 6

1. A solvent is a substance into which a solute can dissolve. The most common solvent on earth is water (which is why it is sometimes referred to as the "universal solvent").

3. Strong electrolytes conduct electricity in solution by the motion of the ions through the solvent. Cations (positively charged ions) move toward the cathode (negatively charged electrode where reduction takes place), and anions (negatively charged ions) move toward the anode (positively charged electrode where oxidation takes place).

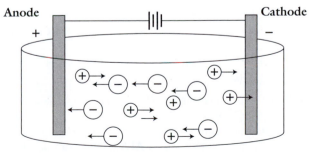

Anode +   Cathode −

5. Electrolytes are substances that conduct electricity when dissolved in solution. Nonelectrolytes do not conduct electricity when they are dissolved. NaCl is an electrolyte, and sugar ($C_{12}H_{22}O_{11}$) is a nonelectrolyte.

7. A supersaturated solution can be prepared by heating the solvent above room temperature. The solubility of most solid substances increases as the temperature increases, so more of the solid will dissolve. Once the solid has dissolved, allowing the solution to cool back to room temperature will cause a supersaturated solution to form. In practice, this can be very difficult to do—sometimes cooling the solution back to room temperature will cause the solute to precipitate from solution instead of remaining in the supersaturated solution.

9. An acid is a substance that produces hydrogen ions (H^+) (sometimes called hydronium ions, H_3O^+) when dissolved in water. A base is a substance that produces hydroxide ions (OH^-) when dissolved in water. Acids dissolve to make solutions with low pH (less than 7), and bases dissolve to make solutions with high pH (greater than 7).

11. Weak acids are not inherently safer than strong acids; safety is dependent on the concentration and type of acid. A highly concentrated weak acid can be more dangerous than a dilute strong acid.

13. Bases can undergo neutralization reactions:
$H^+(aq) + OH^-(aq) \rightarrow H_2O(l)$
Bases can also undergo saponification reactions:
$OH^-(aq) + grease (s) \rightarrow soap (aq)$

15. In everyday language, "salt" refers to table salt, which is sodium chloride (NaCl). When a chemist uses the term *salt*, he or she is referring to one of a number of ionic compounds. Sodium chloride is one particular example of these compounds. They are made up of positively charged cations and negatively charged anions combined to make a compound that has no overall charge.

17. The concept of pH was developed as a way to express the acidity of a solution. It is described as being the "power of hydrogen" and was developed in 1909 by Søren Sørenson.

19. A buffer is a solution that resists changes in pH. It is a combination of a weak acid and the anion produced from neutralizing the weak acid; it works by having the weak acid (or the anion) reacting with the base (or the acid) as it is added to keep it from changing the pH of the solution significantly.

21. Spectator ions do not participate in precipitation reactions. They simply balance the charges of the ions that do participate and are left behind in

solution as the solid precipitate forms. It is possible for a precipitation reaction to occur without spectators, but only if all the ions in solution become part of the precipitate (quite an unusual situation).

23. The total ionic equation is a more accurate depiction of what's occurring in solution; it shows all ions and species that are present and the form in which they are present.

25. A saturated solution is one where the dissolved solute is in equilibrium with undissolved solute. A good example of this is when sugar is added to iced tea and much of the sugar remains at the bottom of the glass. The solution has become saturated with sugar. An unsaturated solution has all the solute completely dissolved. The unsaturated solution is uniform throughout and does not have any undissolved solute present.

27. a. electrolyte
b. nonelectrolyte
c. electrolyte
d. electrolyte

29. a. weak electrolyte
b. strong electrolyte
c. strong electrolyte
d. weak electrolyte

31. a. unsaturated
b. saturated
c. unsaturated
d. supersaturated

33. a. 25.0 g
b. 2.58 g
c. 1,130 g
d. 125 g

35. a. 0.2691 M NaCl
b. 0.05003 M MgS
c. 0.004542 M $CuSO_4$
d. 0.05203 M NaCN

37. a. 0.0748 mol $CaCO_3$
b. 0.0625 mol $CaCO_3$
c. 0.0877 mol $CaCO_3$
d. 2.91 mol $CaCO_3$

39. a. 7.49 g
b. 6.26 g
c. 8.78 g
d. 291 g

41. a. 9.68 M
b. 0.605 M
c. 0.807 M
d. 1.94 M

43. a. 888 mL
b. 289 mL
c. 107 mL
d. 2.07 mL

45. a. H_2CO_3 = carbonic acid, weak acid
b. HNO_3 = nitric acid, strong acid
c. H_3PO_4 = phosphoric acid, weak acid
d. H_2SO_4 = sulfuric acid, strong acid

47. $SO_3(g) + H_2O(l) \rightarrow H_2SO_4(aq)$

49. a. $2 K(s) + 2 H_2O(l) \rightarrow 2 KOH(aq) + H_2(g)$
b. $2 Rb(s) + 2 H_2O(l) \rightarrow 2 RbOH(aq) + H_2(g)$
c. $Sr(s) + 2 H_2O(l) \rightarrow Sr(OH)_2(aq) + H_2(g)$
d. $Mg(s) + 2 H_2O(l) \rightarrow Mg(OH)_2(s) + H_2(g)$

51. a. $K_2O(s) + H_2O(l) \rightarrow 2 KOH(aq)$
b. $Rb_2O(s) + H_2O(l) \rightarrow 2 RbOH(aq)$
c. $SrO(s) + H_2O(l) \rightarrow Sr(OH)_2(aq)$
d. $MgO(s) + H_2O(l) \rightarrow 2 Mg(OH)_2(aq)$

53. $HC_2H_3O_2(aq) + NH_4OH(aq) \rightarrow H_2O(l) + NH_4C_2H_3O_2(aq)$

55. a. Vinegar: acid, acetic acid, $HC_2H_3O_2$, weak electrolyte
b. Toilet cleaner: base, sodium hypochlorite, NaClO, strong electrolyte
c. Car battery: acid, sulfuric acid, H_2SO_4, strong electrolyte
d. Carbonated soft drink: acid, carbonic and phosphoric acids, H_2CO_3 and H_3PO_4, weak electrolytes

57. a. 2.6
b. 1.0
c. 1.1
d. 4.5

59. a. insoluble
b. insoluble
c. insoluble
d. soluble

61. Balanced equation: $Na_2SO_4(aq) + CaCl_2(aq) \rightarrow 2 NaCl(aq) + CaSO_4(s)$
Total ionic equation: $2 Na^+(aq) + SO_4^{2-}(aq) + Ca^{2+}(aq) + 2 Cl^-(aq) + \rightarrow 2 Na^+(aq) + 2 Cl^-(aq) + CaSO_4(s)$
Net ionic equation: $Ca^{2+}(aq) + SO_4^{2-}(aq) \rightarrow CaSO_4(s)$

63. Balanced equation: $2 Al(NO_3)_3(aq) + 3 Na_2S(aq) \rightarrow Al_2S_3(s) + 6 NaNO_3(aq)$
Total ionic equation: $2 Al^{3+}(aq) + 6 NO_3^-(aq) + 6 Na^+(aq) + 3 S^{2-}(aq) \rightarrow Al_2S_3(s) + 6 Na^+(aq) + 6 NO_3^-(aq)$
Net ionic equation: $2 Al^{3+}(aq) + 3 S^{2-}(aq) \rightarrow Al_2S_3(s)$

65. KCN: 50 g/100 mL = 7.7 M
NaCN: 48 g/100 mL = 9.8 M
$Cu(CN)_2$: 0 g/100 mL = 0 M
$Cd(CN)_2$: 1.7 g/100 mL = 0.10 M

67. 0.00341 M

69. 9.6 g $C_6H_8O_7$

71. 0.0150 M

73. $CaO(s) + H_2O(l) \rightarrow Ca(OH)_2(aq)$

The poem says that the man was chained "with fetters on each foot, wrapped in a sheet of flame!" While quicklime does burn the skin, it doesn't do so with a sheet of flame. It slowly dissolves away the flesh off a man's body. Wilde exaggerates its use somewhat, but it still makes an interesting forensic science-related poem.

75. The difference between the two statements is that the presence of Coke doesn't necessarily mean that $TlNO_3$ is present. Saying that "the samples are consistent with thallium nitrate being added" is more specific.

77. 0.617 M $C_{12}H_{22}O_{11}$

Chapter 7

1. Polar solvents have dipole–dipole forces. They also have London dispersion forces (van der Waals forces), but these are much weaker (and therefore less important) than the dipole–dipole forces.

3. Hydrogen bonds form between molecules that have hydrogen bonded to nitrogen, oxygen, or fluorine atoms within them.

5. London dispersion forces are the weakest type of intermolecular forces because the induced dipole is only formed temporarily. Since it only exists for a limited amount of time, the force is weaker.

7. When ionic compounds are dissolved in a polar solvent, they dissociate. This leads to the development of ion–dipole forces within the solution.

9. The dissolution of a nonpolar molecule in a polar solvent takes place because of the dipole–induced dipole force. London dispersion forces also provide a minor contribution to the attractive forces between a nonpolar molecule and a polar solvent. This combined attractive force allows oxygen gas to dissolve in water.

11. In reality, there are many more water molecules than either potassium ions (large dark gray spheres) or sulfide ions (smaller light gray spheres). However, that would make the drawing much too cluttered. The ratio

of two potassium ions for every sulfide ion is required for the ionic compound to be uncharged overall.

13. Gasoline is composed of a mixture of several different nonpolar compounds (e.g., octane, C_8H_{18}). Because grease is composed of various nonpolar compounds, gasoline is a good solvent to dissolve away the grease from mechanics' hands.

15. The rate of dissolving a soluble compound is affected by temperature, surface area of solid, and rate of stirring.

17. Water boils at 100°C only when it is completely pure and the atmospheric pressure is the same as that at sea level. Since many researchers do experiments in locations where the atmospheric pressure is significantly different from that at sea level, it is important to calibrate thermometers using other criteria.

19. The boiling point of a solution is the temperature at which the vapor pressure of the solution is equal to atmospheric pressure. We can observe this when bubbles form spontaneously within the solution and rise to the top of the solution.

21. The solute particles block solvent molecules from moving into the vapor phase by decreasing the number of solvent molecules available to escape into the vapor phase. This decreases the vapor pressure of the solution.

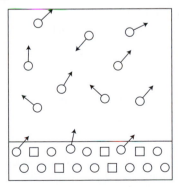

23. If saltwater has a freezing point of −4°C, then lowering the temperature to −5°C will require that the entire sample become a solid. All the liquid water molecules line up around the dissolved salt ions and form a crystal that eventually spreads through the entire sample to make a solid block of frozen saltwater. In the diagram below, circles represent water molecules, squares represent sodium ions, and triangles represent chloride ions. The precise arrangement of ions in the crystal can vary dramatically, depending on concentration, but it is important to note that the number of sodium ions must equal the number of chloride ions in the sample.

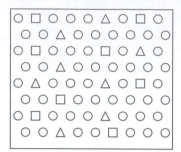

25. Calcium chloride is not more efficient than sodium chloride at lowering the freezing point of water because there are three moles of ions formed for every 110.986 g of $CaCl_2$, while there are only two moles of ions formed for every 58.44277 g of NaCl. That means that more ions per gram are formed from sodium chloride than from calcium chloride, making the sodium chloride more efficient.

27. High salt concentrations used in pickling cause high osmotic pressures within the cells. Any bacteria or mold cells that attempt to survive in this environment will become extremely dehydrated and are unlikely to be successful.

29. HPLC uses the solubility of various components in an eluting solvent. If the solvent and solute have similar intermolecular forces, the solute will be very soluble in the solvent and will elute quickly. If the solute has intermolecular forces similar to the stationary phase of the column, it will elute more slowly. By controlling the composition of the solvent, HPLC controls the rate at which solutes elute and give an indication of the composition of a mixture of liquids.

31. The regular octahedron has 19% more surface area than the sphere with equal volume.

33. Be < Mg < Ca < Sr

35. a. <u>Hydrogen bonding</u>, dipole–dipole forces, dipole–induced dipole forces, and London dispersion forces
   b. <u>Hydrogen bonding</u>, dipole–dipole forces, dipole–induced dipole forces, and London dispersion forces
   c. <u>London dispersion forces</u>

37. a. <u>Dipole–induced dipole</u> forces and London dispersion forces
   b. <u>Hydrogen bonding</u>, dipole–dipole forces, dipole–induced dipole forces, and London dispersion forces
   c. <u>Ion–dipole forces</u>, hydrogen bonding, dipole–dipole forces, dipole–induced dipole forces, and London dispersion forces

39. HI will have a higher boiling point than HBr. Both compounds have dipole-dipole forces and London dispersion forces. Although HBr is slightly more polar than HI, HI has significantly stronger London dispersion forces.

41. KBr and Na_2SO_4 will dissolve, while KCl, NaCl, and KNO_3 will be supersaturated.

43. a. 28 g/100 mL
   b. 35 g/100 mL
   c. 48 g/100 mL
   d. 60 g/100 mL

45. a. 53°C
   b. 13°C
   c. 46°C
   d. 25°C

47. 0.45 M LiI = 0.30 M $CaCl_2$ > 0.25 M Na_3PO_4

49. 0.50 m Na_2S > 0.66 m NaCl > 0.15 m $C_6H_{12}O_6$

51. 0.66 m NaCl: 100.68°C
   0.25 m Na_2S: 100.38°C
   0.15 m $C_6H_{12}O_6$: 100.08°C

53. a. 100.58°C
   b. 101.43°C
   c. 102.81°C

55. a. −2.11°C
   b. −5.21°C
   c. −10.21°C

57. 0.66 m NaCl: −2.46°C
   0.25 m Na_2S: −1.40°C
   0.15 m $C_6H_{12}O_6$: −0.28°C

59. a. −12.11°C
   b. −18.60°C
   c. −6.29°C

61. a. 1.51 m
   b. 5.27 m
   c. 2.54 m

63. a. 272.04 g
   b. 949.44 g
   c. 457.61 g

65. Osmosis naturally takes place to make solute concentrations the same on both sides of a membrane. Water molecules must pass through the membrane from the pure water to the seawater and solute particles must pass through the membrane from the seawater to the pure water.

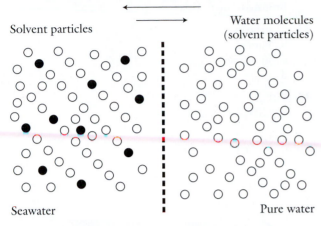

Solvent particles

Water molecules (solvent particles)

Seawater

Pure water

67. a. 1890 Pa (0.0187 atm, 14.2 Torr)
   b. 152 Pa (0.0015 atm, 1.14 Torr)
   c. 453 Pa (0.00447 atm, 3.40 Torr)

69. Alcohol permeates into the vitreous humor as soon as it is present in the bloodstream. If the deceased person consumed an extremely large portion of alcohol, the only way it would be in the bloodstream is if there was enough time for it to be absorbed through the stomach and small intestine. The answer to this question depends on the amount of time that passed between the drinking of the alcohol and the fatal accident. Other physical properties that could be examined to give more information would be freezing point and solubility in water.

71. Having both boiling point and density information about the liquid sample will give a better indication of what is present. It is important to remember that the boiling point can change with the amount of dissolved solvent (boiling point elevation) and that the density can also change depending on the amount of solvent. If the boiling point is measured and it matches the boiling point of propanol, the only way to be sure the liquid sample is propanol is if the density of the liquid is also a match. (In reality, other techniques such as infrared spectroscopy would be used to help identify the liquid. The more points of comparison between a sample and the known substance, the better!)

73. Immunoassay methods rely on intermolecular forces to select for specific antibody/antigen pairs in the assay. The "lock-and-key" interactions between antigens and their associated antibodies are very specific applications of specific intermolecular forces. There are often different regions of the molecules, some of which are polar and some of which are nonpolar, in specific arrangements to give the specific interactions that allow for selective determination of a particular molecule.

75. Trace levels of the compounds will be found in the young man's lungs, even if they were the cause of death, because the petroleum-based cleaners will gradually diffuse out of his lungs as the body was left for 16 hours from the time of death. The hypothesis might still be valid. The medical examiner will also look for petroleum-based compounds in the blood and other vital organs as an indication that the compounds were responsible for the young man's asphyxiation.

Chapter 8

1. Historically (until the early nineteenth century, according to our textbook), organic compounds were those isolated from living or once-living organisms. They typically contained a high percentage of carbon, with some hydrogen, oxygen, and nitrogen (sometimes sulfur and phosphorus as well). Inorganic compounds were considered to be composed of the remaining elements (chiefly non-carbon-based) of the periodic table. Rocks and minerals are classic examples of inorganic materials.

3. All alkanes have the formula C_nH_{2n+2}, where n is any positive integer. Alkanes have found utility as solvents and fuel sources.

5. Alkenes that contain only one double bond have the formula C_nH_{2n}, where n is any positive integer. Alkenes that contain only one triple bond have the formula C_nH_{2n-2}, where n is any positive integer.

7. Line structures are a shorthand notation of the structures of organic molecules. Each line represents a bond, and each corner (that has not been labeled otherwise) represents a carbon atom. Hydrogen atoms are implied; since carbon typically forms four bonds in organic compounds, hydrogen atoms are bonded to each carbon atom to bring its total number of bonds to four.

9. Two isomers have at least one physical property that is different. By rearranging the atoms in a molecule, we change the shape of the molecule. This impacts polarity, which in turn affects solubility, melting/freezing point, boiling/melting point, vapor pressure, and many other properties.

11. A stereoisomer is a compound with the same combination of functional groups but placed in a different arrangement (e.g., cis- and trans- isomers). Two compounds that are constitutional isomers are ethanol (CH_3CH_2OH) and dimethyl ether (CH_3OCH_3).

13. A monomer is a small molecule of which multiple copies can be combined to form a polymer. The polymer is a long chain molecule made up of individual repeating units that are derived from the monomer.

15. A copolymer is a polymer synthesized from more than one monomer simultaneously.

17. Resonance structures occur when more than one equivalent structure (where no atoms are rearranged) can be drawn for a molecule. It means that there is more than one way to arrange the electrons in the molecule and still have the same structure, and typically means that the molecule is more stable than it would be otherwise.

19. Although the Lewis structure of benzene contains C—C (carbon–carbon single bonds) and C=C (carbon–carbon double bonds) as well as C—H single bonds, all the carbon–carbon bond lengths in benzene are actually equal. This is because of resonance—the carbon–carbon bonds are actually intermediate in length between a single and a double bond. They can be thought of as an average between a single and a double bond, or a one-and-a-half bond, because of resonance.

21. Ketones, ethers, and esters all contain oxygen atoms. Ketones have a C=O group (carbon double-bonded to oxygen) near the middle of the molecule (aldehydes have C=O at the end of the molecule, with an H atom bonded to the carbon). Ethers have an O atom within the molecule: C—O—C. Esters have a C=O group in the middle of the molecule that is bonded to an O atom, which is in turn bonded to another carbon atom (carboxylic acids have the C=O bonded to an OH group instead).

23. A plasticizer is a compound added to polymers to make them more flexible.

25. The amine functional group is a nitrogen bonded to the molecule ($-NH_2$ or similar). A primary amine has only one carbon bonded to the nitrogen (CNH_2), while secondary and tertiary amines have two (C_2NH) and three (C_3N) carbons bonded to the nitrogen, respectively.

27. Water is produced in a condensation reaction.

29. Congeners are compounds that have the same functional groups as one another. Methanol (CH_3OH), ethanol (CH_3CH_2OH), 1-propanol ($CH_3CH_2CH_2OH$), 2-propanol ($CH_3CH(OH)CH_3$), and other alcohols are congeners.

31. A Brønsted-Lowry base is a proton acceptor. An Arrhenius base is a compound that produces OH^- in aqueous solution (when dissolved in water). NaOH is a good example of an Arrhenius base. NH_3 is a Brønsted-Lowry base.

33. a. propane
    b. butane
    c. methane
    d. heptane

35. a. C_6H_{14} c. C_5H_{12}
    b. C_4H_{10} d. $C_{10}H_{22}$

37. a. $CH_3CH_2CH_3$
    b. $CH_3CH_2CH_2CH_3$
    c. CH_4
    d. $CH_3CH_2CH_2CH_2CH_2CH_2CH_3$

39. a.
    b.
    c.
    d.

41. a. C_3H_4, CH_3CCH,
    b. C_9H_{18}, $CH_3CH_2CHCHCH_2CH_2CH_2CH_2CH_3$,
    c. C_7H_{14}, $CH_3CHCHCH_2CH_2CH_2CH_3$,
    d. C_5H_8, $HCCCH_2CH_2CH_3$,

43. a. Two unique isomers: 1-butyne and 2-butyne
    b. Four unique isomers: 1-octyne, 2-octyne, 3-octyne, and 4-octyne
    c. Three unique isomers: 1-pentene, 2-pentene, and 3-pentene
    d. One unique isomer

45. a. 2-methylhexane
    b. 3-methylheptane
    c. 3-methylheptane
    d. 2-methylpentane

47. a.
    b.
    c.
    d.

49. a. C_6H_6,

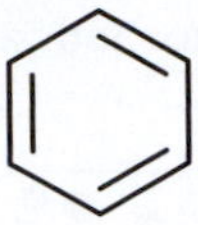

b. C_6H_5OH,

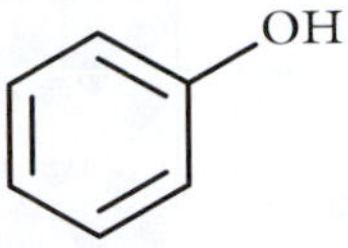

c. C_4H_8,

d. C_6H_{12},

51. a. $CH_3OCH_2CH_2CH_3$
b. $CH_3CH_2OCH_2CH_3$
c. $CH_3CH_2CH_2CH_2OCH_2CH_2CH_2CH_2CH_2CH_3$
d. $CH_3CH_2CH_2OCH_2CH_2CH_2CH_3$

53. a. $CH_3COCH_2CH_2CH_2CH_3$
b. $CH_3COCH_2CH_2CH_3$
c. $CH_3CH_2COCH_2CH_2CH_2CH_3$
d. $CH_3CH_2CH_2COCH_2CH_2CH_2CH_2CH_3$

55. a. $CH_3OCOCH_2CH_3$
b. $CH_3CH_2OCOCH_3$
c. $CH_3CH_2OCOCH_2CH_3$
d. $CH_3CH_2CH_2OCOCH_3$

57. a. 3-hexanone
b. Ethoxybutane
c. Ethyl pentanoate
d. 2-pentanone

59. a. $CH_3CH_2CH_2CH_2CH_2NH_2$
b. $CH_3CH_2NH(CH_2CH_2CH_3)$
c. $(CH_3)_3N$
d. $CH_3CH_2CH_2CH_2N(CH_3)(CH_2CH_2CH_3)$

61. a. 
b. 
c. 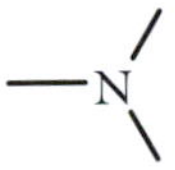
d.

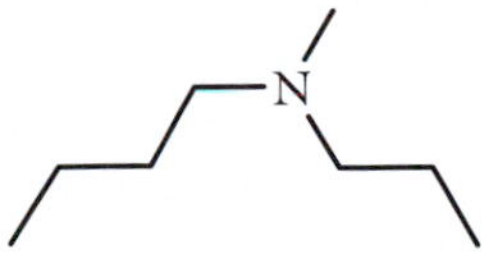

63. a. $CH_3CH(OH)CH_2CH_3$
b. $CH_3CH(OH)CH_3$
c. $CH_3CH_2CH_2CH_2CH_2OH$
d. $CH_3CH(OH)CH_2CH_2CH_2CH_3$

65. a. $CH_3CH_2CH_2CHO$
b. $CH_3CH_2CH_2CH_2CH_2CHO$
c. CH_3CHO
d. $CH_3CH_2CH_2CH_2CH_2CH_2CH_2CHO$

67. a. $CH_3(CO)OH$
b. $CH_3CH_2CH_2CH_2CH_2CH_2CH_2(CO)OH$
c. $CH_3CH_2CH_2CH_2CH_2CH_2CH_2CH_2CH_2(CO)OH$
d. $CH_3CH_2CH_2(CO)OH$

69. a. Pentanal
b. Hexanoic acid
c. 2-pentanol
d. Hexanal

71. $CH_3CH_2CH_2CH_2CH_2CH_2OH$, 1-hexanol;
$CH_3CH_2CH_2CH_2CH(OH)CH_3$, 2-hexanol;
$CH_3CH_2CH(OH)CH_2CH_2CH_3$, 3-hexanol

73. a. Butane
b. Pentane
c. Named correctly
d. Propane

75. a. 1-propanoic acid, or simply propanoic acid, because the carboxylic acid group should receive the lowest possible position number.
b. 2-propanone, or simply propanone (commonly called acetone), because an aldehyde group cannot exist on the second carbon in a propane molecule.
c. Butanal, because a ketone group cannot exist on a terminal carbon in a chain.
d. 2-butanol, because the alcohol group should get the lowest possible position number, and the third carbon becomes the second carbon when the butane is numbered from the opposite group.

77. a. Neutral compound
b. Acid
c. Acid
d. Base

79. High density polyethylene is a straight-chain alkane polymer.

81. Polyester, polystyrene, and polycarbonate polymers all containe benzene.

83. Polyester contains the ester functional group.

85. Nylon contains a secondary amine functional group that would be susceptible to attack by a strong acid.

87. a. CH_3OH
    b. $CH_3CH_2CH_2CH_2OH$
    c. $CH_2(OH)CH_2CH{=}CH_3$
    d. $CH_3CH(OH)CH_2CH_3$
    e. $CH_3CH(CH_3)CH_2OH$
    f. $CH_3CH_2CH(CH_3)CH_2OH$
    g. $CH_3CH(CH_3)CH_2CH_2OH$

89. CH_2O, $CH_3(C{=}O)H$, $CH_3CH_2(C{=}O)H$,
    $CH_3CH_2CH_2(C{=}O)H$, $CH_3CH_2CH_2CH_2(C{=}O)H$,
    $CH_3(CH_2)_4(C{=}O)H$, $CH_3(CH_2)_5(C{=}O)H$,
    $CH_3(CH_2)_6(C{=}O)H$, $CH_3(CH_2)_7(C{=}O)H$,
    $CH_3(CH_2)_8(C{=}O)H$

91. To extract aspirin from the stomach contents, an extraction in a separatory funnel with hexanes should be sufficient. Aspirin is both an ester and a carboxylic acid, and it should be neutral (uncharged) at the low pH found in stomach acid. Amines will be protonated by the strongly acidic solution and so will not be soluble in hexanes. (It is fairly likely that the stomach acid will have decomposed some of the aspirin—acetylsalicylic acid—and removed the ester functionality to leave the combination of carboxylic acid and alcohol—salicylic acid. This is likely to be protonated in the acidic stomach contents, so it will not be extracted into hexanes along with the unreacted aspirin.)

93. Kevlar has London dispersion forces, dipole-dipole forces, and hydrogen bonding as intermolecular forces between chains.

Chapter 9

1. Since the melting point of copper metal is nearly 1083°C, at both 50°C and 500°C, copper is a solid (temperature is *below* the melting point). Solids are composed of atoms that are closely packed together. However, the atoms of copper are *not* stationary. There is enough thermal energy to cause them to "wiggle" about in their packed arrangements. As the temperature is raised from 50°C to 500°C, the atoms have more energy available and "wiggle" more than they did at the lower temperature. Once the two samples are brought together, energy is transferred as the atoms "bump" into each other, until the energy is distributed evenly throughout the sample, thus reaching thermal equilibrium (see figure below). The temperature at this point ($\approx$275°C) is the average of the two initial temperatures since both samples were equal in mass.

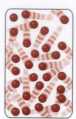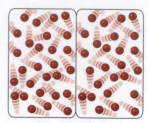

Cu at 50°C   Cu at 500°C   Cu at thermal equilibrium (275°C)

If the 500°C sample were twice the mass of the 50°C sample, then thermal equilibrium would be reached at 350°C because the 500°C mass represents 2/3 of the total mass.

3. Heat is a form of energy that is transferred from hot objects to cold objects when they come in contact with one another. Temperature is a measurement of the average kinetic energy of the particles in a system, not a measurement of how much heat is in a system.

5. When a thermometer is placed in a solution, heat will flow either from the solution into the thermometer or from the thermometer into the solution. When this happens, the temperature of the solution will change. Thermometers are designed to minimize the amount of heat that flows in this case, but there is still a slight change in temperature whenever a thermometer is placed into a solution.

7. A backdraft forms when an excess of fuel is heated in the absence of oxygen. It can be prevented by ventilating the room (providing a source of oxygen from the air surrounding the room).

9. Organic compounds and certain metals and nonmetals can undergo combustion reactions.
   $CH_4(g) + 3\ O_2(g) \rightarrow CO_2(g) + 2\ H_2O(g)$
   $2\ Ca(s) + O_2(g) \rightarrow 2\ CaO(s)$
   $S_8(s) + 12\ O_2(g) \rightarrow 8\ SO_3(g)$

11. We can determine whether an element within a compound is oxidized or reduced in a reaction by comparing the oxidation number of the element on each side of the chemical equation. The precaution to be followed is that oxidation is a loss of electrons and reduction is a gain of electrons. However, electrons have a negative charge, so oxidation results in an oxidation number that is more positive, and reduction results in an oxidation number that is less positive (more negative).

13. An endothermic reaction absorbs energy from its surroundings and an exothermic reaction releases energy into its surroundings. If the products have more energy than the reactants, the system must absorb energy from the surroundings (endothermic). If the products have less energy than the reactants, the system must release energy into the surroundings (exothermic).

15. Flashover takes place when heat from a fire in a small area rises to the ceiling and then spreads out across the room. As the fire progresses, the heated layer of smoke and gases just below the ceiling increases its temperature and thickness. Because the heat is restricted by the ceiling, it radiates downward, heating the remaining combustible materials in the room. This continues until the materials reach the autoignition temperature, where the materials have absorbed enough heat energy from the surroundings to overcome the activation energy of the combustion reaction. Flashover conditions can be prevented by ventilating the room at the ceiling, allowing the heated layer of smoke to escape.

17. Compounds with strong intermolecular forces have larger heat capacities than those with weak intermolecular forces. Compounds with hydrogen bonding (like water) have large heat capacities, and nonpolar compounds (like hexane) have small heat capacities.

19. Tile floor feels colder than carpeting even if they are at the same temperature because of the flow of heat from your bare foot into the material. Tile floor conducts heat more quickly than carpeting, so the heat flows more quickly from your foot into the tile than it does into the carpet. In addition, tile floor has a greater relative heat capacity than carpeting and can absorb more energy as heat before it reaches the same temperature as your foot. The net effect is that the tile floor feels cold and the carpet feels warmer.

21. Accelerants are often not completely consumed in a fire because the vaporization of the liquid fuel is an endothermic process. This lowers the temperature of the region surrounding the fuel to the point that there is insufficient heat for complete combustion to take place. As a result, there is usually some trace amount of accelerant remaining in the residue left after a fire.

23. Crude oil is refined using a process called partial distillation. By heating the crude oil, compounds boil out of the mixture in quantities grouped by boiling point. According to our textbook, gasoline has components with boiling points from approximately 40°C to 220°C, and kerosene has components with boiling points ranging from 175°C to 270°C. Gasoline has more low molecular weight compounds and kerosene has more high molecular weight compounds.

25. a. 272 J
    b. 695,000 J
    c. 74,600 J
    d. 1.31×10^6 J

27. a. 46,000 J/g
    b. 48,100 J/g
    c. 41,800 J/g
    d. 41,500 J/g

29. a. 43,300 J/g
    b. 43,500 J/g
    c. 43,700 J/g
    d. 44,000 J/g

31. a. $CH_4 + 2\,O_2 \rightarrow CO_2 + 2\,H_2O$
    b. $C_5H_{12} + 8\,O_2 \rightarrow 5\,CO_2 + 6\,H_2O$
    c. $C_6H_{14}O + 9\,O_2 \rightarrow 6\,CO_2 + 7\,H_2O$
    d. $2\,C_8H_{16}O + 23\,O_2 \rightarrow 16\,CO_2 + 16\,H_2O$

33. a. $2\,C_4H_{10}(g) + 13\,O_2(g) \rightarrow 8\,CO_2(g) + 10\,H_2O(g)$
    b. $2\,C_8H_{18}(l) + 25\,O_2(g) \rightarrow 16\,CO_2(g) + 18\,H_2O(g)$
    c. $2\,C_5H_{10}(l) + 15\,O_2(g) \rightarrow 10\,CO_2(g) + 10\,H_2O(g)$
    d. $2\,C_6H_{10}(l) + 17\,O_2(g) \rightarrow 12\,CO_2(g) + 10\,H_2O(g)$

35. a. $C_3H_8(g) + 5\,O_2(g) \rightarrow 3\,CO_2(g) + 4\,H_2O(g)$
    b. $C_5H_{12}(l) + 8\,O_2(g) \rightarrow 5\,CO_2(g) + 6\,H_2O(g)$
    c. $2\,CH_3CH_2CH_2OH(l) + 9\,O_2(g) \rightarrow 6\,CO_2(g) + 8\,H_2O(g)$
    d. $2\,CH_3CH_2CH_2CH_2CH_2OH(l) + 15\,O_2(g) \rightarrow 10\,CO_2(g) + 12\,H_2O(g)$

37. a. Al oxidation state +3
    b. Fe oxidation state +2, O oxidation state −2
    c. H oxidation state +1, S oxidation state −2
    d. Fe oxidation state +3, O oxidation state −2

39. a. CO_2 is being reduced and H_2O is being oxidized.
    b. CH_4 is being oxidized and O_2 is being reduced.

41. a. CO is being oxidized and O_2 is being reduced.
    b. Na is being oxidized and H_2O is being reduced.

43. 0.950 J/(g °C)

45. 290,000 J

47. 5.0 g

49. 4.52×10^6 J

51. 0.2817 L H_2O

53. 17.9 kJ

55.

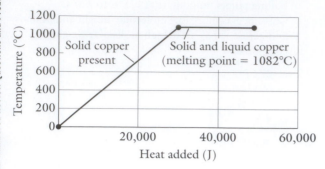

57. 0.551 J/(g °C)

59. 339.2°C

61. 8.62 g Teflon

63. 36.3°C

65. 5.50×10^3 J/g

67. In a fire sprinkler, a solid plug should have a melting temperature that allows it to melt at temperatures typically reached in a fire but not during normal building operation. The liquid in a vial should have a boiling temperature that allows it to boil and shatter the vial in a fire but not in normal conditions.

69. If benzene is found in a blank sample of carpeting that has been analyzed by gas chromatography, the only way the presence of benzene in the suspected sample could still be used to determine whether an accelerant was used to ignite the fire is if there is significantly more benzene in the suspected sample than in the blank. Even still, it would be difficult for a jury to accept that evidence "beyond reasonable doubt."

71. Signs of electrical arcing in an electrical box do not necessarily prove that a structural fire was caused by the arcing. There also must be sufficient evidence to show that the electrical box was the source of the fire.

73. Smoke consists of small particles freed from materials burning in a fire. If flames appear to be forming out of the smoke that is pouring out of a structural fire, that could be entirely possible. If the temperature of the smoke is high enough, the exposure of the smoke to the oxygen-enriched atmosphere outside the structure could allow the particles that make up the smoke to burst into flames once they reach open air.

75. No; cone calorimeters are measuring the decrease of oxygen when being used up in the combustion reaction, so the values will be a bit different.

77. If a large amount of heated smoke is trapped near the ceiling, it is entirely possible that a plastic fire detector might melt during a fire even if it had never been tampered with.

79. Fire requires fuel, heat, and oxygen to burn. Charring between deck boards can easily result if the deck boards are sufficiently close together to limit the amount of oxygen. The charring would take place because there is insufficient oxygen present for complete combustion.

Chapter 10

1. For a compound to be explosive, it must be capable of releasing a large amount of energy. In addition, the compound must react instantaneously and release substantial amounts of gaseous products.

3. Most low explosives serve as propellants for guns and military artillery.

5. It is important for explosive molecules to contain C, H, and O because they have gaseous combustion products (carbon dioxide and water), and the compound already contains some (if not all) of the oxygen necessary to produce those products. This makes them easier to oxidize than a pure hydrocarbon (due to their oxygen balance).

7. The four principles of kinetic-molecular theory: (1) Gas particles are extremely small and have relatively large distances between them. (2) Gas particles act independently of one another; there are no attractive or repulsive forces between gas particles. (3) Gas particles are continuously moving in random, straight-line motion as they collide with each other and the container walls. (4) The average kinetic energy of gas particles is proportional to the temperature of the gas.

9. Gases can be compressed to a much greater extent than liquids and solids because they have much more space between their particles (see premise #1 of kinetic-molecular theory).

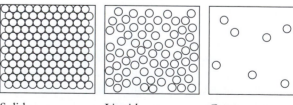

Solid (molecules packed)   Liquid (molecules touching)   Gas (molecules far apart)

11. Above the UEL (upper explosive limit), the fuel-air mixture is too fuel-rich for the fuel to react explosively because the reaction is limited by the amount of oxygen present in the mixture.

13. Avogadro's law says that volume is directly proportional to the number of gas particles if temperature and pressure are held constant. As the number of particles increases, the volume of gas increases. As the number of particles decreases, the volume of gas decreases.

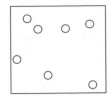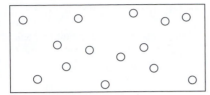

If we double the number of gas molecules, we also double the volume of gas present.

15. Gay-Lussac's law says that pressure is directly proportional to temperature if volume and number of moles are held constant. As the temperature increases, the pressure inside the gas chamber increases. As the temperature decreases, the pressure decreases.

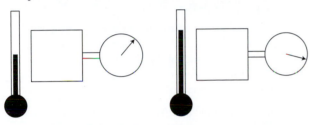

17. STP is an abbreviation for standard temperature and pressure. These are the conditions of 1 atm pressure (760 mmHg, 760 torr, 101.325 kPa) and 273.15 K (0°C, 32°F).

19. Dalton's law of partial pressures states that the total pressure of a gas mixture is the sum of the pressures of each component in the mixture. This requires that each gas behave independently of all other gases in the mixture and that each gas continue to undergo collisions in the same exact manner. This is consistent with the four premises of the kinetic-molecular theory of gases (see answer to Question 7). Premise #2 is most applicable, saying that the gas particles act independently of one another. If particles of one gas act independently of other particles of the same gas, it is a natural extension to say that particles of one gas act independently of particles of a second gas.

21. Chemical markers are high vapor pressure compounds that are included in plastic explosives when manufactured. The high vapor pressure compound forms a vapor surrounding the plastic explosive that is more easily detected than the explosive itself. As a result, it is easier to detect the explosives (using canine or electronic detection systems) and prevent their illegal use.

23. Taggants, markers, and isomers help security officials during all phases of explosive detection and identification. Chemical markers enable security officials to detect explosives before they are used. Taggants and isomers allow the identification of an explosive either before or after detonation. This helps the security officials track the explosive from factory through sale to use and identify any illegal activity along the way.

25. a. 3.5 moles O_2 required
   b. 6.5 moles O_2 required
   c. 3 moles O_2 required
   d. 3 moles O_2 required

27. a. 3.5 molO_2
   b. 3 mol O_2
   c. 6.5 mol O_2
   d. 4.5 mol O_2

29. a. If volume decreases, pressure increases (Boyle's Law).
   b. If temperature increases, volume increases (Charles's Law).
   c. If temperature decreases, pressure decreases (Gay-Lussac's Law).
   d. If the amount of gas decreases, pressure decreases (Avogadro's Law).

31. a. $n_2 = 0.187$ mol
   b. $V_1 = 816.0$ mL
   c. $V_2 = 4.34$ L
   d. $n2 = 1.60$ mol

33. a. $P_2 = 3.06$ atm
   b. $V_2 = 21.2$ mL
   c. $V_1 = 0.642$ L
   d. $P_1 = 2.10$ atm

35. a. $T_2 = 2200$ K
   b. $P_1 = 3.74$ atm
   c. $T_1 = 462$ K
   d. $P_2 = 10.4$ atm

37. a. $T_2 = 3010$ K
   b. $V_1 = 5.46$ L
   c. $V_1 = 2.07$ L
   d. $T_2 = 355$ K

39. a. 140 L
   b. 515 L
   c. 16.6 L
   d. 18.2 L

41. a. 10.0 L
   b. 2.81 L
   c. 45.6 L
   d. 26.8 L

43. a. 0.658 g/L
    b. 1.31 g/L
    c. 1.15 g/L
    d. 1.80 g/L

45. 9.02 L

47. 266 K (−6°C)

49. 52.7 atm

51. 18.87 L

53. 6340 K (6060°C)

55. 7.06 mol Ar

57. 7.29 atm

59. Neon

61. 172 g NaN_3

63. 61.2 atm

65. Negative oxygen balance

67. Explosives with a positive oxygen balance will produce white smoke because the major component will be steam condensing on the small particles produced by the explosion. Explosives with a negative oxygen balance will produce black, sooty smoke because there will be partially combusted particles composed primarily of leftover carbon.

69. Gases are in constant, random motion and collide with other gas molecules. An explosion requires that the fuel, oxygen, and ignition source be present simultaneously. If the partial pressure of fuel is below the LEL, an explosion will not occur; if it is above the UEL, there is insufficient oxygen for the reaction to continue. Because fuels are in constant random motion, a fuel initially present at a partial pressure above the UEL may soon become explosive as the fuel mixes with oxygen gas from the environment, thus lowering its partial pressure.

71. For proper breathing, a person's lungs must be able to expand and contract (via diaphragm movements) to create a pressure difference with respect to the outside atmosphere. To inhale, the lungs expand (increase V, decreasing P), thereby reducing the pressure of the air in the lungs, drawing in outside air. In order to exhale, the lungs contract (decrease V, increasing P), increasing the pressure of the air in the lungs and forcing some outside of the body. If a puncture wound is suffered, then the lungs have great difficulty creating a pressure difference and it is very hard to breathe.

73. The isomers of DNT are 2,3-dinitrotoluene, 2,4-dinitrotoluene, 2,5-dinitrotoluene, 2,6-dinitrotoluene, 3,4-dinitrotoluene, and 3,5-dinitrotoluene.

2,3-dinitrotoluene   2,4-dinitrotoluene   2,5-dinitrotoluene

2,6-dinitrotoluene   3,4-dinitrotoluene   3,5-dinitrotoluene

75. $P_{CO_2} = 3.628$ atm
    $P_{H_2O} = 3.023$ atm
    $P_{N_2} = 1.814$ atm
    $P_{O_2} = 0.3023$ atm
    $P_{total} = 8.767$ atm

77. If the explosion had ruptured the cabin integrity at high altitude, the higher pressure inside the cabin than that surrounding the airplane would have caused the air to rush out through the rupture, depressurizing the entire airplane.

79. The second TLC plate shows the best separation of all five explosive components. The first plate does not adequately separate the TNT, PETN, and NG components from each other. The third plate does not adequately separate the RDX and tetryl components from each other.

Chapter 11

1. The rates of chemical reactions are very important to understand when trying to establish evidence of time in forensic science. Our textbook opens this chapter talking about the case of "Kari" in the Hawaiian islands. A forensic etymologist was able to use his knowledge about the rate of growth of insects and their feeding habits to more accurately estimate the time of Kari's death. Knowing the rates of other chemical reactions can give similar sorts of information.

3. The three principles of collision theory: (1) The reactants must collide for a reaction to take place. (2) Collisions must have high energy. (3) The colliding particles must be properly oriented for a reaction to occur.

5. The initial concentration of the reactants affects most reaction rates because it determines the number of collisions that can potentially take place between

reactant molecules. According to collision theory, reactants must collide for a reaction to take place. Therefore, the initial concentration will often have an effect on reaction rates.

7. Bloody clothes recovered as evidence are immediately dried because the moisture in the clothes is a reactant in the decomposition reactions of the blood. If the clothes are dry, the biological evidence degrades more slowly, and more information can be obtained by the forensic scientists.

9. It is not enough simply for the molecules to have sufficient energy. The collisions must also have the correct orientation for a reaction to take place. As a result, only a fraction of collisions between molecules having sufficient energy will result in the reaction of the reactant molecules.

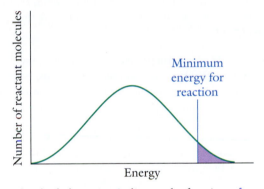

The shaded portion indicates the fraction of reactant molecules that have sufficient energy to react.

11. Increasing the temperature of a reaction system does not change the amount of energy needed for reactants to cross the activation barrier. As the temperature increases, the average kinetic energy of the molecules increases. This causes the collisions to have greater energy and allows the reaction to occur more quickly.

13. The greater the surface area of a heterogeneous catalyst, the more available reaction sites on the surface of the catalyst. Since reaction rate depends on the number of collisions that take place, having more available reaction sites on the surface increases the rate of the reaction.

15. Adding a catalyst is the only method of increasing the reaction rate that works by lowering the activation energy. The catalyst provides an alternative pathway (mechanism) for the reaction to take place, which is lower in energy than the pathway for the uncatalyzed reaction.

17. A catalytic converter is a heterogeneous catalyst. Solid particles are suspended on a porous support to maximize surface area, and the reactants pass over the catalyst in the gas phase.

19. Most enzyme-catalyzed reactions have zero-order kinetics. Doubling the concentration of the enzyme will not have any effect on the rate of the reaction. However, at some point, the enzyme molecules impede the mixing and collision of reactant molecules, so the rate will eventually decrease. Since a very small amount of enzyme is required to catalyze the reaction, this is a highly unusual situation.

21. The analysis of blood alcohol can be done hours after a person has been detained, and an accurate value for the BAC can be determined as of the time a person was arrested. The method for this analysis relies on knowing the rate of oxidation of ethanol to carbon dioxide in the body.

23. If an investigator determines the blood serum levels of both legal and illegal drugs, he or she can use this information coupled with the half-lives of those drugs to estimate the blood serum level of those drugs at some earlier time (such as when the drugs were originally taken into the body). This can help predict the level of intoxication and determine whether a drug could have been the cause of death for a victim.

25. 2.67 hours

27. The injured person claims the headlights were out at the time of the accident, while the driver claims that the broken headlight resulted from the accident. The key lies in the rate of oxidation of the filament. If the headlights were on at the time of the accident, the filament would be at 3000 K as the headlight broke. This would result in the rapid oxidation of the filament. If it was off, it would be at a much lower temperature (approximately 300 K is a normal temperature) and the rate of oxidation would be much slower and less oxidized material would be present. If the headlight broke at a different time (before the accident), there are two possibilities: Either the headlight was on, or it was off. In either case, the same oxidation processes would take place, but there would be additional oxidation over the time between the break and the accident. The investigator would need to carefully measure the amount of oxidized material and compare it to the amount of oxidized material in headlights of the same make and model that are tested in a controlled environment. Then the courts can determine who is telling the truth about the headlights.

29. Lead-210 would not work well for a recent homicide victim because it has a relatively long half-life of 22.3 years. Similarly, polonium-210 would not work on an older homicide because its half-life is only 138 days. If the victim was found buried 30 years after having disappeared, essentially all of the polonium-210 would have decayed and the time since death would not be determinable.

Chapter 12

1. Henri Becquerel originally thought that phosphorescence was related to X-rays. He used a mineral crystal containing salts of uranium on a wrapped photographic plate. When the mineral crystal was exposed to sunlight and then put in the dark, it exhibited phosphorescence and exposed the photographic plate. By chance, he did the same experiment with a crystal that had not been exposed to sunlight (it was left in a drawer). He discovered that the same pattern of exposure was present on the photographic plate. Clearly, the X-rays resulted from something other than phosphorescence. The mineral crystal was phosphorescent, but it also emitted X-rays in the absence of exposure to sunlight. This eventually led to the discovery of radioactivity.

3. If Becquerel had used a mineral ore other than one that contained uranium, it is likely that he would not have discovered radioactivity. He would have found that phosphorescent materials do not emit X-rays and would have tried to find another way to explain that phenomenon, altogether missing the opportunity to shape history by the accidental discovery of the radioactivity of uranium.

5. Neutrons help to stabilize the nucleus of the atom. Since protons have positive charge, they repel one another. The neutrons can be thought of as the "glue" holding the nucleus together, according to our textbook.

7. Alpha decay is the emission of an alpha particle (a helium nucleus). As a result, the atomic mass decreases by 4 atomic mass units (amu). The nucleus loses two protons and two neutrons. Beta decay is the emission of a beta particle (an electron). As a result, the atomic mass decreases by a very small amount (less than 0.001 amu). The number of neutrons decreases by 1 and the number of protons (and the atomic number) increases by 1. Gamma particles have no mass, so gamma emission has no effect on the atomic mass or atomic number of the atom.

9. Alpha particles have the greatest ionization power (+2 charge). Gamma particles have the least ionization power (no charge).

11. Inside the body, alpha particles pose the greatest risk. This is because they have the greatest ionization power and can damage the tissues as a result. They also have the least penetrating power, so they remain in the body until they have impacted a large amount of tissue. Gamma particles have a higher energy, but they have a great penetrating power and will leave the body before doing too much damage.

13. The total mass of protons and neutrons on each side of the equation is conserved, not the identity of the elements.

15. The longer the half-life, the more stable the nucleus. A more stable nucleus will take longer to decay, so its half-life will be longer.

17. Radioisotopes used for medical diagnosis should have an affinity for a particular organ or should target a specific system within the body that is being analyzed. They should also be short-lived within the body and have minimal ionization power. Radioisotopes used for cancer treatment should also have an affinity for a particular organ or should target a specific system within the body. They should also be short-lived within the body and should have minimal penetration power to limit the damage done to organs or systems not intended to receive the treatment.

19. Carbon-14 atoms are produced when cosmic rays interact with nitrogen-14 atoms in the atmosphere. The resulting carbon-14 atoms react with atmospheric oxygen to produce carbon dioxide; this carbon dioxide is taken up by plants during photosynthesis to produce sugars. These are then eaten by animals. As living things continually take in nutrients, the level of carbon-14 in living tissue reaches a steady state.

21. As described in the figure caption for Figure 12.4, uranium fuel rods release large amounts of heat (top right) while undergoing fission reactions in the reactor core. The heat from the reactor core is used to convert water into steam, which is used to turn a steam turbine that generates the electricity. The steam is then cooled and converted back into liquid water by fresh water pumped in from a lake or river. The fresh water is then cooled in the cooling towers before it is returned to the lake or river.

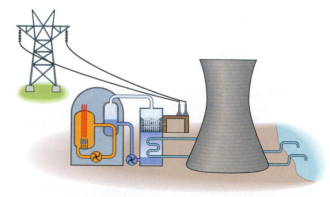

(Illustration adapted from HowStuffWorks.com)

23. A nuclear chain reaction occurs when a fission reaction generates multiple small particles (often neutrons) that in turn initiate more fission reactions. Control rods are used in a nuclear power plant to absorb excess neutrons produced in the fission reaction to avoid a chain reaction.

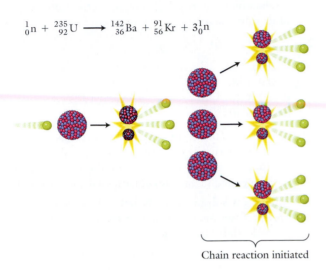

$$_{0}^{1}n + _{92}^{235}U \longrightarrow _{56}^{142}Ba + _{36}^{91}Kr + 3_{0}^{1}n$$

Chain reaction initiated

25. Enriched uranium has a higher percentage of radioactive ^{235}U. Depleted uranium has a higher percentage of less radioactive ^{238}U. Enriched uranium is used for nuclear fuel rods and nuclear fission weapons. Depleted uranium is used as a high density metal in projectile weapons.

27. It is desirable to have nondestructive analysis of samples in forensic science because that leaves open the possibility of analyzing the sample using another technique at a later date. If a destructive technique is used, the sample is lost after that analysis takes place and no subsequent techniques can be used. This makes neutron activation analysis (NAA) advantageous.

29. It is supposed to be used for demoralizing the public by hitting a target that has economic, historic, psychological, or political importance. The scientific issues involved deal primarily with the clean-up process: How can you do it cheaper and faster and get rid of all radioactive substances?

31. Underground testing of nuclear weapons is a safer alternative to above-ground testing because there is less chance that radioactive isotopes will be vaporized and dispersed into the atmosphere. Below ground, the radioactive isotopes are better contained and (theoretically) the long-term impacts of the detonation can be minimized.

33. a. $_{92}^{239}U \rightarrow _{93}^{239}Np + _{-1}^{0}e$

b. $_{84}^{210}Po \rightarrow _{82}^{206}Pb + _{2}^{4}He$

c. $_{38}^{90}Sr \rightarrow _{39}^{90}Y + _{-1}^{0}e$

d. $_{79}^{174}Au \rightarrow _{77}^{170}Ir + _{2}^{4}He$

35. a. $_{79}^{198}Au \rightarrow _{-1}^{0}e + _{80}^{198}Hg$

b. $_{94}^{239}Pu \rightarrow _{2}^{4}He + _{92}^{235}U$

c. $_{1}^{3}H \rightarrow _{-1}^{0}e + _{2}^{3}He$

d. $_{92}^{235}U \rightarrow _{2}^{4}He + _{90}^{231}Th$

37. a. 500.8 g

b. 148 kg

c. 5.89 mg

d. 0.00459 g

39. a. $_{27}^{57}Co + _{-1}^{0}e \rightarrow _{26}^{57}Fe + \gamma$

b. $_{27}^{59}Co + _{0}^{1}n \rightarrow _{27}^{60}Co + \gamma$

c. $_{50}^{111}Sn + _{-1}^{0}e \rightarrow _{49}^{111}In$

d. $_{16}^{32}S + _{2}^{4}He \rightarrow _{18}^{36}Ar$

41. Hair grows slowly and is made from waste protein and other compounds from within the body. Drug metabolites are incorporated as part of the hair. If the rate of hair growth is known (or can be estimated), NAA could be used to detect drug metabolites in the hair, and the distance the metabolites are located above the follicle could determine how long ago the person had ingested the drugs.

43. 724.5 days (approximately 2 years)

45. By using their lips to keep a fine point on their paint brushes, the young women at Westclox were ingesting small amounts of radioactive radium. Madame Curie, on the other hand, was merely exposing the exterior of her body to the radioactive emissions of the radium isotopes. As a result, the young women showed symptoms of radiation poisoning whereas Madame Curie did not.

47. 20%

Chapter 13

1. The reverse reaction of a system can occur as long as product molecules collide with sufficient energy to overcome the activation barrier and the correct orientation to produce the reagents.

3. The rate of forward reaction decreases as a function of reactant time typically because the amount of reagent decreases as a function of time.

5. It appears that a reaction has stopped when it reaches equilibrium because there is no net change in concentration of reactants or products with time. In fact, there is still conversion of reactant to product and product to reactant, but the rate of forward reaction is equal to the rate of reverse reaction.

7. A catalyst has no effect on the equilibrium constant, or the concentrations of the reactants or products at equilibrium. It simply speeds up the initial rates of the forward reactions, allowing it to reach equilibrium more quickly.

9. If an equilibrium system for the reaction A + B → C + D had a very large equilibrium constant, the reaction mixture would contain mostly C and D when the system reached equilibrium.

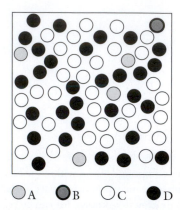

○A ◔B ○C ●D

11. The *K* number will tell you whether it favors the right side of the arrow (large *K* value) or the left side of the arrow (small *K* value).

13. a. 1.2×10^{-5} M
    b. 3.2×10^{-14} M
    c. 1.0×10^{-6} M
    d. 0.12 M

15. c. $HCHO_2(aq)$, $K_a = 1.8 \times 10^{-4}$ will have the highest $[H^+]$ (lowest pH)

17. The four stresses and the response of the system are (1) Add reactant: produce more product. (2) Remove reactant: produce more reactant. (3) Add product: produce more reactant. (4) Remove product: produce more product.

19. Oxyhemoglobin (HbO_2) is transported to the cells in the blood. Once it arrives there, the oxygen is released by the oxyhemoglobin to produce hemoglobin (Hb), which is then circulated back to the lungs to gather more oxygen.

21. Carbon monoxide binds to hemoglobin more strongly than does oxygen. This competitive binding makes carbon monoxide toxic, as it prevents the delivery of oxygen to cells.

23. a. H_2CO_3 concentration will increase and HCO_3^- concentration will decrease
    b. H_2CO_3 concentration will decrease and HCO_3^- concentration will increase

c. H^+ concentration will increase and HCO_3^- concentration will increase
    d. H_2CO_3 concentration will increase and H^+ concentration will decrease

25. Adding a sulfide such as Na_2S to solution would cause the solubility equilibrium to shift toward production of more solid, minimizing the amount of dissolved arsenic.

27. $K_{sp} = 2.1 \times 10^{-8}$

29. a. 0.0035 M
    b. 0.015 M
    c. 8.2×10^{-4} M
    d. 3.4×10^{-6} M

31. According to our textbook, the term *toxin* usually refers to a naturally occurring poisonous substance produced by a living organism. A poison is defined as any compound that injures or harms a living organism.

33. Almost any substance can be lethal if enough is present. Even water, which is necessary for life, can cause death by drowning.

35. The toxicity of a compound is measured according to the lethal dose in 50% of the population of interest. The shorthand notation for this quantity is the LD_{50} for the compound.

37. Chronic poisoning is poisoning that has taken place over a long period of time. In most cases, a toxic substance has accumulated in the body over a long period of time so that the cumulative effect of the toxic substance causes poisoning after that amount of time.

39. Acidosis is a condition characterized by excessive acid in the body fluids. The acid/base status of the body (pH) is regulated by the kidneys and the lungs. Acidosis is caused by an accumulation of acid or a significant loss of bicarbonate. The major categories of acidosis are respiratory acidosis and metabolic acidosis. The human body is programmed to correct for either respiratory or metabolic acidosis to maintain normal pH. For example, if the acidosis was caused by excessive carbon dioxide (which is an acid), then the body will correct the pH by retaining bicarbonate (a base). Respiratory acidosis develops when there is excessive carbon dioxide in the body, primarily caused by decreased breathing. Other names for this include *hypercapnic acidosis* and *carbon dioxide acidosis*.

There are several types of metabolic acidosis. Diabetic acidosis (also called diabetic ketoacidosis and DKA) develops when ketone bodies accumulate during uncontrolled diabetes. Hyperchloremic acidosis results from excessive loss of sodium bicarbonate from the body, as in severe diarrhea, for example. Lactic acidosis is an accumulation of lactic acid. This can be caused by many conditions, including prolonged lack of oxygen (e.g., from shock, heart failure, or severe anemia), prolonged exercise, seizures, hypoglycemia (low blood sugar), alcohol, liver failure, malignancy, or certain medications like salicylates. Other causes of metabolic acidosis include severe dehydration—resulting in decreased tissue perfusion (decreased blood flow), kidney disease (see distal renal tubular acidosis and proximal renal tubular acidosis), and other metabolic diseases.

41. Dimercaprol binds to mercury, making it unable to cause its toxic effects in the body (which require the mercury(II) to be a free ion in solution). This is due to the extremely large binding constant between the mercury(II) ion with the —SH group of the cysteine.

43. The alcohol in your blood is in equilibrium with the alcohol in air in your lungs. The more alcohol in the blood, the more alcohol in the air in your lungs. This is a direct application of equilibrium principles: The ratio of alcohol in the blood to alcohol in the air in your lungs remains constant.

45. Arsenic poisoning is an example of chronic poisoning. The danger level is more substantial for an infant drinking formula prepared using contaminated water than for an adult drinking the same water, because the amount of arsenic present is more substantial in relation to the infant's mass than it is in relation to the mass of an adult drinking the same water.

47. Cyanide poisoning can be "cured" by administering amyl nitrite ampoules that are broken onto a gauze pad and held under the nose. This process should be repeated 30 seconds every minute for 3 minutes and then sodium nitrite infused intravenously, and then finally sodium thiosulfate infused intravenously. Exposure to the nitrites converts a portion of hemoglobin to methemoglobin. Cyanide binds more strongly to methemoglobin than it does to cytochrome C oxidase. Sodium thiosulfate converts the cyanomethemoglobin to hemoglobin, thiocyanate, and sulfite. This allows the cyanide to be removed from the body as thiocyanate.

49. 1.74×10^{-6} M $C_2O_4^{2-}$

Chapter 14

1. Lipid molecules include a variety of compounds such as oils, fats, and waxes. Their shared physical property is that they are not soluble in water but are soluble in a nonpolar solvent.

3. According to our textbook, waxes are made by combining a fatty acid (a long carbon chain carboxylic acid molecule) and a long-chain alcohol to form an ester. Both waxes and fatty acids are long-chain carboxylic compounds, but the waxes are esters and the fatty acids are carboxylic acids.

5. Fats and oils are both examples of triglycerides. The difference between them is that fats are semisolids at room temperature and oils are liquids at room temperature.

7. A fatty acid with a 24-carbon chain that is completely saturated is most likely a solid at room temperature. Stearic acid (an 18-carbon chain that is completely saturated) melts at 69°C, so it is a solid at room temperature. Arachidonic acid (a 20-carbon chain that is completely saturated) melts at 77°C, so it is also a solid at room temperature. As the chain length gets longer, the melting point increases, so it is reasonable to assume that a 24-carbon chain that is completely saturated will also be a solid at room temperature.

9. Starch is a polysaccharide of glucose, which means the starch is the polymer and glucose is the monomer. Cellulose is also a polysaccharide of glucose, which means that cellulose is the polymer and glucose is the monomer.

11. Cellulose:

Starch:

13. Amino acids contain an amino functional group, a carboxylic acid functional group, and an organic group bonded to the middle carbon atom that distinguishes one amino acid from another.

15. The primary structure of a protein is the polypeptide chain that links amino acid monomer units. The secondary structure is how the amino acid chains interact with each other to fold into various superstructures.

17. Secondary structures are held together through hydrogen bonding and disulfide bonds.

19. Tertiary and quaternary structures of proteins are held together by hydrophobic interactions, ionic bridges, disulfide bonds, and hydrogen bonds.

21. Strands of DNA are held together through hydrogen bonding of the DNA base pairs. Covalent bonding between strands would be problematic because DNA replication and other functions require the "unzipping" of the DNA strands, and covalent bonds would not allow this.

23. [CTGAATCCC]

25. [GGCACCAACGAACCCGGCCGC]

27. His-Pro-Arg-Arg-His-Lys-Asn

29. Glu-Lys-Pro-Gln-Asn-Asn-Thr

31. [GAACGCCCCCACGGCGACACC]

33. [ACCCCCCCCAACGACCAACAA]

35. Mitochondrial DNA (mtDNA) comes only from the maternal side of the family. This can be a disadvantage because it makes it impossible to distinguish between relatives on the mother's side of the family. mtDNA does not decompose as easily as nuclear DNA, and there are multiple copies of mtDNA within a single cell, making recovery of a usable sample very simple.

37. Single nucleotide polymorphism (SNP) can be used in genetic genealogy to track a specific line in a genealogy through the Y chromosome. Because SNPs are so rare (once every few hundred generations), they provide a powerful way to differentiate different segments of the population and link individuals to the specific haplogroups that result.

39. 1 in ~5,240,000 people

41. mtDNA can be used to distinguish between two cousins only if they are related through a brother and sister relationship (one cousin's father is the other cousin's sister). That way, they have different maternal lineage and will have different mtDNA.

43. According to our textbook, "even identical twins are believed to have slightly different DNA due to small variations and mutations that inevitably occur over time." However, this is not currently distinguishable using the techniques described in the chapter.

45. As hair grows, waste proteins are being excreted from the body. The primary structure of proteins contains peptide bonds, and the secondary structure involves hydrogen bonding and sulfide bonding between groups along the peptide chain. The interactions responsible for secondary structure are strong enough to trap trace amounts of arsenic or heavy metal poisons and retain some of those materials in the growing hair shaft.

47. If hair grows at a rate of 0.5 mm per day, then a 1.0 cm length (equivalent to 10. mm) represents 10. mm ÷ 0.5 mm/day = 20 days of growth. If the first, fifth, and eighth samples tested positive for arsenic, then the victim had arsenic in his or her system between 0–20 days, 80–100 days, and 140–160 days prior to sample collection.

INDEX

Note: Page numbers followed by f indicate figures; page numbers followed by t indicate tables. **Boldface** entries indicate definitions.

Polyatomic Ions

Polyatomic Ion	Symbol
Ammonium ion	NH_4^+
Nitrate ion	NO_3^-
Nitrite ion	NO_2^-
Hydroxide ion	OH^-
Acetate ion	$C_2H_3O_2^-$
Cyanide ion	CN^-
Permanganate ion	MnO_4^-
Hypochlorite ion	ClO^-
Chlorite ion	ClO_2^-
Chlorate ion	ClO_3^-
Perchlorate ion	ClO_4^-
Hydrogen carbonate ion	HCO_3^-
Carbonate ion	CO_3^{2-}
Sulfate ion	SO_4^{2-}
Chromate ion	CrO_4^{2-}
Dihydrogen phosphate ion	$H_2PO_4^-$
Hydrogen phosphate ion	HPO_4^{2-}
Phosphate ion	PO_4^{3-}

Names and Formulas of Various Acids

Name	Formula
Hydrochloric acid	HCl
Nitric acid	HNO_3
Sulfuric acid	H_2SO_4
Acetic acid	$HC_2H_3O_2$
Carbonic acid	H_2CO_3
Hydrofluoric acid	HF
Phosphoric acid	H_3PO_4

Names and Formulas of Various Bases

Name	Formula
Ammonium hydroxide	NH_4OH
Sodium hydroxide	$NaOH$
Potassium hydroxide	KOH
Calcium hydroxide	$Ca(OH)_2$

Summary of Organic Compounds

Class	Formula	Example	Naming	Example Name
Alkanes	C_nH_{2n+2}	$CH_3CH_2CH_3$	C chain length + -ane	Propane
Alkenes	C_nH_{2n}	$CH_2{=}CH_2CH_3$	C=C location + C chain length + -ene	1-propene (or propene)*
Alkynes	C_nH_{2n-2}	$CH{\equiv}CCH_3$	C≡C location + C chain length + -yne	1-propyne (or propyne)*
Ethers	R—O—R	$CH_3OCH_2CH_3$	Short C chain + -oxy + long C chain + -ane	Methoxyethane
Ketones	$R-\overset{\overset{\displaystyle O}{\|\|}}{C}-R$	$CH_3\overset{\overset{\displaystyle O}{\|\|}}{C}CH_3$	C=O location + C chain length + -one	2-propanone (or propanone)*
Esters	$R-O-\overset{\overset{\displaystyle O}{\|\|}}{C}-R$	$CH_3O\overset{\overset{\displaystyle O}{\|\|}}{C}CH_3$	C chain not touching C=O, C chain length + -oate	Methyl ethanoate
Amines	R_3N	$CH_3NHCH_2CH_3$	Name each C chain (alphabetical order) + -amine	Ethylmethylamine
Alcohols	R—OH	CH_3OH	C chain length + -ol	Methanol
Aldehydes	$R-\overset{\overset{\displaystyle O}{\|\|}}{C}H$	$CH_3\overset{\overset{\displaystyle O}{\|\|}}{C}H$	C chain length + -al	Ethanal
Carboxylic acids	$R-\overset{\overset{\displaystyle O}{\|\|}}{C}-OH$	$CH_3\overset{\overset{\displaystyle O}{\|\|}}{C}OH$	C chain length + -oic + acid	Ethanoic acid

*When there is only one possible location for a functional group, providing the number is unnecessary.